Noise and Nonlinear Phenomena in Nuclear Systems

NATO ASI Series

Advanced Science Institutes Series

A series presenting the results of activities sponsored by the NATO Science Committee, which aims at the dissemination of advanced scientific and technological knowledge, with a view to strengthening links between scientific communities.

The series is published by an international board of publishers in conjunction with the NATO Scientific Affairs Division

A	**Life Sciences**	Plenum Publishing Corporation
B	**Physics**	New York and London
C	**Mathematical and Physical Sciences**	Kluwer Academic Publishers Dordrecht, Boston, and London
D	**Behavioral and Social Sciences**	
E	**Applied Sciences**	
F	**Computer and Systems Sciences**	Springer-Verlag
G	**Ecological Sciences**	Berlin, Heidelberg, New York, London,
H	**Cell Biology**	Paris, and Tokyo

Recent Volumes in this Series

Volume 185—Nonperturbative Quantum Field Theory
edited by G. 't Hooft, A. Jaffe, G. Mack, P. K. Mitter, and R. Stora

Volume 186—Simple Molecular Systems at Very High Density
edited by A. Polian, P. Loubeyre, and N. Boccara

Volume 187—X-Ray Spectroscopy in Atomic and Solid State Physics
edited by J. Gomes Ferreira and M. Teresa Ramos

Volume 188—Reflection High-Energy Electron Diffraction and Reflection Electron Imaging of Surfaces
edited by P. K. Larsen and P. J. Dobson

Volume 189—Band Structure Engineering in Semiconductor Microstructures
edited by R. A. Abram and M. Jaros

Volume 190—Squeezed and Nonclassical Light
edited by P. Tombesi and E. R. Pike

Volume 191—Surface and Interface Characterization by Electron Optical Methods
edited by A. Howie and U. Valdrè

Volume 192—Noise and Nonlinear Phenomena in Nuclear Systems
edited by J. L. Muñoz-Cobo and F. C. Difilippo

Series B: Physics

Noise and Nonlinear Phenomena in Nuclear Systems

Edited by

J. L. Muñoz-Cobo

Polytechnic University of Valencia
Valencia, Spain

and

F. C. Difilippo

Oak Ridge National Laboratory
Oak Ridge, Tennessee

Plenum Press
New York and London
Published in cooperation with NATO Scientific Affairs Division

Proceedings of a NATO Advanced Research Workshop
on Noise and Nonlinear Phenomena in Nuclear Systems,
held May 23-27, 1988,
in Valencia, Spain

Library of Congress Cataloging in Publication Data

NATO Advanced Research Workshop on Noise and Nonlinear Phenomena in Nuclear Systems (1988: Valencia, Spain)

Noise and nonlinear phenomena in nuclear systems / edited by J. L. Muñoz-Cobo and F. C. Difilippo.

p. cm.—(NATO ASI series. Series B. Physics: v. 192)

"Proceedings of a NATO Advanced Research Workshop on Noise and Nonlinear Phenomena in Nuclear Systems, held May 23-27, 1988, in Valencia, Spain."

"Published in cooperation with NATO Scientific Affairs Division."

Includes bilbiographical references and index.

ISBN-13: 978-1-4684-5615-8 e-ISBN-13: 978-1-4684-5613-4

DOI: 10.1007/978-1-4684-5613-4

1. Nuclear engineering—Congresses. 2. Nonlinear theories—Congresses. 3. Noise—Congresses. I. Muñoz-Cobo, J. L. II. Difilippo, F. C. III. North Atlantic Treaty Organization. Scientific Affairs Division. IV. Title. V. Series.

TK9006.N335 1988 88-39498

621.48—dc19 CIP

Softcover reprint of the hardcover 1st edition 1989

A Division of Plenum Publishing Corporation
233 Spring Street, New York, N.Y. 10013

ORGANIZING COMMITTEE

J. L. Muñoz-Cobo
Polytechnic University of Valencia
Valencia, Spain

F. C. Difilippo
Oak Ridge National Laboratory
Oak Ridge, Tennessee

R. B. Perez
University of Tennessee
Knoxville, Tennessee

J. J. Dorning
University of Virginia
Charlottesville, Virginia

J. D. Lewins
University of Cambridge
Cambridge, United Kingdom

This book is dedicated to Professor Vicente Serradell, who founded the Department of Nuclear Engineering at the Polytechnical University of Valencia. His continuous efforts made possible the formation of a generation of professors in the nuclear sciences, who are now working actively in this and other areas of research.

PREFACE

The main goal of the meeting was to facilitate and encourage the application of recent developments in the physical and mathematical sciences to the analysis of deterministic and stochastic processes in nuclear engineering. In contrast with the rapid growth (triggered by computer developments) of nonlinear analysis in other branches of the physical sciences, the theoretical analysis of nuclear reactors is still based on linearized models of the neutronics and thermal-hydraulic feedback loop, an approach that ignores some intrinsic nonlinearities of the real system. The subject of noise was added because of the importance of the noise technique in detecting abnormalities associated with perturbations of sufficient amplitude to generate nonlinear processes.

Consequently the organizers of the meeting invited a group of leading researchers in the field of noise and nonlinear phenomena in nuclear systems to report on recent advances in their area of research. A selected subgroup of researchers in areas outside the reactor field provided enlightenment on new theoretical developments of immediate relevance to nuclear dynamics theory.

Thirty five presentations were discussed by participants from fourteen countries and international organizations in the pleasant environment provided by the Polytechnical University of Valencia. The logistic support and the active help of the University was crucial to the success of the meeting. The organizers were pleased to witness a cross-fertilization of ideas between researchers: for example new stochastic tools to detect nonlinear processes in nuclear power plants are being developed, whereas the paradigm of noise perturbations in a distorted potential well is being applied to analyze jumping processes in the internal vibrations of nuclear reactors. The papers presented at the meeting fall into some of the following categories:

- Basic methodology to analyze nonlinear deterministic and stochastic processes.

- Deterministic analysis of nuclear dynamics, transition to chaos.

- Noise theory and its application to the surveillance and diagnosis of nuclear systems.

The presentations more relevant from an engineering point of view were related to the use of stochastic methods to monitor nuclear systems and the application of recent developments in nonlinear dynamics to xenon oscillations, control theory, limit cycle analysis, breeding of fertile isotopes, analysis of bifurcations, and jumping processes in nuclear reactors.

Several presentations were dedicated to the discussion of the use of the third moment as a stochastic descriptor, with experimental applications to criticality safety, safeguard, and enhancement of the sensitivity of the monitoring system to nonlinear processes. The theoretical background of the noise techniques in both the time and frequency domains were also discussed; a general theory of space- and energy-dependent stochastic processes in zero power systems was presented, illustrating that the Boltzmann equation for neutrons is only the first moment of a more general formalism. Complications associated with this general formalism make very useful the discussion of detailed solutions of simplified models like the explicit and implicit solutions of the forward and backward stochastic equations with and without including the burn-up in the case of "point" systems. Specific noise techniques like the method of the ^{252}Cf source were analyzed with a very general model that includes space and energy effects.

Particularly relevant to the design of high-intensity neutron sources was the application of nonlinear analysis to the problem of xenon oscillations after coupling the kinetic and conservation equations with the equation for the reactivity. The evolution of the nuclear species in breeding assemblies was presented in the light of the theory of dynamical systems. Recent developments in the control of nuclear reactors using the Pontryagin maximum principle to a nonlinear

reactor dynamics model were discussed; because the method allows parameter tracking it could be used for the on-line surveillance of nuclear systems.

One of the finer presentations from a theoretical point of view was the application of the Hartman-Grobman theorem to the analysis of limit cycles observed in boiling water reactors that allows the parametrization of the condition under which a limit cycle appears. Particularly interesting because of its novelty was the presentation about jumping processes in bistable systems subject to noisy perturbations; the paradigm of a double potential well was used to analyze the noise signatures of the vibrations of reactor components such as ,for example, the pressure vessel. Results obtained recently on strongly nonlinear phenomena relevant to nuclear science and technology were summarized.

There was a general concensus that the workshop had been very useful and that the experience should be repeated some years hence. There are already preliminary plans for a further meeting in 1990.

The Organizing Committee gratefully acknowledges the financial support of the following institutions: Scientific Affairs Division of NATO, Direccción General de Investigación Científica y Técnica of Spain, and the Polytechnical University of Valencia. Their contributions made this workshop possible. The editors also acknowledge the help of Ms. Lucía Ferreres in the preparation of this book.

J.L. Muñoz-Cobo
Felix C. Difilippo
Valencia, Spain, July 26, 1988

CONTENTS

I. Introduction: Basic Concepts

II. Stochastic Processes in Linear Nuclear Systems: Theory

III: Stochastic Processes in Linear Nuclear Systems: Applications

V. Stochastic Processes in Nonlinear Nuclear Systems: Applications

VI. Nonlinear Dynamics and Transition to Chaos in Nuclear Systems. Part I

VII. Nonlinear Dynamics and Transition to Chaos in Nuclear Systems. Part II

CHAPTER I

INTRODUCTION: BASIC CONCEPTS

A GLIMPSE INTO THE WORLD OF

RANDOM WALKS

C. Van den Broeck

Limburgs Universitair Centrum

B-3610 Diepenbeek, Belgium

ABSTRACT

Some classical results from random walk theory are reviewed. Asymptotic properties are derived for random walk on a one-dimensional lattice with static disorder. Fractal time properties are illustrated on a simple example.

INTRODUCTION

It would be impossible, both for reasons of space and competence, to write a comprehensive review on random walks. Besides, many excellent reviews and books have been published on this topic [1-10]. Our purpose here will be to collect a few simple illustrative results, which may hopefully inspire ideas for applications in the field of noise in nuclear reactors.

The concept of "random walk" is used in a broad sense of a stochastic walk on a set M of discrete states $\underset{\sim}{m} \in M$. This, for instance, encompasses Markovian walks such as birth and death processes. The walk is characterized by the transition rules between the states and the geometry of the set M. The mathematical structure of the random walk problem is, in fact, closely related to other physical problems, some of which are discussed briefly in §2. In particular, a Fokker-Planck or Langevin description is included as a limiting case.

A central quantity for stationary random walks is the conditional probability $P(\underset{\sim}{m}|\underset{\sim}{m}_o,t)$ to go from site $\underset{\sim}{m}_o$ to site $\underset{\sim}{m}$ during a time t, as well as its Laplace transform, the so-called Greens function of the walk :

$$\tilde{P}(\underset{\sim}{m}|\underset{\sim}{m}_o,s) = \int_o^\infty dt\, e^{-st}\, P(\underset{\sim}{m}|\underset{\sim}{m}_o,t) \tag{1}$$

Some results for $\tilde{P}$ will be reviewed in §3. Furthermore, it is discussed how various other important quantities can be derived from it.

A lot of recent work on random walks is dealing with the study of anomalous behavior. By this we mean a qualitative departure of various random walk properties (such as mean square displacement and spectral properties) from their standard from (as obtained for Markovian walks on Eucledian lattices). In this context, we derive results for random walks with static disorder in §4 and discuss the case of random walks with fractal aspects in §5. Transport properties in random media have been discussed in the context of reactor noise [12]. The concept of fractal structures however, is probably new but may, in view of its universality, find some application in this field.

RANDOM WALK AND RELATED PROBLEMS

In the case of a Markovian random walk, the conditional probability $P(\underset{\sim}{m}|\underset{\sim}{m}_o,t)$ obeys the so-called Master Equation :

$$\partial_t P(\underset{\sim}{m},t) = \sum_{\underset{\sim}{m}} [W(\underset{\sim}{m}|\underset{\sim}{m}') P(m',t) - W(\underset{\sim}{m}'|\underset{\sim}{m}) P(\underset{\sim}{m},t)] \qquad (2)$$

$W(\underset{\sim}{m}|\underset{\sim}{m}')$ stands for the transition probability, per unit time, to go from $\underset{\sim}{m}'$ to $\underset{\sim}{m}$. Note that equation (2) is linear in P : the different random walkers do not interact. "Nonlinearity" or "collisions" can however be included in a mean field way by considering a nonlinear dependence of the transition probability on the state of the system. Nonlinear birth - and death processes are of this type : a neutron, that is produced, is added to the bath of all existing neutrons, and each individual neutron is subsequently interacting with this bath.

Equation (2) has an electrical and mechanical analogue (see Fig. 1) [5]. In the electrical system, the electrical potential $V(\underset{\sim}{m},t)$ obeys the following equation :

$$\partial_t V(\underset{\sim}{m},t) = \sum_{\underset{\sim}{m}'} \frac{(V(\underset{\sim}{m}',t) - V(\underset{\sim}{m},t))}{R(\underset{\sim}{m},\underset{\sim}{m}')C(\underset{\sim}{m})} \qquad (3)$$

where $C(\underset{\sim}{m})$ is the capacitance linking site $\underset{\sim}{m}$ to the earth and $R(\underset{\sim}{m},\underset{\sim}{m}')$ is the resistance between sites $\underset{\sim}{m}$ and $\underset{\sim}{m}'$. In the mechanical analogue, one considers a distribution of harmonically interconnected masses in a plane. The spring constant between sites $\underset{\sim}{m}$ and $\underset{\sim}{m}'$ is denoted by $k(\underset{\sim}{m},\underset{\sim}{m}')$. The small deviations $x(\underset{\sim}{m},t)$, orthogonal to the plain, obey the following set of Newton equations :

$$\partial_t^2 x(\underset{\sim}{m},t) = \sum_{\underset{\sim}{m}'} k(m,m') [x(\underset{\sim}{m}',t) - x(\underset{\sim}{m},t)] \qquad (4)$$

As an illustrative example of a quantum-mechanical problem that is described by an equation of the same form, we cite the Kronig-Penney model of a particle (e.g. an electron) in a one-dimensional periodic delta function potential. If we call $\psi(m)$ the value of an energy eigenfunction for the energy equal to E at the right hand side of the

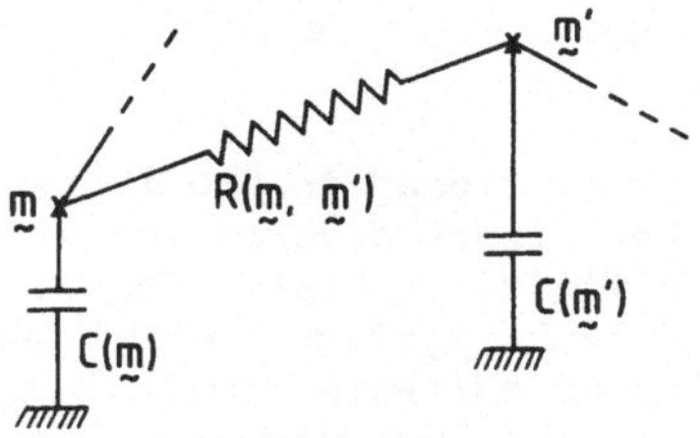

Fig. 1a. Electrical Analogue

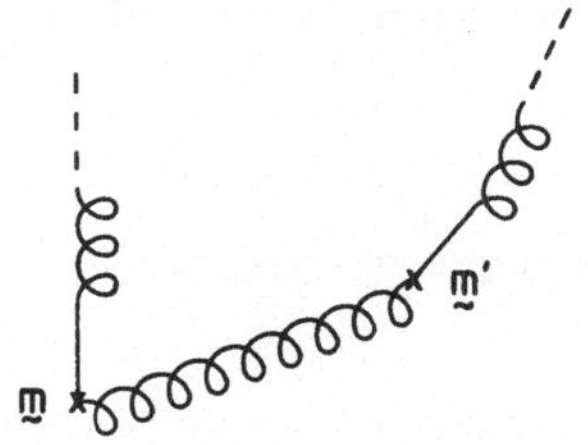

Fig. 1b. Mechanical Analogue

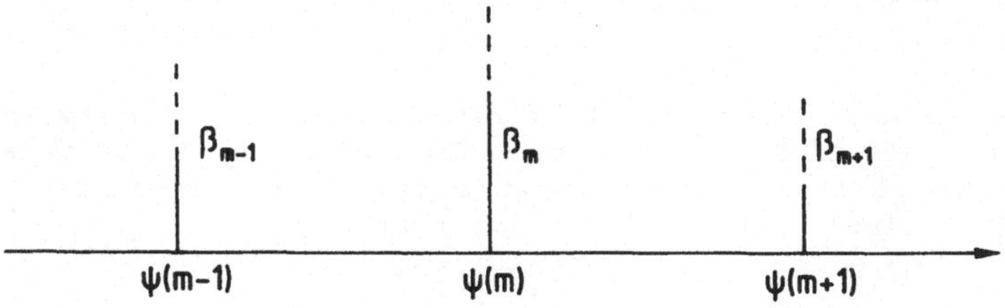

Fig. 1c. Quantum-Mechanical Analogue

Figure 1 . Electrical, Mechanical and Quantum-Mechanical Analogue of the Random Walk Master Equation

potential peak m, then an easy calculation shows that the Schrödinger equation implies the following set of equations [12] :

$$2(\cos k + \beta_m \frac{\sin k}{2k} - 1)\psi(m) = \psi(m+1) + \psi(m-1) - 2\,\psi(m) \tag{5}$$

where β_m is the weight of the potential peak at m, $k = a(2m\,E/\hbar)^{\frac{1}{2}}$ and a is the lattice spacing. We have written the right hand side such that it has the same form as the right hand side of (2) for a symmetric nearest neighbour random walk.

Finally, as already mentioned in the introduction, equation (2) is more general then a Fokker-Planck description since the latter can be obtained in the appropriate limit. A disturbing discrepancy subsists in the thermodynamic limit of systems with extensive transition probabilities in the region of multiple stable states [14]. Moreover, (2) is certainly a more natural description if one is concerned with internal fluctuations, i.e. the fluctuations that arise from the very dynamics of the system. Also, the Ito versus Stratonovich (or other) interpretations of Langevin equations with multiplicative noise reflect the fact that relevant information on the interplay between noise and fast dynamical variables may have been discarded [6,13].

THE CONDITIONAL PROBABILITY AND RELATED QUANTITIES

Once the explicit solution of equation (2) has been found, a lot of interesting random walk properties become available. Usually, it is easier to obtain an explicit compact form for the Laplace transform (1). A list of results for Markovian nearest neighbour random walks in 1 dimension is compiled in table I. In the following, we restrict ourselves to stationary Markov processes (W in equation (2) is time-independent).

Let us call $F(\underset{\sim}{m},\underset{\sim}{m}_o,t)$ the first passage time density to reach $\underset{\sim}{m}$ starting from $\underset{\sim}{m}_o$ after a time t. The following "renewal equation" states that to go from $\underset{\sim}{m}_o$ to $\underset{\sim}{m}$ after a time t, one must have reached $\underset{\sim}{m}$ (for the first time) at an earlier time τ[1] :

$$P(\underset{\sim}{m}|\,\underset{\sim}{m}_o,t) = \int_0^t P(\underset{\sim}{m}|\,\underset{\sim}{m},t-\tau)\;F(\underset{\sim}{m}|\,\underset{\sim}{m}_o,\tau)\;d\tau \tag{6}$$

By Laplace transformation, one finds [1] :

$$\tilde{F}(\underset{\sim}{m}|\,\underset{\sim}{m}_o,s) = \frac{\tilde{P}(\underset{\sim}{m}|\,\underset{\sim}{m}_o,s)}{\tilde{P}(\underset{\sim}{m}|\,\underset{\sim}{m},s)} \tag{7}$$

For a one-dimensional nearest-neighbour random walk, the following stronger result, due to Siegert [15], holds true :

$$\tilde{F}(m|\,m_o,s) = \frac{\tilde{P}(m'\,|\,m_o,s)}{\tilde{P}(m'\,|\,m,s)} \qquad m_o \le m \le m' \tag{8}$$

The moments T_n of the first passage time are given by the following result :

$$T_n(\underset{\sim}{m}|\underset{\sim}{m}_o) = \int_0^\infty t^n F(\underset{\sim}{m}|\underset{\sim}{m}_o,t)\, dt = (-1)^n \frac{\partial^n}{\partial s^n} \tilde{F}(\underset{\sim}{m}|\underset{\sim}{m}_o,s)|_{s=0} \tag{9}$$

This implies that the limiting behavior of $\tilde{F}$ hence of $\tilde{P}$ for $s \to 0$, is sufficient to evaluate the moments T_n. This limiting behavior is available for the case of a nearest neighbour random walk in one dimension and one finds (for $m > m_o$) :

$$T_n (m|m_o) = n \sum_{r=m_o}^{m-1} \sum_{s=-\infty}^{r} \frac{T_{n-1}(m|s)\, P(s)}{W(r|r+1)P(r)} \tag{10.a}$$

with

$$T_o (m|s) = 1 \tag{10.b}$$

and $P(s)$ is the steady state probability for the random walk with transition rates W and a reflecting boundary condition at $s = m$, $W(m|m-1) = 0$.
The result for $n = 1$ is well known [17], and in the limit of Fokker-Planck dynamics, the results of reference [18] are recovered.
These results can also be generalized to the case of first passage time statistics to more then one final destination [19].

The average number of distinct sites (or span) $S(\underset{\sim}{m}_o,t)$ visited in a random walk starting at $\underset{\sim}{m}_o$ after a time t is another important quantity. $S(\underset{\sim}{m}_o,t)$ can be written as follows [1] :

$$S(\underset{\sim}{m}_o,t) = \sum_{\underset{\sim}{m}} < \theta(\underset{\sim}{m},t) > \tag{11}$$

where $\theta(\underset{\sim}{m},t)$ is a random variable which is equal to 1 if site $\underset{\sim}{m}$ has been visited before t, i.e. if the first passage time is smaller then t, and zero otherwise. Obviously, one has :

$$<\theta(\underset{\sim}{m},t)> = \int_0^t F(\underset{\sim}{m}|\underset{\sim}{m}_o,\tau)\, d\tau \tag{12}$$

From (11) and (12), one finds [1], for a translationally invariant system :

$$\tilde{S}(\underset{\sim}{m}_o,s) = \int_0^\infty e^{-st}\, S(\underset{\sim}{m}_o,t)dt$$

$$= \sum_{\underset{\sim}{m}} \frac{\tilde{F}(\underset{\sim}{m}|\underset{\sim}{m}_o,s)}{s} = \frac{1}{s^2\tilde{P}(\underset{\sim}{m}_o|\underset{\sim}{m}_o,s)} \tag{13}$$

where we have made use of equation (7). We conclude that the span is intimately linked to the probability of return to the origin.

A situation occurring in a large variety of physical problems is that of a particle with a random velocity u describing the rate of change of a coordinate x. More precisely, we suppose that u depends on an internal state $\underset{\sim}{m}$, $u = u(\underset{\sim}{m})$, which is undergoing a random walk on a state space M. The equations of motion are thus :

$$\partial_t x = u(\underset{\sim}{m}(t)) \tag{14}$$

combined with equation (2) describing the stochastic dynamics of $\underset{\sim}{m}(t)$. The resulting transport properties such as the average position $<x(t)>$ and dispersion $<\delta x^2(t)>$ can again be expressed in terms of the random walk statistics [16.b]. For example, one has :

$$<\delta\tilde{x}^2(s)> = \int_0^\infty e^{-st} <[x(t) - <x(t)>]^2 > dt$$

$$= \sum_{\underset{\sim}{m}} \sum_{\underset{\sim}{m}_o} [u(\underset{\sim}{m})-\bar{u}] \; [u(\underset{\sim}{m}_o)-\bar{u}] \; \tilde{P}(\underset{\sim}{m}|\underset{\sim}{m}_o,s)P(\underset{\sim}{m}_o) \tag{15}$$

As a final example, we consider the spectral problem in a state space M with a finite number N of states ordered from 1 to N. The master equation (2) can now be written under the following matrix form :

$$\partial_t \underset{\sim}{P} = \underset{\approx}{W} \underset{\sim}{P} \tag{16}$$

while the Greens function (1) is given by :

$$\tilde{P}(m|m_o,s) = \left(\frac{1}{s\underset{\approx}{1} - \underset{\approx}{W}}\right)_{m,m_o} \tag{17}$$

i.e. it is the element (m,m_o) of the inverse of the matrix $s\underset{\approx}{1} - \underset{\approx}{W}$. A quantity of interest, especially in the mechanical and quantum-mechanical analogues discussed in the previous §, is the spectral density $g(\lambda)$ of the operator $\underset{\approx}{W}$:

$$g(\lambda) = \frac{1}{N} \sum_{r=1}^{N} \delta(\lambda-\lambda_r) \tag{18}$$

where $\lambda_r, r = 1,\dots,N$, are the eigenvalues of the $N \times N$ matrix $\underset{\approx}{W}$. If we call $\underset{\approx}{X}$ and $\underset{\approx}{Y}$ the matrix of orthonormalized right and left eigenvectors of $\underset{\approx}{W}$, one has :

$$\tilde{P}(m|m_o,s) = \sum_{r=1}^{N} X_{mr} \frac{1}{s - \lambda_r} Y_{rm_o} \tag{19}$$

For a translational invariant system, one can now write :

$$\tilde{P}_o(s) = \frac{1}{N} \sum_{m_o=1}^{N} \tilde{P}(m_o|m_o,s) = \frac{1}{N} \sum_{r=1}^{N} \frac{1}{s - \lambda_r} \tag{20}$$

where we have used the property $\underset{\approx}{Y} \cdot \underset{\approx}{X} = \underset{\approx}{1}$. By noting that Lorentzians with vanishing width converge to delta functions, we conclude that (see e.g. [5]).

$$g(\lambda) = -\frac{1}{\pi} \lim_{\substack{\epsilon \to 0 \\ >}} \operatorname{Im} \tilde{P}_o(\lambda + i\epsilon) \tag{21}$$

In translational invariant systems, the Laplace transformed probability of return to the origin thus yields the spectral density.
To illustrate this result, consider a symmetric nearest neighbour random walk on a ring of N sites in the limit of $N \to \infty$. One has (cf. Table I):

$$\tilde{P}_o(s) = \lim_{N\to\infty} \frac{\cosh(\xi_o N/2)}{2W \sinh \xi_o \sinh (\xi_o N/2)} = [s(s+4W)]^{-\frac{1}{2}} \tag{22}$$

To obtain a nonvanishing imaginary part, cf. equation (21), one must have $-4W \le s \le 0$. We thus conclude :

$$g(\lambda) = [4|\lambda| W - \lambda^2]^{-\frac{1}{2}}/\pi \qquad -4W \le \lambda \le 0 \tag{23}$$

To transform this result from the random walk to the mechanical problem, we note that an eigenvector of $\underset{\approx}{W}$ corresponds to an eigenmode of the harmonic lattice, while the eigenvalue λ_r is equal to minus the square of the corresponding eigenfrequency $-\omega_r^2$ (with the replacement of W by the spring constant k). The density of eigenfrequencies $g(\omega)$ thus reads

$$\rho(\omega) = 2\omega g(-\omega^2) = \frac{2}{\pi} [4k - \omega^2]^{-\frac{1}{2}} \qquad 0 \le \omega^2 \le 4k \tag{24}$$

For the Kronig-Penney model, described by equation (5) with $\beta_m = \beta$ constant, an eigenvector of $\underset{\approx}{W}$ corresponds to an energy eigenfunction, and the relation :

$$2(\cos k + \beta \frac{\sin k}{2k} - 1) = \lambda \tag{25}$$

It is easy to verify that this relation together with $-4 \le \lambda \le 0$ (note that $W = 1$ here), generates the energy-band structure of the Kronig Penney model [12].

RANDOM WALKS WITH STATIC DISORDER [20]

The study of Fokker-Planck equations with random (or even stochastic) components in its drift or diffusion part has been the object of recent studies [21]. In particular, problems related to noise in nuclear reactors have been tackled [22]. We will address here the related problem of a random walk with static disorder (for reviews, see [5], [11]). Note that these problems are also of paramount importance in the other physical problems discussed in §2 (for example, randomness in a one-dimensional quantum problem induces localization of the wave function [23]).

To have an idea of what is happening, let us start with one of the few, relatively simple dynamical results, that are known for random walks with general transition rates. We consider the moments $T_n(m|m_o)$ of the first passage time to go from m_o to m. Furthermore, we introduce a reflecting boundary at 1, and set $0 < m_o < m$. We then have (cf. (10)) :

$$T_1 = \sum_{r=m_o}^{m-1} \sum_{k=1}^{r} \frac{W(k|k+1)\ W(k+1|k+2)\ \ldots\ W(r-1|r)}{W(k+1|k)\ W(k+2|k+1)\ \ldots\ W(r|r-1)\ W(r+1|r)} \tag{26}$$

and similar, but more complicated expressions for the higher order moments T_n.

We will consider 3 types of randomness. In the case of bond randomness,

$W(k+1|k)$ and $W(k|k+1)$ are correlated random variables, belonging to the bond $(k,k+1)$, but the rates at different bonds are uncorrelated. For "out-site" randomness, $W(k+1|k)$ and $W(k-1|k)$ are correlated, but outgoing rates belonging to different sites are uncorrelated.
For "in-site" randomness, $W(k|k-1)$ and $W(k|k+1)$ are correlated, and incoming rates belonging to different sites are uncorrelated. By perfoming the average over these different types of randomness in (26), one finds :

$$\langle T_1\rangle^{\text{bond}} = \langle T_1\rangle^{\text{out-site}} = \frac{1}{(1-g_1)^2} \langle \frac{1}{W_\rightarrow} \rangle [g_1^m - g_1^{m_o} + (m - m_o)(1-g_1)] \tag{27}$$

with

$$g_1 = \langle \frac{W_\leftarrow}{W_\rightarrow} \rangle \tag{28}$$

and

$$\langle T_1\rangle^{\text{in-site}} = \langle \frac{1}{W_\rightarrow} \rangle \langle W_\leftarrow \rangle \langle \frac{W_\leftarrow}{W_\rightarrow} \rangle^{-1} \langle T_1\rangle^{\text{bond}} \tag{29}$$

Let us discuss some particular cases. For the case of non-random, constant rates $W_\leftarrow = 2qW$ and $W_\rightarrow = 2pW$, one recovers the familiar result for the case without disorder :

$$T_1 = \frac{1}{(1-q/p)^2} \frac{1}{2pW} [(q/p)^m - (q/p)^{m_o} + (m-m_o)(1-q/p)] \tag{30}$$

If $W_\leftarrow$ and $W_\rightarrow$ are independent but identically distributed random variables, (27) - (29) reduce to (for all three cases) :

$$\langle T_1\rangle = \frac{1}{(1-g_1)^2} \langle \frac{1}{W} \rangle [g_1^m - g_1^{m_o} + (m-m_o)(1-g_1)] \tag{31}$$

with

$$g_1 = \langle W\rangle \langle \frac{1}{W} \rangle \geq 1 \tag{32}$$

Note that, even though $W_\leftarrow$ and $W_\rightarrow$ are identically distributed, g_1 is not equal to one (except if W is non-random) so that randomness induces an effective bias $g_1 > 1$ directed from m to m_o.
For perfectly symmetric rates,

$$W_\rightarrow = W_\leftarrow = W \tag{33}$$

with W random, one obtains :

$$\langle T_1(m|m_o)\rangle = \frac{1}{2} \langle \frac{1}{W} \rangle [m(m-1) - m_o(m_o-1)] \tag{34}$$

To compare these results with those obtained in the literature, we consider the asymptotic dependence of $\langle T_1\rangle$ for large values of m and m_o. One finds (for $g_1 < 1$) :

$$\langle T_1\rangle^{\text{bond}} \approx \frac{1}{1-g_1} \langle \frac{1}{W_\rightarrow} \rangle (m-m_o) \tag{35}$$

This result should be compared with the asymptotic result for a process with drift velocity $V > 0$ and diffusion coefficient D :

$$\langle T_1\rangle \approx \frac{m-m_o}{V} \tag{36}$$

By identification of (35) and (36), we conclude

$$V^{bond} = \left(1 - \left< \frac{W_{\leftarrow}}{W_{\rightarrow}} \right>\right) \left< \frac{1}{W_{\rightarrow}} \right>^{-1} \tag{37}$$

This result is in agreement with the one obtained by Derrida ([24], Eq. (67)) from a direct calculation of the average value $<m(t)> \approx Vt$, for large times t. To evaluate the diffusion coefficient, we have to calculate the dispersion of the first passage time :

$$<T_2 - T_1^2> \approx \frac{2D}{V^3} (m-m_o) \tag{38}$$

These averages can be calculated on the basis of the explicit expression given by Eq. (10), along the same lines as those outlined above for the calculation of $<T_1>$. After straightforward but lengthy calculations, one finds, for large values of m and m_o :

$$\frac{< T_2 - T_1^2 >}{2(m-m_o)} \overset{bond}{\approx} \left< \frac{1}{W_{\rightarrow}^2} \right> \frac{1 + g_1}{2(1-g_1)(1-g_2)} + \left< \frac{W_{\leftarrow}}{W_{\rightarrow}^2} \right> \left< \frac{1}{W_{\rightarrow}} \right> \frac{1 + g_1}{(1-g_1)^2(1-g_2)} \tag{39}$$

with

$$g_2 = \left<\left(\frac{W_{\leftarrow}}{W_{\rightarrow}}\right)^2\right> \tag{40}$$

By comparison with (38), one concludes

$$D^{bond} = \frac{1 - <W_{\leftarrow}/W_{\rightarrow}>^2}{1-<(W_{\leftarrow}/W_{\rightarrow})^2>} \left< \frac{1}{W_{\rightarrow}} \right>^{-3} \left[\left< \frac{1}{W_{\rightarrow}} \right> \left< \frac{W_{\leftarrow}}{(W_{\rightarrow})^2} \right> + \frac{1}{2} \left< \frac{1}{W_{\rightarrow}^2} \right> \left(1 - \left< \frac{W_{\leftarrow}}{W_{\rightarrow}} \right>\right)\right] \tag{41}$$

which is again in agreement with Derrida [24] obtained from the identification $<[m(t) - <m(t)>]^2> = 2Dt$, for t large.

We point out three advantages of the above calculation method. Firstly we have bypassed a major effort in calculation by starting from the known expressions for first passage time moments. Secondly, we do not have to start with a periodic lattice like in the work of Derrida, in which, at the end, the limits of time going to infinity and that of an infinite random system have to be commuted. Our present calculation in fact supports the legitimacy of this commutation. Finally, other types of disorder can be easily studied and the approach to the asymptotic results (35) and (39) can be investigated.

The results (35) and (39) indicate that the random walk with static disorder reduces, on large time and length scales, to a regular random walk characterized by an effective drift coefficient (37) and diffusion coefficient (41). In particular, no anomalous behavior is observed. These are, however, somewhat hasty conclusions. Indeed, in deriving (35) and (39), we have supposed that all the averages that appear also exist. For example, the existence of an averge like $<1/W>$ expresses the fact that high barriers between two sites, corresponding to very low values of W, are very unlikely. Secondly, to be meaningfull, D should be a finite, positive quantity, implying that g_1 and g_2 should lie between zero and one. If these conditions are not fulfilled, the properties of a random walk with static disorder become more complicated and other quantities, such as non-integer moments of the first passage time should be considered.

To close this paragraph, we mention the growing interest for lattices with an aperiodic arrangement rather then a random arrangement of the transition frequencies (for example two different transition rates following each other according to a Fibonacci sequence). These lattices are believed to exhibit a behavior (e.g. for localization and transport properties) intermediate between that of regular and random lattices.

RANDOM WALKS AND FRACTALS

One of the oldest examples of a problem, exhibiting the paradoxical peculiarities of a fractal process, is maybe the Petersburg game [26]. It was discussed by Daniel Bernoulli back in the 18th century. A single trial in this game consists in tossing a coin until it falls heads. If this occurs at the rth throw, the player receives 2^r dollars. The probability for such a sequence is 2^{-r}, so that the expected gain diverges, and the law of large numbers is not appliable. The question is to determine entrance fees with which this game will have the properties of a fair game, i.e. the accumulated entrance fees are asymptotically converging for a large number of trials to the accumulated gain. Feller has shown that this can be achieved by introducing variable entrance fees, i.e. the entrance fee per trial should be $\log_2 n$, if n trials are going to be played (in the entire life of the casino !). The intuitive reason for this increasing fee with n is that, as more trials are being performed, the actual realization of a "very expensive trial" (with a large value of r) becomes likely.

To make the connection with "fractal-time" random walks, let us translate the Petersburg game into a waiting time problem. Let us suppose that the time between steps in the random walk are independent random variables, equal to 2^r with a probability 2^{-r}, $r = 1,2,\ldots$. The probability density $\psi(\tau)$ for a waiting time equal to τ thus reads :

$$\psi(\tau) = \sum_{r=1}^{\infty} \frac{\delta(\tau-2^r)}{2^r} \tag{42}$$

Note that the Laplace transform $\tilde{\psi}(s)$ obeys the following scaling equation :

$$\tilde{\psi}(s) = \int_0^{\infty} dt\, e^{-st}\psi(t) = 2^{-m}\,\tilde{\psi}(2^m s) + \sum_{r=1}^{m} \frac{\exp(-2^r s)}{2^r} \tag{43}$$

as noted by Shlesinger and Montroll [27] for some other examples of "fractal" functions.

Now, we start by making a step at time $t = 0$, and ask what will be the average number of steps that have occurred in a time interval $]0,t[$ (we do not count the step at $t = 0$). The probability $P(n,t)$ to take n steps in $]0,t[$, is the sum of the probabilities for all the possible (and mutually exclusive) realizations, namely, a first step at time τ_1, $0 < \tau_1 < t$, a second step at time $\tau_1 + \tau_2$, with $\tau_1 + \tau_2 < t$ and $\tau_2 > 0$, etc ..., and the $n+1^{th}$ step at a time $\tau_1 + \ldots + \tau_{n+1}$ later then t :

$$P(n,t) = \int_0^t d\tau_1\ \psi(\tau_1) \int_0^{t-\tau_1} d\tau_2\ \psi(\tau_2) \ldots \int_0^{t-\tau_1-\ldots-\tau_{n-1}} d\tau_n\ \psi(\tau_n) \int_{t-\tau_1-\ldots-\tau_n}^{\infty} d\tau_{n+1}\ \psi(\tau_{n+1}) \tag{44}$$

By Laplace transformation, one finds :

$$\begin{aligned} \tilde{P}(n,s) &= \int_0^\infty dt\ e^{-st} P(n,t) \\ &= \tilde{\psi}^n(s)\ \frac{1-\tilde{\psi}(s)}{s} \end{aligned} \tag{45}$$

For the Laplace transform af the average number of points, one finds :

$$<\tilde{n}(s)> = \int_0^\infty dt\ e^{-st} <n(t)> = \frac{\tilde{\psi}(s)}{[1-\tilde{\psi}(s)]s} \tag{46}$$

A simple argument to estimate the limiting behavior of $\tilde{\psi}(s)$ for $s \to 0$ goes as follows. Consider the recursive equation (43) for m = 1, in the limit of small s-values :

$$\tilde{\psi}(s) = \frac{1}{2}\ \tilde{\psi}(2s) + \frac{1}{2}\ \exp(-2s) \approx \frac{1}{2}\ \tilde{\psi}(2s) + \frac{1}{2} - s \tag{47}$$

An exact solution of the approximate recursive relation is :

$$\tilde{\psi}(s) = 1 + s\frac{\ell ns}{\ell n2} \tag{48}$$

The general solution of (47) is then equal to the particular solution (48) plus the general solution of the homogenous equation $\tilde{f}(s) = \frac{1}{2}\ \tilde{f}(2s)$. One easily verifies that $\tilde{f}(s) = O(s)$ for s small, so that (48) gives the leading behavior. Inserting this result in (46), one finds :

$$<\tilde{n}(s)> \approx \frac{\ell n2}{s^2\ \ell ns}$$

This result can be verified in a more explicit way by studying the Mellin transform of $\tilde{\psi}(s)$. The corresponding long-time behavior of $<n(t)>$ reads :

$$<n(t)> \approx \frac{t}{\log_2 t} \tag{49}$$

which is very similar to the result of Feller (with the interpretation that "time is money", $<n(t)>$ gives the average number of trials before the accumulated gain exceeds t). (49) predicts that the jumps of the random walk become sparser as a larger time interval is considered, and, incidentally, in the same way as the number of prime numbers becomes sparser [28].

For an object, possessing dilation rather than translation symmetry, a new so-called fractal (or Haussdorf) dimension d_F can be defined as follows : the (average) number of points n(R), inside a sphere of radius R, centered at a point of the fractal, obeys the following relation :

$$d_F = \lim_{R\to\infty} \frac{\ell n\ n(R)}{\ell n\ R} \tag{50}$$

For systems with translation symmetry, one has $N(R) \sim R^d$, and d_F is equal to the Euclidean dimension d. If we generate a series of points on the time-axis on the basis of the waiting distribution (43), one finds that the fractal dimension of this set of points is still equal to one (according to the definition). The Lorentzian distribution :

$$\psi(\tau) = \frac{2\tau_o}{\pi(\tau^2+\tau_o^2)} \tag{51}$$

has a similar divergence of $\bar{\tau}$ and the asymptotic result (48) is also obtained (with much less effort). These distributions are at the borderline between nonfractal and fractal behavior. Truly fractal behavior, with $0 < d_F < 1$, is generated, for example, by the following waiting time distribution :

$$\psi(\tau) = \sum_{n=0}^{\infty} a^n \ (1-a)\ \lambda^n \exp(-\lambda^n\tau)$$

$$\langle n(t)\rangle \sim t^{d_F} \quad ; \quad d_F = \frac{\ell na}{\ell n\lambda} \quad \text{for } \lambda < a < 1 \tag{52}$$

see [27], and by :

$$\psi(\tau) = \left(\frac{\tau_o}{\pi\tau^3}\right)^{\frac{1}{2}} \exp\left(-\frac{\tau_o}{\tau}\right) \tag{54}$$

$$\langle n(t)\rangle \sim t^{d_F} \quad ; \quad d_F = 1/2 \tag{55}$$

Suppose now that a particle is performing a nearest neighbour random walk on a one-dimensional regular lattice $m \in \mathbb{N}$. If it starts out at $m_o = 0$, and its waiting time is given by $\psi(\tau)$, after which it jumps with equal probability to the left or to the right, one finds that :

$$\langle m(t)\rangle = 0 \tag{56}$$

and, for large times, that :

$$\langle m^2(t)\rangle \sim \langle n(t)\rangle \tag{57}$$

In all the considered cases of waiting-time distributions, (42), (51), (52) and (54) one finds sub-diffusive behavior. This anomalous behavior of course results from the possibility of extremely long residence times at a given site, and it will arise in all the other properties of the walk (range, first passage times, etc.).

The property of having an infinite waiting time may, at first, seem surprising or even unphysical. Let me give four arguments why this is not so. Firstly, the waiting time τ is a mathematical object, describing, for example, the escape of a thermally agitated particle out of a potential well with barrier hight U. In an amorphous system, it is expected that U is a random variable. The properties of τ then follow from the Kramers or Arrhenius result :

$$\tau^{-1} \sim \exp(-U/k_BT) \tag{58}$$

It follows that the divergence of $\bar{\tau}$ is tantamount to the divergence of $\overline{\exp(U/k_B T)}$, which is a relatively weak condition on P(U). Secondly, if the waiting time corresponds to a return to the origin on a regular lattice random walk, then it is known that this return occurs with probability one in d = 1 and d = 2, although the mean time until return is infinite. Thirdly, dilation symmetry is effectively realized in many physical processes, at least in a very broad window of time and length scales [29]. Finally, an outstanding example of a physical process to which the above notions have been successfully applied is conductivity and anomalous transport in amorphous semi-conductors [30, 31, 32]. In this study, all the effects of spatial and energetic disorder were lumped into a waiting time distribution for the charge carrier which was further assumed to be hopping on a regular lattice. In order to describe the experimental results, it was necessary that $\bar{\tau}$ be infinite.

Similar notions for fractal properties of the random walk in space can be developed (see e.g. [27]). These notions in fact go back to the seminal work of Levy, who had the name, amongst the mathematicians of this time, to be like a visitor from a strange planet (see quotation in [29]). For example, at a time where everybody was looking for the conditions for the validity of the central limit theorem, he found simple exceptions to it. These exceptions, the Levy stable distributions, are intimately linked with fractal properties [29].

Apart from random walks, with fractal time or space properties, an interesting new field is that of random walks on a fractal object. Indeed, many objects have (or are believed to have) a fractal structure (e.g. the random walk itself, cf. polymer physics [21], the infinite cluster at the percolation threshold [33], strange attractors in dissipative dynamical systems [34], porous media, etc.). The study of "random" fractals is quiet complicated (see e.g. [35]). A popular example of a "deterministic" fractal, for which a lot of analytic results can be obtained [36] - [42], is the Sierpinsky gasket, whose fractal dimension is $d_F = \ell n3/\ell n2$. To study the spectral properties of a walk on this lattice, one can use a very elegant renormalization procedure, see especially [38]. The spectral density is very peculiar. It consists of a superposition of a pure point spectrum, with corresponding localized eigenstates and a continuous spectrum with a Cantor set as fractal support. A similar renormalization procedure can also be used to elucidate some of the dynamical properties [39]. The spectral properties and types of eigenfunctions are of particular interest when considering the mechanical or quantum-mechanical analogue of the Sierpinsky gasket (cf. §2). The results for the spectrum can also be used to calculate, for example, the probability for return to the origin, cf. equation (21) and reference [43] for a recent review.

ACKNOWLEDGEMENTS

We thank Profs V. Balakrishnan, R.M. Mazo and J. Piasecki for helpful discussions and I. Claes for checking the results of Table I.
We thank the N.F.W.O. Belgium and the I.U.A.P. Belgium for financial support.

TABLE I

DEFINITION

Probability/Unit time

$W(m+1|m) = W_{\rightarrow} = W(1+g)$

$W(m|m+1) = W_{\leftarrow} = W(1-g)$

$f^2 = \frac{1+g}{1-g} = \frac{W_{\rightarrow}}{W_{\leftarrow}} = e^{2\beta}$

$\tilde{P}(m|m_o,s) = \int_o^\infty e^{-st} P(m|m_o,t)\, dt$

$m_>$ and $m_<$ is the largest and smallest of m and m_o, respectively.

with bias ($g \neq 0$), $\cosh \xi = (1+s/2W)/(1-g^2)^{1/2}$;

Reflecting boundaries at one and N : $W(0|1) = W(N+1|N) = 0$

$$\frac{f^{(m-m_o)}}{s \sinh\xi \sinh(N\xi)}\{\sinh[(N-m_>+1)\xi]-f\sinh[(N-m_>)\xi]\}\times\{\sinh[m_<\xi]-f^{-1}\sinh[(m_<-1)\xi]\}$$

Absorbing boundaries at zero and N + 1 : $W(1|0) = W(N|N+1) = 0$

$$\frac{f^{(m-m_o)}(f + f^{-1}) \sinh(m_<\xi) \sinh[(N + 1 - m_>)\xi]}{2W \sinh \xi \sinh[(N + 1)\xi]}$$

Periodic boundaries : $W(N|1) = W_{\leftarrow}$, $W(1|N) = W_{\rightarrow}$

$$\frac{f^{(m-m_o)}(f + f^{-1})\{\sinh[\xi(N-m_>+m_<)]-\sinh(N\xi)\sinh[\xi(m-m_o)]+\cosh(N\beta)\sinh[\xi(m_>-m_<)]\}}{4W \sinh \xi [\cosh(N\xi) - \cosh(N\beta)]}$$

Reflecting at one, absorbing at N + 1, $W(0|1) = 0$, $W(N|N+1) = 0$

$$\frac{f^{(m-m_o)}(f + f^{-1}) \sinh[(N+1 - m_>)\xi] \{f \sinh(m_<\xi) - \sinh[(m_< - 1)\xi]\}}{2W \sinh\xi \{f \sinh[(N+1)\xi] - \sinh(N\xi)\}}$$

Half-infinite line m > 0, reflecting at one, $W(0|1) = 0$

$$f^{(m-m_o)} e^{-m_>\xi} \frac{(e^\xi - f)}{s \sinh \xi} \{\sinh(m_<\xi) - f^{-1} \sinh[(m_< - 1)\xi]\}$$

Half-infinite line m > 0, absorbing at zero, $W(1|0) = 0$

$$\frac{f^{(m-m_o)}(f + f^{-1})e^{-m_>\xi} \sinh(m_<\xi)}{2W \sinh \xi}$$

Infinite line $m \in \mathbb{N}$

$$\frac{f^{(m-m_o)}(f + f^{-1})e^{-\xi(m_>-m_<)}}{4W \sinh \xi}$$

REFERENCES

1. E.W. Montroll and G.H. Weiss, J. Math. Phys. 6, 167 (1965).
2. F. Spitzer, Principles of a Random Walk (Van Nostrand, Princeton, NJ, 1964).
3. M.N. Barber and B.W. Ninham, Random and Restricted Walks, Theory and Applications (Gordon and Breach, N.Y., 1970).
4. N.S. Goel and N. Richter-Dyn, Stochastic Models in Biology (Academic Press, N.Y., 1974).
5. S. Alexander, J. Bernasconi, W.R. Schneider and R. Orbach, Rev. Mod. Phys. 53, 175 (1981).
6. N.G. Van Kampen, Stochastic Processes in Physics and Chemistry, (North Holland, Amsterdam, 1981).
7. G.H. Weiss and R.J. Rubin, Adv. Chem. Phys. 52, 363 (1983).
8. M.E. Fisher, J. Stat. Phys. 34, 667 (1983).
9. B.D. Hughes and R. Prager, Lect. Not. Math. 1035, 1 (1983).
10. E.W. Montroll and M.F. Shlesinger, A Wonderful World of Random Walks, in CCNY Physics Symposium : M. Lax Sixtieth Birhtday, Ed. H. Falk (City College of New York Physics Department , New York, 1983).
11. J.W. Haus and K.W. Kehr, Phys. Rep. 150, 264 (1987).
12. J.V. José, in : Stochastic Processes Applied to Physics and other Related Fields, B. Gomez, S.M. Moore, A.M. Rodriguez-Vargas, A. Reuda, Eds. (World Scientific, 1983).
13. R. Graham and A. Schenzle, Phys. Rev. A26, 1676 (1982).
14. G. Nicolis and J.W. Turner, Physica 89A, 326 (1977).
15. A.J.F. Siegert, Phys. Rev. 81, 617 (1951).
16. a) M. Khanta and V. Balakrishnan, Phys. Rev. B29, 4679 (1984).
    b) R.M. Mazo and C. Van den Broeck, J. Chem. Phys. 86, 454 (1986).
17. G.H. Weiss, Adv. Chem. Phys. 13, 1 (1966); V. Seshadri, B.J. West and K. Lindenberg, J. Chem. Phys. 72, 1145 (1980).
18. K. Lindenberg, K.E. Shuler, J. Freeman and T.J. Lie, J. Stat. Phys. 12, 217 (1975).
19. C. Van den Broeck and M. Bouten, J. Stat. Phys. 45, 1031 (1986).
20. C. Van den Broeck and V. Balakrishnan, unpublished.
21. W. Horsthemke and R. Lefever, Noise Induced Transitions (Springer, New York, 1983), see also A. Rodriguez, L. Pesquèra, M. San Miguel and J.M. Sancho, J. Stat. Phys. 40, 669 (1985).
22. M.A. Rodriguez, M. San Miguel and J.M. Sancho, Ann. Nucl. Energy 10, 263 (1983).
23. For a review, see P.A. Lee and T.V. Ramakrishnan, Rev. Mod. Phys. 57, 287 (1985).
24. B. Derrida, J. Stat. Phys. 31, 433 (1982) see also : R. Zwanzig, J. Stat. Phys. 28, 127 (1982).
25. B.D. Hughes and M. Sahimi, J. Stat. Phys. 29, 781 (1982); B.D. Hughes, M. Sahimi and T. Davis, Physica 120A, 515 (1983).

26. W. Feller, An Introduction to Probability Theory and its Applications (New York, Wiley, 1966).

27. M.F. Shlesinger and E.W. Montroll, Lect. Not. Math. 1035, 138 (1983).

28. A.E. Ingham, The distribution of prime numbers, (Cambridge, 1983).

29. B.B. Mandelbrot, The fractal geometry of nature (W.W. Freeman, San Francisco, 1982).

30. H. Scher and M. Lax, Phys. Rev. B7, 4491, (1973).

31. H. Scher and E.W. Montroll, Phys. Rev. B12, 2455 (1975).

32. M. Shlesinger, J. Stat. Phys. 10, 421 (1974).

33. D. Stauffer, Phys. Rep. 54, 1 (1979).

34. A.J. Lichtenberg and M.A. Lieberman, Appl. Math. Sci. 38, (1983).

35. Y. Meir and A. Aharony, Phys. Rev. A37, 596 (1988).

36. S. Alexander and R. Orbach, J. Phys. Lett. 43, L625 (1982).

37. R. Rammal and G. Toulouse, Phys. Rev. lett. 49, 1194 (1982).

38. E. Domany, S. Alexander, D. Bensimon and L.P. Kadanoff, Phys. Rev. B28, 3110 (1983).

39. R. Rammal, J. Phys. 45, 191 (1984).

40. O'Shaugnessy and I. Procaccia, Phys. Rev. Lett. 54, 455 (1985).Phys. Rev. A32, 3073 (1985).

41. A. Blumen, G. Zumofen and J. Klafter, Phys. Rev. B28, 6112 (1983).

42. S. Havlin and D. Ben-Avraham, Adv. Phys. 36, 695 (1987).

43. T. Schneider, A. Politi and M.P. Sörensen, Phys. Rev. A37, 948 (1988).

STOCHASTIC BIFURCATIONS IN A GENERIC DYNAMICAL SYSTEM: A QUALITATIVE ANALYSIS

F.J.de la Rubia (#) and W.Kliemann (*)

(#)Department of Chemistry,B-040 and
Institute for Nonlinear Science
University of California,San Diego
La Jolla,CA 92093, U.S.A.

(*)Department of Mathematics
Iowa State University
Ames,Iowa 50011, U.S.A.

INTRODUCTION

One of the main problems in the analysis of nonlinear dynamical systems is the study of the number and stability of the stationary solutions. In a first approach one assumes deterministic conditions, such that all the parameters involved in the problem have fixed values. If this is the case, bifurcation theory[1-3] provides the necessary mathematical tools to handle the problem. In many practical situations, however, the restriction of constant parameters is difficult to maintain, and one would like to be able to extend the precise results of bifurcation theory to, for instance, the case in which the bifurcation parameter fluctuates around a well defined mean value.

While the concepts and methods of deterministic bifurcation theory are well established and accepted, the situation is not the same in the stochastic case. Different authors use different methods and concepts to define stochastic bifurcation[4-7].

In this work we use recently developed methods for the qualitative study of stochastic systems[8,9] to give precise results on the stability of the stationary solutions of a generic nonlinear model with stochastic control parameter. We show that, when the stochastic bifurcation parameter satisfies certain conditions, in particular the requirement of being of bounded variation, we can draw a modified stochastic bifurcation diagram that gives the exact long-time behavior of the system in a completely similar way as in the deterministic case.

Since some of the mathematical concepts required to make the theory rigourous are quite involved, we concentrate on more intuitive arguments, using as much as possible ideas well known from the underlying deterministic system to extract useful information about the behavior of our more general stochastic model.

THE DETERMINISTIC MODEL

Let us consider a generic nonlinear deterministic system whose evolution is given by the following ordinary differential equation

$$\dot{x} = f(x,q) = -\frac{1}{2}x^3 + bx^2 + c(q-q_c)x \tag{1}$$

where $\dot{x} = dx/dt$, the parameters b, c, q_c are nonnegative, and q is the bifurcation parameter. To analyze the long time behavior of (1) we have to:
(i) Find the steady states, i.e., the values of x such that $\dot{x} = 0$.
(ii) Analyze the stability of the steady states.
(iii) Study the change of stability of the steady states as a function of q.

For system (1) the steady states are given by

$$x_0 = 0 \tag{2.1}$$

$$x_1(q) = b + \sqrt{b^2 + 2c(q-q_c)} \tag{2.2}$$

$$x_2(q) = b - \sqrt{b^2 + 2c(q-q_c)} \tag{2.3}$$

Obviously, (2.1) is a steady state for all parameter values, whereas (2.2) and (2.3) are physically meaningful states only if

$$q \geq q_{hc} \equiv q_c - \frac{b^2}{2c} \tag{3}$$

To analyze the stability of the steady states (2) we linearize (1) around these states and find:
x_0 is globally asymptotically stable for all $x > 0$ if $q < q_{hc}$; asymptotically stable for $x < x_2$ if $q_{hc} \leq q \leq q_c$; and unstable if $q > q_c$.
x_1 is asymptotically stable for $x > x_2$ if $q_{hc} \leq q \leq q_c$ and for all $x > 0$ if $q > q_c$.
x_2 is unstable if $q_{hc} \leq q \leq q_c$, and asymptotically stable for $x < 0$ if $q > q_c$.

Figure 1 summarizes this behavior for b= 2, c= 5 and q_c= 0.6, showing a region of bistability for values of q in the interval [q_{hc} , q_c]. This bifurcation diagram is known in the literature as an imperfect pitchfork bifurcation[2], since we can obtain the familiar symmetrical pitchfork by putting b= 0 in (1).

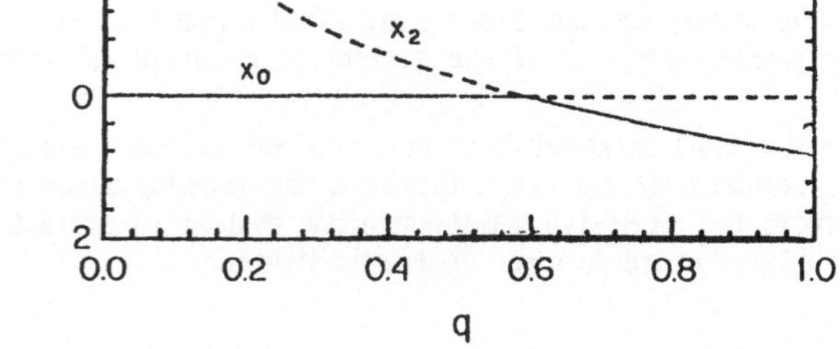

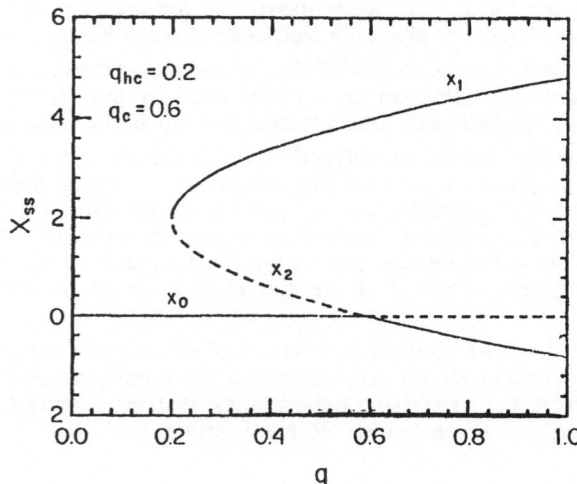

Fig. 1. Bifurcation diagram of system (1) for b= 2, c= 5 and q_c= 0.6. With these values we have q_{hc}= 0.2. Solid (broken) lines indicate stable (unstable) states.

THE STOCHASTIC MODEL

Suppose now that the bifurcation parameter is a stochastic process q_t. It is customary to assume that the noise is gaussian (colored or not) but this implies that it may take values in an infinite range. In a practical situation, however, one never has parameters of unbounded variation, and with this in mind we consider q_t as a noisy parameter taking values in a finite interval $[\alpha,\beta]$. As will become clear soon, this requirement is also necessary to obtain our results on stability. We therefore take q_t as a bounded diffusion process, with values in an interval $[\alpha,\beta]$ and satisfying the Stratonovich stochastic differential equation[10] (Ssde)

$$dq_t = a(q_t)\,dt + \sigma(q_t)\,dW \tag{4}$$

where W is the Wiener process (Brownian motion), $a(\cdot)$ and $\sigma(\cdot)$ are general continuous functions that for our present analysis need not be specified, and q_t lives in an interval

$$\alpha \leq q_t \leq \beta \qquad \text{for all } t \tag{5}$$

With these assumptions, the pair process (x_t,q_t) satisfying equations (1) and (4) is a Markov process with joint probability distribution $P(x,q;t)$ such that the marginal distribution $p(x;t)$ for x alone is obtained by integrating over q.

Our aim is to study the asymptotic ($t\to\infty$) behavior of x_t depending on the values of α and β, and to redraw the bifurcation diagram as a function of the mean value of the stochastic bifurcation parameter. To do this we will use results of the qualitative theory of stochastic systems[8,9]. Without going into the details what we do is to study the behavior of the *stochastic* system $\dot{x} = f(x_t,q_t)$ by using the control theory of an associated *deterministic* system $\dot{x} = f(x,u)$ in order to find invariant control sets such that once the system gets into them it remains inside (see Ref.[9] for details). As stated in the introduction, we concentrate on intuitive arguments, using the vector field associated with the deterministic system (1) to analyze the qualitative behavior of our stochastic model depending upon the different locations of the interval of variation of the noise (Fig.2).

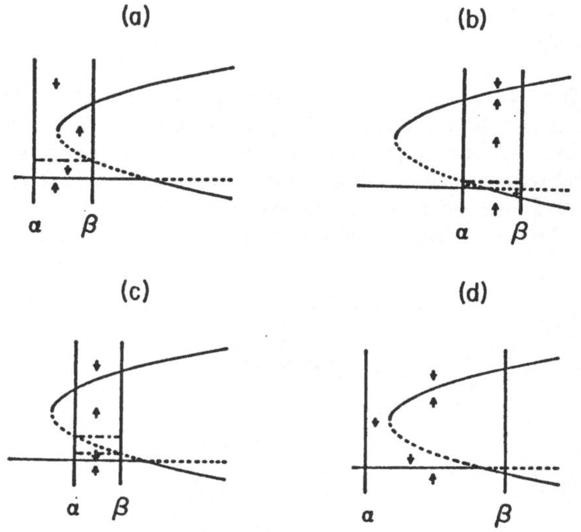

Fig. 2. Different situations considered in the text depending on the location of the interval of variation of the noise q_t. The arrows indicate the vector field associated with the deterministic system (1).

In the following we analyze in some detail the behavior of the system in the much more interesting upper part of the bifurcation diagram, i.e., for values of $x > 0$, although the behavior in the lower part will also be contained in our final stochastic bifurcation diagram.

We consider the following cases:

(A): $\alpha < q_{hc}$ and $q_{hc} < \beta < q_c$

This is case (a) of Fig.2. Define the points $A_1 = x_2(\beta)$, $A_2 = x_1(\beta)$ where the functions x_1 and x_2 are defined in (2.2) and (2.3) respectively. We have

$$0 = x_2(q_c) < A_1 < x_1(q_{hc}) = b \qquad (6.1)$$

$$b = x_2(q_{hc}) < A_2 < x_1(q_c) = 2b \qquad (6.2)$$

Then[9], **D** = $(A_1 , A_2]$ is the unique control set in this case, but it is clearly not invariant since the flow drives the system outside of this domain with probability one (w.p.1). It is on the other hand obvious that when the system crosses A_1 it can not come back. Therefore, we have that, for *all initial conditions* $x(0) > 0$,

$$\text{Prob} \left[\lim_{t \to \infty} x_t[x(0)] = 0 \right] = 1 \qquad (7)$$

and the steady state $x_0 = 0$ is globally asymptotically stable w.p.1.

(B): $q_{hc} \leq \alpha < q_c$ and $\beta \geq q_c$

This is case (b) in Fig.2. Now we define $A_1 = x_2(\alpha)$, $A_2 = x_1(\alpha)$, $A_3 = x_1(\beta)$ with

$$0 < A_1 < b < A_2 < 2b \leq A_3 \qquad (8)$$

We find that[9] **D** = $(0 , A_1)$ is an open, hence variant control set but **C** = $[A_2 , A_3]$ is the invariant set we are looking for. Using the same vector field arguments as in the first case we easily see that for *all initial conditions* $x(0) > 0$ the system is going to enter, w.p. 1, the domain **C** where it will remain from then on (as a matter of fact it is enough for the system to cross just once the point A_1). Once inside that domain the stationary behavior of x_t will be determined by the stationary marginal probability distribution $p_s(x) = p(x;t \to \infty)$.

In conclusion, for all initial conditions $x(0) > 0$, and w.p.1

$$x_t[x(0)] \to \text{(in distribution) } p_s(x) \text{ in } \mathbf{C} \qquad (9)$$

(C): $\alpha \geq q_{hc}$ and $\beta < q_c$

Since we always consider $\alpha < \beta$, this is case (c) of Fig.2. Again, define $A_1 = x_2(\beta)$, $A_2 = x_2(\alpha)$, $A_3 = x_1(\alpha)$, $A_4 = x_1(\beta)$ with

$$0 < A_1 < A_2 \leq b \leq A_3 < A_4 < 2b \qquad (10)$$

Now we have the following sets: $\mathbf{D} = (A_1, A_2)$ is an open, variant set and $\mathbf{C} = [A_3, A_4]$ is the compact invariant set. Furthermore, the sets $\mathbf{B}_1 = (0, A_1]$, $\mathbf{B}_2 = [A_2, A_3]$ and $\mathbf{B}_3 = (A_4, \infty)$ consist of points without control properties, where the system, for all controls u, moves in the same direction, either upwards (in $\mathbf{B}_1$ and $\mathbf{B}_3$) or downwards (in $\mathbf{B}_2$).

In this case the long time behavior depends on the initial conditions in the following manner:
(i) $A_2 \leq x(0) < \infty$. With the same arguments as in the previous situation we have the result (9) above.
(ii) $0 \leq x(0) \leq A_1$. Then clearly we have convergence towards the stationary solution $x_0 = 0$, w.p.1.
(iii) $A_1 < x(0) < A_2$. There is a probability of absorption $s[x(0)]$ such that

$$\text{Prob}[\lim_{t\to\infty} x_t[x(0)] = 0] = s[x(0)] \tag{11}$$

and therefore

$$x_t[x(0)] \to s[x(0)]\,\delta(x) + (1 - s[x(0)])\,p_s(x) \tag{12}$$

where $\delta(x)$ is the Dirac delta function and the convergence is in distribution, as usual.

It is important to realize that although at a qualitative level the probability $s[x(0)]$ must always be a decreasing (from 1) function of the initial condition, its actual shape, e.g., how fast it decreases and in which interval it differs from one or zero, depends strongly on the characteristics of the noise, i.e., the interval of variation, correlation time, variance, etc.

(D): $\alpha < q_{hc}$ and $\beta \geq q_c$

This is the final case (d) of Fig.2. Now $A_1 = x_1(\beta)$, $A_1 \geq 2b$ and $\mathbf{C} = (0, A_1]$ is the unique invariant control set but it is *not compact.* Unfortunately there is no general theory to work with this situation of non compact invariant control sets from the form of the vector fields f(x,u). The problem here is that both branches in the bifurcation diagram are conected, and the long time behavior will depend on the initial conditions and the characteristics of the noise. We will comment more on this later.

STOCHASTIC BIFURCATION DIAGRAM

In order to draw the stochastic bifurcation diagram using the mean value r of q_t as the new bifurcation parameter, we introduce the following notation

$$q_t = r + \xi_t \tag{13}$$

where ξ_t is a symmetrical zero-mean noise with values in the interval $[-\gamma, +\gamma]$. Then q_t is also symmetrical and $\alpha = r - \gamma$ and $\beta = r + \gamma$ define its interval of variation. According to the analysis of the previous section and depending upon the size of the interval, we can distinguish three situations:

(A): $\beta - \alpha < q_c - q_{hc}$

Then we know that if $\alpha < q_{hc}$ we have convergence towards the state $x_0 \equiv 0$ [see (7)]. However, the symmetry of the distribution of q_t and $\alpha < q_{hc}$ imply

$$r = \alpha + \gamma < q_{hc} + \gamma \tag{14}$$

and as a consequence, the point

$$A = q_{hc} + \gamma \tag{15}$$

may be considered as a bifurcation point for the stochastic system.

On the other hand, if $\beta > q_c$ the system converges towards the marginal distribution $p_s(x)$, i.e., we have the result (9), and now we get

$$r = \beta - \gamma \geq q_c - \gamma \tag{16}$$

and the point

$$B = q_c - \gamma \tag{17}$$

is a second bifurcation point.

With this information we can draw the stochastic bifurcation diagram that looks like shown in Fig.3. In that picture we can separate several regions: (i) in region **I** the stationary solution $x_0 \equiv 0$ is asymptotically stable w.p.1, i.e., for any initial condition (positive or negative) the system converges to that solution (note that for $x < 0$ the region of stability of the stationary solution x_0 extends to $q_c = 0.6$, whereas for $x > 0$ the solution $x_0 = 0$ loses its stability at $B = q_c - \gamma$); (ii) in region **II** the stationary solution $x_1(q_t)$ is asymptotically stable, in the sense that all systems starting in this area converge to the interior of the invariant set $C_r = [y_2(r), y_1(r)]$ where

$$y_1(r) = x_1(r+\gamma) \tag{17.1}$$

$$y_2(r) = x_1(r-\gamma) \tag{17.2}$$

and this set C_r is also the support of the invariant marginal distribution $p_s(x)$; (iii) in region **III**, and for negative values only, the stationary solution $x_2(q_t)$ is stable w.p.1, with the set $D_r = [z_2(r), z_1(r)]$ as the support of the invariant distribution. The boundaries of this set are given by

$$z_1(r) = x_2(r+\gamma) \tag{18.1}$$

$$z_2(r) = x_2(r-\gamma) \tag{18.2}$$

(iv) finally in region **IV** and for each $x(0)$ in this area, there is a probability $s[x(0)]$ such that x_0 is attractive with probability $s[x(0)]$ and $x_1(q_t)$ is attractive with probability $1 - s[x(0)]$. The curves z_1 and z_2 given by (18.1) and (18.2) indicate the boundaries of this region with regions **II** and **I** respectively. It is important to note that in this case the region of bistability for $x > 0$ is smaller than in the deterministic situation and that the bifurcation takes place at different values, whereas for $x < 0$ the diagram is pretty much the same as the deterministic one.

(B): $\beta - \alpha = q_c - q_{hc}$

As can be seen from the previous discussion, the larger γ the smaller the bistability region, and as soon as $\gamma = \gamma_c \equiv b^2/4c$ the two points A and B coincide. Figure 4 depicts the bifurcation diagram for this limit. The behavior in regions **I-III** and the meaning of the different curves is the same as explained earlier.

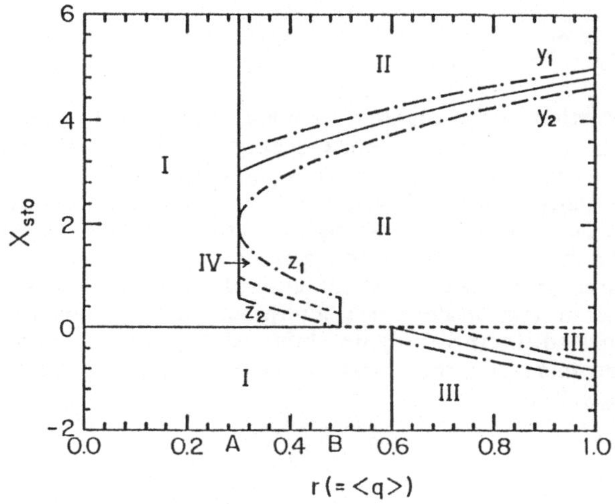

Fig. 3. Stochastic bifurcation diagram for the same values as in Fig.1. and $\gamma=0.1$, showing the regions with different asymptotic behaviors. Notice the region of bistability between A= $q_{hc}+\gamma= 0.3$ and B= $q_c-\gamma= 0.5$. Solid and broken lines indicate stable and unstable solutions respectively (the meaning of the dot-dashed lines is explained in the text).

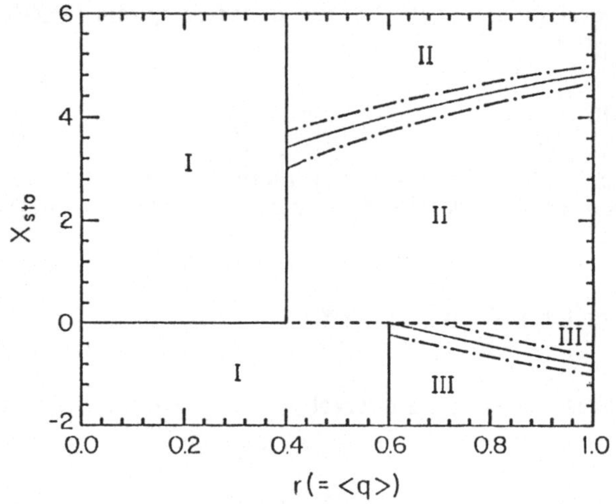

Fig. 4. Stochastic bifurcation diagram for the same values as in Fig.1 and $\gamma=\gamma_c=0.2$. There is no region of bistability and the unique bifurcation point is located at A= $q_c-\gamma=q_{hc}+\gamma= 0.4$.

The important point to note here is that the intermediate region **IV** disappears, and, therefore, there is no bistability region. As a consequence there is only one bifurcation point located at $A = q_{hc} + \gamma = q_c - \gamma$ for $x > 0$, and at q_c for $x < 0$.

(C): $\beta - \alpha > q_c - q_{hc}$

The only situation not covered previously is when $\alpha < q_{hc}$, $\beta > q_c$ and $q_{hc} \leq r \leq q_c$. In this case we do not know precisely how to draw the bifurcation diagram, but based on the preceding discussion we can argue the following:
(i) There is no bistability region and only one bifurcation point will be present. This is a consequence of the fact that $\gamma > \gamma_c$. Although we can not locate the bifurcation point we know that it must lie inside the interval $[q_{hc}, q_c]$.
(ii) If we further increase the interval of variation of the noise, a plausible limit would be that of white noise in the Stratonovich interpretation[10]. In this limit one can easily prove[11] that, as expected, there is only one bifurcation point located at q_c. Therefore we conjecture that upon increasing the length of the interval $[\alpha, \beta]$ the critical point should move from $(q_c + q_{hc})/2$ to q_c.

DISCUSSION

In the last two sections we have analyzed the changes in the bifurcation diagram when the bifurcation parameter is perturbed by a noise with bounded variation. We would like to stress that to obtain the quite detailed picture presented in Figs.3 and 4 we only used general arguments without resorting to any particular characteristic of the noise or of the deterministic system. This clearly indicates that the results reported here should remain valid for any system with a bifurcation diagram similar to the one depicted in Fig.1.

However, for these results to be of any practical use, the time taken for the system to actually bifurcate must not only be finite, but also relatively small, at least small enough to be in the accesible region for an experimentalist. To check this we have performed some numerical simulations[12,13] of the stochastic system (1),(4). For the noise in the simulations we write

$$q_t = r + \gamma \eta_t \tag{19}$$

where r (mean value) and γ (length of the interval of variation) are given parameters and η_t is a Pearson process described in detail in Ref.[14], obeying the stochastic differential equation

$$d\eta_t = -\frac{1}{\tau}\eta_t dt + D\sqrt{2(1-\eta_t^2)}\,dW \tag{20}$$

This process takes values in the interval [-1 , + 1] and has the following characteristics[14]

$$< \eta_t > = 0 \tag{21.1}$$

$$< \eta_t \eta_{t'} > = \left[\frac{D^2\tau}{1 + D^2\tau}\right] e^{-|t-t'|/\tau} \tag{21.2}$$

and the parameters D, τ, must fulfill the condition[14] $D^2\tau \leq 1/2$ in order to obtain a zero-mean stationary symmetrical distribution for η_t.

Figs. 5 and 6 depict the time-dependent solution of the set of equations (1),(19),(20) for different values of the parameters and initial conditions. Since the time scale of the deterministic system (γ=0) is of the order of two (in dimensionless units), the stochastic system takes longer to reach the final stationary state, but, nevertheless, the time is clearly finite and small enough to be observed in an actual experimental situation.

Fig. 5 is for the parameters b= 2, c= 5, q_c= 0.6, r= 0.24, γ= 0.12 and different values of τ (D is varied accordingly to keep the variance of η_t fixed). The initial condition is x(0)= 2, hence $x_1(r)< x(0)< x_2(r)$. With these values the system starts in region I of Fig. 3 but well inside the bistability region for the deterministic system (Fig.1). It should be stressed that Fig. 5 merely indicates that different trajectories take different times to bifurcate and that trajectory (a) bifurcates in a finite time as well.

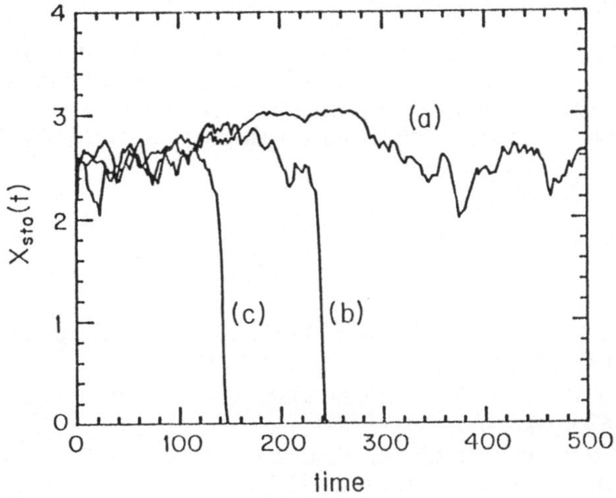

Fig. 5. Time evolution of the stochastic system for the parameter values and initial conditions indicated in the text and three different correlation times of the noise: (a) τ= 0.05;(b) τ= 0.1;(c) τ= 0.5

In Fig. 6 we have the same values for b, c, q_c and now r= 0.52, γ= 0.16, τ= 0.5, D= 0.5, x(0)= 0.1 ($x(0) < x_2(r)$). We are in region II of Fig.3 but within the basin of attraction of x_0= 0 for the deterministic system.

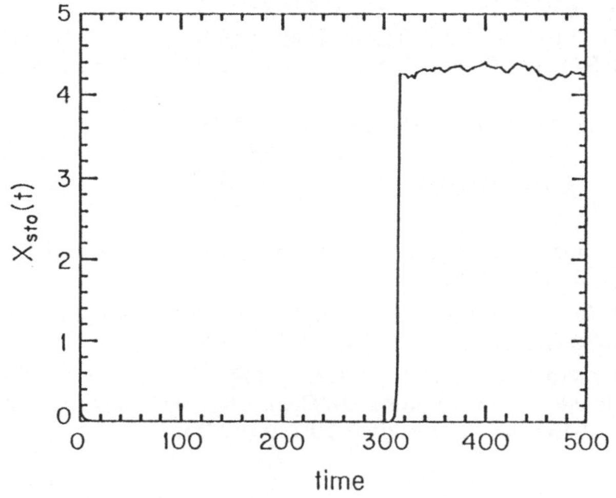

Fig. 6. Time evolution of the stochastic system for the parameter values and initial condition indicated in the text.

To summarize, in this work we have shown how the bifurcation diagram can be modified in the presence of external fluctuations. In particular we have seen that, with some restrictions on the stochastic forces, it is possible to provide precise results to characterize stochastic bifurcations. Since it is conceivable that some of the noises that are actually present in an experiment fulfill the conditions required in this work, we believe that our results may be of some interest in quite different fields, provided that the underlying deterministic model shows a bifurcation diagram similar to the one considered here.

ACKNOWLEDGMENTS

We thank Prof.K.Lindenberg and Dr.B.West for many stimulating discussions. This work has been partially supported by DOE grant DE-FG03-86ER13606. One of us (FJR) also acknowledges financial support by a grant of the Scientific Committee of the NATO organization and from Direccion General de Investigacion Cientifica y Tecnica (Spain), project PB-85-0024.

REFERENCES

1. J.Guckenheimer and P.Holmes, "Nonlinear Oscillations, Dynamical Systems and Bifurcation of Vector Fields", Springer-Verlag, New York(1983).
2. M.Golubitsky and D.G.Schaeffer, "Singularities and Groups in Bifurcation Theory, Vol.1", Springer-Verlag, New York(1985).
3. D.W.Jordan and P.Smith, "Nonlinear Ordinary Differential Equations (second edition)", Oxford University Press, Oxford(1987).
4. E.Knobloch and K.A.Wiesenfeld, Bifurcations in Fluctuating Systems:The Center-Manifold Approach, J.Stat.Phys. 33:611 (1983).
5. M.Lucke and F.Schank, Response to Parametric Modulation near an Instability, Phys.Rev.Lett. 54:1465 (1985).
6. L.Arnold, Lyapunov Exponents of Nonlinear Stochastic Systems, in: "Nonlinear Stochastic Dynamic Engineering Systems", G.I.Schueller and F.Ziegler,eds., Springer-Verlag, Berlin (1987).
7. C.Meunier and A.D.Verga, Noise and Bifurcations, J.Stat.Phys. 50:345 (1988).
8. W.Kliemann, Qualitative Theory of Stochastic Dynamical Systems: Aplications to Life Sciences, Bull.Math.Biology 45:483 (1983).
9. L.Arnold and W.Kliemann, Qualitative Theory of Stochastic Systems, in: "Probabilistic Analysis and Related Topics, Vol.3", A.T.Barucha-Reid, ed., Academic Press, New York(1983).
10. L.Arnold, "Stochastic Differential Equations: Theory and Applications", Wiley, New York(1974).
11. I.I.Gihman and A.V.Skorohod, "Stochastic Differential Equations", Springer-Verlag, Berlin(1972).
12. W.Rumelin, Numerical Treatment of Stochastic Differential Equations, SIAM J.Numer.Anal. 19:604 (1982).
13. J.M.Sancho,M.San Miguel,S.L.Katz and J.D.Gunton, Analytical and Numerical Studies of Multiplicative Noise, Phys.Rev. A26: 1589 (1982).
14. F.J.de la Rubia and M.G.Velarde, White versus Non-White Stochastic Environment of Intrinsically Stochastic Systems: A Computational Approach to the Evolutionary Problem, J.Non-Equilib.Thermodyn. 10:219 (1985).

UNIFIED THEORY OF FLUCTUATIONS AND PARAMETRIC NOISE

M. San Miguel[+] and M.A. Rodríguez[*]

+ Dpto. de Física, Univ. de las Islas Baleares, E-07071 Palma de Mallorca, Spain
* Dpto. de Física Moderna, Univ. de Cantabria, E-39005 Santander, Spain

1.- INTRODUCTION

Generally speaking a system composed of many particles which are being created and annihilated has an intrinsic stochasticity associate with birth and death processes. Fluctuations associated with this process are usually called internal fluctuations. They are described by a master equation or probability balance equation[1]. In the context of stochastic nuclear reactor models the birth and death process is the stochastic fission process with associated emission and absorption probabilities per unit time[2]. In the limit in which the number of particles N goes to infinity or system size goes to infinity (thermodynamic limit) internal fluctuations become negligeable. In this limit the master equation leads to a deterministic or rate equation. For nuclear reactors this is the point kinetic reactor equation.

In addition of the internal fluctuations it is generally necessary to consider parametric or external noise due to the fact that the control parameters of the system have not well defined values but rather they are fluctuating quantities. An example of parametric noise in nuclear reactors is given by fluctuations of the reactivity parameter. The usual way to describe parametric noise is through stochastic differential equations. Such equations are obtained from the deterministic rate equations replacing the control parameters by stochastic processes. This procedure leads to stochastic reactor kinetic equations[2]. It is important to note that in this usual description of parametric noise, internal fluctuations are completely neglected.

We are then in the situation in which well known techniques exist for the separate description of internal fluctuations and parametric noise. The question we wish to address here is the possibility of having a framework in which internal fluctuations and parametric noise are considered simultaneously. The possible joint treatment depends, of course, on the level of description. Keeping the description at the phenomenological level used for the separate description of internal and parametric noise we propose[3,4,5] a unified theory of fluctuations based on a stochastic master equation (SME). Such equation is obtained introducing parametric noise sources in the original master equation describing the stochastic fission process. The transition probabilities for unit time of the master equation become stochastic quantities and the probability P(N,t) of having N neutrons at time t becomes a functional of the noise sources. Averaging the SME over the parametric noise sources one generally obtains an effective master equation which gives the desired joint description of internal and parametric noises. The general framework of the methodology can be summarized in the following equation: Internal fluctuations are described by a birth and death master equation.

$$\partial_t P(N,t) = W_+(N-1)\, P(N-1,t) + W_-(N+1)\, P(N+1,t) - (W_+(N) + W_-(N))\, P(N,t) \tag{1}$$

where $W_\pm(N)$ are respectively, the birth and death transition probabilities per unit time. They are assumed to be proportional to the system size V,

$$W_+(N) = V\, q(x) \qquad W_-(N) = V\, r(x) \tag{2}$$

where x=N/v. In the limit $V \to \infty$, eq(1) becomes the deterministic rate equation

$$\partial_t x = q(x) - r(x) \tag{3}$$

The functions q(x) and r(x) have a general form

$$q(x) = q_0^c(x) + \alpha\, q_1(x) \tag{4}$$

$$r(x) = r_0^c(x) + \beta\, r_1(x) \tag{5}$$

where α and β are the relevant control parameters. Parametric noise is described in the limit $V \to \infty$ by replacing α by $\alpha + \xi_\alpha(t)$ and β by $\beta + \xi_\beta(t)$ in (3), where ξ_α and ξ_β are the parametric noise sources. These replacements lead to a stochastic differential equation

$$\partial_t x = q_c(x) - r_c(x) + \xi_\alpha(t)\, q_1(x) - \xi_R(t)\, r_1(x) \tag{6}$$

The joint description of internal and parametric noise is given by the effective master equation obtained making the replacement of α and β by stochastic quantities in (1) instead of doing it in (3)

$$\partial_t P(N,t) = \Big[(E^{-1}-1)\, W_+^c(N) + (E^{+}-1)\, W_-^c(N) + (E^{-1}-1)\, W_+^1(N)\, \xi_\alpha(t) + (E^{+1}-1)\, W_-^1(N)\, \xi_R(t) \Big] P(N,t) \tag{7}$$

where $E^{\pm} = e^{\pm \partial_N}$ and $W_\pm^1$ are associated respectively with $q_1(x)$ and $r_1(x)$.

In Sections 2 and 3 of this paper we give some results of this scheme for the joint treatment of internal and parametric noise in two stochastic point nuclear reactor models. For clarity of presentation we consider in Sect. 2 a simple point reactor model which incorporates fluctuations of the fission rate, capture rate and source-event rate. In this model delayed neutrons are neglected and only two neutrons are emitted in any fission event. Results for the counting process are given in Sect. 3 for a more general model which includes delayed neutrons and a detector. There are three main novel physical effects that appear in the joint treatment of fluctuations presented here. The first one is the occurrence of "crossed-fluctuation" terms which couple internal and parametric noise. These terms give contributions to the neutron variance which are not obtained without a joint treatment of fluctuations. Second is that this "crossed-fluctuation" terms permit to distinguish between reactivity fluctuations due to fission-rate fluctuations and those due to capture-rate fluctuations. Thirdly, our results for the counting process show that information on parametric multiplicative noise can be obtained from measurements associated with the counting process and the stationary neutron correlation function. Hence, it is possible to extend the standard counting methods used in internal noise measurements to more general situations including measurement of external noise.

As a separate matter of methodological interest we discuss in Sect. 4 difficulties and possible pitfalls that appear in other apparently natural approaches to the problem of a joint description of internal and parametric noise. In particular we show that it is not a valid procedure to introduce parametric noise in the Langevin approximation to the original master equation which appear in the formalism of equivalent noise sources[6].

2.- SIMPLE POINT REACTOR MODEL

In order to gain insight in our ideas explained above, we apply them in this section to the most simple model of stochastic nuclear reactor. This model takes into account the fission process by considering that two neutrons are liberated with probability one. The absorption and source processes are taken as in the standard formulation. Hence, if Λ and λ are respectively the fission and absorption probability rates per neutron and s is the source probability rate per volume V, the probability balance equation for this model reads

$$\begin{aligned} \partial_t P(N,t) = & \left[\Lambda(N-1) + sV\right] P(N-1,t) + \\ & + \lambda(N+1)\, P(N+1,t) - \\ & - \left[(\Lambda+\lambda)N + sV\right] P(N,t) \end{aligned} \tag{8}$$

where P(N,t) is the probability of finding N neutrons at time t. Comparing this equation with the more general one considered in the introduction (1) it is possible to identify the creation term

$$W^{+}(N) = \Lambda(N-1) + Vs$$

and the destruction one

$$w_-(N) = \lambda (N+1)$$

The first includes the fission and independent source contributions and the second only the neutron capture proccess.

We obtain the usual deterministic point kinetic equation by taking the thermodynamic limit $V \to \infty$ in the equation (8). Defining the density of neutrons $n = N/V$ and taking this limit we obtain

$$\lim_{V \to \infty} V P(N,t) = \delta(N/V - n(t)) \quad (9)$$

with $n(t)$ obeying the above mentioned kinetic equation

$$\partial_t n(t) = (\Lambda - \lambda) n + s = \frac{\rho}{\ell} n + s \quad (10)$$

where we identify $(\Lambda - \lambda)$ with the ratio between the reactivity ρ and the neutron fifetime ℓ.

At this point we introduce the external noise in the parameter of fision absorption and source ξ_Λ, ξ_λ and ξ_s .

$$\Lambda \Rightarrow \Lambda + \xi_\Lambda(t) \quad ; \quad \lambda \Rightarrow \lambda + \xi_\lambda(t) \quad ; \quad s \Rightarrow s + \xi_s(t) \quad (11)$$

We remark that in standard theories[7] these parametric noise sources are introduced at the level of the deterministic equation (10) obtaining a stochastic differential equation as

$$\partial_t n(t) = (\Lambda - \lambda) n + s + (\xi_\Lambda(t) - \xi_\lambda(t)) n + \xi_s(t) \quad (12)$$

which completely neglects the internal fluctuations. In our theory we introduce the external noise at the level of the master equation (8) in such a manner that both, the internal and external fluctuations are taken into account in a natural way. The explicit calculation of statistical properties within this theory can be done using standard methods, with no additional complications. Let us introduce the generating function as usually:

$$G(x,t) = \sum_{n=0}^{\infty} x^N P(N,t) \quad (13)$$

From the master equation (8) with stochastic rates (11) we obtain

$$\begin{aligned} \partial_t G(x,t) = {} & \left[\lambda(1-x) + \Lambda x(x-1)\right] \partial_x G(x,t) + \\ & + sV(x-1) G(x,t) + \left[\xi_\lambda(t)(1-x) + \xi_\Lambda(t) x(x-1)\right] \partial_x G(x,t) \quad (14) \\ & + \xi_s(t) V(x-1) G(x,t) \end{aligned}$$

Now the generating function becomes a functional of the noise sources $G(x,t;[\xi_\lambda, \xi_\mu, \xi_s])$. (In the following we omit this explicit notation for simplicity). The average of $G(x,t)$ over noise sources is denoted by $\overline{G}(x,t)$. It gives an equivalent generating function which incorporates the effects of external fluctuations. We consider external Gaussian white noise for ξ_λ, ξ_μ and ξ_s with intensity D

$$\langle \xi_\alpha(t)\, \xi_\alpha(t')\rangle = D_\alpha\, \delta(t-t') \quad , \quad (\alpha = \mu, \lambda, s) \tag{15}$$

The average of (14) can be straight-forwardly done bearing in mind the functional formula valid for Gaussian white noise

$$\overline{\xi_\alpha\, G(x,t)} = D_\alpha \frac{\overline{\delta G(x,t)}}{\delta \xi_\alpha(t)} \tag{16}$$

where $\delta/\delta\xi(t)$ denotes a functional derivative. Operating in this way we obtain the averaged generating function

$$\begin{aligned} \partial_t \overline{G}(x,t) &= [\lambda(1-x) + \mu x(x-1)]\, \partial_x \overline{G}(x,t) + s v(x-1)\, \overline{G}(x,t) \\ &+ [D_\lambda (1-x)\, \partial_x (1-x)]\, \partial_x \overline{G}(x,t) \\ &+ D_\mu\, x(x-1)\, \partial_x\, x(x-1)\, \partial_x \overline{G}(x,t) \\ &+ D_s\, v^2 (x-1)^2\, \overline{G}(x,t) \end{aligned} \tag{17}$$

Equation (17) corresponds to a well defined master equation. This master equation has some interesting differences with respect to the original one (8). Indeed, in the master equation description of (17) we obtain terms with two step transitions[3] $W(N \rightarrow N+2)$ due to the average of the external white noise. We must remark that memory effects do not appear in (17) due to the white character of the noise. Such terms appear when considering non white noise[8].

The calculation of moments from (17) follows also in a standard way. Using the formula for the factorial moments $\phi_m = \langle N(N-1)\ldots(N-m+1)\rangle$

$$\phi_m = \left.\frac{\partial^m \overline{G}}{\partial x^m}\right|_{x=1}$$

it is inmediate to obtain statistical moments up to desired order. The two first stationary moments for the density $n = N/v$ are

$$\langle n\rangle = \frac{s}{\lambda-\mu} + \frac{s(D_\mu + D_\lambda)}{(\lambda-\mu)(\lambda-\mu-D_\lambda-D_\mu)} \tag{18}$$

$$\begin{aligned} \langle n^2\rangle &= \frac{D_s}{(\lambda-\mu+2D_\lambda+2D_\mu)} + \\ &+ \frac{s[s + \lambda/v + D_\mu/v - 2D_\lambda/v]}{[\lambda-(\mu+2D_\lambda+2D_\mu)][\lambda-(\mu+D_\lambda+D_\mu)]} \end{aligned} \tag{19}$$

In these formulas we have simoultaneously effects of external and internal fluctuations. For example we can identify changes on the stability of the system which is a genuine effect of multiplicative external noise. Conditions for stability now reads:

$$(\lambda - 1) > 2 (D_\lambda + D_1)$$

With expressions (18,19) we can now explicitly show the consistency of the theory in the sense that both limits, the thermodynamic $V \to \infty$, that neglects internal fluctuations, and the limit of no external noise $D_\rho = D_1 = D_s = 0$ reproduces the correct results obtained from (12) in the first limit and from (8) in the second. Also it is possible to see novel physical effects that appear as a result of the more fundamental level of description used in this theory. One is the existence of crossed fluctuation terms D_α / V that are a consequence of the interplay between the internal and the external fluctuations. They can only be found with a theory that takes into account both kinds of noise. Other novel feature is the fact that noise sources afecting the reactivity, ξ_λ and ξ_1 , have in our formulation an assimetry that does not exist in the standard equation (12). Namely (19) is not invariant under the interchange of D_λ and D_1 . The terms that break this symmetry are precisely those which couple internal and external noise. This indicates the greater level of information hold in the master equation with relation to the deterministic description. On the other hand this result opens the possibility, at least theoretically of differentiating between noise sources due to absorption or fission causes. Experiments concerning this point would be certainly very interesting.

To finish this section we note that the Gaussian noise assumption taken for the external sources of noise can not be strictly correct because the transition probability entering in a master equation has to be always positive and realizations of Gaussian noise are unbounded for both positive and negative values. However, in order to avoid complexities it is convenient to use Gaussian noise with a careful interpretation of the results. This limitation does not apply only in this theory but it is usual in most of the cases in which a Gaussian noise is used to modelize a physical process. The use of bounded processes such as Dichotomic and Poisson noise is strictly correct and as can be seen in refs. 8 and 5 provides the same basic results than with Gaussian noise.

3.- MODEL WITH DELAYED NEUTRONS AND COUNTING PROCESS

In this section we consider a more general model of nuclear reactor. Keeping the point assumption we are able to deal with a general zero power reactor with an arbitrary number of groups of delayed neutrons and a detector. The relevant variables are the neutron number, N , precursor number of the ith group c_i and counts R. The only variable observed will be R so we must reduce our results to time dependent moments of this variable R and relate them with the value of parameters of the model.

Following the probability balance method it is easy to write the master equation for this model[2]:

$$
\begin{aligned}
\partial_t P(N, c_i, R, t) = {} & \lambda (N+1) P(N+1, c_i, R, t) \\
& + \Lambda \sum_{m_0} \sum_{m_1} \cdots \sum_{m_I} p(m_0, m_1 \ldots m_I)(N+1-m_0) P(N+1-m_0, c_i - m_i, R, t) \\
& + \sum_{i=1}^{I} \lambda_i (c_i+1) P(N-1, c_1, \ldots c_i+1, \ldots c_I, R, t) + \\
& + \lambda_d (N+1) P(N+1, c_i, R-1, t) + sV P(N-1, c_i, R, t) \\
& - \left[sV + (\lambda + \Lambda + \lambda_d) N + \sum \lambda_i c_i \right] P(N, c_i, R, t)
\end{aligned}
\qquad (20)
$$

where $p(m_0, m_1, \ldots m_I)$ is the probability of finding m_0 instantaneous neutrons and m_i precursors of the ith group as result of a fission. The meaning of s, V, λ and Λ are the same than in the previous model, and now λ_d is the detection rate per neutron and λ_i the decay constant of precursors of the ith group.

The deterministic description obtained in the thermodynamic limit of (20) as

$$
\lim_{V \to \infty} V^{I+2} P(n, c_i, r) = \delta\left(\frac{N}{V} - n(t)\right) \delta\left(\frac{c_i}{V} - c_i(t)\right) \delta\left(\frac{R}{V} - r(t)\right)
$$

provides for $n(t)$, $c_i(t)$ and $r(t)$ the standard point-reactor kinetic equations:

$$
\begin{aligned}
\frac{dn}{dt} &= -\left[\lambda + \lambda_d - (\nu_0 - 1)\Lambda\right] n + \sum_{i=1}^{I} \lambda_i c_i + s \\
\frac{dc_i}{dt} &= \Lambda \nu_i n - \lambda_i c_i \\
\frac{dr}{dt} &= \lambda_d n
\end{aligned}
\qquad (21)
$$

with $\nu_i = \sum_{m_0} \cdots \sum_{m_I} m_i \, p(m_0, \ldots m_I)$. These equations are the starting point for the consideration of the external noise in standard formulations. We consider fluctuations of each parameter entering in the model

$$
\begin{aligned}
\lambda &\to \lambda + \xi_\lambda(t) \\
\Lambda &\to \Lambda + \xi_\Lambda(t) \\
\lambda_d &\to \lambda_d + \xi_{\lambda d}(t) \\
s &\to s + \xi_s(t)
\end{aligned}
\qquad (22)
$$

We take ξ_α to be a Gaussian white noise with intensity D_α as in the previous section. The treatment of (21) with (22) leads to well known results for the mean, variances and correlations. When introducing (22) into the master equations (20) and repeating the same method than in section 2 via the generating function $G(x, y_i, z)$ it is possible to obtain stationary moments and correlations. The reader interested in a detailed description of these calculations is refered to ref. 4. Here we only remark the points we have yet obtained in section 2 namely that it is possible to recover the usual descriptions by taking respectively the $D_\alpha = 0$ and $V \to \infty$ limits. One also find here crossed fluctuations terms which only appear in this theory.

We focus our attention on observable quantities that is, on the manner in which it is possible to extract information by means of a counting process. We define the modified variance of the counting variable $\mathcal{N}_{RR}$ as

$$\mathcal{N}_{RR}(t) = \langle R^2 \rangle - \langle R \rangle^2 - \langle R \rangle$$

This quantity can be bound experimentally with standard counting techniques.

After a tedious but straight forward calculation it is possible to find the expression for $\mathcal{N}_{RR}(t)$.

$$\mathcal{N}_{RR}(t) = 2 D_{\lambda d} (\langle N^2 \rangle - \langle N \rangle) t + 2(\lambda d - D_{\lambda d}) \sum_{i=1}^{I} y_i \left\{ t - \frac{[1 - \exp(-\tilde{\alpha}_i t)]}{\tilde{\alpha}_i} \right\} \tag{23}$$

where

$$A = (\lambda d - 3 D_{\lambda d}) (\langle N^2 \rangle - \langle N \rangle) -$$

$$- (\lambda d - D_{\lambda d}) \langle N \rangle^2$$

$$y_j = A + \sum_{i=1}^{I} \frac{\lambda_i B_i}{\lambda_i - \tilde{\alpha}_j} \Big/ \left\{ \tilde{\alpha}_j \left(1 + \tilde{\Lambda}_1 \sum_{i=1}^{I} \frac{\nu_i \lambda_i}{\lambda_i - \tilde{\alpha}_j} \right) \right\}$$

$$B_i = (\lambda d - D_{\lambda d}) \mathcal{N}_{NC_i}$$

and $\tilde{\alpha}_i$ are the roots of an equivalent inhour equation

$$\tilde{\rho} = \frac{\tilde{k} - 1}{\tilde{k}} = \frac{\tilde{\ell} u}{\tilde{k}} + u \sum_{i=1}^{I} \frac{\beta_i}{\lambda_i + u} \tag{24}$$

with

$$\tilde{\ell} = \frac{1}{\tilde{\lambda}_1 + \tilde{\Lambda}_1} \qquad \tilde{\lambda}_i = \lambda + \lambda d - i D_\lambda - i D_{\lambda d}$$

$$(1 - \beta) \tilde{k} = \frac{\nu_1 \tilde{\Lambda}_1}{\tilde{\Lambda}_1 + \tilde{\lambda}_1} \qquad \tilde{\Lambda}_i = \Lambda + i(\nu_i - 1) D_\Lambda$$

From $\mathcal{N}_{RR}(t)$ and using Pluta's formula we have inmediatly the correlation function:

$$C(z) = \frac{1}{2} \frac{d^2 \mathcal{N}_{RR}(z)}{d z^2} = (\lambda d - D_{\lambda d}) \sum_{i=1}^{I} y_i \tilde{\alpha}_i \exp(-\tilde{\alpha}_i z) \tag{25}$$

This expression is formally identical to the one obtained in the absence of external fluctuations, but now with effective coefficients. The simplicity observed in the correlation formula (25) is due to the consideration of Gaussian white sources of noise. The modeling with noise sources with more complicated statistics would lead to difficulties in the calculation of correlations. These difficulties are also encountered in the formulation (21) and (22) in which internal fluctuations are neglected. At this stage of the theory the key idea is to show that it is possible to obtain information on both external and internal fluctuations using standard counting methods. This is a conclusion of practical importance since counting methods are not often regarded as useful to describe effects of external fluctuations.

4.- PARAMETRIC NOISE AND LANGEVIN APROXIMATIONS

It is a common practice to use a Langevin equation instead of a master equation for modeling internal sources of noise. The idea, as formulated by Saito[6] and Williams[2] in the context of Stochastic Nuclear Reactor Theory is to use an equivalent aditive, Gaussian and white noise source of noise. The equivalence has been discused by several authors[2,6,9]. The equivalent source of noise in the Langevin equation is given by Schottky's formula[10]. We discuss the problem of parametric noise in the Langevin equation in the context of the simple model of section 2. The Langevin equation for the neutron density is given by

$$\partial_t n = -[\lambda - \Lambda] n + S + V^{-\frac{1}{2}} \xi(t) \tag{26}$$

$$\langle \xi(t) \rangle = 0$$

where Scottky's formula is in this case

$$\langle \xi(t)\, \xi(t') \rangle = \left[(\Lambda + \lambda) \{ n \}_{av} + S \right] \delta(t - t') \tag{27}$$

Here $\{ \ \}_{av}$ means the average over the internal fluctuations. It is easy to see that the Langevin equation is equivalent to the master equation (8) in the sense that both equations reproduce the same two first stationary moments and correlations. These are the requirements of equivalence between the master and Laugevin equations formulated by Ackasu[9].

The master equation formulation has a more fundamental level of description than the Langevin description. However it is simpler to deal with the Langevin description. This last point suggest to introduce the external parametric noise at the level of the Langevin description[6,11]. Indeed, in this way apparently we take into account both kinds of fluctuations. It also seems to be an easier manner to proceed than in our previous formulation. The question now is the equivalence of these two approaches. In order to show the difficulties encountered when using a Langevin description let us introduce parametric noise in the capture parameter λ in the Langevin equation (26).

$$\lambda \longrightarrow \lambda + \xi_\lambda(t)$$

The external and internal sources of noise are in principle independent

$$\langle \xi_\lambda(t)\, \xi_n(t') \rangle = 0 \tag{28}$$

The Langevin equation with external noise reads

$$\partial_t n = -[\lambda - \Lambda] n + S + V^{-\frac{1}{2}} \xi(t) + \xi_\lambda n \tag{29}$$

Being $\xi(t)$ and $\xi_\lambda(t)$ independent it is straight forward to find the equation satisfied by the moments. For example, the equation for $\langle n^2 \rangle$ is

$$\partial_t \langle n^2 \rangle = -2[\lambda - \Lambda]\langle n^2 \rangle + [2S + V^{-1}(\lambda + \Lambda)]\langle n \rangle + \\ + V^{-1}S + 4D_\lambda \langle n^2 \rangle \tag{30}$$

Comparing this equation with the one which follows from (17) one realizes that a crossed fluctuation term $-3(D_\lambda/V)\langle n \rangle$ appearing with the master equation formulation does not appear in eq. (30) obtained in the Langevin aproach. We note that this fact is not surprising since in the definition of the equivalent source by means of the Schottky formula the parameter λ appears implicitly. It is in this formula where the cross fluctuation term is by-passed.

Following the idea of the equivalent source of noise we can try to solve this problem by defining a Langevin equation as

$$\partial_t n = -[\lambda - \Lambda]n + S + V^{-1}[(\lambda + \Lambda)n + S]^{1/2}\eta(t) \tag{31}$$

with $\langle \eta(t)\eta(t') \rangle = \delta(t-t')$. With the Ito prescription this equation is equivalent, in the sense of Ackasu to (26)[12] and we can introduce the external noise $\lambda \to \lambda + \xi_\lambda(t)$ in the systematic and stochastic parts. This seems to solve the problem mentioned above. However this is not a good procedure because it deals with mathematical objects whith are difficult to interpret. For instance a term like $[\xi_\lambda n]^{1/2}\eta(t)$ would appear in the stochastic equation. As a consequence we believe that our original theory explained in sect. 2 is a better way to deal simultaneously with internal and external fluctuations. Likewise, the Langevin equation is not a useful starting point for a joint description of internal and parametric noise.

5.- CONCLUSIONS

We have presented a method that gives a joint description of external and internal fluctuations. This method permits us to obtain theoretical and practical consequences. Here we have discussed some of these consequences mainly in connection with a linear model of nuclear reactor. From the expression of moments of the neutron density probability one can identify terms that are a consequence of the interplay between external and internal sources of noise. These terms do not appear when introducing the external moise at the level of a Langevin equation.

The significance of the consideration of external and internal fluctuations can be described by the importance of these crossed fluctuation terms. Tipically a zero power reactor is dominated by internal fluctuations and a power reactor by external noise. There would be situations in which both sources of noise are of the same magnitude, for instance in some frecuency region of the power spectral density. This last case has not been experimentally investigated. Hence the simultaneous treatment of internal and external fluctuations seem not to be necessary in a standard application of diagnosis of a power plant or parameter measurement of a zero power reactor. However there are experimental situations in which information about an isolated phenomena is required and hence it is necessary to separate it from the rest. For example, in ref. 13 the effect of bubles is analyzed by injecting air in a reactor at zero power, isolating in this way this effects of temperature and presion effects, always present in a power reactor. Obviously internal fluctuations act, so a joint analysis of both fluctuations are necessary in these cases.

On the other hand and as general matter our formulation would be very useful for analyzing parameters and for diagnosis purposes when using counting statistics techniques, either in zero power or at power reactors. Indeed, in this case one needs to deal with a probability of the number of counts and this is just what our treatment provides including the effect of external noise.

ACKNOWLEDGEMENT

Financial support of Dirección General de Investigación Científica y Técnica (Spain), Proyect nº PB 86-0534 and of the CAICYT, Proyect Nº 361/84 are acknowledged. Part of the original work on which this paper is based was done in collaboration with J.M. Sancho.

REFERENCES

1.- N.G. Van Kampen, "Stochastic Processes in Physics and Chemistry", North-Holland, Amsterdam (1983).
2.- M.M.R. Williams, "Random Processes in Nuclear Reactors", Pergamon Press, Oxford (1974).
3.- M.A. Rodríguez, M. San Miguel and J.M. Sancho, Ann. nucl. Energy, 10, 263 (1983).
4.- M.A. Rodriguez, M. San Miguel and J.M. Sancho, Ann. nucl. Energy, 11, 321 (1984).
5.- M.A. Rodríguez, L. Pesquera, M. San Miguel and J.M. Sancho, J. Stat. Phys. 40, 669 (1985).
6.- K. Saito, Prog. nucl. Energy, 3:157 (1979).
7.- W. Horsthemke and R. Lefever, "Noise Induced Transitions", Springer, New York (1983).
8.- J.M. Sancho and M. San Miguel, J. Stat. Phys. 37: 151 (1984).
9.- Z. Ackasu, J. Stat. Phys. 16:33 (1977).
10.-K. Saito, Ann. nucl. Energy 1:31, 107, 209 (1974).
11.-W. Seifritz, Atomkernenergie 16:29 (1970).
12.-N.G. Van Kampen, Physica 25:3 (1981).
13.-S.A. Wright, R.W. Albrecht and M.F. Edelmann, Ann. nucl. Energy 2:367 (1975).

AN INFLUENCE FUNCTIONAL APPROACH FOR THE ELIMINATION OF VARIABLES IN STOCHASTIC PROCESSES

Horacio S. Wio[1]

Departament de Fisica, Facultat de Ciences
Universitat de les Illes Balears
E - 07071 Palma de Mallorca, Spain

ABSTRACT

We present a non-adiabatic scheme for the elimination of variables in stochastic processes, based on the "influence functional" method of Feynman. Particularly, the case of multivariate Fokker-Planck equations, or equivalently a set of coupled Langevin equations driven by white noises, is analyzed, and applications to the non-white noise problem and Kramers equation are discussed.

INTRODUCTION

A large number of physical systems show an interplay of mechanisms that evolve on different time scales. The equations of motion of such systems can often be simplified by eliminating the rapid variations. When there is a well defined separation of time scales, with some of the variables varying over a time scale which is characteristically very much shorter than that of the rest, and if dissipation is present, it is possible for the fast variables to relax to a quasistationary state, in which their values follow the values of the slow variables (Haken`s slaving principle [1]).

(1) Member of CONICET, ARGENTINA. On leave from Centro Atomico Bariloche, 8400 - S.C.Bariloche, ARGENTINA

This problem has been recently treated by methods of adiabatic elimination of variables [2,3] . However, if the above mentioned separation in time scales is not strictly fulfilled, such an elimination procedure is not so obvious.

On the other hand, in the context of the kinetic or transport description of physical phenomena, there is a wide range of situations, in physics as well as in other sciences, where it is desirable (and sometimes necessary) to describe the problem not in terms of a full set of (N) macroscopic variables, but only via a small set (K<N) of "relevant" variables. A few examples include the kinetics of multicomponent chemical reactions, cluster growth in homogeneous nucleation, kinetics of phase transitions, heavy ion reactions at low energies, etc [4] .

One of the problems to be faced if one starts from the `full` description, is to find the form of the reduced one, that is the reduced kinetic equation and/or its solution.

In this communication we present a new approach for eliminating irrelevant variables based on Feynman`s influence functional method [5] . This approach , not resorting to adiabatic arguments, is then able to eliminate fast or slow variables. The starting point of the presentation is the assumption that the complete description ,that means in terms of the full set of N-macroscopic variables, is Markovian. This assumption is not strictly necessary, but simplifies the argument of having a path integral representation of the solution or the conditional probability, for instance by a reiterative use of the Chapman-Kolmogorov relation, employing some adequate representation of the short time propagator. Here we discuss the case of multivariate Fokker-Planck equations, although the method can be applied to the equivalent set of coupled Langevin equations (with additive or multiplicative) white noise sources. As a matter of fact the case of colored noise can also be handled [6] .

We put the emphasis on the form of the solution of the problem, after the elimination of variables has been carried out. The finding of the form of the reduced equation is briefly addressed, a complete discussion will be given elsewhere.

We expect that this approach could be also applied to other kinetic schemes, at least in those cases where the Markovianicity previously refered to is satisfied.

THE METHOD

We will follow the notation, and use some of the results of ref.[7,8] . We start with the multivariate Fokker-Planck equation given by :

$$\frac{\partial}{\partial t} P(\mathbb{r},t|\mathbb{r}_o,t_o) = \sum_j \frac{\partial}{\partial r_j}\left[h_j(\mathbb{r})\, P\right] + \tfrac{1}{2}\sum_{jl} D_{jl} \frac{\partial^2}{\partial r_j \partial r_l} P \qquad (1a)$$

where $\mathbb{r} = (r_1, r_2, \ldots\ldots, r_N)$. In an obvious short hand notation we have :

$$\frac{\partial}{\partial t} P(\mathbb{r},t|\mathbb{r}_o,t_o) = -\frac{\partial}{\partial \mathbb{r}}\left[\mathbb{h}(\mathbb{r})\, P\right] + \tfrac{1}{2} \frac{\partial}{\partial \mathbb{r}} \hat{\mathbb{D}} \frac{\partial}{\partial \mathbb{r}} P \qquad (1b)$$

$P(\mathbb{r},t|\mathbb{r}_o,t_o)$ being the conditional probability for reaching the `point` $\mathbb{r}$ at time t, if we started from $\mathbb{r}_o$ at t_o ; and $(\hat{\mathbb{D}})_{ij} = D_{ij}$. The set of equivalent Langevin equations can be written as :

$$\dot{r}_j = h_j(\mathbb{r}) + \Gamma_j(t) \qquad (1c)$$

with $\Gamma(t)$ white noises of zero mean and correlation functions

$$\langle\langle \Gamma_i(t)\, \Gamma_j(t') \rangle\rangle = 2\, D_{ij}\, \delta(t-t') \qquad (1d)$$

Following ref. [7,8], we arrive at the full propagator :

$$P(\mathbb{r}_f,t_f|\mathbb{r}_o,t_o) = \int \mathcal{D}\mathbb{r}\ \exp\left\{-\tfrac{1}{2}\int dt\ (\dot{\mathbb{r}}(t) - \mathbb{h}(\mathbb{r}))^t \cdot \hat{\mathbb{D}}^{-1} \cdot (\dot{\mathbb{r}}(t) - \mathbb{h}(\mathbb{r})) + \frac{\partial}{\partial \mathbb{r}} \mathbb{h}(\mathbb{r})\right\} \qquad (2)$$

$\mathcal{D}\mathbb{r}$ is the measure as usual.

We introduce now the idea of the subset of relevant variables, assuming that :

$$\{\mathbb{r}\} = \{\{\mathbb{x}\}, \{\mathbb{y}\}\} \qquad (3a)$$

$\{\mathbb{x}\}$ being the subset of the relevant variables, and $\{\mathbb{y}\}$ the subset of the remaining variables. Similarly, we write:

$$\{\mathbb{h}(\mathbb{r})\} = \{\{\mathbb{h}_x(\mathbb{r})\}, \{\mathbb{h}_y(\mathbb{r})\}\} \qquad (3b)$$

and also adopting the separation :

$$\begin{aligned} \mathbb{h}_x(\mathbb{r}) &= \mathbb{h}_{o,x}(\mathbb{x}) + \mathbb{h}_{1,x}(\mathbb{x},\mathbb{y}) \\ \mathbb{h}_y(\mathbb{r}) &= \mathbb{h}_{o,y}(\mathbb{y}) + \mathbb{h}_{1,y}(\mathbb{x},\mathbb{y}) \end{aligned} \qquad (3c)$$

In order to simplify the notation, in what follows we will consider only one relevant variable : x ; the extension to several relevant variables is straightforward.

The full Lagrangian :

$$L(\mathbb{r},\dot{\mathbb{r}}) = \tfrac{1}{2}(\dot{\mathbb{r}} - \mathbb{h}(\mathbb{r}))^{t}\, \hat{\mathbb{D}}^{-1}\, (\dot{\mathbb{r}} - \mathbb{h}(\mathbb{r})) + \tfrac{1}{2}\frac{\partial}{\partial \mathbb{r}}\mathbb{h}(\mathbb{r}) \quad (4)$$

can be writen as the sum of three contributions :

$$L(\mathbb{r},\dot{\mathbb{r}}) = L_x(x,\dot{x}) + L_y(y,\dot{y}) + L_{int}(x,\dot{x},y,\dot{y}) \quad (5)$$

which are given by :

$L_x(x,\dot{x})$ = containing all the dependence on x and $\dot{x}$ alone.

$L_y(y,\dot{y})$ = same with y and $\dot{y}$ alone.

L_{int} = including the interaction terms.

The definition of L_x and L_y implies the possibility that some of them (or both) could not be "bona fide" thermodynamic Lagrangians. That means that them would not have the structure (4).

We now define the marginal or inclusive conditional probability by means of :

$$P_{incl}(x_f,t_f|x_o,t_o) = \int dy_f \int dy_o P_y(y_o)\; P(x_f,y_f,t_f|x_o,y_o,t_o) \quad (6)$$

where we have taken a weighted average on the initial distribution of the subset $\{y_o\}$ (assuming that the initial distribution at time t_o, is separable : $P_o(\mathbb{r}_o) = P_x(x_o)\; P_y(y_o))$, and is summed over all possible final values $\{y_f\}$ as usual.

In order to make the connection with the Feynman approach, the marginal probability can be written as :

$$P_{incl}(x_f,t_f|x_o,t_o) = \int \mathcal{D}x\, \exp\{-\int dt\; L_x(x,\dot{x})\}\; F[x,\dot{x}] \quad (7)$$

where the functional $F[x,\dot{x}]$ is given by :

$$F[x,\dot{x}] = \int dy_f \int dy_o \, P_y(y_o) \int \mathcal{D} y \, \exp\{-\int dt (L_y + L_{int})\} \quad (8a)$$

This has the form of Feynman`s influence functional [5], although the latter arose in a different context (it must be remarked that Feynman`s influence functional is, in fact, a double path integral).

In the same spirit of Feynman`s scheme, we write :

$$F[x,\dot{x}] = \exp\{-\int dt \; \Psi[x(t),\dot{x}(t)] \} \quad (8b)$$

introducing the following Ansatz for $\Psi[x,\dot{x}]$:

$$\Psi[x,\dot{x}] = f_o(t) \, x(t) - \int d\tau \, x(t) \, C(t,\tau) \, x(\tau) \quad (9a)$$

Where $f_o(t)$ will have the form :

$$f_o(t) = f_{oo}(x_f,x_o) \, \delta(t-t_o) + \int d\tau \, f_{0,1}(t,\tau) \, \dot{x}(\tau) \quad (9b)$$

f_{oo} being some function of the initial and final coordinates x_o and x_f .

We will show that such an Ansatz can be justified, or that a sound interpretation can be assigned to it, for a broad class of situations.

In order to see how such a result could arise in a practical case we will assume the simplified situation :

$$h_{0,x}(x) = A\,x \quad ; \quad h_{1,x}(x,y) = \varepsilon\, g(y)$$

$$h_{0,y}(y) = h(y) \quad ; \quad h_{1,y}(x,y) = \varepsilon\, B\, x$$

$$\mathbb{D} = D\,\mathbb{I} \quad (10a)$$

$\mathbb{I}$ being the identity matrix, then the interaction Lagrangian will become :

$$\begin{aligned} L_{int} &= (\varepsilon/D) \{[Ag(y) + Bh(y) - B\dot{y}]x(t) - g(y)\dot{x}(t)\} \\ &= \varepsilon \{ \mu(y,\dot{y}) \, x(t) - \Phi(y) \, \dot{x}(t) \} \end{aligned} \quad (10b)$$

$\mu(y,\dot{y})$ being a function of y and $\dot{y}$, and $\Phi(y)$ of y only.

Replacing (10b) in (8a) we obtain :

$$F[x,\dot{x}] = \int dy_f \int dy_o \, P_y(y_o) \int \mathcal{D}y \, \exp\{-\int dt \, L_y\} \exp\{-\varepsilon \int dt(\mu x - \phi \dot{x})\} \quad (11)$$

In the case of weak coupling ($\varepsilon <<1$) , we can expand the second exponential in powers of the coupling constant and perform the path integral, obtaining :

$$\begin{aligned} F[x,\dot{x}] = F_o\{1 &- \varepsilon \int dt [\langle\langle \mu(y(t),y(t)) \rangle\rangle \, x(t) - \langle\langle \phi(y(t)) \rangle\rangle \, \dot{x}(t)] \\ &+ \varepsilon^2 \int dt \int dt' \, [x(t) \langle\langle \mu(y(t)y(t)) \mu(y(t')y(t')) \rangle\rangle \, x(t') \\ &+ x(t) \langle\langle \mu(y(t)y(t)) \phi(y(t')) \rangle\rangle \, \dot{x}(t') \\ &+ \dot{x}(t) \langle\langle \phi(y(t)) \mu(y(t')y(t')) \rangle\rangle \, x(t') \\ &+ \dot{x}(t) \langle\langle \phi(y(t)) \phi(y(t')) \rangle\rangle \, \dot{x}(t') \,] \} \end{aligned} \quad (12)$$

where the angular brackets denotes the average

$$\langle\langle \Theta(y(t)) \rangle\rangle = F_o^{-1} \int dy_f \int dy_o P_y(y_o) \int \mathcal{D}y \, \Theta(y(t)) e^{-\int dt L} \quad (13)$$

and F_o will be one only if L_y is a "bona fide" thermodynamic Lagrangian.

Transforming the perturbational expansion (12) into a cumulant expansion, and restricting ourselves to the case in which second order perturbation leads to reasonably accuracy, the influence functional becomes :

$$\begin{aligned} F[x,\dot{x}] = F_o \exp\{-\varepsilon \int dt \, [x(t) \, \mu_o(t) - \dot{x}(t) \, \phi_o(t)] \\ + \varepsilon^2 \int dt \int dt' \, [x(t) \, \mu_1(t,t') \, x(t') \\ + \ldots\ldots \quad + \dot{x}(t) \, \phi_1(t,t') \, \dot{x}(t')]\} \end{aligned} \quad (14a)$$

with

$$\begin{aligned} &\mu_o(t) = \langle\langle \mu(y(t)y(t)) \rangle\rangle \quad ; \quad \phi_o(t) = \langle\langle \phi(y(t)) \rangle\rangle \\ &\mu_1(t,t') = \langle\langle \mu(y(t)y(t)) \mu(y(t')y(t')) \rangle\rangle \quad ; \\ &\phi_1(t,t') = \langle\langle \phi(y(t)) \phi(y(t')) \rangle\rangle \quad ; \quad \text{etc.} \end{aligned} \quad (14b)$$

We see that (14a,b) have the desired structure (9).

What we intent to do next is to give some arguments to reach a meaningful interpretation of the above Ansatz. We return to eq.(8b) and start asking about the stationary path for the action associated with the effective Lagrangian. It is given by the solution of the Euler-Lagrange equation :

$$\frac{d}{dt}\frac{\partial}{\partial \dot{x}} L_x - \frac{\partial}{\partial x} L_x = -\left\{\frac{d}{dt}\frac{\partial}{\partial \dot{x}} \Psi - \frac{\partial}{\partial x} \Psi\right\} = -\Omega[x,\dot{x}] \quad (15)$$

which can be rewritten as ($\dot{v} = \ddot{x}$) :

$$\dot{v} = -\frac{\partial}{\partial x} U_{eff}(x) + \tilde{\Omega}[x,\dot{x}] \tag{16}$$

It is now necessary to calculate or to make reasonable models for the influence (or in this case for $\Omega[x,\dot{x}]$). In order to give a plausible interpretation of the Ansatz (9) or the result (14), we will speculate by exploiting the underlying stochasticity of the processes at hand. Following previous authors [9], we assume that Ω can be written as :

$$-\int d\tau\, \alpha(t-\tau)\, \dot{x}(\tau) + f_{fl}(t) \tag{17 }$$

where the first term is a systematic known effect, and f_{fl} is not known exactly, but instead we know the probability distribution functional $P[f_{fl}(t)]$ for different realizations of $f_{fl}(t)$. Then in order to get sensible results we must consider the average of the influence functional which leads to a form as (17a), over such a distribution :

$$F_{av}[x,\dot{x}] = \int \mathcal{D} f_{fl}\;\; P[f_{fl}]\;\; F_{fl}[x,\dot{x}] \tag{18a}$$

When $P[f_{fl}]$ is a Gaussian distribution of zero mean, such an average gives the result [5] :

$$F[x,\dot{x}] = \exp\left\{-\int dt f_o(t)x(t)+\int dt\int dt' x(t)C(t,t')x(t')\right\} \tag{18b}$$

where $C(t,t') = \langle f_{fl}(t), f_{fl}(t')\rangle$ is the correlation function of the "stochastic force". Then we find that the above Ansatz can be interpreted as the effect of a Gaussian colored noise, coming from the eliminated irrelevant variables, on the relevant one. If the correlation function $C(t,t')$ (or $\alpha(t-t')$ as it must be proportional to the other) has a very short correlation time ("short memory"), the systematic contribution becomes :

$$\alpha_o\, \dot{x}(t)$$

and the above effect would correspond to a Gaussian white noise.

A few inital examples that we want to mention here are : (i) colored noise, (ii) Kramers equation in the high friction limit, (iii) a simplified Lotka-Volterra model. For the first case we consider an Ornstein-Uhlenbeck noise, corresponding the coupled set of equations :

$$\begin{aligned} \dot{x} &= h(x) + g(x)\, u \\ \dot{u} &= -\sigma\, u + \sigma\, \Gamma(t) \end{aligned} \tag{19a}$$

$\Gamma(t)$ being a Gaussian white noise of zero mean and correlation function « $\Gamma(t)\Gamma(t')$» = $2D\,\delta(t-t')$, and σ is the inverse of the correlation time. The corresponding Fokker-Planck equation has a singular diffusion tensor, making more adequate a "phase space-like" picture for the path integral. After integration over u and its associate conjugate variable, we reach a form of the influence functional (in "phase space") which has the same structure than the proposed Ansatz. At this stage we can expand the exponents in powers of σ^{-1} (we must remark that in the original problem the limit $\sigma \rightarrow \infty$ corresponds to the white noise case). Up to zero order in σ^{-1} we find a marginal conditional probability corresponding to the Stratonovich prescription (that means the middle point discretization in which we have been working) of a multiplicative white noise [7,8] . The next order in σ^{-1}, in the case of additive noise (i.e.: $g(x)=1$) coincides with the lowest order of the "exact" result of Pesquera et al 6 . The analysis of higher order terms, as well as a comparison with the work of other authors is under way.

The second case, corresponding to the Kramers problem is described by the following set of equations :

$$\dot{x} = u \qquad\qquad \dot{u} = -\sigma u - V(x)' + \Gamma(t) \tag{19b}$$

with « $\Gamma(t)\Gamma(t')$» = $\sigma\,\delta(t-t')$, and σ is the friction parameter. The form of this set of equations is similar to the previous one. That means that we must resort to the same "phase-space" description, and also to expand the exponents in powers of σ^{-1} . In this case the lowest order in σ^{-1} (corresponding to the high friction limit) gives the Smoluchowski's result [10] . Higher order contributions are under study.

Finally, for the simplified Lotka-Volterra model we have:

$$\dot{x} = b\,y + \Gamma_1(t) \qquad\qquad \dot{y} = -b\,x + \Gamma_2(t) \tag{19c}$$

where, as before, the $\Gamma_j(t)$ are white noise with zero mean and correlation function « $\Gamma_i(t)\Gamma_j(t')$» = $2\,D\delta_{ij}\,\delta(t-t')$. As the involved Lagrangians results to be cuadratic in coordinates and velocities, the problem can be exactly integrated. Particularly the influence functional has an exact expresion with the same structure as the Ansatz (here we like to remark that in this case both variables clearly evolve on the same time scale, then an adiabatic elimination procedure is not applicable). In this case is also possible to find the equation that governs the evolution of the reduced or marginal conditional probability. It has the form of a Fokker-Planck equation but with a drift coefficient explicitly dependent on time and on the initial conditions, making evident the non-Markovian character of the reduced problem.

As a final point we want to briefly discuss the way to find out what kind of equation governs the evolution of the marginal conditional probability (7). If we follow Feynman's approach[5], we must analyze the following limit :

$$\lim_{\tau \to 0} (1/\tau)\{P_{incl}(x_f,t_f+\tau|x_o,t_o) - P_{incl}(x_f,t_f|x_o,t_o)\} =$$

$$= \lim (1/\tau)\{\int \mathcal{D}x\, e^{-\int^{t_f+\tau} dt..} - \int \mathcal{D}x\, e^{-\int^{t_f} dt..}\} \qquad (20)$$

Expanding the r.h.s. up to terms of first order in τ, and after some lengthly algebra, we would obtain the desired equation. What we expect to get is, in general, a kind of "generalized" Fokker-Planck equation [11], including non-Markovian terms, such as kernels non-local in time for the drift and diffusion coeffients. It is also possible to find some inhomogenity which will work as a probability "source" or "sink", depending on its sign. The former case will imply that the remainder variables will contribute to the process, while the latter implies that their effect has a dissipative character. In practical cases much care must be exerted in order to avoid problems with the probability normalization. It must be remarked that a result such as (14), at least in the case of weak coupling, comes from a perturbation like expansion, which will be valid only on a restricted range.

DISCUSSION

In this communication we have presented a scheme of elimination of variables from multivariable Fokker-Planck equations, based on Feynman`s influence functional method and the Ansatz for such a functional. The starting point of the scheme is that the full description is Markovian. This assumption is not strictly necessary, but simplifies the argument of having a path integral representation of the solution, by a reiterative use of the Chapman-Kolmogorov relation, employing an adequate form of the short time propagator.

We have shown that the proposed Ansatz can be justified, at least in those cases with weak coupling, via a perturbational expansion. Here we would like to remark, that this is true as far as the interaction Lagrangian is linear in the relevant variables and/or its velocities. For the more general case it is easy to see that the form of the Ansatz could be the same, but with the functions of the relevant variable or its velocity replacing them. Also, it can be interpreted as the effect of colored but still Gaussian noise coming fron the eliminated variables. However, if it is necessary to consider higher perturbational orders, we can generalize the Ansatz in order to include non-Gaussian effects. The examples examined show the possibilities of this scheme, in particular in situations where we have not a clear separation of time scales, as is the case for the simplified Lotka-Volterra model.

The emphasis of the presentation lies on the form of the solution, as at the very end this is the quantity of interest. In principle, we can evaluate it in a saddle-point, or eventually in a uniform approximation, or even numerically. Nevertheless, there could be interest in knowing the form of the reduced equation, this point has been briefly discussed here. A deeper analysis is still lacking as it is a subject under investigation.

As a test, the present scheme must be applied to well known problems which have been solved by other means. One of such cases is the reduction of the Kramers-Chandrasekhar equation, in the high friction limit, to the Smoluchowski equation [11] , as well as the higher order corrections. This case, together with the simplified Lotka-Volterra model as well as applications to the non-white noise problem, will be analyzed elsewhere [12]. Also, a comparison of this scheme with those presented in the literature is in progress.

ACKNOWLEDGEMENTS

The analysis of the Lotka-Volterra model and Kramers problem has been done in collaboration with C.Budde and C.Briozzo from the IMAF, Universidad Nacional de Cordoba, Argentina.

The author greatly acknowledges fruitful discussions with C.Budde, C.Briozzo, M.Caceres, and Profs. M.San Miguel and E.Tirapegui, as well as the kind hospitality at the Department of Physics of the Universitat de les Illes Balears, Palma de Mallorca, Spain, and to the Physik Department der TUM, Munich, W.Germany.

This work was partially performed under grants N.12091/84, and PID 3-012000/85, CONICET, Argentina.

REFERENCES

1. H.Haken, Rev.Mod.Phys. 47, 67 (1975).

2. Among others : C.W.Gardiner, Phys.Rev. A29,2814 and 2823 (1984); W.Theiss and U.M.Titulauer; Physica 130A,123 and 143 (1985); N.G.Van Kampen and I.Oppenheim; Physica138A, 231 (1986).

3. N.G.Van Kampen; Phys.Rep. 124, 69 (1985). In this review there is an extense list of references to resent papers on adiabatic elimination procedures.

4. L.E.Reichel and W.C.Scieve (Eds.), "Instabilities, Bifurcations and Fluctuations in Chemical Systems" (Univ. Texas Press, Austin, 1982); K.Kitahara, H.Metiu and J.Ross, J.Chem.Phys. 64, 292(1976) and 65, 393(1976); H.A.Weidenmuller, Prog. Partic. and Nucl. Phys. 3, 49 (1980).

5. R.P.Feynman and A.R.Hibbs, "Quantum Mechanics and Path Integrals" (McGraw Hill, New York, 1965); R.P.Feynman and F.L.Vernon, Ann.Phys. (N.Y.) 24, 118(1963).

6. L.Pesquera, M.Rodriguez and E.Santos, Phys.Lett.94A, 287 (1983); R.Phytian, J.Phys. A10, 777(1977).

7. R.Graham, Z.Phys. B26, 281 (1977).

8. F.Langouche, D.Roekaerts and E.Tirapegui, "Functional Integration and Semiclassical Expansions" (D.Reidel Pub. Co., Dordrecht, 1982).

9. D.M.Brink, Prog. Partic. and Nucl. Phys. 4, 323 (1980); K.Moehring and U.Smilansky, Nucl.Phys. A338, 227 (1980).

10. N.G.Van Kampen, "Stochastic Processes in Physics and Chemistry"(North Holland, Amsterdam, 1981); C.W.Gardiner "Handbook of Stochastic Methods"(Springer-Verlag, Berlin , 1983).

11. J.Nordholm and P.Zwanzig, J. Stat. Phys. 13, 347 (1975); H.Risken, "The Fokker-Planck Equation" (Springer-Verlag, Berlin, 1983).

12. H.S.Wio, C.Budde and C.Briozzo, to be submitted for publication.

MIXING PROPERTIES AND RESONANCES IN CHAOTIC DYNAMICAL SYSTEMS

by Stefano Isola

Dipartimento di Fisica, Universitá di Firenze

Largo E. Fermi 2 I-50125, Firenze, Italy

We are interested here in discussing the *mixing* properties of chaotic time evolution, namely the rate at which a chaotic dynamical system looses memory about its past history.

A (differentiable) dynamical system is a time evolution $x(t) = f^t x(0)$ on a manifold M, where f is a differentiable function and the *time* t may be an integer or a real number. A *chaotic* time evolution is defined by the presence of sensitive dependence on initial conditions; namely any small noise (always present in physical as well as in computer experiments) will be exponentially amplified as the time goes on. Due to this feature, a suitable framework of investigation is a statistical one.

In fact, we assume that the time averages of a given continuous function $A : M \to R$ (i.e. an *observable*):

$$< A >= \lim_{T\to\infty} \frac{1}{T} \int_0^T A(f^t x_0) dt = \rho(A)$$

exists for some x_0, defining a probability measure ρ, which is invariant under the time evolution and ergodic (see refs. 1, 2).

Then, given two observables $A, B : M \to R$, the *correlation function* $C_{AB}(t)$ is defined as follows:

$$C_{AB} = \rho[(A \circ f^t)B] - \rho(A)\rho(B)$$

In the simplest case where the observables are the "position variable" x one gets the *autocorrelation function* of the x-coordinate. Picking an initial point $x(0)$ and another point $x(t)$ on the time evolution, this function tells us how much the deviations of these two points from their common mean value know about each other, on the average over all initial points. Therefore the correlation function yields a measure of the irregularity of the motion which is quite different from the one provided by measuring characteristic exponents. A time evolution for which

$C_{AB}(t) \to_{t\to\infty} 0$, for all choices of A, B, is called *mixing*. In fact, one formulation of the mixing property is just that:

$$\lim_{t\to\infty} \rho[(A \circ f^t)B] = \rho(A)\rho(B)$$

Then, the problem is to know *how*, for a mixing system, the correlation function decays to zero. This is important because it gives the rate at which the time evolution becomes independent of where the system began. However, it seems that at present there are no successful techniques to measure the decay rate directly from the correlation function. An hopeful direction is rather to study the properties of its Fourier transform:

$$S_{AB}(\omega) = \int_{-\infty}^{+\infty} dt \exp(i\omega t) C_{AB}(t)$$

Note that $S_{AA}(\omega)$ is the *power spectrum* of the signal $A(x(t))$. Usually a chaotic time evolution exhibits a continuous power spectrum and for this property it has been largely utilized as an indicator to distinguish between regular and chaotic motions. Nevertheless the function S_{AB} turns out to be useful also in order to investigate mixing properties, namely to distinguish between different chaotic regimes. In fact the decay rate of the correlation function is strictly related to the analytic properties of its Fourier transform S_{AB}. More precisely the location of poles of S_{AB} in the complex frequency plane (interpreted as *resonances*) may provide an adequate description of the behaviour of $C_{AB}(t)$.

For example, if $C(t) \sim \exp(-\lambda|t|)\cos(\alpha|t| + \phi)$, then $S(\omega)$ has complex poles located at $\omega = \pm\alpha \pm i\lambda$.

Such problem has been studied recently for a particular class of differentiable dynamical systems, called Axiom-A systems, for which the invariant probability measure is naturally chosen as a *Gibbs state* (see refs. 3, 4). In this case, its was shown that the Fourier transform of $C(t)$ is meromorphic in a strip $|\mathrm{Im}\,\omega| < \alpha_0$, and the position of the poles does not depend on the observable monitored. Furthermore, for a discrete time mixing system (i.e. an Axiom-A diffeomorphism), the poles stay at a finite distance from the real axis, and this corresponds to a correlation function which behaves like an exponentially damped oscillation (if the poles are not purely imaginary, otherwise only an exponential decay would be observed).

Let us see what happens for more general (non-Axiom-A) dynamical systems. In this case, numerical studies are often the only accessible tool of investigation, so that the invariant probability measure is given directly by the time average:

$$< A >= \frac{1}{T}\sum_{t=0}^{T-1} A(f^t x)$$

where we have a record of finite length T. Here, an important fact to take into account is the presence of *noise* (in the case of computer-generated time evolution this is due to the floating point truncation), which introduces some imprecision in evaluating $f^t x$. Moreover, for a chaotic system the largest characteristic exponent λ_1 is positive so that, if the noise level is ν, the errors will grow as $\nu\exp(\lambda_1 t)$. Due to this feature the noise (it does'nt matter how small it is) will swamp completely the time evolution after a time

$$t_c \sim \frac{|\log \nu|}{\lambda_1}$$

if we assume the attractor to have size one.

This provides a heuristic estimate for computing correlation functions from a chaotic time evolution: in order to observe only the intrinsic decay properties of $C(t)$, one has to satisfy the inequality $t < t_c$. Conversely, in order to ensure the selection of the right *physical* measure (see refs. 1, 2), the total observation time T must be much greater than t_c.

The behaviour of some interesting dynamical systems has been recently investigated and it turns out that a modulated decay of correlation functions (suggesting the presence of resonances with non-zero real part) is a quite well shared feature [5,6]. Moreover, for the case of the two-dimensional map of the plane (the Hènon map):

$$x_1(t+1) = x_2(t) + 1 - \alpha[x_1(t)]^2$$
$$x_2(t+1) = \beta x_1(t)$$

with $\beta = 0.3$ and α ranging from 1.3 to 1.4, the introduction of a numerical technique based on Padè approximants has made possible to successfully locate the position of the leading resonance (the nearest to the real axis), thus providing an estimate of the decay rate as well as the modulation frequency of correlations [5]. When $\alpha = 1.3$, the time evolution lies on a periodic orbit of period 7 and a resonance located at $\omega = \frac{2\pi}{7}$ is observed. Increasing α toward the value 1.4 (where the system has a well known strange attractor [8]) the resonance moves away from the real axis and accelerates the exponential decay of correlations.

An analogous result has been obtained for the logistic map of the interval $[0,1]$ into itself $x(t+1) = rx(t)[1-x(t)]$, for a set of values of the parameter r for which the system exhibits intermittent behaviour.

Very recently, the hypothesis that modulated correlation functions occur for intermittent systems has been developed with encouraging results [6]. A probabilistic model of the correlation function for one-dimensional discrete time dynamical systems with Type I intermittency [7], provides a good description (with close agreement with numerical experiments) of the typical situation occurring in an intermittent dynamical regime: two basic frequencies modulate the correlation function, one coming from the "ghost" of a periodic orbit (which is becomed unstable through a saddle-node bifurcation, for example), and the other, slower, representing the mean return time in the laminar region.

Numerical studies in this direction are in progress with dynamical systems like the Lorenz system, which are more interesting from a physical point of view.

References

1. J. -P. Eckmann and D. Ruelle: Rev. Mod. Phys. **53**, 643 (1985)
2. D. Ruelle: *Chaotic evolutions and strange attractors*,
Notes prepared by S. Isola from the "Lezioni Lincee" (Rome, 1987), Cambridge University Press, Cambridge UK, 1988

3. D. Ruelle: Phys. Rev. Lett. **56**, 405 (1985);
J. Stat. Phys. **44**, 281 (1986);
J. Differ. Geom. **99** (1987);
J. Differ. Geom. **117** (1987)
4. M. Pollicott: Invent. Math. **81**, 413 (1985)
5. S. Isola: Commun. Math. Phys. **116**, 343 (1988)
6. V. Baladi, J. -P. Eckmann and D. Ruelle: *Resonances for intermittent systems*, Preprint (1988), to appear in Nonlinearity.
7. Y. Pomeau and P. Manneville: Comm. Math. Phys. **74**, 189 (1980)
8. M. Hénon: Commun. Math. Phys. **50**, 69, (1976)

CHAPTER II

STOCHASTIC PROCESSES IN LINEAR NUCLEAR SYSTEMS : THEORY

APPLICATION OF STOCHASTIC TRANSPORT THEORY TO THE DETERMINATION OF THE STATISTICAL DESCRIPTORS USED IN NOISE ANALYSIS

J.L. Muñoz-Cobo and G. Verdú

Departamento Ingeniería Química y Nuclear
Universidad Politécnica de Valencia. P.O. Box 22012
46071. Valencia

Abstract

In this paper we develop the theory of the covariance and cross correlation between two detectors, when delayed neutrons are considered, and we extend this previous theory to the calculation of the Cross-Power-Spectral-Density. This paper includes also the single mode approximation and the reduction to a lumped parameter model. Finally we develop calculable methods and we apply them to the interpretation of neutron noise experimental data.

1. Introduction

Recently a renewed interest has emerged for the use of stochastic methods in the areas of nuclear subcriticality safety [1-3], and nondestructive fuel assay techniques [4]. In both applications one utilizes the fluctuations of the neutron field arising from the inherent stochasticity of the nuclear processes and/or those excited by the insertion of Cf-252 source into the system.

In this paper we follow up on the line of Pal [5] and Bell [6] fundamental work, and we have developed the theory of the covariance and the cross-correlation function between two detectors, when delayed neutrons are included. Also we will perform the reduction to a lumped parameter model in the unimodal approximation which provides a good test of the theory. Finally we want to provide a rigurous formalism for the calculation of the statistical descriptors used in zero power noise analysis and to develop some computational tools which will serve us to interpret some experiments [1] recently performed.

This paper has been organized as follows:

In sections 2, and 3, we review the theory of the interaction of one single detector with the stochastic neutron field [7,8], and from this theory we obtain calculable expressions for the variance and the Feynman Y-function. In section 4 we review the unimodal approximation and the reduction to a lumped parameter model [8,9].

In section 5 we develop the theory of the covariance and cross correlation between two detectors when delayed neutrons are included.

In section 6 we extend this previous theory to the calculation of the Cross-Power-Spectral Density (C.P.S.D.), and we develop the single mode aproximation and the reduction to a lumped parameter model. Finally in section 7 we develop calculable methods and we apply them to the interpretation of neutron noise experimental data.

2. The Interaction of a Neutron Detector with the Stochastic Neutron Field

The detector counts are produced as result of neutron collision events inside the detector sensitive volume. For instance in a boron detector, the counts are produced through (n,α) reactions; while in a fission chamber, counts are triggered by fissions.

The detector will be considered as a part of the multiplicative system with total collision macroscopic cross section $\Sigma(\vec{r},v,t)$ and macroscopic detection cross section $\Sigma_D(\vec{r},v,t)$.

To study the counting statistics of a neutron detector, which is turned on for Δt_c sec, we define the following probabilities :

i) $P_{Nc}(d)$, which is the probability of observing N_c counts at a final time t_f, upon insertion of a neutron source into the system. d, denotes the counting interval $(t_f-\Delta t_c, t_f)$ defined by the couple of variables $(t_f, \Delta t_c)$.

ii) $K_{nc}(1,t,d)$, which is the probability of observing n_c, counts upon injection of one neutron at the phase space point and time $(1,t)$. The phase space point $(\vec{r},v,\vec{\Omega})$ is denoted by 1.

iii) $K^{(i)}_{nc}(\vec{r},t,d)$ which is the probability of observing n_c counts, upon injection into the system of one delayed neutron precursor of i-th kind at the position, and time $(\vec{r},t)$.

The associated probability generating functions are:

$$G_S(Z,d)=\sum_{N_c=0}^{\infty} Z^{Nc} \, . \, P_{Nc}(d) \tag{1}$$

$$G_K(Z,1,t,d)=\sum_{n_c=0}^{\infty} Z^{nc} K_{nc}(1,t,d) \tag{2}$$

$$G^{(i)}_K(Z,\vec{r},t,d)=\sum_{n_c=0}^{\infty} Z^{nc} K^{(i)}_{nc}(\vec{r},t,d) \tag{3}$$

The next step is to define the so called C-probabilities as follows [8]:

i) $C_{0c}(\vec{r},v,t)=$ Probability that a neutron collision leads to a capture event without neutron emission, this probability can be expressed in terms of nuclear parameters as follows [8]:

$$C_{0c}(\vec{r},v,t)=\Sigma_c(\vec{r},v,t)/\Sigma(\vec{r},v,t) \tag{4}$$

where $\Sigma_c(\vec{r},v,t)$ and $\Sigma(\vec{r},v,t)$ are the capture and total macroscopic cross-sections respectively.

ii) $C_{1S}(\vec{r},v,\vec{\Omega},t/v',\vec{\Omega}')\,dv'd\Omega'$ = Probability that a collision event with a neutron of speed and direction $v,\vec{\Omega}$, leads to one neutron scattered with velocity within the interval $(v',v'+dv')$, and direction within a $d\Omega'$ around $\vec{\Omega}'$. This probability is given in terms of nuclear parameters by:

$$C_{1S}(\vec{r},v,\vec{\Omega},t/v',\vec{\Omega}') = \Sigma_S(\vec{r},v,t)\, f_S(\vec{r},v,\vec{\Omega}/v',\vec{\Omega}')/\Sigma(\vec{r},v,t) \tag{5}$$

where $\Sigma_S(\vec{r},v,t)$ is the macroscopic scattering cross-section, and $f_S(\vec{r},v,\vec{\Omega}/v',\vec{\Omega}')$ the scattering Kernel.

iii) $C_{jl1l2\ldots l6}(\vec{r},v,t)$ ≡ Probability that a neutron collision with a neutron of speed v, leads to j prompt neutrons, and $l_1,l_2,\ldots l_6$ delayed neutron precursors of groups 1,2,...,6 respectively. This probability is given in terms of nuclear parameters by:

$$C_{jl1l2\ldots l6}(\vec{r},v,t) = \Sigma_f(\vec{r},v,t).\varepsilon_{jl1l2\ldots l6}(v)/\Sigma(\vec{r},v,t) \tag{6}$$

where $\varepsilon_{jl1l2l6}(v)$, is the probability that a fission event with a neutron of speed v, leads to j prompt neutrons and $l_1,l_2,\ldots l_6$ delayed neutron precursors of group 1,2,...6 respectively; and $\Sigma_f(\vec{r},v,t)$ is the macroscopic fission cross section.

Because not all neutron collisions events within the sensitive detector volume V_D, necessarily lead to detector counts, we need to define for each collision type, the probability to have one count or not. To this end, one must redefine the C_{0c} probability as follows:

$$C_{0c}(\vec{r},v,t) = C^{(0)}_{0c}(\vec{r},v,t) + C^{(1)}_{0c}(\vec{r},v,t) \tag{7}$$

where $C^{(i)}_{0c}$, (i=0,1), is the probability that a neutron collision leads to a neutron capture event and to i(0 or 1) counts in the detector. Obviously $C^{(1)}_{0c}(\vec{r},v,t)$ is different from zero only when t belongs to the counting interval, i.e. $t \in [t_f-\Delta t_c, t_f]$. The second condition that must hold in order to have $C^{(1)}_{0c}$ 0 is that r belongs to the detector sensitive volume, i.e. $\vec{r}\in V_D$. In the same way, we introduce similar definitions for the C_{1S} and $C_{jl1l2\ldots l6}$ probabilities.

$$C_{1S}(1,t/v',\vec{\Omega}') = C^{(0)}_{1S}(1,t/v',\vec{\Omega}') + C^{(1)}_{1S}(1,t/v',\vec{\Omega}') \tag{8}$$

$$C_{jl1l2\ldots l6}(\vec{r},v,t) = C^{(0)}_{jl1l2\ldots l6}(\vec{r},v,t) + C^{(1)}_{jl1l2\ldots l6}(\vec{r},v,t) \tag{9}$$

where $C^{(i)}_{1S}(1,t/v',\vec{\Omega}')\,dv'd\Omega'$, is the probability to have in a collision at (1,t) a scattering event leading to i (0 or 1) counts in the detector and to one scattered neutron with speed within $v',v'+dv'$ and direction within a $d\Omega'$ around $\vec{\Omega}'$. $C^{(i)}_{jl1l2\ldots l6}(\vec{r},v,t)$, with i(0 or 1), is the probability to have in a collision a neutron fission event, leading to: j prompt neutrons, $l_1,l_2,\ldots l_6$ delayed neutron precursors and to i (0 or 1) counts. We now write down the probability balances for the functions $K^{(i)}_{nc}(\vec{r},t,d)$ and $K_{nc}(1,t,d)$ in terms of $K_{nc}(1,t,d)$. We start with the probability balance for $K^{(i)}_{nc}(\vec{r},t,d)$ with i ranging from 1 to 6:

$$K^{(i)}_{nc}(\vec{r},t,d) = \delta_{nc0}.\exp(-\lambda_i(t_f-t)) + \int_t^{t_f} dt' \int dv' \int d\Omega'$$

$$\exp(-\lambda_i(t'-t)).\lambda_i.X^i(v').K_{nc}(\vec{r},v',\vec{\Omega}',t',d)/4\pi \tag{10}$$

where λ_i and $X^i(v)$ are the decay constant and the normalized neutron spectrum respectively of the i-th kind of precursors. The first term in equation (10), takes on account the probability of no precursor disintegration in the time interval (t,t_f); if this happens we do not inject the neutron into the system and therefore will be zero the number of counts. The second term is formed by:

i) $\exp(-\lambda_i\ (t'-t)).\lambda_i dt'$, which is the probability that the i-precursor disintegrate in the time interval $(t',t'+dt')$. It is assumed, that each precursor disintegration gives one neutron, this is the standard convention followed by many authors [10].

ii) $X^{(i)}(v')dv'd\Omega'/4\pi$, probability of the neutron to be isotropically emitted with velocity and direction within the intervals $(v',v'+dv')$, and $d\Omega'$ around $\vec{\Omega}'$.

iii) $K_{nc}(\vec{r},v',\vec{\Omega}',t',d)$, which has been defined previously.

We now, set up the probability balance for the function $K_{nc}(1,t,d)$ [8]:

$$K_{nc}(1,t,d)=\int_0^{S_v} dS\ P(1,t,S)\left[\sum_{q=0}^{1} C^{(q)}{}_{0c}\ (\vec{r}+s\vec{\Omega},\ v,t\ +s/v).\delta_{ncq}\right.$$

$$+\int dv'\int d\Omega' \sum_{q=0}^{1} C^{(q)}{}_{1S}\ (\vec{r}+s\vec{\Omega},v,\vec{\Omega},t+s/v/v',\vec{\Omega}')K_{nc-q}\ (\vec{r}+s\vec{\Omega},v',\vec{\Omega}',t+s/v,d)$$

$$+\sum_{q=0}^{1}\sum_{j=0}^{I}\left[\sum_{m_1(q)=0}^{n_c-q}\cdots\sum_{m_j(q)=0}^{n_c-q}\right]\sum_{l_1=0}^{L_1}\left[\sum_{m'_1(q)=0}^{n_c-q}\cdots\sum_{m'l_1(q)=0}^{n_c-q}\right]\cdots$$

$$\sum_{l_6=0}^{L_6}\left[\sum_{m''_1(q)=0}^{n_c-q}\cdots\sum_{m''l_6(q)=0}^{n_c-q}\right] C^{(q)}{}_{jl1\ldots l6}\ (\vec{r}+s\vec{\Omega},v,t+s/v)\prod_{k=1}^{j}\int dv'_K\int d\Omega'_K$$

$$\frac{X(v'k)}{4\pi}\ .\ K_{m_k(q)}\ (\vec{r}+s\vec{\Omega},v'_k,\ \vec{\Omega}'_k,t+s/v,d)\prod_{p=1}^{l_1} K^{(1)}{}_{m'_p(q)}\ (\vec{r}+s\vec{\Omega},t+s/v,d)$$

$$\left.\ldots\prod_{r=1}^{l_6} K^{(6)}{}_{m''_r(q)}\ (\vec{r}+s\vec{\Omega},t+s/v,d)\right] +A_d \qquad (11)$$

where the convention $\prod_{k=1}^{0}(\quad)=1$, has been assumed.

In equation (11) we have the constraints:

$$\sum_{k=1}^{j} m_k(q) + \sum_{p=1}^{l_1} m'_p(q) + \ldots + \sum_{r=1}^{l_6} m''_r(q) = n_c - q \quad \text{with } q=0,1 \tag{12}$$

Also in writing equation (11), we have introduced the following definitions:

$$S_v = \text{minimum } \{S_B, (t_f - t)v\} \tag{13}$$

where S_B is the distance between the neutron birthplace, $\vec{r}$, and the convex boundary of the system along the direction $\vec{\Omega}$.

$P(1,t,S)dS$= Probability that a neutron born at (1,t) interacts with the system within the distance interval $[\vec{r}+s\vec{\Omega}, \vec{r}+(s+ds)\vec{\Omega}]$ and within the time interval $[t+s/v, t+(s+ds)/v]$.

$$P(1,t,S) = \Sigma(\vec{r}+S\vec{\Omega}, v, t+S/v).\exp\left(-\int_0^S ds' \, \Sigma(\vec{r}+s'\vec{\Omega}, v, t+s'/v)\right) \tag{14}$$

and A_d is given by:

$$A_d = \begin{cases} \delta_{nc0}.\exp\left(-\int_0^{S_B} .ds.\Sigma(\vec{r}+s\vec{\Omega}, v, t+s/v)\right) & \text{if } v(t_f-t) \geq S_B \\ \delta_{nc0}.\exp\left(-\int_0^{v(t_f-t)} ds \, \Sigma(\vec{r}+s\vec{\Omega}, v, t+s/v)\right) & \text{if } v(t_f-t) < S_B \end{cases} \tag{15}$$

The physical interpretation of equation (11) in spite of its complicated appearance is quite obvious. The first term within the curly bracket gives the probability that an interaction in the path interval dS, should lead to a neutron capture event and to n_c= 0, or n_c = 1 counts in the detector.

The second term arises from the probability that after a scattering event one neutron emerges with speed and direction v', $\vec{\Omega}'$ and we have considered the possibility that, if the detector is a proton-recoil one, the scattering event within the sensitive volume could lead to zero or one count.

The third term arises from the probability to have in a neutron collision event a fission which yields j prompt neutrons and $l_1, l_2 \ldots l_6$ delayed neutron precursors. I and L_1, $L_2, \ldots L_6$ are the maximum number of neutrons and delayed precursors in a fission event.

Finally the extraterm A_d, accounts for the probability of a collisionless neutron leaving the system through the boundary, or which has not interacted at time t_f.

The next step is to multiply equations (10) and (11) by Z^{nc}, and then to sum up from $n_c=0$ to $n_c=\infty$. In this way it is obtained the following system of integral equations [8]:

$$G_K(Z,1,t,d)=\int_0^{S_V} ds\ P(1,t,s)\left[\sum_{q=0}^{1} Z^q\ C^{(q)}{}_{0C}\ (\vec{r}+s\vec{\Omega},v,t+s/v)\right.$$

$$+\int dv'\int d\Omega'.\sum_{q=0}^{1} Z^q\ C^{(q)}{}_{1S}\ (\vec{r}+s\vec{\Omega},v,\vec{\Omega},t+s/v/v',\vec{\Omega}')G_K(Z,\vec{r}+s\vec{\Omega},v',\vec{\Omega}',t+s/v,d)$$

$$+\ldots+\sum_{j=0}^{I}\sum_{l_1=0}^{L_1}\ldots\sum_{l_6=0}^{L_6}\sum_{q=0}^{1} Z^q\ C^{(q)}{}_{jl1l2\ldots l6}\ (\vec{r}+s\vec{\Omega},v,t+s/v).$$

$$.\left[\int dv'\int d\Omega'\frac{X(v')}{4\pi}.\ G_K\ (Z,\vec{r}+s\vec{\Omega},v',\vec{\Omega}',t+s/v,d)\right]^j.$$

$$\left.\cdot\prod_{i=1}^{6}\left[G^{(i)}{}_K\ (Z,\vec{r}+s\vec{\Omega},t+s/v,d)\right]^{l_i}\right]+A'_d \qquad (16)$$

$$G^{(i)}{}_K\ (Z,\vec{r},t,d)=\exp(-\lambda_i\,(t_f-t))+\int_t^{t_f} dt'\int dv'\int d\Omega'.\exp(-\lambda_i(t'-t)).\lambda_i.$$

$$X^i\ (v').G_K\ (Z,\vec{r},v',\vec{\Omega}',t',d)/4\pi \qquad i=1,2\ldots6 \qquad (17)$$

where

$$A'_d=\begin{cases}\exp(-\int_0^{S_B} ds.\Sigma(\vec{r}+s\vec{\Omega},v,\ t+s/v)) & \text{if } v(t_f-t)\geq S_B\\[2ex] \exp(-\int_0^{(t_f-t)} ds.\Sigma(\vec{r}+s\vec{\Omega},v,t+s/v)) & \text{if } v(t_f-t)<S_B\end{cases}$$

Now, we apply Pal's methodology [5], to equation (16), i.e we compute the following limit

$$\lim_{\frac{\delta S}{v}\to 0}\frac{G_K(Z,\vec{r}+\delta S\vec{\Omega},v,\vec{\Omega},t+\delta S/v,d)-G_K(Z,\vec{r},v,\vec{\Omega},t,d)}{\frac{\delta S}{v}}$$

In this way it is obtained the following result:

$$\left[1/v\partial/\partial t+\vec{\Omega}.\vec{\nabla}\ -\Sigma\ (\vec{r},v,t)\right]G_K\ (Z,1,t,d)=\Sigma\ (\vec{r},v,t)\ \left[\sum_{q=0}^{1} Z^q\ C^{(q)}{}_{0C}(\vec{r},v,t)\right.$$

$$+\int dv'\int d\Omega' \sum_{q=o} Z^q\ C^{(q)}{}_{1S}\ (\vec{r},v,\vec{\Omega},t/v',\vec{\Omega}')\ G_K\ (Z,\vec{r},v',\vec{\Omega}',t,d)+$$

$$+\sum_{j=0}^{I}\sum_{l_1=0}^{L_1}\cdots\sum_{l_6=0}^{L_6}\sum_{q=0}^{1} Z^q\ C^{(q)}{}_{j1112\ldots16}(\vec{r},v,t).$$

$$\left.\cdot\left[\int dv'\int d\Omega' X(v')G_K(Z,\vec{r},v',\vec{\Omega}',t,d)/4\pi\right]^j \cdot\prod_{i=1}^{6}\left[G^{(i)}{}_K(Z,\vec{r},t,d)\right]^{l_i}\right] \tag{18}$$

Next, we derivate the equation for $G^{(i)}{}_K(Z,\vec{r},t,d)$ with respect to t which yields:

$$\frac{\partial G^{(i)}{}_K(Z,\vec{r},t,d)}{\partial t} = \lambda_i G^{(i)}{}_K(Z,\vec{r},t,d)-\lambda i\int dv'\int d\Omega'.\ \frac{X^i(v')}{4\pi}\ G_K(Z,\vec{r},v'\vec{\Omega}',t,d) \qquad i=1,2\ldots,6 \tag{19}$$

The boundary conditions associated with equations (18) and (19) are easily deduced. For a neutron injected into the system at a time $t > t_f$, the function $K_{nc}(1,t,d)$ is equal to zero according to the causality condition, and we have

$$G_K(Z,1,t,d) = 1,\ t > t_f \tag{20}$$

A similar reasonning lead us to:

$$G^{(i)}{}_K\ (Z,\vec{r},t,d) = 1,\ t > t_f \tag{21}$$

Clearly for an outwardly directed neutron injected at the convex boundary of the system one must have $K_{nc}\ (1,t,d)=\delta_{nc0}$; from this result it follows

$$G_K\ (Z,1,t,d) = 1 \qquad \vec{n}_B\ .\ \vec{\Omega} > 0 \text{ and } \vec{r} \in S_B$$

where S_B denotes the convex boundary of the system.

The average number of detector counts, the variance and so on..., can be obtained from the generating function $G_S(Z,d)$, defined in terms of the probability $P_{Nc}(d)$ of observing N_c counts at a final time t_f.

The connection between the generating function $G_S(Z,d)$ and $G_K(Z,1,t,d)$ is derived on the basis of a probability balance for $P_{Nc}(d)$ in terms of $K_{nc}(1,t,d)$, after some calculations one arrives to [7,8]:

$$G_S(Z,d)= \exp \left[\sum_{j=0}^{I_S} \varepsilon_{s,j} \int_{t_0}^{t_f} dt \int dr\, S_0(\vec{r},t) \left[\left[\int d\Omega \right.\right.\right.$$

$$\left.\left.\left. \int dv\, X_S(v)/4\pi G_K(Z,1,t,d) \right]^j -1 \right] \right] \tag{22}$$

where $S_0(\vec{r},t)$ is the source disintegration rate per unit volume at time t, $\varepsilon_{S,j}$ gives the probability of emission of j neutrons per source disintegration, $X_S(v)$ is the normalized source spectrum, and finally Is gives the maximum number of neutrons released per source disintegration.

3. The Average Number of Counts, the Variance and the Feynman Y. function

To get an expression for the average number of counts $\bar{N}_c(d)$, we operate on equation (22), with $\partial/\partial Z|_{Z=1}$, to obtain after some trivial calculus:

$$\bar{N}_c(d)= \int_{t_0}^{t_f} dt \int dr\, \bar{\nu}_S\, S_0(\vec{r},t) \int dv \int d\Omega\, X_S(v)/4\pi . \bar{n}_c(1,t,d) \tag{23}$$

where:

$\bar{n}_c(1,t,d)= \partial/\partial Z\; G_K(Z,1,t,d)|_{Z=1}$, gives the average number of counts upon injection of one single neutron at (1,t), and $\bar{\nu}_S$ is the average number of neutrons released per source disintegration. Expresion (23), can be written in a more compact form, defining the neutron source strength per phase space and time units S(1,t) as follows:

$$S(1,t) = \bar{\nu}_S\, S_0\, (\vec{r},t)\, X_S\, (v)/4\pi \tag{24}$$

with this definition we recast equation (23) in the form:

$$\bar{N}_c(d)= < \bar{n}_c\, (1,t,d)\, S(1,t)> \tag{25}$$

where the symbol < > means integration over the phase space variables and time. The next step is to compute $\bar{n}_c(1,t,d)$, to this end we operate on equations (18) and (19), with $\partial/\partial Z|_{Z=1}$, to obtain after some lengthly calculations, the following result:

$$K^+\, \psi^+{}_C= \begin{bmatrix} \Sigma_D(\vec{r},v,t,d) \\ 0 \\ . \\ . \\ . \\ 0 \end{bmatrix} \tag{26}$$

where, K^+, is the time dependent adjoint Boltzmann matrix operator, which is given by:

$$K^{+} = -1/v\ \partial/\partial t . I + \begin{bmatrix} H^{+} & -M^{+}_{1} & \dots & -M^{+}_{6} \\ -\lambda_1/v & \lambda_1/v & \dots & 0 \\ \cdot & \cdot & & \cdot \\ \cdot & \cdot & & \cdot \\ \cdot & \cdot & & \cdot \\ -\lambda_6/v & 0 & & \lambda_6/v \end{bmatrix} \tag{27}$$

where I is unit matrix and the operators H^{+}, $M^{+}_{i}(i=1,2...6)$, are given by

$$H^{+} = -\vec{\Omega}.\vec{\nabla} + \Sigma(\vec{r},v,t) - M^{+}_{0} - S^{+}_{0} \tag{28}$$

$$M^{+}_{i} = \int dv' \int d\Omega'\ \overline{\beta}_i\ (v).\nu(v).\Sigma_f\ (\vec{r},v,t)X^{i}(v')/4\pi \tag{29}$$

and where S^{+}_{0} and M^{+}_{0} are the scattering and fission adjoint operators respectively, which are given by:

$$S^{+}_{0} = \int dv' \int d\Omega'\ \Sigma_S\ (\vec{r},v,t)\ f_S\ (\vec{r},v,\vec{\Omega}/v',\vec{\Omega}') \tag{30}$$

$$M^{+}_{0} = \int dv' \int d\Omega'[1-\overline{\beta}(v)]\ \overline{\nu}(v).\Sigma_f(\vec{r},v,t).X(v')/4\pi \tag{31}$$

where $\overline{\beta}_i(v)$ is the fraction of fission neutrons due to the decay of the i-th kind of precursors and the rest of the symbols is well known or has been defined previously, Σ_D $(\vec{r},v,t,d)$ is the macroscopic detection cross section. Finally the column vector $\psi^{+}_{C}(t)$, is given by:

$$\psi^{+}_{C}(t) = \begin{bmatrix} \overline{n}_C(1,t,d) \\ \overline{C}^{+}_{C1}(1,t,d) \\ \cdot \\ \cdot \\ \cdot \\ \overline{C}^{+}_{C6}(1,t,d) \end{bmatrix} \tag{32}$$

where $\overline{C}^{+}_{Ci}(1,t,d)$ is defined by

$$\overline{C}^{+}_{Ci}(1,t,d) = \int_{t}^{t_f} dt'\lambda_i\ \exp(-\lambda_i(t'-t))\overline{n}_C(1,t',d) \tag{33}$$

Now from equations (25), (26) and the standard Boltzmann equation, it can be shown that $\overline{N}_C(d)$ is also given by [8]:

$$\overline{N}_C\ (d) = \langle \Sigma_D\ \phi \rangle \tag{34}$$

which expresses as expected that the average number of counts is given by the product of the average flux and the macroscopic detection cross section, integrated over the phase space variables and the time.

It is very easy to check that the variance of neutron detector counts, denoted by VAR $[N_c\ (d)]$, is given by:

$$\mathrm{VAR}\ [Nc(d)] = \left.\frac{\partial^2\ G_S(Z,d)}{\partial Z^2}\right|_{Z=1} + \left.\frac{\partial G_S(Z,d)}{\partial Z}\right|_{Z=1} - \left[\left.\frac{\partial G_S(Z,d)}{\partial Z}\right|_{Z=1}\right]^2 \tag{35}$$

Now, on account of equations (22), (24), and (25), it is obtained:

$$\mathrm{VAR}\ [N_c(d)] = \langle\ \overline{n^2}_C(1,t,d)\ S(1,t)\ \rangle + Q_S(d) \tag{36}$$

where S(1,t) is the source strength per unit of phase space and time, defined in (24), and we have defined:

$$Q_S(d) = \overline{\nu_S(\nu_S-1)} \int_{t_0}^{t_f} dt \int dr\ S_0(\vec{r},t)\ I^2_S(\vec{r},t,d) \tag{37}$$

with

$$\overline{\nu_S(\nu_S-1)} = \sum_{j=0}^{I_S} \varepsilon_{S,j}.j.(j-1) \tag{38}$$

$$I_S(\vec{r},t,d) = \int dv \int d\Omega\ X_S(v)/4\pi.\overline{n}_c(1,t,d) \tag{39}$$

$Q_S(d)$ gives the contribution to the variance coming from the fluctuations in the number of neutron released per source disintegration, this contribution is proportional to $\overline{\nu_S(\nu_S-1)}$, and to the square of the weight function $I_S(\vec{r},t)$, which is the source fission spectrum weighted adjoint function.

We now must find the function $\overline{n^2}_C(1,t,d)$ which enter into equation (91). To this end we operate on equations (18) and (19) with $\partial^2/\partial Z^2|_{Z=1}$ and after a lengthly calculation it is obtained [8]:

$$K^+ \begin{bmatrix} \overline{n^2}_c(1,t,d) \\ \overline{C^{+2}}_{C1}(1,t,d) \\ \cdot \\ \cdot \\ \overline{C^{+2}}_{C6}(1,t,d) \end{bmatrix} = \begin{bmatrix} Q(1,t,d) \\ 0 \\ \cdot \\ \cdot \\ 0 \end{bmatrix} \tag{40}$$

where K^+ is given by (27), the functions $\overline{C^{+2}}_{Ci}(1,t,d)$, i= 1, to 6 are defined by

$$\overline{C^{+2}}_{Ci}(1,t,d)= \int_t^{t_f} dt'\exp(-\lambda_i\ (t'-t)).\lambda_i.\overline{n^2}_c(1,t',d) \tag{41}$$

Finally Q(1,t,d), plays the role of a source term in equation (40) and is given by:

$$Q(1,t,d)= \Sigma_D\ (\vec{r},v,t,d)\ +\ Q_1(1,t,d)\ +\ \sum_{i=1}^{6} Q^{i(d)}(1,t,d)\ +\ 2\ Q_{1D}(1,t,d)$$

$$+\ \sum_{i,j(i\neq j)}^{6} Q^{ij(d)}(1,t,d)\ +\ 2\ \sum_{i=1}^{6}\ Q^{i(P)}(1,t,d) \tag{42}$$

It is interesting to analyze the physical meaning of the various terms which enter into (42). These terms are:

i) $\Sigma_D(\vec{r},v,t,d)$ is the well known macroscopic detection cross-section.

ii) $Q_1(1,t,d)$ is a source term defined by:

$$Q_1(1,t,d)= \overline{\nu_p\ (\nu_p-1)}.\Sigma_f(\vec{r},v,t).I^2{}_1(\vec{r},t,d) \tag{43}$$

with

$$I_1(\vec{r},t,d)= \int dv \int d\Omega\ X(v)/4\pi\ \overline{n}_C(1,t,d) \tag{44}$$

This source term arises from the fluctuations in the number of prompt neutrons $\overline{\nu}_p$, released per fission. We note that this term is proportional to $\overline{\nu_p(\nu_p-1)}$, and to the macroscopic fission cross section of the reactive assembly, multiplied by the square of $I_1(\vec{r},t,d)$, which is the adjoint function weighted with the prompt neutron spectrum.

iii) $\sum_{i=1}^{6} Q^{i(d)}$ (1,t,d), is a source term that is negligible for practical purposes,

$Q^{i(d)}$ is given by:

$$Q^{i(d)}\ (1,t,d)= \overline{l_i(l_i-1)}\Sigma_f(\vec{r},v,t)][I^{i(d)}\ (\vec{r},t,d)]^2 \tag{45}$$

This term arises from the fluctuations in the number l_i, of delayed neutron precursors of group i, released per fission. We note that $\overline{l_i(l_i-1)}$ is very small, because it is very improbable to have a fission where the two fragments belong to the same delayed neutron precursor group. $I^{i(d)}$ is an importance weight function for the i-delayed neutron precursor group given by:

$$I^{i(d)}(\vec{r},t,d)= \int dv \int d\Omega\ X^i(v)/4\pi\ \overline{C^+}_{Ci}(1,t,d) \tag{46}$$

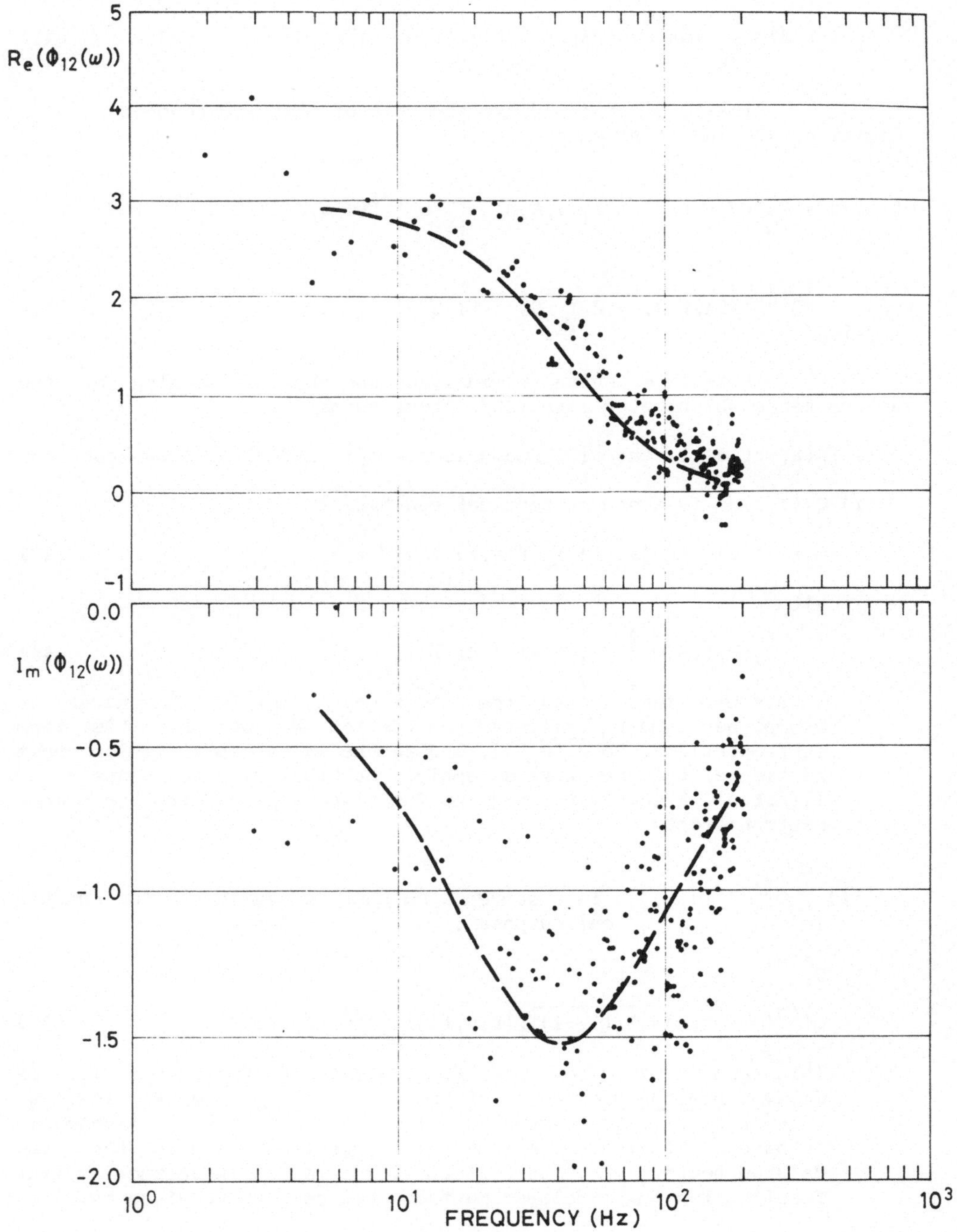

Fig. 1. Real and imaginary parts of $\Phi_{12}(\omega)$ (arbitrary units), versus frequency, for run AB.

where $X^i(v)$ is the neutron spectrum for the (i) group of neutron precursors and $C^+_{Ci}(1,t,d)$ has been defined in (33).

iv) $\sum_{i,j\,(i\neq j)} Q^{ij(d)}(1,t,d)$

this source term arises from the correlations between delayed neutron precursors of different groups, i.e. the contributions to this term come from fissions where one fragment belongs to group, i, and the other one to group j. $Q^{ij(d)}(1,t,d)$ is defined by:

$$Q^{ij(d)}(1,t,d)=\overline{l_i l_j}\ \Sigma_f(\vec{r},v,t)\ I^{i(d)}(\vec{r},t,d)I^{j(d)}(\vec{r},t,d) \tag{47}$$

where

$$\overline{l_i l_j}=\sum_{j=0}^{I}\sum_{l_1}^{L_1}\cdots\sum_{l_6}^{L_6} l_i l_j\ \varepsilon_{jl1\ldots l6} \tag{48}$$

v) $2\sum_{i=0}^{I} Q^{i(p)}(1,t,d)$, this term arises from the correlations between prompt and delayed neutrons, where $Q^{i(p)}(1,t,d)$ is obviously given by:

$$Q^{i(p)}=\overline{l_i\nu_p}\ \Sigma_f(\vec{r},v,t)\ I^{i(d)}(\vec{r},t,d)\ I_1(\vec{r},t,d) \tag{49}$$

This term is produced because in one fission, it is possible to have one or several prompt neutrons and one fragment which is a delayed neutron precursor; obviously these prompt neutrons are correlated with this delayed neutron.

vi) The source term $2Q_{1D}(1,t,d)$, arises from the detector and for a fission chamber $Q_{1D}(1,t,d)$ is given by:

$$Q_{1D}(1,t,d)=\overline{\nu}_D\cdot\Sigma_{fD}(\vec{r},v,t,d)\left[(1-\overline{\beta}_D)I_D(\vec{r},t,d)+\sum_{i=1}^{6}\overline{\beta}^i{}_D\ I^{i(d)}{}_D(\vec{r},t,d)\right] \tag{50}$$

where, $\overline{\nu}_D$ is the average number of neutrons produced per fission inside the chamber, $\overline{\beta}_D$ is the delayed neutron fraction, $\Sigma_{fD}(\vec{r},v,t,d)$ is the macroscopic detection cross section into the fission chamber and finally $I_D(\vec{r},t,d)$ and $I^{i(d)}{}_D(\vec{r},t,d)$ are the spectral weighted importance functions for the prompt and delayed neutrons released into the chamber respectively. These functions are given by:

$$I_D(\vec{r},t,d)=\int dv\int d\Omega\ .\ X_D(v)/4\pi\ .\ \overline{n}_c(1,t,d) \tag{51}$$

$$I^{i(d)}{}_D(\vec{r},t,d)=\int dv\int d\Omega\ .\ X^i{}_D(v)/4\pi\ \overline{C}^+{}_{Ci}(1,t,d) \tag{52}$$

where $X_D(v)$ and $X^i{}_D(v)$ are the normalized neutron spectrum of the prompt and delayed neutrons respectively.

In order to calculate the first term, we need some previous definitions, the scalar product of two real or complex vectors:

$$\psi(1,t)= \text{col } [\psi_1(1,t), \psi_2(1,t)...\psi_n(1,t)]$$

and

$$\varphi(1,t)= \text{col } [\varphi_1(1,t), \varphi_2(1,t)... \varphi_n(1,t)]$$

is defined by:

$$<\psi|\varphi> = \int dt \int d^3r \int d^3v \sum \psi^*{}_i(1,t) \varphi_i(1,t) \tag{53}$$

Now we remind the standard Boltzmann equation [8,10]

$$K \psi(t)= S \tag{54}$$

where $\psi(t)$ is the column vector

$$\psi(t)= \begin{bmatrix} \bar{\phi}(1,t) \\ vC_1(\vec{r},t)X^1(v)/4\pi \\ \cdot \\ \cdot \\ \cdot \\ vC_6(\vec{r},t)X^6(v)/4\pi \end{bmatrix} \tag{55}$$

S is the column vector:

$$S= \text{col } [S(1,t), 0, ...,0] \tag{56}$$

$\bar{\phi}(1,t)$ is the neutron angular flux, K is the standard Boltzmann matrix operator, and $C_1(\vec{r},t)...C_6(\vec{r},t)$ are the concentrations of delayed neutron precursors.

To calculate the first term of the right hand side of equation (36), we form the inner product of $\psi(t)$, given by (55), with equation (40), which on account of the commutation relation between a matrix operator, and its adjoint [10] yields:

$$<\psi(t)|K^+\psi_{2C}>=<K\psi(t)|\psi^+{}_{2C}>=<S(1,t)\overline{n^2{}_C(1,t)}>=<\bar{\phi}(1,t)Q(1,t,d)> \tag{57}$$

To calculate the relative variance RVAR $[N_C(d)]$, we divide equation (36), by the average number of counts $\bar{N}_C(d)$, thus we have:

$$\text{RVAR } [N_C(d)]= 1 + Y(d) \tag{58}$$

where Y(d) is the excess of relative variance over one, i.e. over the characteristic relative variance of a Poissonian process.

Y(d) is known as the Feynman Y-Function and on account of (36), (42), (57) and (58) it is obtained that Y(d) is given by:

$$Y(d) = \frac{<\bar{\phi}\, Q_1>}{\bar{N}_c(d)} + \sum_{i=1}^{6} \frac{<\bar{\phi}\, Q^{i(d)}>}{\bar{N}_c(d)} + 2\sum_{i=1}^{6} \frac{<\bar{\phi}\, Q^{i(p)}>}{\bar{N}_c(d)} + \sum_{i,j(i\neq j)}^{6} \frac{<\bar{\phi}\, Q^{ij(d)}>}{\bar{N}_c}$$

$$+ 2\,\frac{<\bar{\phi}\, Q_{1D}>}{\bar{N}_c(d)} + \frac{Q_S(d)}{\bar{N}_c(d)} \qquad (59)$$

Now we are going to discuss the more important terms in equation (59)

(i) $Y_{sys}(d) = <\bar{\phi}\, Q_1>/N_C$ (60)

This term gives the excess of relative variance over one due to the fluctuations in the number of prompt neutrons released per fission in the system.

(ii) $Y_S(d) = Q_S(d)/\bar{N}_c$ (61)

This term gives the excess of relative variance over one due to the fluctuations in the number of neutrons released per disintegration in the neutron source. We note that this contribution is proportional to $\overline{\nu_S(\nu_S-1)}$ and to the integral over the system volume and time of the product of $S_0(\vec{r},t)$ and the square of the importance function $I_S(\vec{r},t,d)$. This means that for source neutrons produced far away from the detector, the contribution to $Y_S(d)$ will be smaller than for source neutrons produced close to the neutron detector.

(iii) $Y_D = 2<\bar{\phi}\, Q_{1D}>/N_C$ (62)

This term gives the excess of relative variance over one produced by the detector. We have already examined the physical meaning of Q_{1D} for the case of a fission chamber.

(iv) $Y_{pd} = 2\sum_{i=6}^{6} <\bar{\phi}\, Q^{i(p)}>/N_C$ (63)

This term gives the excess of relative variance over one due to the correlation between prompt and delayed neutrons, where $Q^{i(p)}$ is given by equation (49).

Inspection of equation (59), reveals that to calculate the variance and the Feynman Y(d) function, one needs the results of two deterministic calculations. First we need to solve equation (54), which is the standard Boltzmann transport equation for the average flux, and then to solve equation (32), which is the adjoint Boltzmann transport equation with the macroscopic neutron detection cross section as a source.

4. The Single Mode Approximation for a Subcritical Assembly

In this approximation we write $n_c(1,t,d)$ as follows

$$\bar{n}_c(1,t,d)= T(t)\ \phi^+_0(1) \tag{64}$$

where $\phi^+_0(1)$ is the solution of the adjoint eigenvalue equation:

$$L^+\ \psi^+_0(1)= -\ \alpha/v\ \psi^+_0(1) \tag{65}$$

where $\psi^+_0(1)$ is the column vector col $(\phi^+_0(1),\ C^+_{10}(1),\ldots,C^+_{60}(1))$, L^+ is the matrix operator:

$$L^+= \begin{bmatrix} H^+ & -M^+_1 & \ldots & -M^+_6 \\ -\lambda_1/v & \lambda_1/v & \ldots & 0 \\ -\lambda_6/v & 0 & \ldots & \lambda_6/v \end{bmatrix} \tag{66}$$

and α is the well known α-eigenvalue. From (65) it follows after some manipulations that $\phi^+_0(1)$ is solution of the following equation:

$$(H^+ - \sum_j M^+_j)\ \phi^+_0 = -\ \alpha/v\ \phi^+_0 - \sum_j\ [\alpha/\lambda_j+\alpha]\ M^+_j\ \phi^+_0 \tag{67}$$

If equation (67) is multiplied by the stationary average flux of the subcritical assembly $\phi(1)$, and then integrated over the phase space variables (r,Ω,v), we have the Inhour equation:

$$\rho/\Lambda + \alpha\ [1 + \sum_j \beta_j/\Lambda\ \ .\ 1/(\alpha+\lambda_j)]= 0 \tag{68}$$

where we have defined:

$$\rho \text{ (subcritical reactivity) } = (\phi^+_0,S)/F \tag{69}$$

$$\Lambda \text{ (mean generation time)}= (1/v\ \phi^+_0,\bar{\phi})/F \tag{70.a}$$

$$\beta_j \text{ (j-group effective delayed neutron fraction)}= (\phi,M^+_j\phi^+_o)/F \tag{70.b}$$

$$F \text{ (weighted neutron production) } = \bar{\nu}(I_0(\vec{r}),\Sigma_f\bar{\phi}) \tag{70.c}$$

$$I_0(\vec{r}) \text{ (neutron spectrum weighted importance)}= \int dv \int d\Omega\ X(v)/4\pi\ \phi^+_0(1) \tag{71}$$

and the symbol (,) means integration over the phase space variables. The roots of equation (69) will be denoted by α_k.

Now, if we consider that the detector gate is open during the time interval $(t_f-\Delta tc,t_f)$, we can write:

$$\Sigma_D(\vec{r},v,t,d)= \Sigma_D(\vec{r},v)\ \{H(t-(t_f-\Delta tc)) - H(t-t_f)\} \tag{72}$$

where H( ), are Heaviside functions and $\sum_D(r,v)$ can be assumed as time independent during the counting interval. Now from equations (26), (64) and (72) it is obtained after some trivial calculus that T(t) is given by:

$$T(t)=\frac{(\sum_D \bar{\phi})}{(\frac{1}{v}\ \phi^+{}_0\bar{\phi})}\sum_{k=1}^{7} V(\alpha_k)\ \exp\left[-\alpha_k(t-t_f)\right].\{1-\exp(-\alpha_k\ \Delta tc)\} \quad \text{for } t < t_f-\Delta tc \tag{73}$$

$$T(t)=\frac{(\sum_D \bar{\phi})}{(\frac{1}{v}\ \phi^+{}_0\bar{\phi})}\sum_{k=1}^{7} V(\alpha_k)\ \{\exp\left[-\alpha_k(t-t_f)\right]-1\} \quad \text{for } t_f-\Delta tc < t < t_f \tag{74}$$

where we have defined the function $V(\alpha_k)$ by:

$$V(\alpha_k)=\frac{1}{\alpha_k\left[1+\left(\sum_{j=1}^{6}\bar{\beta}_j\ \lambda_j/[\Lambda\ (\alpha_k+\lambda_j)^2]\right)\right]} \tag{75}$$

and $\{\alpha_k\}\Big|_{k=1}^{7}$ are the roots of the inhour equation [10].

Now we have all the ingredients that are needed to compute the Feynman Y(d) function in the unimodal approximation. We shall deduce in this approximation the two more important contributions to Y(d), i.e. $Y_{sys0}(d)$ and $Y_{s0}(d)$, which have been obtained from equations (60) and (61) respectively.

From (61), (37), and (64) we have that the source contribution is given by:

$$Y_{s0}(d)=\frac{\overline{\nu_s(\nu_s-1)}}{\bar{N}_c}\ .\ \int_{-\infty}^{t_f} dt \int dr.S_0(\vec{r})\ I^2{}_{s0}(\vec{r})\ T^2(t) \tag{76}$$

where the additional subscript 0, means unimodal (76) approximation and $I_{s0}(r)$ is a spectral weighted importance for the source neutrons given by:

$$I_{s0}(\vec{r})=\int dv \int d\Omega\ X_s(v)\ \phi^+{}_0(1)/4\pi \tag{77}$$

Finally we have considered the neutron source as time independent.

Now expression (76) for $Y_{s0}(d)$ can be integrated over time on account of expressions (73) and (74); and after a little calculus it is obtained:

$$Y_{s0}(d)=\varepsilon\ .\ \bar{\nu}_s/\bar{\nu}\ .\ D_s/\rho\ .\ R_s\ .\ Z(\Delta tc) \tag{78}$$

where ε, is the detector efficiency in counts per fissions, D_s is the

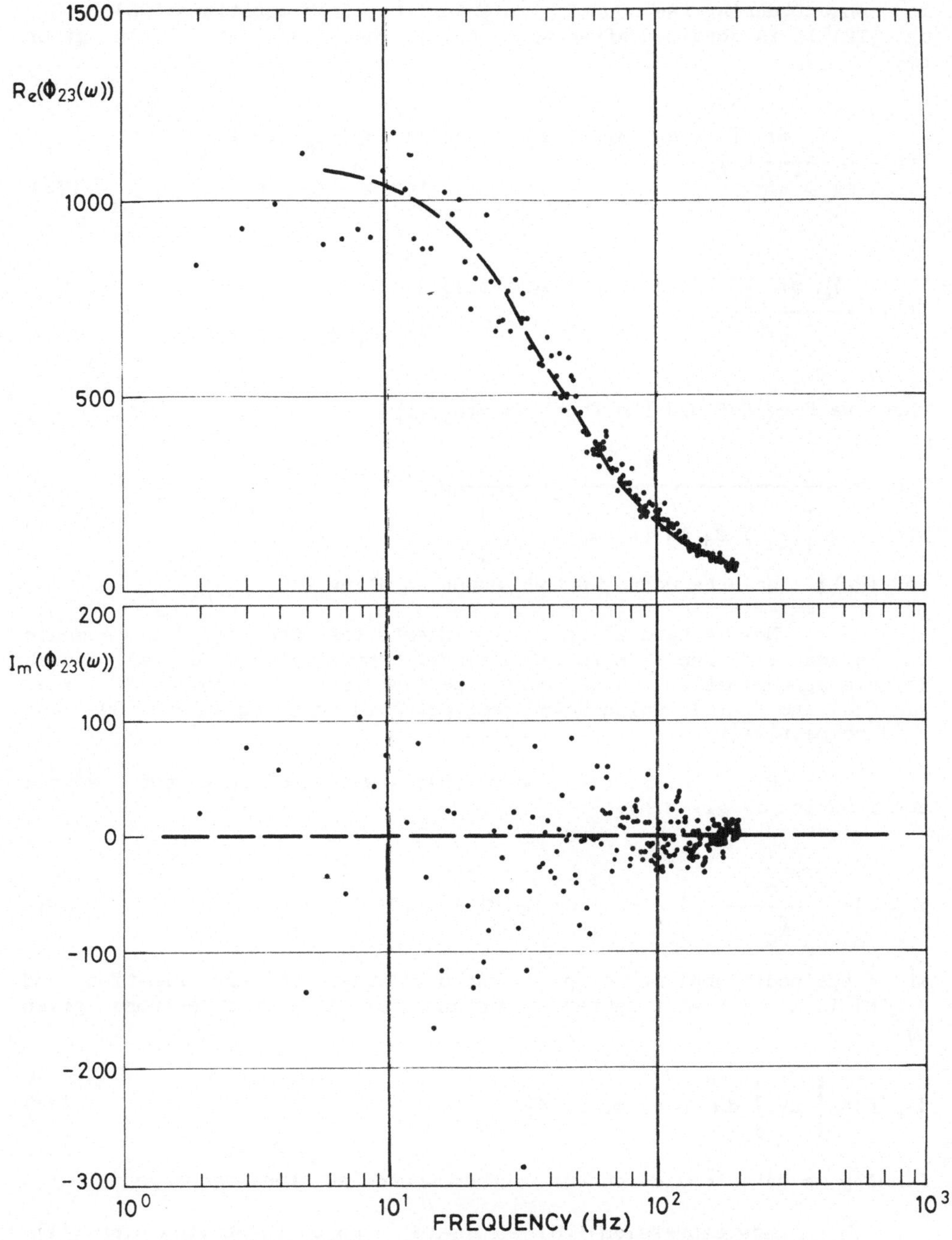

Fig. 2. Real and imaginary parts of $\Phi_{23}(\omega)$ (arbitrary units), versus frequency, for run AB.

source Diven factor [11], the subcritical reactivity, R_S is a factor which accounts for the neutron source spatial effects and is equal to:

$$R_S = \frac{F_T \int dr . \vec{\nu}_S . S_0(\vec{r}) . I^2_{S0}(r)}{(\Phi^+_0, S) . (I_0, \Sigma_f \bar{\phi})} \tag{79}$$

where F_T is the fission rate of the subcritical assembly. Finally $Z(\Delta tc)$ is a function defined by:

$$Z(\Delta tc) = \sum_{k=1}^{7} \alpha^2_0/\alpha^2_k . F(\alpha_k \Delta tc) . \alpha_k . V(\alpha_k) \tag{80}$$

where $\alpha_0 = (-\rho/\Lambda)$ and the $F(\alpha_k \Delta tc)$ function is defined by:

$$F(\alpha_k \Delta tc) = 1 + \frac{[1-\exp(\alpha_k . \Delta tc)]}{\alpha_k . \Delta tc} \tag{81}$$

Now from (60), (43), and (64) it is obtained that the excess of relative variance produced by the fluctuations in the number of prompt neutrons released per fission in the system is given in the unimodal approximation by:

$$Y_{sys0}(d) = \varepsilon . D . R . Z(\Delta tc)/\rho^2 \tag{82}$$

where D is the prompt neutron Diven factor and R is a factor which account for the system spatial effects and is given by:

$$R = \frac{F_T . \int F_T(\vec{r}) . I^2_0(\vec{r}) dr}{[\int F_T(\vec{r}) I_0(\vec{r}) dr]^2} \tag{83}$$

where $F_T(r)$ is the fission rate per unit volume at point r and $I_0(r)$ is the spectral weighted importance for prompt neutrons, given by:

$$I_0(\vec{r}) = \int dv \int d\Omega \; X(v) \; \phi^+_0 \; (1)/4\pi \tag{84}$$

It is interesting to observe, that when delayed neutrons are not considered, and spatial and source effects are not included, the Feynman Y function of this theory reduces to the standard expresion quoted by Williams and Lewins [13]:

$$Y(d) = \varepsilon . D/ \rho^2 \left[1 + \frac{[1-\exp(\alpha_0 . \Delta tc)]}{\alpha_0 \Delta tc} \right] \tag{85}$$

5. Covariance and cros correlation between two detectors

In this section we shall extend the discussion of section 2, to the two detector case and give the two master equations necessary for the derivation of the covariance function relating the counts in both detectors.

We introduce the probability $K_{ncmc}(1,t,d_1,d_2)$, that upon injection of one neutron at (1,t) one registers n_c counts in detector D_1 and m_c in detector D_2, in the time counting intervals $d_1(t_{f1}-\Delta tc_1, t_{f1})$ and $d_2(t_{f2}-\Delta tc_2)$. Also let, $K^{(i)}{}_{ncmc}(\vec{r},t,d_1,d_2)$ the probability that upon injection of one delayed neutron precursor of the i-th group at $(\vec{r},t)$ one registers n_c counts in D_1 and m_c in D_2. Finally let $P_{NcMc}(d_1,d_2)$ be the probability that after the insertion of the neutron source at to, there will arise N_c counts in D_1 and M_c counts in D_2. The generating functions associated with these three probabilities are denoted by $G_K(Z_1,Z_2,1,t,d_1,d_2)$, $G^{(i)}{}_K(Z_1,Z_2,\vec{r},t,d_1,d_2)$, $G_S(Z_1,Z_2,d_1,d_2)$ respectively.

Their definitions are the standard ones; for instance $G_S(Z_1,Z_2,d_1,d_2)$ is defined by:

$$G_S(Z_1,Z_2,d_1,d_2) = \sum_{N_c=0}^{\infty} \sum_{M_c=0}^{\infty} Z_1^{N_c} Z_2^{M_c} P_{NcMc}(d_1,d_2) \tag{86}$$

similar definitions hold for $G_K(Z_1,Z_2,1,t,d_1,d_2)$ and $G^{(i)}{}_K(Z_1,Z_2,\vec{r},t,d_1,d_2)$.

In terms of the above probability generating functions we define the following quantities:

$COV(d_1,d_2)$= Covariance function= $\overline{N_cM_c}(d_1,d_2)-\overline{N}_c(d_1).\overline{M}_c(d_2)$

where $\overline{NcMc}(d1,d2)= \left[\frac{\partial^2 \quad G_S}{\partial Z_1,\partial Z_2}\right]_{Z_1=Z_2=1}$, $\overline{N}_c(d_1)= \left[\frac{\partial \quad G_S}{\partial Z_1}\right]_{Z_1=Z_2=1}$,

$$\overline{M}_c(d_2)= \left[\frac{\partial \quad G_S}{\partial Z_2}\right]_{Z_1=Z_2=1}$$

$$\overline{n}_c(1,t,d_1)= [\partial/\partial Z_1 \; G_K(Z_1,Z_2,1,t,d_1,d_2)|Z_1=Z_2=1 \tag{87}$$

$\overline{n}_c(1,t,d_1)$ gives the average number of counts in D_1, upon injection of one neutron at (1,t). A similar definition holds for $\overline{m}_c(1,t,d_2)$.

$$\overline{n}^{(i)}{}_c(\vec{r},t,d_1)= [\partial/\partial Z_1 \; G^{(i)}{}_K(Z_1,Z_2,\vec{r},t,d_1,d_2)]\,|Z_1=Z_2=1 \tag{88}$$

$\overline{n}^{(i)}{}_c(\vec{r},t,d_1)$ gives the average number of counts in D_1, upon injection of one delayed neutron precursor of group i at $(\vec{r},t)$ a similar definition holds for $\overline{m}^{(i)}{}_c(\vec{r},t,d_2)$.

$\overline{n_c m_c}(1,t,d_1,d_2) = [\partial^2/\partial Z_1,\partial Z_2\ G_K]\,|Z_1=Z_2=1$ = average number of "doublets" upon injection of one neutron at (1,t).

We redefine now the C-probabilities to account for the presence of two detectors in the system.

$$C_{0c}(\vec{r},v,t) = C^{(0,0)}{}_{0C}(\vec{r},v,t) + C^{(1,0)}{}_{0C}(\vec{r},v,t) + C^{(0,1)}(\vec{r},v,t) \tag{89}$$

where:

$$C^{(1,0)}{}_{0C}(\vec{r},v,t) = \frac{\Sigma_{CD1}(\vec{r},v)}{\Sigma(\vec{r},v,t)} H_0(t,d_1) = \frac{\Sigma_{CD1}(\vec{r},v,t,d_1)}{\Sigma(\vec{r},v,t)} \qquad \vec{r} \in V_{D1} \text{ and } t \in d_1 \tag{90}$$

$$C^{(0,1)}{}_{0C}(\vec{r},v,t) = \frac{\Sigma_{CD2}(\vec{r},v)}{\Sigma(\vec{r},v,t)} H_0(t,d_2) = \frac{\Sigma_{CD2}(\vec{r},v,t,d_2)}{\Sigma(\vec{r},v,t)} \qquad \vec{r} \in V_{D2} \text{ and } t \in d2 \tag{91}$$

with the following definition of $H_0(t,d_i)$ (i=1, or 2) in terms of Heaviside functions:

$$H_0(t,d_i) = H[t-(t_{fi} - \Delta tc_i)] - H(t-t_{fi}) \qquad i = 1 \text{ or } 2 \tag{92}$$

we note that $C^{(1,0)}{}_{0C}(\vec{r},v,t)$ or $C^{(0,1)}{}_{0C}(\vec{r},v,t)$ are equal to zero when $\vec{r} \notin V_{D1}$ or $\vec{r} \notin V_{D2}$ respectively. Where V_{D1} and V_{D2} denote the sensitive volume of detectors D_1 and D_2 respectively. $\Sigma_{CD1}(\vec{r},v)$ is the macroscopic capture detection cross section for detector D_1, during the counting interval. Expresions (90) and (91) hold for capture detectors.

We note that with this definition $C^{(i,j)}{}_{0C}(\vec{r},v,t)$ is the probability to have in a neutron collission a neutron capture event leading to i(0 or 1) counts in detector D_1 and j(0 or 1) counts in detector D_2, i and j can not be simultaneously equal to one.

In the same way we redefine the $C_{jl1...l6}(\vec{r},v,t)$ probability as follows

$$C_{jl1l2...l6}(\vec{r},v,t) = C^{(0,0)}{}_{jl1l2...l6}(\vec{r},v,t) + C^{(1,0)}{}_{jl1l2...l6}(\vec{r},v,t) + C^{(0,1)}{}_{jl1l2...l6}(\vec{r},v,t) \tag{93}$$

where $C^{(1,0)}{}_{jl1l2...l6}$ is defined as the probability to have in a neutron collission a fission event leading to one count in detector D_1, zero in detector D_2, j prompt neutrons and $l_1 \ldots l_6$ delayed neutrons precursors. Obviously $C^{(1,0)}{}_{jl1l2...l6}$ is equal to zero for $\vec{r} \notin V_{D1}$ or $t \notin d_1$. A similar definition holds por $C^{(0,1)}{}_{jl1l2...l6}$ but interchanging D_1 by D_2.

For neutron detectors such as fission chambers we write:

$$C^{(1,0)}{}_{jl1l2...l6}(\vec{r},v,t) = \frac{\Sigma_{fD1}(\vec{r},v)}{\Sigma(\vec{r},v,t)} \cdot \varepsilon_{jl1l2...l6}(v) \cdot H_0(t,d_1) \cdot H_1(t,d_1,d_2) \quad \text{for } r \in V_{D1} \tag{94}$$

$$C^{(0,1)}{}_{jl1l2...l6}(\vec{r},v,t) = \frac{\Sigma_{fD2}(\vec{r},v)}{\Sigma(\vec{r},v,t)} \cdot \varepsilon_{jl1l2...l6}(v) \cdot H_0(t,d_2) \cdot H_2(t,d_1,d_2) \quad \text{for } r \in V_{D2} \tag{95}$$

where $\Sigma_{fD1}(\vec{r},v)$, $\Sigma_{fD2}(\vec{r},v)$, are the macroscopic fission detection cross section for detectors D_1, and D_2, during the counting intervals d_1 and d_2 respectively.

The factors H_1, are Heaviside functions which account for the detector gate closings. For instance, if $t_{f2} > t_{f1}$, those fission events taken place at the detector fission chamber D_2, at times larger than t_{f1}, will not affect the count rate at the detector D_1.

$$H_1(t,d_1,d_2) = H\{t + \tau - (t_{f2} - \Delta tc_2)\} \tag{96}$$

$$H_2(t,d_1,d_2) = H\{t + \tau - (t_{f1} - \Delta tc_1)\} \tag{97}$$

here τ refers to the time delay between correlated counts.

In the same way we introduce a similar definition for the C_{1S} probability. With the above redefinitions of the C-probabilities, the integrodifferential equation system satisfied by the generating function $G_K(Z_1,Z_2,1,t,d_1,d_2)$ is obtained by the procedure shown in section 2, and after a lenghtly calculus we have obtained that:

$$\left[1/v\ \partial/\partial t + \bar{\Omega}.\bar{\nabla}_{\bar{r}} - \Sigma(\vec{r},v,t)\right] G_K(Z_1,Z_2,1,t,d_1,d_2)$$

$$= - \Sigma(\vec{r},v,t) \left[C^{(0,0)}{}_{0C}(\vec{r},v,t) + Z_1\ C^{(1,0)}{}_{0C}(\vec{r},v,t) + Z_2\ C^{(0,1)}{}_{0C}(\vec{r},v,t)\right.$$

$$+ \int dv' \int d\Omega' \left[C^{(0,0)}{}_{1S}(1,t|v',\vec{\Omega}') + Z_1\ C^{(1,0)}{}_{1S}(1,t|v',\vec{\Omega}') +\right.$$

$$\left. +Z_2\ C^{(0,1)}{}_{1S}(1,t|v',\vec{\Omega}')\right] .\ G_K(Z_1,Z_2,\vec{r},v',\Omega',t,d_1,d_2) +$$

$$+ \sum_{j=0}^{I} \sum_{l_1=0}^{L_1} \cdots \sum_{l_6=0}^{L_6} (C^{(0,0)}{}_{jl1l2\ldots l6}(\vec{r},v,t) +$$

$$+ Z_1.C^{(1,0)}{}_{jl1l2\ldots l6}(\vec{r},v,t) + Z_2.\ C^{(0,1)}{}_{jl1l2\ldots l6}(\vec{r},v,t)).$$

$$.\left[\int dv' \int d\Omega' .\ X(v')/4\pi\ .\ G_K(Z_1,Z_2,\vec{r},v',\vec{\Omega}',t,d_1,d_2\right]^j .$$

$$\left. .\ \prod_{i=1}^{6} \left[G^{(i)}{}_K(Z_1,Z_2,\vec{r},t,d_1,d_2)\right]^{l_i}\right] \tag{98}$$

$$\partial G^{(i)}{}_K\ (Z_1,Z_2,\vec{r},t,d_1,d_2)/\partial t = \lambda_i G^{(i)}_K(Z_1,Z_2,\vec{r},t,d_1,d_2) -$$

$$- \lambda_i \int dv' \int d\Omega' X^i/4\pi(v').\ G_K(Z_1,Z_2,\vec{r},v',\vec{\Omega}',t,d_1,d_2) \tag{99}$$

Finally the relationship between $G_S(Z_1,Z_2,d_1,d_2)$ and $G_K(Z_1,Z_2,1,t,d_1,d_2)$ is:

$$G_S(Z_1,Z_2,d_1,d_2)= \exp \left[\sum_{j=0}^{I_S} \varepsilon_{sj} \int_{t_0}^{t_f} dt \int dr.S_0(\vec{r},t).\right.$$

$$\left.\left[\left[\int d\Omega \int dv\, X_S(v)/4\pi\, G_K(Z_1,Z_2,1,t,d_1,d_2)\right]^j -1\right]\right]$$

$$\text{where } t_f= \max(t_{f1},\ t_{f2}) \tag{100}$$

The next step is to obtain a calculable expression for the covariance function $COV(d_1,d_2)$, to this end first we operate on eq (100) with $[\partial^2/\partial Z_1\, \partial Z_2]\ Z_1=Z_2=1$, to obtain after some manipulations:

$$COV(d_1,d_2)= \overline{\langle n_c m_c S\rangle + \nu_S(\nu_S-1)} \int_{t_0}^{t_f} dt \int dr\ S_0(\vec{r},t).I^{(1)}{}_S(\vec{r},t,d_1)I^{(2)}{}_S(\vec{r},t,d_2) \tag{101}$$

where $I^{(1)}{}_S(\vec{r},t,d_1)$ and $I^{(2)}{}_S(\vec{r},t,d_2)$ are the spectral weighted source neutron importances for detectors D_1 and D_2 respectively, given by:

$$I^{(1)}{}_S(\vec{r},t,d_1)= \int dv \int d\Omega\ X_S(v)/4\pi\ \bar{n}_c(1,t,d_1) \tag{102}$$

$$I^{(2)}{}_S(\vec{r},t,d_2)= \int dv \int d\Omega\ X_S(v)/4\pi\ \bar{m}_c(1,t,d_2) \tag{103}$$

$\langle\ \rangle$ means integration over the phase space variables and time, and S is the neutron source strength defined in (24).

Operating on eq (98) and (99), with $[\partial/\partial Z_1]|Z_1=Z_2=1$ and $[\partial/\partial Z_2],\ Z_1=Z_2=1$ it is obtained that $\bar{n}_c(1,t,d_1)$ and $\bar{m}_c(1,t,d)$ are solutions of the adjoint Boltzmann equations:

$$K^+\psi_{C1}(t)= \begin{bmatrix} \Sigma_{D1}(\vec{r},v,t,d_1) \\ 0 \\ . \\ . \\ . \\ 0 \end{bmatrix} \tag{104}$$

$$K^+\psi_{C2}(t)= \begin{bmatrix} \Sigma_{D2}(\vec{r},v,t,d_2) \\ 0 \\ . \\ . \\ . \\ 0 \end{bmatrix} \tag{105}$$

where K^+ is the time dependent adjoint Boltzmann matrix operator, given by (27). $\Sigma_{D1}(\vec{r},v,t,d_1)$ and $\Sigma_{D2}(\vec{r},v,t,d_2)$ are the macroscopic detection cross sections for detectors D_1 and D_2 respectively. Finally ψ_{C1} and $\psi_{C2}(t)$ are the column vectors

$$\psi_{C1}(t)= \mathrm{col}[\bar{n}_c(1,t,d_1),\ \bar{C}^+{}_{C1}(q,t,d_1),\ldots,\bar{C}^+{}_{C6}(1,t,d_1)] \tag{106}$$

$$\psi_{C2}(t)= \mathrm{col}[\bar{m}_c(1,t,d_2),\ \bar{C}^+{}_{C1}(1,t,d_2),\ldots,\bar{C}^+{}_{C6}(1,t,d_2)] \tag{107}$$

with $\bar{C}^+{}_{Ci}(1,t,d_1)$ defined by (33) and $\bar{C}^+{}_{Ci}(1,t,d_2)$ defined also by eq (33) but substituting $\bar{n}_c(1,t,d_1)$ by $\bar{m}_c(1,t,d_2)$.

To calculate the covariance function, we need to calculate the first term of the right hand side of equation (101), to this end first we operate on equations (98) and (99), with the operator $[\partial^2/\partial Z_1\partial Z_2]\ Z_1= Z_2= 1$, and after some calculus it is obtained:

$$K^+\ \psi_{cov}\ (1,t,d_1,d_2) = Q_{cov} \tag{108}$$

where K^+ has been defined previously and the column vectors $\psi_{cov}(1,t,d_1,d_2)$ and Q_{cov}, are defined by:

$$\psi_{cov}(1,t,d_1,d_2) \equiv \mathrm{col}(\overline{n_c m_c}(1,t,d_1,d_2),\overline{C^{+(1)}{}_{C1}\ C^{+(2)}{}_{C1}}(1,t,d_1,d_2),\ldots,\overline{C^{+(1)}{}_{C6}\ C^{+(2)}{}_{C6}}(1,t,d_1,d_2)) \tag{109}$$

$$Q_{cov} \equiv \mathrm{col}\ (Q_{cov}(1,t,d_1,d_2),0,\ldots 0) \tag{110}$$

where:

$$\overline{C^{+(1)}{}_{Ci}\ C^{+(2)}{}_{Ci}}(1,t,d_1,d_2)= \int_t^{t_f} dt'\exp(-\lambda_i(t'-t))\ \lambda_i.\overline{n_c m_c}(1,t',d_1,d_2) \tag{111}$$

and $Q_{cov}(1,t,d_1,d_2)$ is a source term equal to:

$$Q_{cov}(1,t,d_1,d_2) = Q_{12}(1,t,d_1,d_2) + Q_{12D}(1,t,d_1,d_2)$$

$$+ \sum_{i=1}^{6} Q^{i(p)}{}_{12}(1,t,d_1,d_2) + \sum_{i=1}^{6} Q^{(p)i}{}_{12}(1,t,d_1,d_2) + \sum_{i=1}^{6} Q^{i(d)}{}_{12}(1,t,d_1,d_2)$$

$$+ \sum_{\substack{i=1\\ i\neq j}}^{6} \sum_{j=1}^{6} Q^{ij(d)}{}_{12}(1,t,d_1,d_2) \tag{112}$$

It is convenient to analyze the expressions and the physical meaning of the various terms which enter into (112). This source term drives the correlated counts $\overline{n_c m_c}(1,t,d_1,d_2)$, produced upon injection one single neutron at (1,t)

i) $Q_{12}(1,t,d_1,d_2)$, this source term is equal to

$$Q_{12}(1,t,d_1,d_2)= \overline{\nu_p(\nu_p-1)}.\Sigma_f(\vec{r},v,t).I^{(1)}(\vec{r},t,d_1)\ I^{(2)}(\vec{r},t,d_2) \tag{113}$$

where $I^{(1)}(\vec{r},t,d_1)$ and $I^{(2)}(\vec{r},t,d_2)$ are the spectral weighted prompt neutron importances for detectors D_1 and D_2 respectively. These importances are obtained from expressions (102) and (103) changing $X_S(v)$ by the prompt neutron spectrum $X(v)$.

ii) $Q_{12D}(1,t,d_1,d_2)$

This source term arises from the neutrons that produced in a detector such as a fission chamber give correlated counts in the other detector. If detector D_1 and D_2 are both fission chambers this source term is equal to:

$$Q^f_{12D}(1,t,d_1,d_2) = \bar{\nu}_{D1}(v)(1-\bar{\beta}_{D1})\Sigma_{fD1}(\vec{r},v,t,d)I^{(2)}_D(\vec{r},t,d_2) + $$
$$+ \bar{\nu}_{D2}(v)(1-\bar{\beta}_{D2})\Sigma_{fD2}(\vec{r},v,t,d_2)I^{(1)}_D(\vec{r},t,d_1) \quad (114)$$

where $\Sigma_{fD1}(\vec{r},v,t,d_1)$, $\Sigma_{fD2}(\vec{r},v,t,d_2)$ are the fission detection cross-section of detectors D_1 and D_2 respectively; $I^{(1)}_D(\vec{r},t,d_1)$, $I^{(2)}_D(\vec{r},t,d_2)$ are the spectral weighted prompt neutron importances for neutrons produced at D_1 and D_2 respectively.

iii) $$\sum_{i=1}^{6} Q^{i(d)}_{12}(1,t,d_1,d_2) + \sum_{\substack{i=1\\i\neq j}}^{6} \sum_{j=1}^{6} Q^{ij(d)}(1,t,d_1,d_2) =$$

$$= \sum_{i=1}^{6} \overline{l_i(l_i-1)}\ \Sigma_f(\vec{r},v,t)\ I^{i(d)}(\vec{r},t,d_1)\ I^{i(d)}\ (\vec{r},t,d_2)$$

$$+ \sum_{\substack{i=1\\i\neq j}}^{6} \sum_{j=1}^{6} \overline{l_i\ l_j}\ \Sigma_f(\vec{r},v,t)\ I^{i(d)}\ (\vec{r},t,d_1)I^{j(d)}(\vec{r},t,d_2) \quad (115)$$

where

$$I^{i(d)}\ (\vec{r},t,d_k) = \int dv \int d\Omega\ X^i(v)/4\pi\ \overline{C^+}_{ci}(1,t,d_k) \quad (116)$$

k=1 or 2

The first term arises from those fissions where it is produced more than one delayed neutron precursor of the same group. The second one arise from those fissions where two delayed neutron precursors belonging to different groups are produced.

iv) $$\sum_{i=1}^{6} (Q^{i(p)}_{12}\ (1,t,d_1,d_2) + Q^{(p)i}_{12}(1,t,d_1,d_2) =$$

$$= \sum_{i=1}^{6} \overline{l_i\ \nu_p}\ \Sigma_f(\vec{r},v,t)\ (I^{(1)}(\vec{r},t,d_1)\ I^{i(d)}(\vec{r},t,d_2) +$$

$$+ I^{i(d)}(\vec{r},t,d_1)\ I^{(2)}(\vec{r},t,d_2)) \quad (117)$$

This source term of correlated counts, arises from those fission events where prompt neutrons and delayed neutrons precursor are produced.

To get a calculable expression for the covariance function $COV(d_1,d_2)$, we proceed as follows, first we form the inner product of equation (108) and the vector $\psi(t)$, then we use the commutation relation and finally we take on account equation (54), in this way it is obtained:

$$< K^+ \psi_{cov}|\psi > = < \psi_{cov}|K \psi > = < \overline{n_c\ m_c}\ S > \qquad (118)$$

and

$$< K^+ \psi_{cov}|\psi > = < Q_{cov}|\psi > = < Q_{cov}(1,t,d_1,d_2)\ \phi > \qquad (119)$$

Neglecting (ii) given by (114) and (iii) given by (115) we have on account of (101), (118) and (119) that the covariance function is given by:

$$COV(d_1,d_2)= \overline{\nu_S(\nu_S-1)} \int_{t_0}^{t_f} dt \int dr.S_0(\vec{r},t)\ I^{(1)}{}_S(\vec{r},t,d_1)\ I^{(2)}{}_S(\vec{r},t,d_2)$$

$$+ \int_{t_0}^{t_f} dt \int dv \int dr\ \overline{\nu_p(\nu_p-1)}.F_T(\vec{r},v,t)\ I^{(1)}(\vec{r},t,d_1)\ I^{(2)}(\vec{r},t,d_2)$$

$$+ \sum_{i=1}^{6} \int_{t_0}^{t_f} dt \int dv \int dr\ \overline{l_i\ \nu_p}.F_T(\vec{r},v,t)\ (I^{(1)}(\vec{r},t,d_1)\ I^{i(d)}(\vec{r},t,d_2)\ +$$

$$+ I^{i(d)}(\vec{r},t,d_1)\ I^{(2)}(\vec{r},t,d_2)) \qquad (120)$$

where:

$$F_T(\vec{r},v,t) = \Sigma_f(\vec{r},v,t)\ \bar{\phi}(\vec{r},v,t) \qquad (121)$$

$F_T(\vec{r},v,t)$ is obviously the fission rate per unit volume and velocity at time t. The first term in equation (120) gives the contribution to the covariance coming from the source neutrons, the second one the contribution that arises from the prompt fission neutrons, and the third one the contribution coming from the correlations between prompt and delayed neutrons.

In actual practice the descriptor utilized in noise analysis is the cross-correlation function [7], which is defined in terms of instantaneous count rates, as the following limit of the covariance function for infinitely small gate length Δtc_1, Δtc_2

$$\phi_{12}(t_{f1}\ t_{f2}) = \lim_{\substack{\Delta tc_1 \to 0 \\ \Delta tc_2 \to 0}} \frac{COV(d_1,d_2)}{\Delta tc_1\ \Delta tc_2} \qquad (122)$$

This limit procedure yields the following result:

$$\phi_{12}(t_{f1}, t_{f2}) = \phi_{S12}(t_{f1}, t_{f2}) + \phi_{sys12}(t_{f1}, t_{f2}) \tag{123}$$

where $\phi_{S12}(t_{f1}, t_{f2})$ comes from the first term (source term) of the right hand side of equation (120) and is equal, to:

$$\phi_{S12}(t_{f1}, t_{f2}) = \overline{\nu_S(\nu_S-1)} \int_{t_0}^{t_f} dt \int dr\, S_0(\vec{r}, t)\, I^{(1)}{}_S(\vec{r}, t-t_{f1})\, I^{(2)}{}_S(\vec{r}, t-t_{f2}) \tag{124}$$

where:

$$I^{(1)}{}_S(\vec{r}, t-t_{f1}) = \int dv \int d\Omega\, X_S(v)/4\pi\, \bar{n}_c(1, t-t_{f1}) \tag{125}$$

and:

$$I^{(2)}{}_S(\vec{r}, t-t_{f2}) = \int dv \int d\Omega\, X_S(v)/4\pi\, \bar{m}_c(1, t-t_{f2}) \tag{126}$$

To find the equations satisfied by the rates $\bar{n}_c(1, t-t_{f1})$ and $\bar{m}_c(1, t-t_{f2})$, we divide equations (104) and (105) by Δtc_1 and Δtc_2 respectively and then we take on account the following limit:

$$\lim_{\Delta tc_1 \to 0} \frac{\Sigma_{D1}(\vec{r}, v, t, d_1)}{\Delta tc_1} = \Sigma_D(\vec{r}, v) \lim_{\Delta tc_1 \to 0} \frac{H(t-(t_{f1}-\Delta tc_1))-H(t-t_{f1})}{\Delta tc_1} = \tag{127}$$

$$= \Sigma_{D1}(\vec{r}, v)\, .\, \delta(t-t_{f1})$$

with this limit procedure equation (104) gives:

$$K^+ \begin{bmatrix} \bar{n}_c(1, t, t_{f1}) \\ \overline{C^+}_{C1}(1, t, t_{f1}) \\ \cdot \\ \cdot \\ \overline{C^+}_{C6}(1, t, t_{f1}) \end{bmatrix} = \begin{bmatrix} \Sigma_{D1}(\vec{r}, v)\delta(t-t_{f1}) \\ 0 \\ \cdot \\ \cdot \\ 0 \end{bmatrix} \tag{128}$$

a similar equation holds for $\bar{m}_c(1, t, t_{f2})$. In view of equation (128), we conclude that the count rates $\bar{n}_c(1, t, t_{f1})$ and $\bar{m}_c(1, t, t_{f2})$ are displacement kernels in $t-t_{f1}$ and $t-t_{f2}$ respectively. So we write them as $\bar{n}_c(1, t-t_{f1})$, $\bar{m}_c(1, t-t_{f2})$.

The second term in equation (123), i.e., $\phi_{sys}(t_{f1}, t_{f2})$ (system contribution) comes from the second and third term of the right hand side of equation (120); after performing the limiting procedure of eq (122), it is obtained the following result:

$$\phi_{sys}(t_{f1}, t_{f2}) = \int_{t_0}^{t_f} dt \int dv \int dr. \overline{\nu_p(\nu_p-1)}. F_T(\vec{r}, v, t). I^{(1)}(\vec{r}, t-t_{f1}).$$

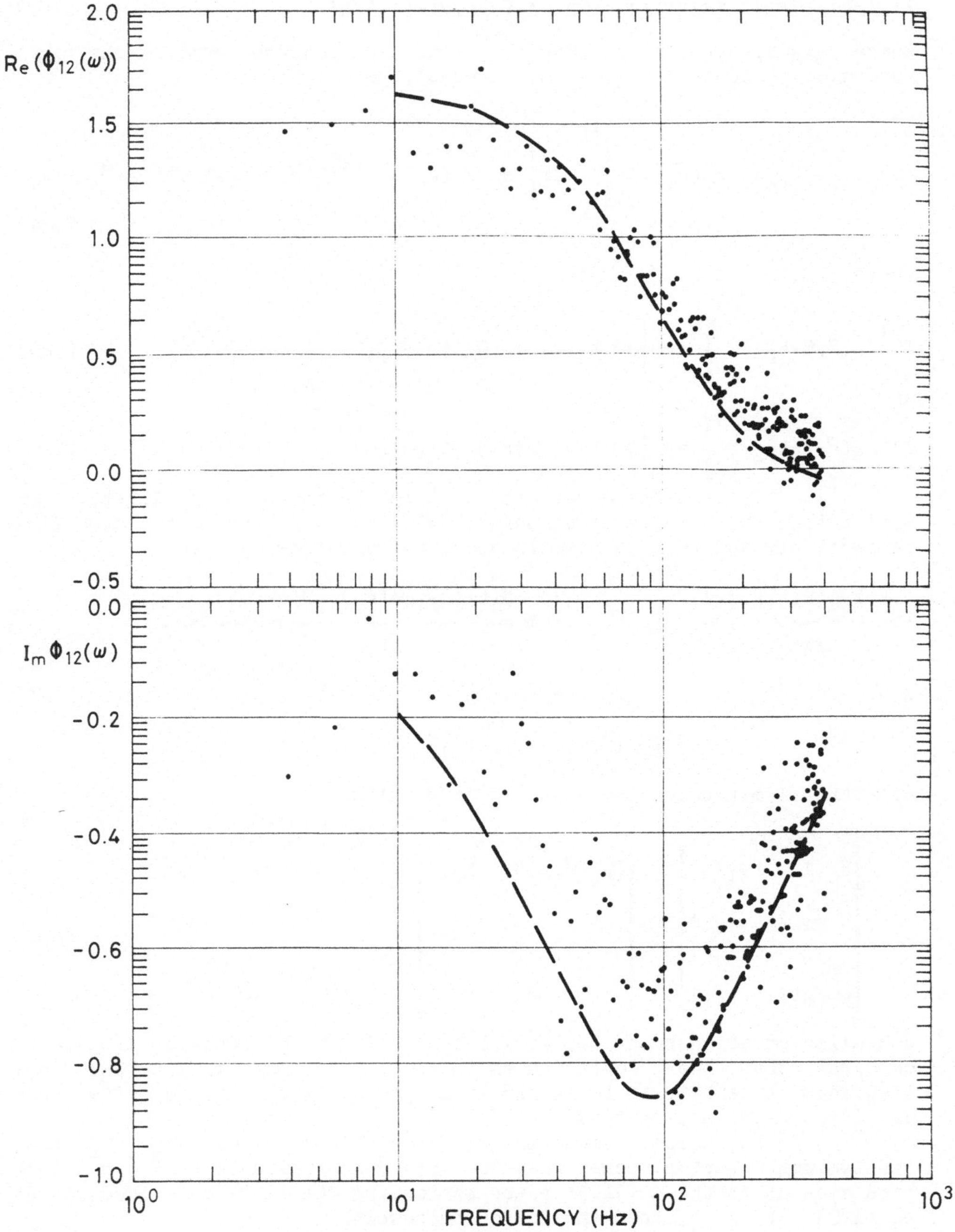

Fig. 3. Real and imaginary parts of $\Phi_{12}(\omega)$ (arbitrary units), versus frequency, for run AC.

$$. I^{(2)}(\vec{r},t-t_{f2}) + \sum_{i=1}^{6} \int_{t_0}^{t_f} dt \int dv \int dr\ \overline{l_i\ \nu_p}\ .\ F_T(\vec{r},v,t).$$

$$.(I^{(1)}(\vec{r},t-t_{f1}).I^{i(d)}(\vec{r},t-t_{f2})+I^{i(d)}(\vec{r},t-t_{f1}).I^{(2)}(\vec{r},t-t_{f2})) \qquad (129)$$

where

$$I^{(1)}(\vec{r},t-t_{f1}) = \int dv \int d\Omega\ X(v)/4\pi\ \bar{n}_c(1,t-t_{f1}) \qquad (130)$$

$$I^{i(d)}(\vec{r},t-t_{f1}) = \int dv \int d\Omega\ X^i(v)/4\pi\ \overline{C^+}_{Ci}(1,t-t_{f1}) \qquad (131)$$

where on account of eq (33) $\overline{C^+}_{Ci}(1,t-t_{f1})$ is given by:

$$\overline{C^+}_{Ci}(1,t-t_{f1}) = \int_t^{t_{f1}} dt'\lambda_i.\exp(-\lambda_i(t'-t))\ \bar{n}_c(1,t'-t_{f1}) \qquad (132)$$

Similar definitions to (130), (131) and (132), hold for $I^{(2)}(\vec{r},t-t_{f2})$, $I^{i(d)}(\vec{r},t-t_{f2})$ and $\overline{C^+}_{Ci}(1,t-t_{f2})$.

6. The Cross Power Spectral Density (C.P.S.D.)

6.a The C.P.S.D. for a subcritical assembly with constant source introduced in the remote past

In this particular case we have assumed $S_0(\vec{r})$ time independent and $t_0= -\infty$, and on account that for $t>t_{f1}$, $\bar{n}_c(1,t-t_{f1})= 0$ and for $t>t_{f2}$, $\bar{m}_c(1,t-t_{f2})= 0$, we can extend the upper limits of (129) and (124) up to ∞. Making $t_{f1}= t_{f2}+\tau$, where τ is the time delay between the two detectors, and applying to both sides of equation (123) the operator $\int_{-\infty}^{\infty} d\tau\ \exp(-iw\tau)$, it is obtained the C.P.S.D, or Fourier transform of the cross-correlation function:

$$\phi_{12}(w)= \phi_{S12}(w) + \phi_{sys12}(w) \qquad (133)$$

where $\phi_{S12}(w)$ gives the source contribution to the C.P.S.D. and ϕ_{sys12}, gives the system contribution. To obtain $\phi_{S12}(w)$, we have made the following calculations, first we express $I^{(1)}{}_S(\vec{r},t-t_{f1})$ and $I^{(2)}{}_S(\vec{r},t-t_{f2})$ in terms of their respective Fourier transforms; this gives:

$$\phi_{S12}(w)= \int_{-\infty}^{\infty} d\tau \int_{-\infty}^{\infty} dt \int_{-\infty}^{\infty} dw_1/2\pi \int_{-\infty}^{\infty} dw_2/2\pi \int dr.\exp(-iw\tau)\exp(iw_1(t-t_{f2}-\tau))$$

$$\exp(iw_2(t-t_{f2})).\ S_0(r).\ I_S(\vec{r},w_1).\ I_S(\vec{r},w_2) \qquad (134)$$

on account of the development of the Dirac-δ distribution in Fourier integral, we arrive to the expression:

$$\phi_{S12}(w)= \overline{\nu_S(\nu_S-1)} \int dr\ S_0(r).I^{(1)}{}_S(\vec{r},-w).I^{(2)}{}_S(\vec{r},w) \qquad (135)$$

where $I^{(2)}{}_S(\vec{r},w)$ is the Fourier transform of $I^{(2)}{}_S(\vec{r},t-t_{f2})$ and $I^{(1)}{}_S(\vec{r},-w)$ is the complex conjugate of $I^{(1)}{}_S(\vec{r},w)$ these Fourier transforms are given from (125) and (126) by:

$$I^{(1)}{}_S(\vec{r},w)= \int dv \int d\Omega\ X_S(v)/4\pi\ \bar{n}_c(1,w) \qquad (136)$$

$$I^{(2)}{}_S(\vec{r},w)= \int dv \int d\Omega\ X_S(v)/4\pi\ \bar{m}_c(1,w) \qquad (137)$$

The Fourier transformed average count rates, $\bar{n}_c(1,w)$, and $\bar{m}_c(1,w)$ satisfy, in view of Eq (128), the following adjoint transport equations:

$$K^+(w)\ \psi_{C1}(w)= \begin{bmatrix} \Sigma_{D1}(\vec{r},v) \\ 0 \\ . \\ . \\ . \\ 0 \end{bmatrix} \qquad (138)$$

$$K^+(w)\ \psi_{C2}(w)= \begin{bmatrix} \Sigma_{D2}(\vec{r},v) \\ 0 \\ . \\ . \\ . \\ 0 \end{bmatrix} \qquad (139)$$

where the adjoint Boltzmann operator in the frequency domain is given by:

$$K^+(w)= iw/v\ I + L^+ \qquad (140)$$

where I is the unit matrix and L^+ is the matrix operator given by eq (66). $\psi_{C1}(w)$ are the column vectors:

$$\psi_{C1}(w)= \text{col}\ [\bar{n}_c(1,w),\overline{C^{+(1)}}{}_{C1}(1,w),\ldots\overline{C^{+(1)}}{}_{C6}(1,w)] \qquad (141)$$

$$\psi_{C2}(w)= \text{col}\ [\bar{m}_c(1,w),\overline{C^{+(2)}}{}_{C1}(1,w),\ldots\overline{C^{+(2)}}{}_{C6}(1,w)] \qquad (142)$$

where the supraindexes(1) and (2) refers to detector D_1 and D_2 respectively. Finally from (138) and (139) it follows that

$$\overline{C^{+(1)}}_{Ci}(1,t)= \frac{\lambda_i}{-iw+\lambda_i} \, \overline{n}_c(1,w) \tag{143}$$

$$\overline{C^{+(2)}}_{Ci}(1,t)= \frac{\lambda_i}{-iw+\lambda_i} \, \overline{m}_c(1,w) \tag{144}$$

To obtain the system contribution $\phi_{sys12}(w)$ to the C.P.S.D., we have followed the same steps that were used to deduce the source contribution. Now from (129), we have obtained for a subcritical assembly the result:

$$\phi_{sys12}(w)= \int dr \int dv \, \overline{\nu_p(\nu_p-1)} \; F_T(\vec{r},v) \; I^{(1)}(\vec{r},-w). \; I^{(2)}(\vec{r},w)$$

$$+ \sum_{i=1}^{6} \int dr \int dv \, \overline{l_i \; \nu}. F_T(\vec{r},v). (I^{(1)}(\vec{r},-w) I^{i(2)}(\vec{r},w) +$$

$$+ I^{i(1)}(\vec{r},-w) \; I^{(2)}(\vec{r},w)) \tag{145}$$

where the fourier transforms of the spectral factors are given by

$$I^{(1)}(\vec{r},w)= \int dv \int d\Omega \; . \; X(v)/4\pi \; . \; \overline{n}_c(1,w) \tag{146}$$

$$I^{(2)}(\vec{r},w)= \int dv \int d\Omega \; X(v)/4\pi \; . \; \overline{m}_c(1,w) \tag{147}$$

$$I^{i(1)}(\vec{r},w)= \int dv \int d\Omega \; X^i(v)/4\pi \; . \; (\lambda_i/\lambda_i-iw) \; \overline{n}_c(1,w) \tag{148}$$

$$I^{i(2)}(\vec{r},w)= \int dv \int d\Omega \; X^i(v)/4\pi \; . \; (\lambda_i/\lambda_i-iw) \; \overline{m}_c(1,w) \tag{149}$$

and

$$\overline{l_i \; \nu} = \sum_{j=0}^{I} \; \sum_{L_1=0}^{L_1} \; .. \sum_{L_6=6}^{L_6} l_i . j \; \varepsilon_{j1112..16} \tag{150}$$

we note that the spectral factors $I^{i(1)}(\vec{r},w)$ decrease with frequency.

6.b The unimodal approximation for a subcritical assembly, with source introduced in the remote past

In this case we make

$$\bar{n}_{cj}(1,w) = T_j(w).\ \phi^+_o(1) \tag{151}$$

where the subindex j(1 or 2) refers to detector D_1 or D_2 respectively.

Then we substitute (151) into (138) for detector D_1, and into (139) for detector D_2, and then we multiply by $\bar{\phi}(1)$ with the scalar product (,) which means integration over the phase space variables $(\vec{r}, v, \vec{\Omega})$. After a litle algebra it is obtained:

$$T_j(w) = \frac{(\Sigma_{Dj}, \bar{\phi})}{F}\ .\ Y\rho(-iw) \tag{152}$$

where F is the weighted neutron production, given by (70.c), and $Y\rho(-iw)$ is the transfer function of the subcritical assembly,

$$Y\rho(-iw) = \frac{1}{\rho + (1/Y_z(-iw))} \tag{153}$$

with ρ the subcritical reactivity and $Y_z(S)$ equal to:

$$Y_z(S) = \frac{1}{S(\Lambda + \sum_{j=1}^{6} \beta_j.\ \frac{1}{S+\lambda_j})} \tag{154}$$

Expresión (147) holds for a detector working in pulse mode, however if one works in current mode it is necesary to account for the transfer function $h_j(w)$ of the electronic set up, and for the response of the chamber to fissions (If we work with a fission chamber), denoted by U_j.

From (133), it follows that, the C.P.S.D. between detector D_1 and D_2, is given in the unimodal approximation by:

$$\phi_{120}(w) = \varepsilon_1.\varepsilon_2.Y\rho(-iw).Y\rho(iw).F_T.\left[D_S.R_S.\ \frac{(\phi^+_0, S)}{F}\ .\ \frac{\overline{\nu^2}_S}{\bar{\nu}^2} + D.R\right] \tag{155}$$

where D_S and D are the source and System Diven's [11] factors, $\varepsilon_1, \varepsilon_2$ the detector efficiencies in counts per reactor fission, F_T the total fission rate, and finally R_S and R denote the factors which account for source and system spatial effects, given by eqs (79) and (83) respectively. S is the strength source defined by eq (24).

It is interesting to analyze the source and system contributions in the unimodal approximation, to this end we divide both contributions obtaining:

$$\frac{\phi_{S120}(w)}{\phi_{sys120}(w)} = \frac{D_S}{D} \cdot \left[\frac{\bar{\nu}_S}{\bar{\nu}}\right]^2 \cdot \frac{(\phi^+_0, S)}{F} \cdot \frac{R_S}{R} \tag{156}$$

or

$$\frac{\phi_{S120}(w)}{\phi_{sys120}(w)} = \frac{D_S}{D} \cdot \left[\frac{\bar{\nu}_S}{\bar{\nu}}\right]^2 \cdot \frac{\int dr. S_0(\vec{r}). I^2_{S0}(\vec{r})}{\int dr\ F_T(\vec{r}) I^2_0(\vec{r})} \tag{157}$$

where $F_T(\vec{r})$ is the fission rate per unit volume at point r, and the spectral weighted factors $I_{S0}(\vec{r})$, and $I_0(\vec{r})$ are defined by eqs (77) and (84) respectively.

We observe that the ratio of source and system contributions is proportional to:

i) The ratio of source and system Diven's factors, ii) The square of $\bar{\nu}_S/\bar{\nu}$ iii) The ratio of weighted production rates of source and fission neutrons, iv) The ratio of the source and system spatial effects factors.

6.c <u>C.P.S.D. between the count rates of a ionization chamber with a californium source inside and a neutron detector</u>

In this case we have a ionization chamber which contains a californium-252 source inside. The C_f-252 chamber as reported by Mihalczo [13,14], serves as a source of neutron noise for the reactor system, and as a detector provides a highly correlated electrical representation of that noise for use in the analysis.

In the randomly pulsed neutron measurements [13-16], a C_f source in an ionization chamber was placed in a multiplicative assembly which also contained a neutron counter. The neutrons from the spontaneous fission of C_f-252 initiate fission chains in the assembly, and the neutron detector, counted events from the interactions of neutrons from these fission chains with the detector.

From stochastic transport theory we have deduced that the cross correlation function between the count rates of a ionization chamber with a Californium-252 source inside and a neutron detector is given by:

$$\phi_{cf12}(w) = h^*_1(w)\ U^*_1 \int dr\ S_0(\vec{r}).\bar{\nu}_{cf}. I_{cf}(\vec{r}, w) \tag{158}$$

where $S_0(\vec{r})$ is the Californium source disintegration rate per unit volume, $\bar{\nu}_{cf}$ the average number of neutrons released per Californium disintegration, $h^*_1(w)$ is the complex conjugate of the transfer function of the electronic set up, and U_1 the response of the ionization chamber to fissions, finally $I_{cf}(\vec{r}, w)$ is given by:

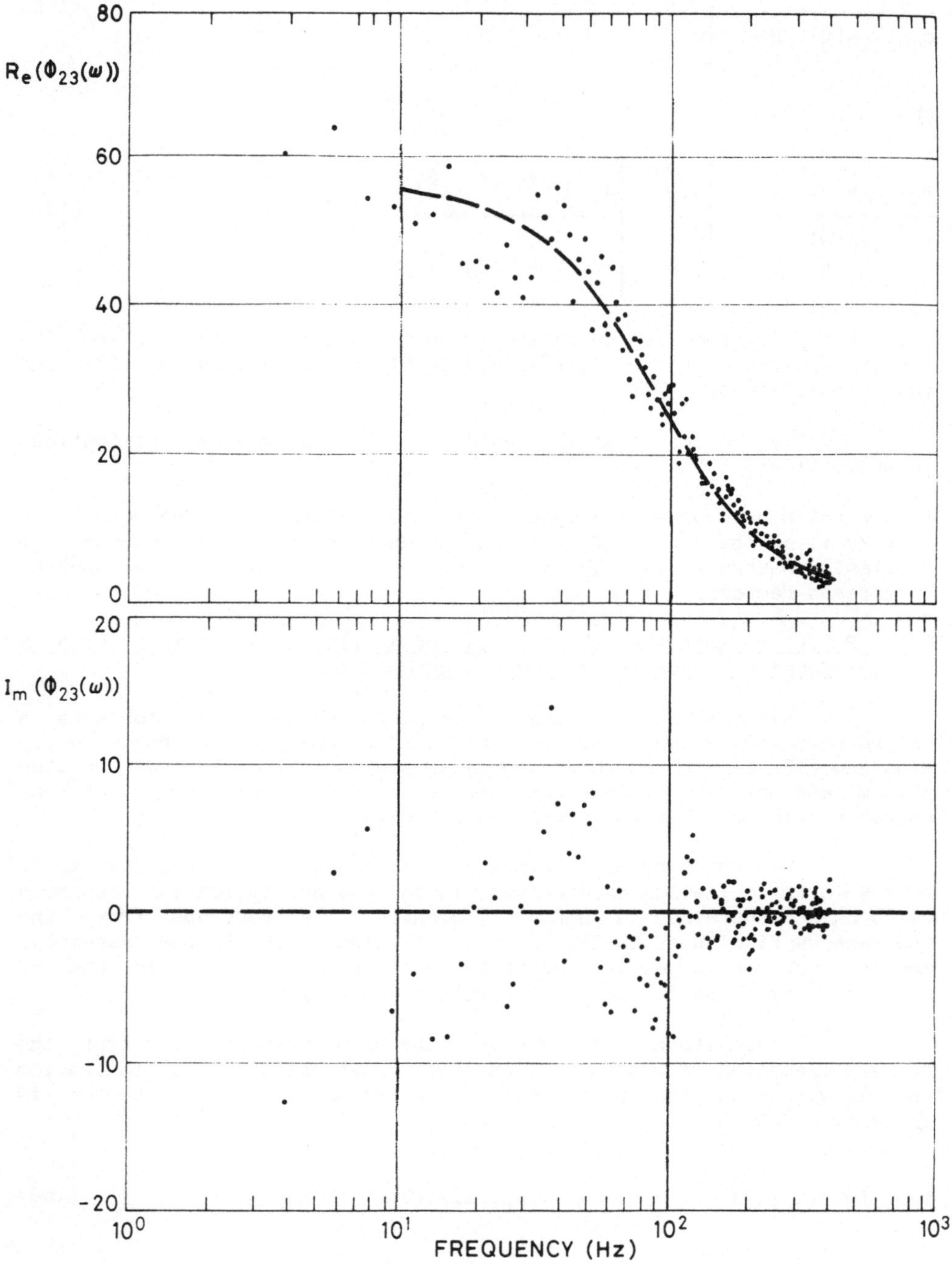

Fig. 4. Real and imaginary parts of $\Phi_{23}(\omega)$ (arbitrary units), versus frequency, for run AC.

$$I_{cf}(\vec{r},w)= \int dv \int d\Omega\ X_{cf}/4\pi\ \bar{n}_c(1,w) \qquad (159)$$

where $\bar{n}_c(1,w)$ is the Fourier transformed average count rate of the neutron detector, which satisfy the adjoint transport equation (138), and $X_{cf}(v)$ the normalized spectrum of C_f-252.

7. C.P.S.D. calculations and comparison with experiments.

Now we are going to analyze theoretically some features of the experiment performed by Mihalczo et al [1], in this experiment a stainless steel cylinder was filled with a UO_2F_2 solution in water, in the cylinder axis was placed a stainless steel duct. Inside this duct, was placed a ionization chamber with a Californium-252 source inside. Close to the cylinder tank wall containing the UO_2F_2 solution were located two neutron detectors.

We denote by D_1, the ionization chamber containing the Californium source, and by D_2, D_3 the two neutron detectors. In order to compute the various C.P.S.D., we developed a code called BESMAC3, that used a modal expansion, to compute $n_c(1,w)$.

In order to calculate $\bar{n}^{(j)}{}_c(1,w)$, where j= 2 or j= 3, for detectors D_2 and D_3 respectively, we worked out equation (138) and after a litle algebra, it was obtained:

$$\left[+\frac{iw}{v} + H^+ - \sum_i \frac{\lambda_i M^+{}_i}{\lambda_i - iw}\right] \bar{n}^{(j)}{}_c(1,w) = \Sigma^{(j)}{}_{CD}(\vec{r},v) \qquad (160)$$

where H^+, and $M^+{}_i$(i= 1..6) are given by eqs (28) and (29) respectively. $\Sigma^{(j)}{}_{CD}(\vec{r},v)$ gives the macroscopic detection cross-section, for the capture detector D_j. The operator within [ ] brackets will be denoted by $U^+(w)$.

To calculate eq (160) we have used the P_1, spherical harmonic approximation, with three energy groups, and we have assumed that all fission neutrons are injected into the fast group; this means that the neutron fractions injected into each group are, X_1= 1 (fast group) and X_2= X_3 = 0.

Now we define $n^{(j)}{}_c(1,\vec{r}_d,w)$ in such a way that its integral over the detector sensitive volume V_{Dj}, of detector D_j, gives $\bar{n}_c(1,w)$. It is quite obvious that $\bar{n}^{(j)}{}_c(1,\vec{r}_d,w)$ satisfies the equation:

$$U^+(w).\bar{n}^{(j)}{}_c(1,\vec{r}_d,w)= \Sigma^{(j)}{}_{CD}(\vec{r},v)\ \delta(\vec{r}-\vec{r}_D) \qquad (161)$$

The spectral function $I^{(j)}(r\text{-->}r_D,w)$ defined by:

$$I^{(j)}(\vec{r}\text{-->}\vec{r}_D,w)= \int dv \int d\Omega\ X(v)/4\pi\ \bar{n}^{(j)}{}_c(1,\vec{r}_D,w) \qquad (162)$$

reduces in the P_1, spherical harmonic apoproximation (Telegraphist equation) with three energy groups to:

$$I^{(j)}(\vec{r}\text{-->}\vec{r}_D, w) = \bar{n}^{(j)}{}_{C1}(\vec{r}\text{-->}\vec{r}_D, w) \tag{163}$$

where the subindex g= 1, means that the fission neutrons are injected into the energy group 1 (fast group). It becomes evident that the integral of $I^{(j)}(\vec{r}\text{-->}\vec{r}_D, w)$ over the detector volume gives the spectral weighted importance function $I^{(j)}(\vec{r}, w)$.

Now, we want to compute $\bar{n}_{c1}(\vec{r}\text{-->}\vec{r}_D, w)$, which is the Fourier transform of the count rate per unit of sensitive volume at detector point $\bar{r}_D$, and time t_{fj} upon injection of one fast neutron at point $\vec{r}$, and time t.

$n^{(j)}{}_{c1}(\vec{r}\text{-->}\vec{r}_D, w)$ can be calculated by means of the expression

$$\bar{n}^{(j)}{}_{c1}(\vec{r}\text{-->}\vec{r}_D, w) = \sum_{g=1}^{3} \Sigma^{(j)}{}_{CDg}\ \bar{\phi}_g(\vec{r}\text{-->}\vec{r}_D, w) \tag{164}$$

where $\Sigma^{(j)}{}_{CDk}$ is the detection cross section for the k-group and j-detector. $\bar{\phi}_g(\vec{r}\text{-->}\vec{r}_D, w)$ is the Fourier transform of the neutron flux at r_D, energy group g, and time t_{fj}, upon injection of one fast neutron at point r and time t. In general, to calculate the functions $\bar{\phi}(r'\text{--->}r, w)$, we solve the equation:

$$U(w) \begin{bmatrix} \bar{\phi}_1(\vec{r}'\text{-->}\vec{r}, w) \\ \bar{\phi}_2(\vec{r}'\text{-->}\vec{r}, w \\ \bar{\phi}_3(\vec{r}'\text{-->}\vec{r}, w \end{bmatrix} = \begin{bmatrix} \delta(\vec{r}-\vec{r}') \\ 0 \\ 0 \end{bmatrix} \tag{165}$$

where the U(w) operator is given in the spherical harmonic P_1 approximation and three energy groups, with X_1= 1 and X_2= X_3= 0, by:

$$U(w) = \begin{bmatrix} H_1(w)-F_1(w) & -F_2(w) & -F_3(w) \\ -\Sigma_{S12} & H_2(w) & -\Sigma_{S32} \\ -\Sigma_{S13} & -\Sigma_{S23} & H_3(w) \end{bmatrix} \tag{166}$$

with:

$$H_g(w) = \frac{iw}{v_g} - D_g(w)\ \nabla^2 + \Sigma_{Rg} \tag{167}$$

$$F_g(w) = (1-\beta)\ \bar{\nu}_g\ \Sigma_{fg} + \sum_i \frac{\lambda_i \beta_i \bar{\nu}_g \Sigma_{fg}}{iw+\lambda_i} \tag{168}$$

$$D_g(w)= \frac{D_g}{1+ \frac{3iwD_g}{v_g}} \tag{169}$$

where D_g is the usual diffusion coefficient for group g, that we have assumed constant throughout the tank solution; Σ_{Rg} is the removal cross sections for group g, Σ_{sgm} the group transfer cross section from group g to group m, Σ_{fg} the fission cross section for group g, v_g the neutron velocity characterizing group g, and the rest of the symbols are well known. To solve equation (169) we perform a modal expansion of $\bar{\phi}_g(\vec{r}' \to \vec{r}, w)$ in Helmholtz modes [17] $N^g_\mu(\vec{r})$, satisfying the equation:

$$\Delta N^g_\mu(\vec{r}) + B^2_{\mu,g} N^g_\mu(\vec{r}) = 0 \tag{170}$$

with the appropiate boundary conditions. The superscript is introduced because the extrapolated length, changes slightly from one group to another one, and for this reason the Helmholtz modes are slightly differents from group to group. The index μ, is used to label the different modes within each group.

The expansion of $\bar{\phi}_g(\vec{r}' \to \vec{r}, w)$ in Helmholtz [17] modes is achieved as follows, first we write:

$$\bar{\phi}_g(\vec{r}' \to \vec{r}, w) = \sum_\mu A^g_\mu(w, \vec{r}').N^g_\mu(\vec{r}) \qquad g=1,2,3 \tag{171}$$

then we substitute eq (171) into eq (169) and we use equation (165) and the closure property [17]:

$$\sum_\mu N^1_\mu(\vec{r}) N^1_\mu(\vec{r}) = \delta(\vec{r}-\vec{r}') \tag{172}$$

The $A^g_\mu(w,\vec{r}')$ coefficients are obtained multiplying each one of the three equations that result from the previous steps by $N^1_\mu(\vec{r})$ and then integrating over the system volumen. In this way it is obtained on account of the approximation:

$$\int_{V_g} d^3r\, N^1_\mu(\vec{r})\, N^g_{\mu'}(\vec{r}) \approx \delta_{\mu\mu'} \tag{173}$$

the following result:

$$\bar{\phi}_1(\vec{r}' \to \vec{r}, w)= \sum_\mu \frac{(H_{2\mu}(w).H_{3\mu}(w)-\Sigma_{S23}\cdot\Sigma_{S32})}{\Delta_\mu(w)} . N^1_\mu(\vec{r}')N^1_\mu(\vec{r})$$

$$\bar{\phi}_2(\vec{r}' \to \vec{r}, w)= \sum_\mu \frac{(\Sigma_{S12}\cdot H_{3\mu}(w)+\Sigma_{S32}\Sigma_{S13})}{\Delta_\mu(w)} . N^1_\mu(\vec{r}')N^2_\mu(\vec{r}) \tag{174}$$

$$\bar{\phi}_3(\vec{r}' \to \vec{r}, w)= \sum_\mu \frac{(\Sigma_{S13}\cdot H_{2\mu}(w)+\Sigma_{S12}\cdot\Sigma_{S23})}{\Delta_\mu(w)} . N^1_\mu(\vec{r}').N^3_\mu(\vec{r})$$

where

$$\Delta_\mu(w) = \text{Det}\ [U_\mu\ (w)] \tag{175}$$

the matrix $U_\mu(w)$ is obtained from $U(w)$, changing ∇^2 for $- B^2_{\mu,g}$. $H_{g\mu}(w)$, is given by:

$$H_{g\mu}(w)= iw/v_g + D_g(w)\ B^2_{\mu,g} + \Sigma_{Rg} \tag{176}$$

For a puntual Californium source, located at r_0 we have:

$$S_0(\vec{r})= S_0\ .\ \delta(\vec{r}-\vec{r}_0) \tag{177}$$

and on account of equations (153), (154), (158), and (159) we have

$$\phi_{Cf12}(w)= h^*{}_1(w)U_1\bar{\nu}_{cf}S_0 \int_{V_{D2}} \bar{n}^{(2)}{}_{c1}(\vec{r}\text{-->}\vec{r}_{D2},w)d^3\vec{r}_{D2} \tag{178}$$

$$\phi_{Cf13}(w)= h^*{}_1(w)U_1\bar{\nu}_{cf}S_0 \int_{V_{D3}} \bar{n}^{(3)}{}_{c1}(\vec{r}\text{-->}\vec{r}_{D3},w)d^3\vec{r}_{D3} \tag{179}$$

where all the symbols that appears in (178) and (179) has been defined previously.

If the only source present in the subcritical system, is the C_f-252 source, and this is assumed to be given by (175). Then the C.P.S.D., $\phi_{23}(w)$ between detectors D_2 and D_3, deduced from (132) yields the following result:

$$\phi_{23}(w)= \overline{\nu_{cf}(\nu_{cf}-1)}.S_0.I^{(2)}{}_{cf}(\vec{r}_0,-w).I^{(3)}{}_{cf}(\vec{r}_0,w) +$$

$$+ \frac{\overline{\nu(\nu-1)}}{\bar{\nu}} \sum_\mu \frac{K_\mu}{1-K_\mu} N^1{}_\mu(\vec{r}_0)\bar{\nu}_{cf}S_0. \int_V dr N^1{}_\mu(\vec{r})I^{(2)}(\vec{r},-w)I^{(3)}(\vec{r},w) \tag{180}$$

where K_μ are the K-eigenvalues. These eigenvalues are obtained from $\Delta_\mu(w= 0)$, substituting $\nu_g\ \Sigma_{fg}$ by $\bar{\nu}_g\ \Sigma_{fg}/K_\mu$, and then solving the equation that results when we equal to zero the resulting determinant. $I^{(2)}{}_{cf}(\vec{r}_0,-w)$ is the complex conjugate of the spectral weighted factor (154), for detector D_2. S_o is the spontaneous fission rate of the Californium-252 neutron source. $I^{(2)}(\vec{r},-w)$ and $I^{(3)}(\vec{r},w)$, have been defined in eqs (141) and (142), and can be calculated from (158), (159) and (169). The point $\vec{r}_0$, is the Californium-252 source position. The second term in equation (180) contains an integral over the system volume.

The BESMAC3 code computes the C.P.S.D $\phi_{cf12}(w)$, $\phi_{cf13}(w)$, and $\phi_{23}(w)$, in cylindrical geometry, and homogeneous media using the formalism developed in this section. The detectors D_1, D_2, and D_3, can be located at any position and can be point (small detectors), or long detector (cylindrical detectors in which the length is not negligible compared with the system dimensions).

If the detectors are located outside the system, the BESMAC3 code, computes the detector response from the neutron current leaving the system instead of use the neutron flux.

In order to see how the theoretical C.P.S.D. look compared with the experimental ones, we run the SHEBA [1], cases AB and AC, for several frequencies. Because in this two cases the Californium-252 neutron source was located on the cylinder axis, we have only radial and axial modes for $\phi_{cf12}(w)$ and $\phi_{cf13}(w)$.

The calculation was performed with 20 radial modes and 20 axial modes. In figure (1) are shown with dashed line the theoretical plots of Real $(\phi_{cf12}(w))$ and Imaginary $(\phi_{cf12}(w))$, versus frequency, for RUN AB 1 (Solution height 35.9 cm, source position at 18 cm from the bottom, detector D_2 heigh 18 cm, detector D_3 height 18 cm, angle between detectors D_2 and D_3, 120°).

Then we performed a calculation of $\phi_{23}(w)$ for run AB, with 8 radial, 8 axial and 8 angular modes to calculate each importance function which enters into equation (178). In figure (2), we have plotted $Re(\phi_{23}(w))$ and $Im(\phi_{23}(w))$ versus w, the theoretical results agree pretty well with the experimental data.

In figures (3) and (4) we have plotted $Re(\phi_{cf13}(w))$, Im $\phi_{cf13}(w))$, and $Re(\phi_{23}(w))$, $Im(\phi_{23}(w))$ versus frequency for RUN AC1, (solution height 34 cm, source and detector positions are the same that in RUN AB). The agreement is much better for ϕ_{23}, that for ϕ_{12}.

8. Conclusions

In this paper first we review the neutron counting statistic of one single detector using stochastic transport theory [7], [8], and including delayed neutrons. Then we present the theory of the covariance and cross-correlation between two detectors, when delayed neutrons are included.

The main contributions to the covariance between two detectors are given by equation (120). The first term gives the source contribution, the second one the contribution that arises from the prompt fission neutrons, and the third one the contribution coming from the correlations between prompt and delayed neutrons.

Then we have deduced the unimodal approximation of this theory. Equation (156) shows that the ratio of source and system contributions to the C.P.S.D., is equal to the product of the following factors:

i) The ratio of source and system Diven's factor.

ii) The square of $\bar{\nu}_S/\bar{\nu}$.

iii) The ratio of the weighted production rates of source and fission neutrons.

iv) The ratio of the source and system spatial effect factors. These factor are the sameones that account for spatial effects in the Feynman Y(d) function. We remark that the factor R, which account for spatial effects in point reactor Kinetics is the same factor that was reported by Otsuka and Ijima [18].

It is interesting to observe that to calculate the C.P.S.D between two detectors, we need to perform a sequence of deterministic calculations, in this case the fourier transformed eqs (138) and (139), for the detector "Fields of View", $\bar{n}_c(1,w)$ and $\bar{m}_c(1,w)$, and the equation for the average flux $\bar{\phi}$.

This theory reduces in the lumped parameter model to the same expressions reported by Difilippo [19], which confirms the soundness of the present formalism.

Finally we present at section 7 the comparison between the results of this theory and a set of experiments performed by J.T. Mihalczo,[1] the agreement can be considered as good, which again confirms that probably we are in the right direction.

References

1: J.T. Mihalczo, R.C. Kryter, W.T. King, E.D. Blakeman. Ann. nucl. Energy Vol 13, No 7. p.p. 351-362 (1986).

2: J.T. Mihalczo, R.C. Kryter, W.T. King, Trans. Am Nucl. Soc., 38 ,359 (1981).

3: J.T. Mihalczo, W.T. King and J.P. Renier. Trans. Am. Nucl. Soc., 41, 619 (1982).

4: N. Shenhav, Y. Segal and A. Notea, Nucl. Sci. Eng., 80, 61 (1982).

5: L. Pál, Il Nuovo Cimento, Suppl. VII, 25 (1958).

6: G.I. Bell, Nucl. Sci. Eng., 21, 390 (1965).

7: J.L. Muñoz-Cobo, R.B.Pérez, G.Verdú, Nucl. Sci. Eng, 95, 83-105 (1987).

8: J.L. Muñoz-Cobo, G. Verdú, Ann nucl. Energy, Vol 14, No 7, pp 327-350 (1987).

9: G. Verdú, J.L. Muñoz-Cobo, Energía Nuclear, Vol 27, No 143 pp 211 (1983).

10: A. Akcasu, G.S. Lellouche, L.M. Shotkin. Mathematical Methods in Nuclear Reactor Dynamic. Academic Press. New York (1971).

11: M.M.R. Williams, Random Processes in Nuclear Reactors, Pergamon Press, oxford (1974).

12. J. Lewins Nuclear Reactor Kinetics and Control, Pergamon Press, Oxford, (1978).

13: J.T. Mihalczo. Nucl. Sci. Engng 56, 271 (1975).

14: J.T. Mihalczo, M.V. Mathis, V.K. Paré, Nucl. Sci. Engng 59, 350 (1976).

16: J. T. Mihalczo, V.K. Pare, G.L. Ragan, M.V. Mathis, G.C. Tillett Nucl. Sci. Engng 60, 29 (1978).

17: G. Th. Analytis. Ann Nucl. Energy. Vol 9 pp 53 to 77 (1982).

18: M. Otsuka and T. Ijima, Nucleonik, 7, 448 (1965).

19: F.C. Difilippo, Nucl. Sci. Engng 90, 13-18 (1985).

DETERMINATION OF KINETICS PARAMETERS USING STOCHASTIC METHODS IN A ^{252}CF DRIVEN SYSTEM

Felix C. Difilippo

Engineering Physics & Mathematics Division
Oak Ridge National Laboratory
P.O. Box 2008
Oak Ridge, Tennessee 37831

INTRODUCTION

Safety analysis and control system design of nuclear systems require the knowledge of neutron kinetics related parameters like effective delayed neutron fraction, neutron lifetime, time between neutron generations and subcriticality margins. Many methods, deterministic and stochastic, are being used, some since the beginning of nuclear power, to measure these important parameters. The method based on the use of the ^{252}Cf neutron source has been under intense study at the Oak Ridge National Laboratory, both experimentally[1] and theoretically,[2] during the last years. The increasing demand for this isotope in industrial and medical applications and new designs of advanced high flux reactors to produce it make the isotope available as neutron source (only few micrograms are necessary). A thin layer of ^{252}Cf is deposited in one of the electrodes of a fission chamber which produces pulses each time the ^{252}Cf disintegrates via α or spontaneous fission decay; the smaller pulses associated with the α decay can be easily discriminated with the important result that we know the time when ν_c neutrons are injected into the system (number of neutrons per fission of ^{252}Cf). Thus, a small (few cm^3) and nonintrusive device can be used as a random pulsed neutron source with known natural properties that do not depend on biases associated with more complex interrogating devices like accelerators. This paper presents a general formalism that relates the kinetics parameters with stochastic descriptors that naturally appear because of the random nature of the production and transport of neutrons.

MEASUREMENT OF THE EFFECTIVE DELAYED FRACTION

In this proposed experiment the subcritical system is driven by a neutron source of multiplicity ν_s, (i.e., ν_s neutrons are emitted per disintegration event in the source) and it is monitored with a neutron detector. The stochastic variables are n, the number of neutrons in the system, and r, the number of counts in the detector at time t. The probability generating function

$$F(x,z,t) = \sum_{n,r} P(n,r,t)x^n z^r \tag{1}$$

satisfies the following partial differential equation that can be derived using the detailed balance of probabilities (see Ref. 3 for details)

$$\frac{\partial F}{\partial t} = C[f_s(x) - 1]F + \{\alpha_1(1-x) + \Lambda_f[f_F(x) - x] + \epsilon\Lambda_d(z-1)\}\frac{\partial F}{\partial x} \quad (2)$$

where

$$f_s = \sum_m p_s(m)x^m \quad (3)$$

and

$$f_F = \sum_i p_F(i)x^i. \quad (4)$$

In the previous equations, $C(1/\text{sec})$ is the disintegration rate of the source, $\alpha_1(1/\text{sec})$ is the fundamental mode decay constant of the neutron population, $\Lambda_f(1/\text{sec})$ and $\Lambda_d(1/\text{sec})$ are the probabilities per unit time that a neutron induces a fission or a detection event, ϵ is the efficiency of the detector and $p_s(m)$ and $p_F(i)$ are the probabilities for the emission of m and i neutrons after, respectively, a disintegration event in the source and a fission in the fuel. Because all neutron noise techniques are interrelated[4] throughout the function F, it would be sufficient to study one technique, for example, the reduced variance or Feynman method,[5] to see the effect of the multiplicity of the source. The departure of the counter statistics with respect to the Poisson distribution is defined by ψ, the reduced variance minus one; evaluating F and its derivatives one arrives at the following equation for ψ

$$\psi(t) \equiv \frac{\bar{r^2} - \bar{r}^2}{\bar{r}} - 1 = E\left[1 - \frac{1 - \exp(-\alpha_1 t)}{\alpha_1 t}\right] \quad (5)$$

where

$$E = \frac{eD_f}{\beta_{ef}^2(1 + |\$|^2)} f \quad (6)$$

and

$$f = 1 + \frac{\bar{\nu}_s D_s}{\bar{\nu}_f D_f}\beta_{ef}(1 + |\$|). \quad (7)$$

In the previous equations, e is the customary efficiency of the detector with respect to the total fission rate; $\bar{\nu}_s, D_s$ and $\bar{\nu}_f, D_f$ are the average numbers of neutrons and the Diven factors for, respectively, a disintegrating event in the source and a fission event in the fuel. The factor f is equal to one when we use a one-neutron-per-event source, thus f can be measured as the ratio of the correlations measured first in the presence of a ^{252}Cf source and then in the presence of a photoneutron source. Because the reactivity in dollar units can be measured independently, a fit of f as a function of \$ allows a direct measurement of the delayed fraction, β_{ef}.

CALCULATION OF THE CROSS POWER SPECTRAL DENSITY (CPSD)

In this section we calculate the CPSD's relevant to the ratio method to measure subcriticality. For this case, the ^{252}Cf is within a fission chamber (labeled detector 1) that measures the spontaneous fission in the ^{252}Cf deposit, at the same time the system is monitored by two neutron detectors (2 and 3). The CPSD's (G) between the fluctuations of the responses of different pairs of detectors are then measured and the ratio R $= \frac{G_{12}^* G_{13}}{G_{11} G_{23}}$ is formed (the star denotes complex conjugation) which has to be related to the multiplication constant. The method has been successfully applied by Mihalczo et al.[1] to a large variety of multiplicative systems, ranging from fast to thermal, heterogeneous to homogeneous, and continuous to a collection of fissile lumps. In this section we summarize the theoretical background of the method, essential for a correct analysis of the abundant experimental data. We will use a Langevin-Schottky approach that reproduces results, shown in Ref. 6, based on a more basic theory. The Fourier component of the neutron flux, ϕ_w, satifies the equation

$$\hat{H}\phi_w + S_w = j\frac{w}{v}\phi_w \tag{8}$$

where $\hat{H}$ is the Boltzmann operator, S_w the Fourier component of a generic source, j the imaginary unit, w is 2π times the frequency and v the velocity of the neutrons.

Because our system is monitored with detectors (labeled i) with cross section Σ_{di}, it is convenient to define the detector field of view η_i throughout the adjoint equation

$$\hat{H}^+\eta_i + \Sigma_{di} = j\frac{w}{v}\eta_i \tag{9}$$

From Eqs. (8) and (9) and using the property of the adjoint operator, we obtain

$$R_{wi} \equiv < \Sigma d_i \phi_w > = < \eta_i S_w > \tag{10}$$

where R_{wi} is the Fourier component of the reaction rate of detector i and the bracket indicates integration in phase space $\vec{\mu} = (\vec{r}, \vec{v})$. The detector field of view can be expanded in adjoint kinetic eigenfunctions

$$\eta_i = \sum_{n=1}^{\infty} b_{ni}\chi_n^+(\vec{\mu}) \tag{11}$$

where χ_n^+ satisfies

$$\hat{H}^+\chi_n^+ = -\frac{\alpha_n}{v}\chi_n^+ \tag{12}$$

where α_n is the decay constant of mode n. After introducing Eq. (11) in Eq. (9) and multiplying scalarly by $\chi_m(\vec{\mu})$, the direct kinetic eigenfunction, we obtain

$$b_{mi} = \frac{d_{mi}}{\alpha_m + jw}; \tag{13}$$

where $d_{mi} = < v\Sigma_{di}\chi_m >$, in the process we used the property $< \chi_m \chi_n^+ > = \delta_{nm}$. The contribution to the fluctuation of the signal from a source of noise around $\vec{\mu}$ is

$$\delta R_{wi}(\vec{\mu}) = \eta_i \delta S_w(\vec{\mu}) \tag{14}$$

and the cross power spectral density (CPSD) between detectors i and ℓ is

$$Gi_\ell = | << \eta_i^*(\vec{\mu})\eta_\ell(\vec{\mu}\,')\delta S_w^*(\vec{\mu})\delta S_w(\vec{\mu}\,') >> | \tag{15}$$

where the double bracket indicates integration in phase spaces $\vec{\mu}$ and $\vec{\mu}\,'$ and the vertical lines indicate ensemble average; because the η's are deterministic the ensemble average can be made before the integration. The ensemble average of the source of noise can be calculated using the Schottky presciption. Considering that the detectors remove the neutrons the result is[7]

$$|\delta S_w^*(\vec{\mu})\delta S_w(\vec{\mu}\,')| = G_{ss}\delta(\vec{\mu} - \vec{\mu}\,') \tag{16}$$

where

$$G_{ss}(\vec{\mu}) = q(\vec{r})\chi_f(\vec{v})\overline{\nu_p(\nu_p - 1)} + C(\vec{r})\chi_c(\vec{v})\overline{\nu_c(\nu_c - 1)} \tag{17}$$

and $q(\vec{r})$ and $C(\vec{r})$ are, respectively, the distribution of the fission rate induced by the neutrons in the system and the distribution of spontaneous fission rate in the ^{252}Cf source, χ_f and χ_c are the respective spectra of the fission neutrons, ν_c is the number of neutrons per fission in the ^{252}Cf and the average over ν_c is made over the distribution of the number of neutrons per fission. The index p in Eq. (17) refers to the prompt approximation, i.e., we assumed the frequency is well above the typical frequencies associated to delayed neutrons, thus the average over ν_p in Eq. (17) refers to the distribution of prompt fission neutrons in the fuel. After substituting Eq. (16) into Eq. (15), the CPSD between neutron detectors is

$$G_{i\ell} = < \eta_i^* \eta_\ell G_{ss}(\vec{\mu}) > . \tag{18}$$

After expanding the η's according to Eqs. (11) and (13), we have the modal expansion for the $G_{i\ell}$

$$G_{i\ell} = A \sum_{n=1}^{\infty} \frac{a_{i\ell}^{(n)}}{\alpha_n^2 + w^2} + jwA \sum_{n=1}^{\infty} \frac{b_{i\ell}^{(n)}}{\alpha_n^2 + w^2} \tag{19}$$

where the common multiplicative constant A is

$$A = d_{1i} d_{1\ell} < f\chi_1^+ >^2 / F \tag{20}$$

with $f = f(\vec{\mu}) = q(\vec{r})\chi_f(\vec{v})$ and F the total fission rate $F =< f(\vec{\mu}) >$. The other parameters of Eq. (19) are given by the following set of equations.

$$a_{i\ell}^{(n)} = \sum_{m=1}^{\infty} \frac{\alpha_n}{(\alpha_n + \alpha_m)} (r_{in} r_{\ell m} + r_{\ell n} r_{im}) g_{nm} \tag{21}$$

$$b_{i\ell}^{(n)} = \sum_{m=1}^{\infty} \frac{1}{(\alpha_n + \alpha_m)} (r_{in} r_{\ell m} - r_{\ell n} r_{im}) g_{nm} \tag{22}$$

$$g_{nm} = \overline{\overline{\nu_p(\nu_p - 1)}} \frac{f_{nm}}{\epsilon_n \epsilon_m} + \frac{\overline{\nu_s(\nu_s - 1)}}{\overline{\nu}_s} \rho \overline{\overline{\nu}} \gamma_k \frac{S_{nm}}{\epsilon_n \epsilon_m I_1} \tag{23}$$

$$f_{nm} = \frac{F < \chi_n^+ \chi_m^+ f(\vec{\mu}) \overline{\nu_p(\nu_p - 1)} >}{< f(\vec{\mu}) \chi_1^+ >^2 \, \overline{\overline{\nu_p(\nu_p - 1)}}} \tag{24}$$

$$\overline{\overline{\nu_p(\nu_p - 1)}} = \frac{< \nu_p(\nu_p - 1) f(\vec{\mu}) >}{F}; \; \overline{\overline{\nu}} = \frac{< \overline{\nu} f(\vec{\mu}) >}{F} \tag{25}$$

$$\epsilon_n = \frac{< S(\vec{\mu}) \chi_n^+ >}{< S(\vec{\mu}) \chi_1^+ >}; \; S(\vec{\mu}) = C(\vec{r}) \chi_c(\vec{v}) \tag{26}$$

$$S_{nm} = \frac{F^2}{F_c} \frac{< \chi_n^+ \chi_m^+ S(\vec{\mu}) >}{< \chi_1^+ f(\vec{\mu}) >^2}; \; F_c =< S(\vec{\mu}) > \tag{27}$$

$$I_n = \frac{F < S(\vec{\mu}) \chi_n^+ >}{F_c < f(\vec{\mu}) \chi_1^+ >} \tag{28}$$

$$r_{in} = \epsilon_n d_{ni} / d_{1i} \tag{29}$$

Because we have introduced in Eq. (23) the static reactivity

$$\rho = (1 - k_1)/k_1 \tag{30}$$

where k_1 is the multiplication constant associated with the static fundamental mode, it appears the kinetic distortion factor γ_k defined as

$$\gamma_k = \frac{<\chi_1^+ S>}{<\phi_1^+ S>} \frac{<\phi_1^+ f>}{<\chi_1^+ f>} \tag{31}$$

where ϕ_1^+ is the adjoint static fundamental mode.

The previous equations for $G_{i\ell}$ are valid for the neutron detectors i and ℓ which have in common the fluctuations induced by the source and the neutron field. By contrast, the CPSD G_{1i} between the source and the detector depends only on the fluctuations of the source. The fluctuations of the signal in detectors 1 and i ($\neq 1$) are given by

$$\delta R_{w1}(\vec{r}) = \epsilon_c \delta C(\vec{r}) \tag{32}$$

and

$$\delta R_{wi}(\vec{\mu}') = \eta_i \delta S_w(\vec{\mu}') \tag{33}$$

where ϵ_c is the efficiency of the fission chamber. The only fluctuation in common between detector 1 and i is the fluctuation of the number of fissions, δC, in the ^{252}Cf which is spatially uncorrelated, then

$$G_{1i} = \epsilon_c \bar{\nu}_c < \eta_i S(\vec{\mu}) > \tag{34}$$

or expanding η_i in eigenfunctions

$$G_{1i} = \bar{\nu}_c \epsilon_c d_{1i} \frac{F_c}{F} < f(\vec{\mu}) \chi_1^+ > I_1 \sum_{n=1}^{\infty} \frac{r_{in}}{\alpha_n + jw}. \tag{35}$$

With similar arguments

$$G_{11} = \epsilon_c^2 F_c. \tag{36}$$

MEASUREMENT OF KINETIC PARAMETERS: THE RATIO METHOD

The calculation of the CPSDs in the previous section shows that the ratio R is, in general, a frequency-dependent complex number whose relationship with the reactivity is not obvious. In this section we propose a method to make this relationship clear.

The fit of the measured CPSDs according to Eqs. (19), (35), and (36) allow the simultaneous determination of the decay constants and modal amplitudes like $A_{12}^{(1)}$ and $A_{13}^{(1)}$, the fundamental modal amplitudes of, respectively, G_{12} and G_{13}, and $A_{23}^{(1)}$, the fundamental mode amplitude of the real part of G_{23}. It is then possible to measure the fundamental mode ratio defined as

$$R_{FM} \equiv \frac{A_{12}^{(1)} A_{13}^{(1)}}{G_{11} A_{23}^{(1)}}. \tag{37}$$

The relationship between the observables of Eq. (37) and the parameters of the system are given by Eqs. (35), (36), (19) and related equations of the previous section. Thus, the reactivity can be inferred from the experiment with the following set of equations:

$$\rho = \frac{R^* Y_1}{1 - Y_1 Y_2 R^*} \tag{38}$$

$$Y_1 = \frac{f_{11}}{\gamma_k I_1} \frac{\overline{\overline{\nu_p(\nu_p - 1)}}}{\bar{\nu}_s \bar{\bar{\nu}}} \tag{39}$$

$$Y_2 = \frac{S_{11}\gamma_k}{f_{11}I_1} \frac{\overline{\nu_c(\nu_c - 1)}}{\overline{\overline{\nu_p(\nu_p - 1)}}} \frac{\overline{\overline{\nu}}}{\overline{\nu}_c} \tag{40}$$

$$R^* = R_{FM}h \tag{41}$$

$$h = 1 + \frac{1}{g_{11}} \sum_{n=2}^{\infty} \frac{\alpha_1}{\alpha_1 + \alpha_n}(r_{2n} + r_{3n})g_{1n} \tag{42}$$

where R^*, the ideal ratio, would be equal to the fundamental mode ratio (an observable) if $h = 1$, i.e., if the detectors or the source are distributed according to the fundamental mode. The equations for Y_1, Y_2 and h contain known physical parameter like ν_p and ν_c, observables like α_n, r_{2n} and r_{3n} and functionals of the flux and source distributions that have to be calculated; all the functionals are defined as adimentional numbers. The expressions for Y_1, Y_2 and g_{11} contains functionals that depend on the fundamental mode, they can be computed by standard transport codes. Extensive calculations by Mihalczo et al.[1] show that because of its definition, as ratio of functionals, Y_1, for example, does not depend on details of the numerical model. On the contrary, h depends, through g_{1n}, on functionals of the higher harmonic that cannot be computed by standard codes.

The sequence to measure ρ is then the following, after computing Y_1 and Y_2 a first estimation of ρ using the approximation $R^* \sim R_{FM}$ can be made, this value of ρ can be entered into the definition of h (Eq. (42)) together with the measured α_n, r_{2n} and r_{3n} and some estimation of the higher harmonic functionals, which would produce a second iteration value for R^* and so on. To minimize errors it is then better to keep h as close to one as possible, either by minimizing the harmonic content of the source and detector field of view, and/or by placing the detectors in symmetric positions in such a way that $r_{2n} = -r_{3n}$ for the first relevant harmonics.

EXAMPLE OF APPLICATION: INTERACTION BETWEEN TWO EQUAL MULTIPLICATIVE SYSTEMS

The simplest parameterization in this case is a two-point kinetics model. The steady state solution satisfies, in general, the equation

$$\hat{H}\vec{N} + \bar{\nu}_c\vec{S} = O \tag{43}$$

where the Boltzmann operator has, in this case, the simple form of a two by two matrix. It can be written as $\hat{H} = \hat{P} - \hat{D}$ where the production and destruction operators are

$$\hat{P} = \frac{k}{\ell}\begin{bmatrix} 1 & 0 \\ 0 & 1 \end{bmatrix} ; \hat{D} = \frac{1}{\ell}\begin{bmatrix} 1 & -k_{in} \\ -k_{in} & 1 \end{bmatrix} \tag{44}$$

where k and ℓ are the multiplication constant and the mean life of a neutron in one-half of the systems and k_{in} is the coupling constant. The static eigenfunctions are defined by

$$\left(\frac{\hat{P}}{k_{ef}} - \hat{D}\right)\vec{\phi} = 0, \tag{45}$$

solving Eq. (45) we obtain two eigenvalues and eigenfunctions subject to the normalization $< \phi_m^+ \hat{P}\phi_n >= \delta_{nm}$ they are

$$k_1 = \frac{k}{1 - k_{in}}; \ \phi_1 = \sqrt{\frac{\ell}{2k}}\begin{bmatrix} 1 \\ 1 \end{bmatrix} \tag{46}$$

and

$$k_2 = \frac{k}{1+k_{in}}; \ \phi_2 = \sqrt{\frac{\ell}{2k}} \begin{bmatrix} 1 \\ -1 \end{bmatrix}, \tag{47}$$

because the operators are self-adjoint $\phi_1^+ = \phi_1$ and $\phi_2^+ = \phi_2$.

The kinetic eigenfunctions are, in the prompt approximation,

$$\hat{H}_p \vec{\chi}_n = -\alpha_n \vec{\chi}_n \tag{48}$$

where $\hat{H}_p$ is similar to $\hat{H}$ with the factor (1-β) multiplying k. In this approximation we have again only two modes subject to the normalization $< \chi_n^+ \chi_m >= \delta_{nm}$, they are

$$\alpha_1 = \alpha - \gamma; \quad \chi_1 = \frac{1}{\sqrt{2}} \begin{bmatrix} 1 \\ 1 \end{bmatrix} \tag{49}$$

$$\alpha_2 = \alpha + \gamma; \quad \chi_2 = \frac{1}{\sqrt{2}} \begin{bmatrix} 1 \\ -1 \end{bmatrix} \tag{50}$$

and because the operators are self-adjoint, $\chi_1^+ = \chi_1$ and $\chi_2^+ = \chi_2$; in Eqs. (49) and (50) $\alpha = [1 - k_1(1-\beta)]/\ell$ and $\gamma = k_{in}/\ell$. The eigenfunctions and eigenvectors were already defined so it is possible to apply the theory of the previous sections. If the detectors i are distributed in region 1 and 2 with detector cross sections $\Sigma d_i^{(1)}$ and $\Sigma d_i^{(2)}$, the coefficients defined in Eq. (13) are $d_{1i} = (\Sigma d_i^{(1)} + \Sigma d_i^{(2)})/\sqrt{2}$ and $d_{2i} = (\Sigma d_i^{(1)} - \Sigma d_i^{(2)})/\sqrt{2}$. After specification of the intensity of the source in region 1, $S_1 = \frac{F_c}{2}(1+\epsilon)E$ and region 2, $S_2 = \frac{F_c}{2}(1-\epsilon)E$, the steady state distribution can be calculated with Eq. (43). The factor ϵ is related to the distribution of the source, $\epsilon = 1$ implies $S_2 = 0$, and E is indicative of the effective source (note that F_c is a direct observable). After solving Eq. (43) and using Eqs. (23) to (29) it is possible to calculate the factor h of Eq. (41); the necessary g_{nm} are

$$g_{11} = \overline{\nu_p(\nu_p - 1)} + \frac{\overline{\nu_s(\nu_s - 1)}}{\bar{\nu}_s} \rho \bar{\nu} \tag{51}$$

$$g_{12} = \overline{\nu_p(\nu_p - 1)} \frac{\rho}{\rho + \frac{2\gamma}{\alpha_1}(\beta + \rho)} + \frac{\overline{\nu_s(\nu_s - 1)}}{\overline{\nu_s}} \rho \bar{\nu} \tag{52}$$

where γ can be measured as $(\alpha_2 - \alpha_1)/2$. This way of approximating h was successfully applied to analyze the interaction between fissile systems.[8]

CONCLUSIONS

A general theoretical background was presented to analyze experiments designed to measure kinetic parameters using the ^{252}Cf neutron source. The nonintrusive type of experiment involved make the method ideally suited for a continuous monitoring of nuclear systems.

Acknowledgement

Research sponsored by Office of Nuclear Safety, U.S. Department of Energy under contract number DE-AC05-84OR21400 with Martin Marietta Energy Systems, Inc.

REFERENCES

1. J. T. Mihalczo, W. T. King and E. D. Blakeman, "^{252}Cf-Source-Driven Neutron Noise Analysis Method," *Proc. of the Workshop on Subcritical Reactivity Measurement*, Albuquerque, New Mexico, August 26-29, 1985, the University of New Mexico.

2. F. C. Difilippo, "Interpretation of Spatial Effects in Reactivity Measurement with the ^{252}Cf Source Noise Method," *Trans. Am. Nucl. Soc.* **52**, 642 (1986).

3. F. C. Difilippo, "Stochastic Processes in a Subcritical Nuclear Reactor in the Presence of a Fission Source," *Nucl. Sci. Eng.* **90**, 13 (1985).

4. N. Pacilio et al., "Toward a Unified Theory of Reactor Neutron Noise Analysis Techniques" *Ann. of Nucl. Energy* **3**, 239 (1976).

5. R. P. Feynman et al., "Dispersion of the Neutron Emission in U-235 Fission" *J. of Nucl. Energy* **3**, 64 (1956).

6. J. L. Muñoz-Cobo et al., "Stochastic Neutron Transport Theory: Neutron Counting Statistics in Nuclear Assemblies" *Nucl. Sci. Eng.*, **95**, 83 (1987).

7. F. C. Difilippo, "Correlation of the Signals from Detectors in the Presence of a Stochastic Neutron Field" *Nucl. Sci. Eng.* **99**, 28 (1988).

8. F. C. Difilippo and C. March-Leuba, *9th Intnl. Conf. on Noise in Physical Systems*, p. 579, World Scientific Publishing Co., 1987.

SOLVING LOW POWER STOCHASTIC EQUATIONS: NONLINEAR STUDIES

Jeffery D. Lewins

University of Cambridge
Engineering Department
Trumpington Street
Cambridge CB2 1PZ
England

INTRODUCTION

We address here the stochastic equations for neutrons, precursors and detected events ('detectrons') in a conventional lumped model of a reactor at low power. Working in the backward Kolmogorov formulation we have the usual first-order partial differential equations in time with nonlinear terms.

It is not generally well known that this form of the equations permits explicit solutions starting with the extinction probability and, once this is known, recursively through higher probabilities. Therefore we illustrate this for the general model.

To do so, we seek the extinction probability following the introduction of one neutron or of one precursor (or indeed of one detectron) into a source-free system. From these elementary results (Green's functions of the nonlinear equations) the full solution can be constructed.

The starting point is to follow up a well known statement (Bartlett, 1978) that the extinction probability is given by the lowest (real) root in the range zero to unity of the polynomial formed by equating the backward equation to zero. Feller (1950) gives, for the equivalent discontinuous time problem, an answer to the obvious consequential questions: firstly, when is there an internal root between 0 and 1 (there is always an end root at 1) and, secondly, how many internal roots are there?

It turns out that the discussion of the general problem with precursors is aided by first discussing the elementary problem with no precursors and, since this serves to introduce the notation etc., the plan of the paper is to deal with the neutron-only model first, followed by an extension to one precursor group and finally to many precursor groups.

Readers might be amused to ask themselves now whether they know the answer to the problems. The purpose of the paper, however, is to illustrate the advantages of the backward equations.

The derivation in the combined 1+I+J groups of neutrons, precursors and 'detectrons' is explicit. We have drawn on earlier work by Salmi and Lewins (1980) for the vector notation in a multi-group neutron model. As we remarked at the time, that model could be specialised to the present case but it is likely to be clearer if we derive the equations explicitly.

THE NEUTRON-ONLY MODEL

There is a single group of neutrons at low power. This means that burn-up or feedback effects are ignored. Nevertheless, the model allows for the possibility of having an unbounded number of neutrons. Should that occur, the model will be faulty.

It is supposed that neutrons have a mean interaction rate $1/l$, a removal that can either lead to capture or to fission producing more neutrons (Ruby and McSwine, 1986). By defining a general fission probability p_υ to include capture, the notation can be compacted at the cost, of course, of not having universal distribution functions. Thus it will not follow that the mean yield is 2.45 for U-235. Suppose also there is a mean source rate S yielding υ neutrons per source event at probability q_υ. In principle, these coefficients can be time dependent.

Let $P_{nm}(t|s)$ be the probability that given m neutrons at an earlier time s, there shall be n neutrons at a later time t. Thus m is the original state leading to n, the final state.

The backward equations are obtained by following the m original neutrons for a short interval δs during which either nothing happens (with suitable probability) or something happens to change the state. From the low power, independence model, the probability balance can be written as

$$P_{mn}(t|s) = [1-\frac{m}{l}\delta s - S\delta s]P_{nm}(t|s+\delta s) + \frac{m}{l}\delta s\sum_{\upsilon=0}^{\infty} p_\upsilon P_{nm-1+\upsilon} + S\delta s\sum_{\upsilon=0}^{\infty} q_\upsilon P_{nm+\upsilon} \tag{1}$$

which, together with the obvious boundary condition at s = t, yields

$$-\frac{\partial P_{nm}}{\partial s} = \frac{m}{l}\sum_{\upsilon=0}^{\infty} p_\upsilon P_{nm-1+\upsilon} - \frac{m}{l}P_{nm} + S\sum_{\upsilon=0}^{\infty} q_\upsilon P_{nm+\upsilon} - SP_{nm} \ ; \ P_{nm}(t|t) = \delta_{nm} \ . \tag{2}$$

The generating function $F_m(t|s;x)$ is introduced together with auxiliary generating functions as

$$F_m(t|s;x) = \sum_{n=0}^{\infty} x^n P_{nm} \ , \ g(x) = \sum_{\upsilon=0}^{\infty} x^\upsilon p_\upsilon \ , \ f(x) = \sum_{\upsilon=0}^{\infty} x^\upsilon q_\upsilon \ . \tag{3}$$

The auxiliary generating functions are such that $g(1) = f(1) = 1$.

As usual, probabilities can be regenerated from these z-transforms and they also serve as factorial moment generating functions.

For the particular case of no source, write the probabilities as Q_{nm} and generating function

$$G_m(t|s;x) = \sum_{n=0}^{\infty} x^n Q_{nm} \ . \tag{4}$$

Thus the backward equations become

$$-\frac{\partial F_m}{\partial s} = \frac{m}{l}[\sum_{\upsilon=0}^{\infty} p_\upsilon F_{m-1+\upsilon} - F_m] + S[\sum_{\upsilon=0}^{\infty} q_\upsilon F_{m+\upsilon} - F_m] \ ; \ F_m(t|t) = x^m \ . \tag{5}$$

The low power model implies the independence of neutrons such that

$$F_m = G_1^m F_0 \ . \tag{6}$$

Therefore if we may find G_1 and F_0, we can say we have a complete solution available. Indeed F_0 itself may be found from G_1 and hence G_1 is essentially the 'Green's function' for the problem.

The special cases are as follows:

(a) $m = 1$, $S = 0$ (noting that $F_0 = 1$),

$$-\frac{\partial G_1}{\partial s} = \frac{1}{l}[\sum_{\upsilon=0}^{\infty} p_\upsilon G_1^\upsilon - G_1] = \frac{1}{l}[g(G_1) - G_1] \equiv H(G_1)\ ;\ \ G_1(t|t) = x\ , \tag{7}$$

(b) $m = 0$, $S \neq 0$,

$$-\frac{\partial F_0}{\partial s} = S[\sum_{\upsilon=0}^{\infty} q_\upsilon G_1^\upsilon - 1]F_0 = S[f(G_1) - 1]F_0\ ;\ \ F_0(t|t) = 1\ . \tag{8}$$

Hence indeed

$$F_0(t|s) = \exp\{-\int_s^t S[1 - f(G_1(t|s'))]ds'\}\ , \tag{9}$$

and F_0 is available by quadratures over G_1.

The equation for G_1 is seen to be a first-order partial differential equation with s-dependent l and g. Rather than solve it as such, observe that we may solve recursively for the probabilities, starting with the extinction probability Q_{01}. That is, put $x = 0$ in the equations and

$$-\frac{\partial Q_{01}}{\partial s} = \frac{1}{l}[g(Q_{01}) - Q_{01}] = H(Q_{01})\ ;\ \ Q_{01}(t|t) = 0\ . \tag{10}$$

This may be solved backwards in s-time from $t-s = 0$.

With Q_{01} found, $\frac{\partial}{\partial x}(\ )_0$ yields

$$-\frac{\partial Q_{11}}{\partial s} = \frac{1}{l}[g'(0)G_1'(0) - G_1'(0)]\ ;\ \ Q_{11}(t|t) = 1\ , \tag{11}$$

i.e.

$$-\frac{\partial Q_{11}}{\partial s} = H'(0) = \frac{1}{l}[\sum_{\upsilon=0}^{\infty} \upsilon p_\upsilon Q_{01}^{\upsilon-1} - 1]Q_{11}\ . \tag{12}$$

And, in principle, we may continue to find higher probabilities. Of course, if our interest were in finding say the probability of $n = 1000$, this would be a formidable task. We might be willing to settle for a few moments, particularly the mean and variance instead. The equations for moments are as readily obtained from the backward formulation as the forward. But the extinction probability is likely to be the single most significant individual probability and hence we shall focus on it alone.

DYNAMICS OF EXTINCTION

Bartlett (1978) discusses the extinction probability for such a case as this in terms of the polynomial of generating functions $H(G_1)$. More directly, consider Q_{01} itself, i.e. put $x = 0$ in equation (7).

The equation for the extinction probability shows that it is originally growing (at $t-s = 0$) from its initial value of zero (if there is one neutron at $s = t$, the system is certainly not extinct). If the corresponding polynomial $H(Q_{01}) = 0$ is studied for its roots, certain cases can be anticipated corresponding to a steady state for Q_{01} (Fig. 1).

(a) There is always a root at $Q_{01} = 1$ since $H(1)$ is composed of proper auxiliaries where $g(1) = 1$. We refer to this as an end root. There is no root at $Q_{01} = 0$ unless there is no removal mechanism (i.e. unless $p_0 = 0$), a trivial case.

(b) If there are one or more real roots internally, in $(0,1)$, then Q_{01} rises to the lowest such root and there stops.

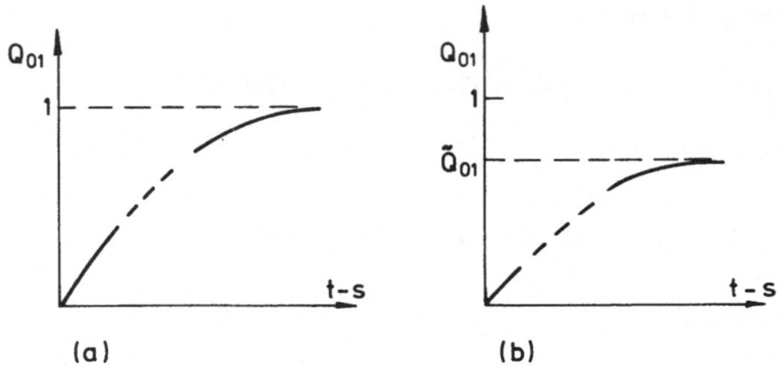

Fig. 1.

Thus after long enough time, the state of the system is extinguished with probability corresponding to the lowest internal root or with certainty if there is no internal root. The interpretation of the probability Q_{0m} for m initial neutrons, no source, is simply that

$$Q_{0m} = Q_{01}^{m} .$$

If there is a source, we may obtain P_{0m} through equation (9), putting G_1 to Q_{01}. The interpretation in this case needs some care. That is, with a source present, there can be no extinction as such but rather the long term solution gives the proportion of time we could sample the system and find no neutron.

Thus the questions to be answered in this model are (i) when is there an internal root, and (ii) is there more than one real internal root?

To answer these questions, we exploit the properties of H(x) as found in the mean, Fig. 2. Consider

$$\frac{\partial H(x)}{\partial x}\Big|_{x=1} = \frac{1}{l}[\sum_{\upsilon=0}^{\infty} \upsilon x^{\upsilon-1} p_\upsilon - 1]\Big|_{x=1} = \frac{1}{l}[<\upsilon> - 1] = H'(1) . \tag{13}$$

Then $H'(1)$ is $\gtrless 0$ according to $k_{ex} \gtrless 0$, the conventional kinetics parameter in the mean, the excess multiplication factor. Indeed the equation for the mean number of neutrons is found, after partial differentiation with respect to x, to be

$$-\frac{\partial}{\partial s}\frac{\partial G_1}{\partial x}\Big|_{x=1} = H'(1)G_1'(1) ,$$

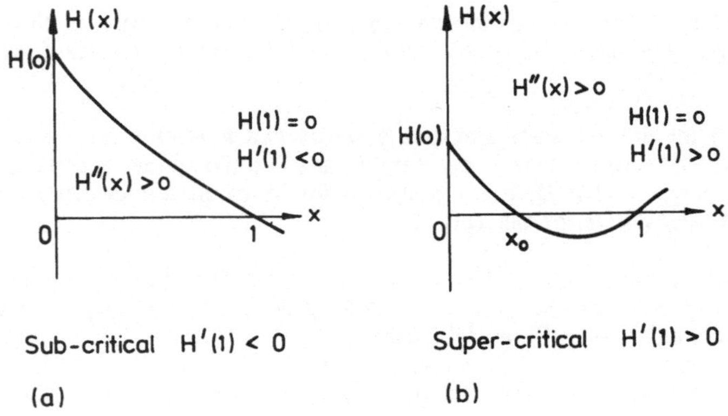

Fig. 2.

or

$$-\frac{\partial}{\partial s}\langle n\rangle = H'(1)\langle n\rangle \,. \tag{14}$$

If the problem is non-trivial, then not only is H(0) > 0 but also H″(x) > 0 for all x within (0,1), since

$$\frac{\partial^2 H}{\partial x^2}\Big|_{x=1} = \frac{1}{l}[\sum_{\upsilon=0}^{\bar{\upsilon}} \upsilon(\upsilon-1)x^{\upsilon-2}p_\upsilon]\Big|_{x=1} = \frac{1}{l}\langle\upsilon(\upsilon-1)\rangle = H''(1) \,. \tag{15}$$

Thus, if (Fig. 2(a)) the system is sub-critical, H(x) follows the curve with no internal root. If the system is super-critical (Fig. 2(b)), it follows a curve that has one and one only internal root. The questions are answered.

GENERAL MODEL

We now add a representation of precursors, up to I groups, and detectors, up to J in number. λ_i is the mean rate of decay of a precursor in the ith group and ε_j/l the mean reaction rate of the jth detector. This is taken to be included in the overall mean reaction rate $1/l$. For simplicity detectors are neutron capturing but not fissioning. (It makes no difference to the subsequent discussion.)

The distribution in fission over neutrons and precursors is to be represented. For simplicity we assume that there is not more than one precursor per fission and that the source does not yield precursors. (The relaxation of these simplificatons is straightforward.) Then take $p_{\upsilon 0}$ to be the probability of a removal giving υ neutrons and no precursors, and $p_{\upsilon i}$ to be the probability of producing υ neutrons and 1 precursor of the ith type. The corresponding auxiliary generating functions,

$$g_0(x,t) = \sum_{\upsilon=0}^{\bar{\upsilon}} p_{\upsilon 0}x^\upsilon, \quad g_i(x,t) = \sum_{\upsilon=0}^{\bar{\upsilon}} p_{\upsilon i}x^\upsilon$$

are not individually 'proper', but in combination we have

$$g_0(1) + \sum_{i=1}^{I} g_i(1) = 1 \,,$$

a 'complete' stochastic generating function.

The state of the system is written as vector

$$\underline{n} = (n, c_1, \ldots c_i, \ldots c_I, d_1, \ldots d_j, \ldots d_J)$$

the number of neutrons, precursors and 'detectrons', i.e. the counts in the detectors. Similarly write

$$\underline{m} = (m, m_1, \ldots m_i, \ldots m_I, o_1, \ldots o_j, \ldots o_J)$$

for the initial state. It is no loss of generality to start with the counters cleared, $o_j = 0$. Now we may write the state probabilities as $P_{\underline{n}\underline{m}}(t|s)$ as before.

Define (Salmi and Lewins, 1980) suitable vector Kronecker deltas:

$$\underline{\delta}_0 = (1, 0, \ldots 0, 0, \ldots 0) \tag{16a}$$

in the neutron position,

$$\underline{\delta}_i = (0, 0, \ldots 1, 0, \ldots 0, 0, \ldots 0) \tag{16b}$$

in the "ith" position,

$$\underline{\delta}_j = (0, 0, \ldots 0, 0, \ldots 1, 0, \ldots 0) \tag{16c}$$

in the "jth" position.

The backward equation may be derived via the probability balance

$$P_{\underline{nm}}(t|s) = [1-\frac{m}{l}\delta s-\sum_{i=1}^{I}m_i\lambda_i\delta s-S\delta s]P_{\underline{nm}}(t|s+\delta s) \qquad \text{no change}$$

$$+\frac{m}{l}\delta s\sum_{\upsilon=0}^{\infty}p_{\upsilon 0}P_{\underline{nm}+(\upsilon-1)\underline{\delta_0}} \qquad \text{fission, no precursor}$$

$$+\frac{m}{l}\delta s\sum_{i=1}^{I}\sum_{\upsilon=0}^{\infty}p_{\upsilon i}P_{\underline{nm}+(\upsilon-1)\underline{\delta_0}+\underline{\delta_i}} \qquad \text{fission with precursor} \qquad (17)$$

$$+\sum_{i=1}^{I}m_i\lambda_i\delta sP_{\underline{nm}+\underline{\delta_0}-\underline{\delta_i}} \qquad \text{precursor decay}$$

$$+S\delta s\sum_{\upsilon=0}^{\infty}q_\upsilon P_{\underline{nm}+\upsilon\underline{\delta_0}} \qquad \text{source event}$$

$$+\frac{m}{l}\delta s\sum_{j=1}^{J}\varepsilon_jP_{\underline{nm}-\underline{\delta_0}+\underline{\delta_j}}\,, \qquad \text{detector event}$$

which gives

$$-\frac{\partial P_{nm}}{\partial s} = -[\frac{m}{l}+\sum_{i=1}^{I}m_i\lambda_i+S]P_{\underline{nm}}+\frac{m}{l}\sum_{\upsilon=0}^{\infty}p_{\upsilon 0}P_{\underline{nm}+(\upsilon-1)\underline{\delta_0}}+\frac{m}{l}\sum_{i=1}^{I}\sum_{\upsilon=0}^{\infty}p_{\upsilon i}P_{\underline{nm}+(\upsilon-1)\underline{\delta_0}+\underline{\delta_i}}$$
$$+\sum_{i=1}^{I}m_i\lambda_iP_{\underline{nm}+\underline{\delta_0}-\underline{\delta_i}}+S\sum_{\upsilon=0}^{\infty}q_\upsilon P_{\underline{nm}+\upsilon\underline{\delta_0}}+\frac{m}{l}\sum_{j=1}^{J}\varepsilon_jP_{\underline{nm}-\underline{\delta_0}+\underline{\delta_j}}\;;\; P_{\underline{nm}}(t|t)=\delta_{\underline{nm}}\,. \qquad (18)$$

Generating functions are defined using the Lewins symbol

$$\underline{x}^{\cdot\underline{n}} = x^n y_1^{c_1} y_2^{c_2} \ldots y_I^{c_I} z_1^{d_1} z_2^{d_2} \ldots z_J^{d_J} \qquad (19)$$

over auxiliary variables $\underline{x} = (x, \ldots y_i, \ldots z_j, \ldots z_J)$. A summation covention over repeated $\underline{n}$-indices allows us to write

$$F_{\underline{m}}(t|s;\underline{x}) = \underline{x}^{\cdot\underline{n}}P_{\underline{nm}}(t|s)\,. \qquad (20)$$

Again we may exploit the independence of neutrons, precursors and detectrons in this low power model. Let $G_{\underline{m}}$ again be $F_{\underline{m}}$ in the absence of a source. Then we have

$$F_{\underline{m}} = G_{\underline{\delta_0}}^{m}\ldots G_{\underline{\delta_i}}^{m_i}\ldots G_{\underline{\delta_j}}^{m_j}\ldots F_{\underline{0}} = \underline{G}^{\cdot\underline{m}}F_{\underline{0}}\,. \qquad (21)$$

It is seen that in this model we seek 1+I+J elementary generating functions, or 'Green's functions', and the source function.

Then the backward probability equation becomes

$$-\frac{\partial F_{\underline{m}}}{\partial s} = -[\frac{m}{l}+\sum_{i=1}^{I}m_i\lambda_i+S]F_{\underline{m}}+\frac{m}{l}\sum_{\upsilon=0}^{\infty}p_{\upsilon 0}F_{\underline{m}+(\upsilon-1)\underline{\delta_0}}+\frac{m}{l}\sum_{i=1}^{I}\sum_{\upsilon=0}^{\infty}p_{\upsilon i}F_{\underline{m}+(\upsilon-1)\underline{\delta_0}+\underline{\delta_i}}$$
$$+\sum_{i=1}^{I}m_i\lambda_iF_{\underline{m}+\underline{\delta_0}-\underline{\delta_i}}+S\sum_{\upsilon=0}^{\infty}q_\upsilon F_{\underline{m}+\upsilon\underline{\delta_0}}+\frac{m}{l}\sum_{j=1}^{J}\varepsilon_jF_{\underline{m}-\underline{\delta_0}+\underline{\delta_j}}\;;\; F_{\underline{m}}(t|t)=\underline{x}^{\cdot\underline{m}}\,. \qquad (22)$$

There are four special cases of interest:

(1) $S = 0$, $\underline{m} = \underline{\delta_0}$,

$$-\frac{\partial G_{\delta_0}}{\partial s} = \frac{1}{l}[\sum_{\upsilon=0}^{\infty}p_{\upsilon 0}G_{\underline{\delta_0}}^{\upsilon}+\sum_{i=1}^{I}\sum_{\upsilon=0}^{\infty}p_{\upsilon i}G_{\underline{\delta_0}}^{\upsilon}G_{\underline{\delta_i}}-G_{\underline{\delta_0}}]+\frac{1}{l}\sum_{j=1}^{J}\varepsilon_jG_{\underline{\delta_j}}\,,$$

subject to $G_{\underline{\delta_0}}(t|t) = x$, or

$$-\frac{\partial G_{\delta_0}}{\partial s} = \frac{1}{l}[g_0(G_{\underline{\delta_0}})+\sum_{i=1}^{I}g_i(G_{\underline{\delta_0}})G_{\underline{\delta_i}}-G_{\underline{\delta_0}}]+\frac{1}{l}\sum_{j=1}^{J}\varepsilon_jG_{\underline{\delta_j}}\,. \qquad (23)$$

(2) $S = 0$, $\underline{m} = \underline{\delta}_i$,

$$-\frac{\partial G_{\underline{\delta}_i}}{\partial s} = \sum_{i=1}^{I} \lambda_i [G_{\underline{\delta}_0} - G_{\underline{\delta}_i}] , \tag{24}$$

subject to $G_{\underline{\delta}_i}(t|t) = y_i$ for all I.

(3) $S = 0$, $\underline{m} = \underline{\delta}_j$,

$$-\frac{\partial G_{\underline{\delta}_j}}{\partial s} = 0 , \tag{25}$$

subject to $G_{\underline{\delta}_j}(t|t) = z_j$ for all J.

(4) $S \neq 0$, $\underline{m} = 0$,

$$-\frac{\partial F_{\underline{0}}}{\partial s} = S[\sum_{\upsilon=0}^{\infty} q_\upsilon G_{\underline{\delta}_0}^{\upsilon} - 1]F_{\underline{0}} = S[f(G_{\underline{\delta}_0}) - 1]F_{\underline{0}} , \tag{26}$$

subject to $F_{\underline{0}}(t|t) = 1$.

From this final result we have a quadrature for the source result over the first of the 1+I+J Green's functions:

$$F_{\underline{0}}(t|s) = \exp\{-\int_s^t S[1 - f(G_{\underline{\delta}_0}(s'))]ds'\} . \tag{27}$$

These special cases therefore give us a complete representation of the state dynamics in the backward formulation.

EXTINCTION PROBABILITIES

Extinction probabilities in the long term are again going to be determined by putting the generating function equations, now 1+I+J in number, to zero. However, it is apparent that the detectron equations are superfluous to this question; we really have only a I+1-dimensional problem. Of course, we should really be considering the G-values at x = 0, i.e the extinction probabilities $Q_{0\underline{\delta}_0}$ and $Q_{0\underline{\delta}_i}$ starting with one neutron or one precursor respectively. It is again evident that there is always a root at $Q_{0\underline{\delta}_0} = Q_{0\underline{\delta}_i} = 1$ from the proper nature of the stochastic variables.

That is, writing the polynomial equations as

$$0 = H_0(G_{\underline{\delta}_0}, G_{\underline{\delta}_i}) = \frac{1}{l}[g_0(G_{\underline{\delta}_0}) + \sum_{i=1}^{I} G_{\underline{\delta}_i} g_i(G_{\underline{\delta}_0}) - G_{\underline{\delta}_0}] ,$$

and

$$0 = H_i(G_{\underline{\delta}_0}, G_{\underline{\delta}_i}) = \lambda_i [G_{\underline{\delta}_0} - G_{\underline{\delta}_i}] ,$$

then

$$H(G_{\underline{\delta}_0}) = \frac{1}{l}[g_0(G_{\underline{\delta}_0}) + \sum_{i=1}^{I} G_{\underline{\delta}_0} g_i(G_{\underline{\delta}_0}) - G_{\underline{\delta}_0}] , \tag{28}$$

and $H(G_{\underline{\delta}_0})$ is just the expression that would be used for a model in which the delay engendered by delayed neutron precursors was ignored but the production of neutrons included. At this stationary condition, the delay is of course immaterial. Correspondingly, $H(G_{\underline{\delta}_0})$ will relate to the reactivity of the system. We see that the roots, if any, lie on the diagonal of the unit hypercube.

The natural questions now are three-fold:

(1) If there is no internal root the system can only evolve to the end root at $\underline{1}$. Does it indeed converge to this root?

(2) If there are internal roots, how many are there and when will they occur?

(3) Will the system necessarily converge to (the lowest) of any internal roots in $(\underline{0},\underline{1})$?

We examine the two-dimensional problem, I = 1, first. H(x) behaves as in the precursor-free case. Thus $H(0) > 0$, $H'(1) \gtrless 0$ as $k_{ex} \gtrless 0$ and $H''(x)$ is positive throughout. Hence the construction of Fig. 2 still applies and we conclude that there is no internal root for a sub-critical system, only one internal root for a super-critical system.

CONVERGENCE

Put $Q_{0010} = x$, $Q_{0001} = y$ and examine in turn the sides of the unit square (0,0) to (1,1).

Case (A): No internal root

$$-\frac{\partial x}{\partial s} = \frac{1}{l}[g_0(x) + yg_1(x) - x] ,$$

and

$$-\frac{\partial y}{\partial s} = \lambda[x - y] ,$$

so

$$\frac{\partial y}{\partial x} = l\lambda[x - y]/[g_0(x) + yg_1(x) - x] . \tag{29}$$

When y = 0

$$\frac{\partial y}{\partial x} = l\lambda x/[g_0(x)-x] , \tag{30}$$

which changes sign. When x = 1

$$\frac{\partial y}{\partial x} = \text{constant} < 0 . \tag{31}$$

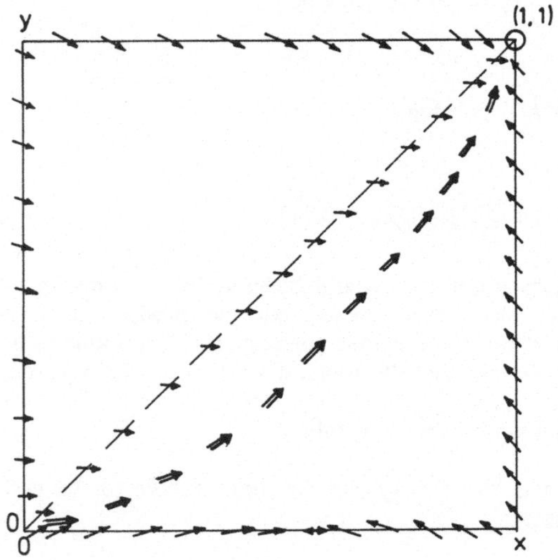

convergence on $\underline{1}$

Fig. 3. No internal root.

Along the diagonal

$$\frac{\partial y}{\partial x} = 0 . \qquad (32)$$

The trajectories, Fig. 3, rise from the origin in the lower triangle and certainly appear to move to the root at (1,1), i.e. to certain extinction. The nature of convergence to this root will be examined shortly.

Case (B): One internal root at (x_0, x_0)

The results of examining the sides and diagonal in turn are shown in Fig. 4. Again they give the appearance of convergence.

Linearisation:

Detail of the approach to the root is obtained by linearising around the root as a potential operating centre. The issue is whether the linearised solutions, which will display eigenvector-eigenvalue behaviour, allow for true convergence or whether there is perhaps a rotation (limit cycle) around the root, a possibility with the internal root.

On linearisation about the internal root (x_0, x_0), one obtains

$$H_0(x,y) \approx \frac{1}{l}[g_0'(x_0)\delta x + g_1(x_0)\delta y + x_0 g_1'(x_0)\delta x - \delta x] , \qquad (33)$$

$$H_1(x,y) = \lambda[\delta x - \delta y] . \qquad (34)$$

Eigenvalues are sought in the form $H_0 = p\delta x$, $H_1 = p\delta y$ with characteristic equation

$$lp = g_0'(x_0) + x_0 g_1'(x_0) - 1 + \lambda g_1(x_0)/(p+\lambda) . \qquad (35)$$

But since

$$H(x) = g_0(x) + x g_1(x) - x ,$$

$$H'(x_0) = g_0'(x_0) + g_1(x_0) + x_0 g_1'(x_0) - 1 , \qquad (36)$$

and

$$lp = H'(x_0) - g_1(x_0) + \lambda g_1(x_0)/(p+\lambda) , \qquad (37)$$

where $H'(x_0) < 0$ for an internal root.

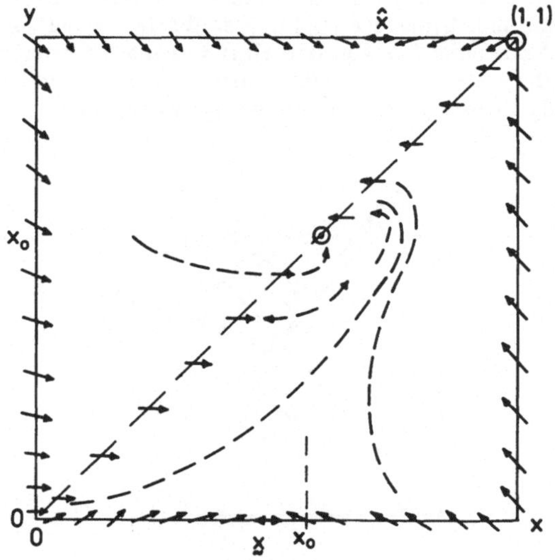

Fig. 4. Internal root.

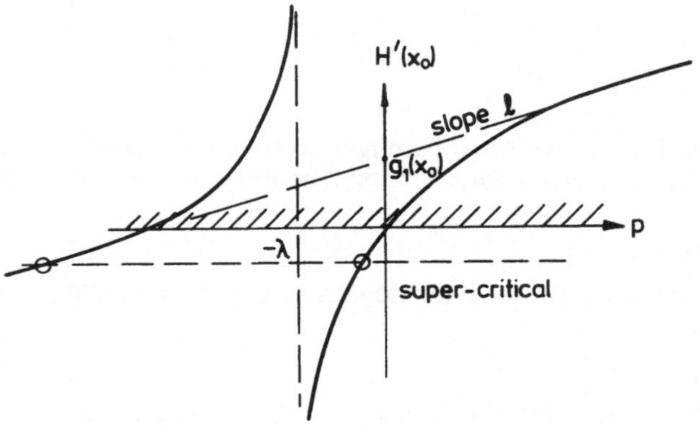

$$\ell p = H'(x_0) - g_1(x_0) + \frac{\lambda}{p+\lambda}\, g_1(x_0)$$

Fig. 5.

For the present case, one neutron group, one precursor group, the two eigenvalues could be found by solving the quadratic characteristic equations. But it suffices to sketch the unknown value of $H'(x_o)$ against the roots, Fig. 5, to see that both roots are real and, since $H'(x_o)$ is known to be negative, lie to the left of $p = 0$. (Examine the asymptotes of the characteristic equation and its singularities.) Thus there are two eigenvalues of the linearised equation at the internal root which correspond to exponentially converging solutions. Indeed in this second-order case (a phase plane rather than a phase volume) we can go on to say that the corresponding eigenvectors are of the form

$$\frac{\delta y}{\delta x} = 1/(\lambda+p) \,, \tag{38}$$

and therefore provide sepatrices at the slopes shown in Fig. 6. Finally, as the system moves towards the critical state, x_o tends to 1 and the linearised solution at the end point is obtained. Here there is one exponentially decreasing eigenvector and one stationary eigenvector, enough, however, for convergence to the external root (in the absence of an internal root) to be guaranteed.

DYNAMICS WITH I GROUPS OF PRECURSORS

The generalisation to several groups of precursors is straightforward. Again, any root must lie on the diagonal of the unit hypercube ($\underline{0},\underline{1}$). Then the I+1 equations of the form $H_o = H_i = 0$ reduce to the equivalent precursor-free equation $H(x) = 0$ and the proof of the occurrence of roots proceeds as before. If the reactor is sub-critical, there is only the end root at $\underline{1}$. If super-critical, there is one and one only internal root, lying on the hyper-diagonal.

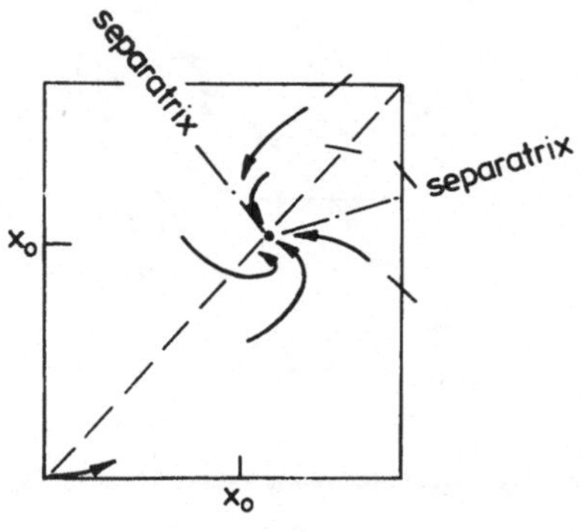

Fig. 6.

The trajectories across faces of the hypercube can be sketched as before and show the general nature of the case I = 1. Trajectories must leave $\underline{1}$.

Linearisation about $\underline{x_0}$ produces the following:

$$H_0(x,\underline{y}) \approx \frac{1}{l}[g_0'(x_0)\delta x + \sum_{i=1}^{I}[g_i(x_0)\delta y_{i.} + x_0 g_i'(x_0)\delta x - \delta x] , \tag{39}$$

and

$$H_i(x,\underline{y}) = \lambda_i[\delta x - \delta y_{i.}] , \tag{40}$$

for which we seek eigenvalues given by

$$lp = g_0'(x_0) + \sum_{i=1}^{I} x_0 g_i'(x_0) - 1 + \sum_{i=1}^{I}\lambda_i g_i(x_0)/(p+\lambda_i) . \tag{41}$$

Recollecting that

$$H(x) = \frac{1}{l}[g_0(x) + \sum_{i=1}^{I} x g_i(x) - x] , \tag{42}$$

then

$$lp = H'(x_0) - \sum_{i=1}^{I} g_i(x_0) + \sum_{i=1}^{I}\lambda_i g_i(x_0)/(p+\lambda_i) , \tag{43}$$

where $H'(x_0) < 0$ for an internal root.

As before, Fig. 7, the nature of the characteristic equation can be found graphically (see for example, Lewins 1978). From the condition $H'(x_0) < 0$, it is found that all I+1 roots are real and negative, providing for convergence to the internal root. Again, in the limit of criticality, I roots are real and negative, one stationary. Really the only difference to the I = 1 case is that the eigenvectors no longer provide sepatrices in the hypercube. However, the faces do provide that trajectories can only enter the hypercube and, in the sub-critical case, the diagonal faces limit the trajectory starting at $\underline{0}$ to the corresponding hyper-tetrahedron.

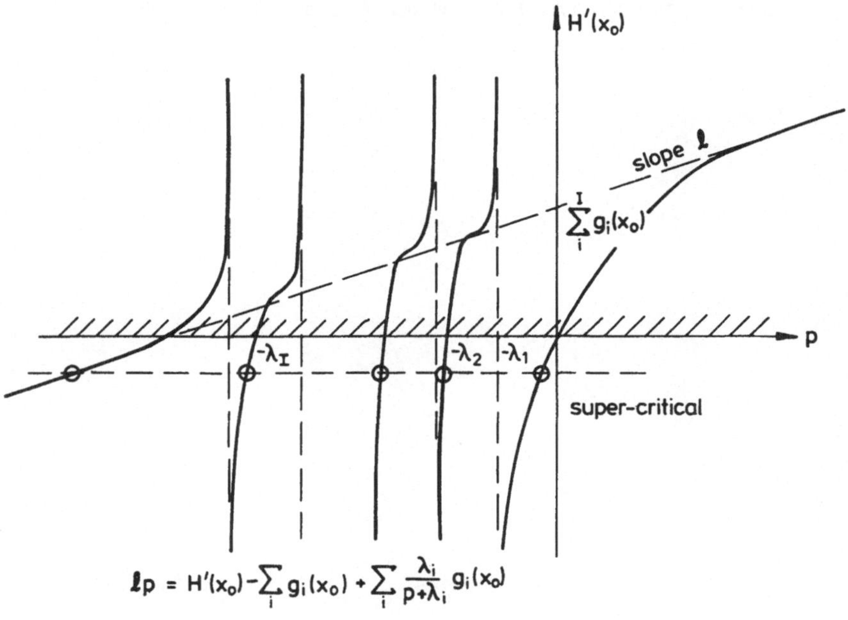

Fig. 7.

CONCLUSIONS

A proof has been given that for arbitrary fission distributions, both the model with and without precursors provides for a single internal root as extinction probability less than unity in a super-critical reactor. The trajectories can be followed with some conviction as converging to that root. In a sub-critical reactor, the convergence is to the end root, i.e. with certainty. The limits of the model, particularly to low power and hence the assumed independence of neutrons, must be recollected.

The result in the prompt and with delayed neutrons models that there exists at most one internal root, dependent on the mean yield, is readily adapted from simple branching process theory (Feller, 1950). The proof that this root actually converges in the I+1-dimensional model is, as we have seen, rather more complicated.

Figs. 5 and 7 were obtained by the author many years ago (Lewins, 1960) in a discussion of reactor kinetics in the mean and differ slightly from the 'in-hour' equation based on excess multiplication and lifetime. That is, instead of interpreting $H'(1)$ as the expression k_{ex}/l, it is desirable to express it as the reactivity equivalent ρ/Λ. Here Λ is the generating or reproduction time, the reciprocal of the production rate rather than the removal rate. With this interpretation, the values of the $g_i'(1)$ become directly related to the delayed neutron fractions β_i and the equations correspondingly simpler, as they are in the mean kinetics expression. There was no idea thirty years ago that this work in the mean would be relevant to the present stochastic study. We are convinced however that the backward Kolmogorov equations have considerable advantages over the forward equations in teasing out the system behaviour. These equations in particular give information on the extinction probability directly.

REFERENCES

Bartlett M. S., 1978, "An Introduction To Stochastic Processes," Cambridge University Press.
Feller W., 1950, "An Introduction To Probability Theory With Applications, 1," Wiley, New York.
Lewins J. D., 1960, The Use Of The Generating Time, Nucl. Sci. Eng., 7:122.
Lewins J. D., 1978, "Nuclear Reactor Kinetics And Control," Pergamon, Oxford.
Ruby L., and McSwine T. L., 1986, Approximate Solution To The Kolmogorov Equation For A Fission Chain-Reacting System, Nucl. Sci. Eng., 94:271.
Salmi U., and Lewins J. D., 1980, Multigroup Energy Formalism For Reactor Stochastic Equations In The Space-Independent Low Power Model, Ann. nucl. Energy, 7:99.

CONTRIBUTION TO THE DISCUSSION

Jeffery D. Lewins

University of Cambridge
Engineering Department
Trumpington Street
Cambridge CB2 1PZ
England

LOW POWER NONLINEAR STOCHASTIC EQUATIONS

Professor Williams sought a physical explanation of the mathematical result, that, in a low power model of a nuclear reactor, the extinction probability is certainty in a critical system whereas the expected number of neutrons remains constant.

A general argument might be to compare this condition with the classical problem of the Gambler's Ruin. In a fair game, the gambler expects to maintain a constant stake. Nevertheless, if the model allows for a bank with infinite resources against the finite resources of the gambler (the 'low power' model), the gambler must anticipate ruin when his stake is reduced to zero but the bank cannot be ruined. Although the game is fair, the play is biased against the gambler.

For a numerical approach, suppose the reactor is the usual lumped model with one group of neutrons only and each fission produces two neutrons. A critical (fair) reactor will have a neutron captured with probability of 0.5 (leading to immediate extinction) or fission (to two neutrons) with probability of 0.5. Then these two neutrons with probability of 0.25 will both be captured (another route to extinction), with probability of 0.5 one will fission and one be captured, and with probability of 0.25 both will fission (leading to four neutrons). The middle case leads to no change of state. At this level a further 0.25 is added to the accumulated extinction probability which now shows the sequence:

$$1/2 + 1/4 + \ldots$$

We can continue. The problem is one of combinatorial algebra and can be shown to yield the infinite geometric progression of ratio one half, whose sum converges to unity - certain extinction.

But there is a neater way to develop the answer employing the spirit of the backward Kolmogorov equations. Let P_1 be the extinction probability for one neutron. The low power model implies the independence of neutrons (already employed above) so that the extinction probability for n neutrons is given by

$$P_n = P_1^n \,. \tag{1}$$

Then at the second stage we can write a conservation equation for the probability of extinction

$$P_1 = 0.5 + 0.5P_2 = 0.5 + 0.5P_1^2 \,. \tag{2}$$

This quadratic equation has two simultaneous roots at $P_1 = 1$ and thus shows the certainty of extinction in the critical reactor with constant expected population.

EXACT AND INEXACT, EXPLICIT AND IMPLICIT SOLUTION TECHNIQUES FOR THE FORWARD AND BACKWARD KOLMOGOROV EQUATIONS

Geoffrey T. Parks

University of Cambridge
Engineering Department
Trumpington Street
Cambridge CB2 1PZ
England

INTRODUCTION

This paper considers possible solution methods for both the forward and backward Kolmogorov equations of a fission chain-reacting system. The notation, based on that used by Ruby and McSwine (1986), is identical to that defined in the preceding paper (Lewins, 1988). Thus the forward equation for a system with one group of neutrons, I groups of precursors and J groups of 'detectrons' is

$$\frac{\partial F_{\underline{m}}}{\partial t} = \frac{1}{l}[-x+g_0(x)+\sum_{i=1}^{I} g_i(x)y_i+\sum_{j=1}^{J}\varepsilon_j(z_j-x)]\frac{\partial F_{\underline{m}}}{\partial x} + \sum_{i=1}^{I}\lambda_i(x-y_i)\frac{\partial F_{\underline{m}}}{\partial y_i} + S(f(x)-1)F_{\underline{m}} \; , \tag{1}$$

while the backward equation for the same system is

$$\begin{aligned} -\frac{\partial F_{\underline{m}}}{\partial s} = &- [\frac{m}{l}+\sum_{i=1}^{I} m_i\lambda_i+S]F_{\underline{m}} + \frac{m}{l}\sum_{\upsilon=0}^{\infty} p_{\upsilon 0}F_{\underline{m}+(\upsilon-1)\underline{\delta}_0} + \frac{m}{l}\sum_{i=1}^{I}\sum_{\upsilon=0}^{\infty} p_{\upsilon i}F_{\underline{m}+(\upsilon-1)\underline{\delta}_0+\underline{\delta}_i} \\ &+ \sum_{i=1}^{I} m_i\lambda_i F_{\underline{m}+\underline{\delta}_0-\underline{\delta}_i} + S\sum_{\upsilon=0}^{\infty} q_\upsilon F_{\underline{m}+\upsilon\underline{\delta}_0} + \frac{m}{l}\sum_{j=1}^{J}\varepsilon_j F_{\underline{m}-\underline{\delta}_0+\underline{\delta}_j} \; . \end{aligned} \tag{2}$$

There is no known exact solution to either equation (1) or (2), but we describe here some techniques which will yield solutions to simplified forms of both equations.

The method employed on the forward equation is based on that of Parks and Lewins (1985) and allows the probability distributions of neutrons, precursors or 'detectrons' to be found as exact but implicit functions of time. To introduce the technique we first consider a simple precursor-, detector- and source-free system.

We go on to show that not only is the same information available from the backward equation but that this formalism also provides a means of calculating the same probabilities as explicit functions of time through numerical integration.

Returning to the forward equation we demonstrate that our implicit solution technique will allow an exact solution of the generalised forward equation under the 'zero prompt lifetime' assumption, originally used by Hurwitz et al. (1964) as the basis for a numerical solution.

THE SIMPLIFIED FORWARD EQUATION

In the absence of precursors and detectors, the forward equation becomes

$$\frac{\partial F_{\underline{m}}}{\partial t} = \frac{1}{l}[g_0(x) - x]\frac{\partial F_{\underline{m}}}{\partial x} + S(f(x)-1)F_{\underline{m}} \ . \tag{3}$$

The independence of neutron behaviour in a low power reactor means that the solutions for differing initial conditions can be constructed by superposition from two fundamental solutions:

$$G_{mn} = G_1^m F_0^n \ , \tag{4}$$

where m is the number of neutrons initially present, n is the number of independent sources, G_1 is the solution for m = 1, n = 0, and F_0 is the solution for m = 0, n = 1.

Furthermore F_0 may be derived from G_1 by convolution:

$$F_0 = \exp\{\int_0^t S[f(G_1(x,t-\tau)) - 1]d\tau\} \ . \tag{5}$$

Thus we seek G_1, the solution of

$$\frac{\partial G_1}{\partial t} = \frac{1}{l}[g_0(x) - x]\frac{\partial G_1}{\partial x} = H(x)\frac{\partial G_1}{\partial x} \ . \tag{6}$$

Using the classic method of Lagrange, we can write

$$dG_1/0 = dt/1 = -dx/H(x) \ . \tag{7}$$

Whence

$$G_1 = \Phi \quad \text{a constant,} \tag{8}$$

and

$$t = \Psi - u(x) \ , \tag{9}$$

where

$$u(x) = \int H^{-1}(x)dx \ . \tag{10}$$

Now

$$\Psi = Fn(\Phi) \ , \tag{11}$$

so

$$t + u(x) = Fn(G_1) \ . \tag{12}$$

But we know that initially, when t = 0, there is one neutron present, $G_1 = x$. Thus

$$Fn(x) = u(x) \ , \tag{13}$$

and hence

$$t + u(x) = u(G_1) \ . \tag{14}$$

If H(x) is of sufficiently simple form, e.g. for doublet fission, this relation may be inverted to give G_1 as an *explicit* function of x and t. However, with a realistic representation of auxiliary generating function $g_0(x)$, e.g. as a polynomial of order five, inversion is not possible.

IMPLICIT PROBABILITIES FROM THE FORWARD EQUATION

Even if equation (14) cannot be inverted, the neutron probabilities may nevertheless be found as *exact, implicit* functions of time. Remarking that

$$\frac{\partial^n G_1}{\partial x^n}\Big|_{x=0} = n!P_n(t) , \tag{15}$$

and setting x = 0 in equation (14), the extinction probability, P_0, is available from

$$u(P_0) = t + u(0) , \tag{16}$$

and other probabilities are found by successive differentiation of equation (14) with respect to x:

$$\frac{\partial^n}{\partial x^n}\left[u(x)\right] = \frac{\partial^n}{\partial x^n}\left[u(G_1)\right] \quad \text{for } n > 0 . \tag{17}$$

On setting x = 0, one obtains relationships such as

$$P_1 = u'(0)/u'(P_0) , \tag{18}$$

and

$$P_2 = [u''(0)-u''(P_0)P_1^2]/2u'(P_0) \quad \text{etc.,} \tag{19}$$

where, since u(x) may be written as

$$u(x) = \sum_{i=1}^{N} A_i \,ln(x-a_i) , \tag{20}$$

$$\frac{\partial^n u}{\partial x^n} = (-1)^{n-1}(n-1)!\sum_{i=1}^{N} A_i(x-a_i)^{-n} . \tag{21}$$

This process may, in theory, be continued to find any probability P_n as an exact, implicit function of time.

EXPLICIT PROBABILITIES FROM THE BACKWARD EQUATION

For the same source-free system the equivalent backward equation is

$$-\frac{\partial G_1}{\partial s} = \frac{\partial G_1}{\partial t} = H(G_1) . \tag{22}$$

Whence, setting x = 0, we obtain

$$\frac{\partial G_1}{\partial t}\Big|_{x=0} = H(G_1)\Big|_{x=0} \tag{23}$$

and

$$\frac{\partial P_0}{\partial t} = \frac{dP_0}{dt} = H(P_0) . \tag{24}$$

Equation (24) may be integrated numerically from $P_0 = 0$ at $t = 0$ to generate P_0 as an *inexact, explicit* function of time.

Now, differentiating equation (22) with respect to x gives

$$\frac{\partial}{\partial x}\left[\frac{\partial G_1}{\partial t}\right] = \frac{\partial H}{\partial G_1}\frac{\partial G_1}{\partial x} . \tag{25}$$

Hence

$$\frac{\partial}{\partial t}\left[\frac{\partial G_1}{\partial x}\right]\Big|_{x=0} = H'(G_1)\frac{\partial G_1}{\partial x}\Big|_{x=0} \tag{26}$$

and

$$\frac{\partial P_1}{\partial t} = \frac{dP_1}{dt} = H'(P_0)P_1 \; . \tag{27}$$

This may also be solved numerically from $P_1 = 1$ at $t = 0$, as $P_0(t)$ is now known, and this process may, in principle, be continued to find any P_n by numerical integration.

IMPLICIT PROBABILITIES FROM THE BACKWARD EQUATION

Alternatively equation (24) may be rewritten as

$$dt = dP_0/H(P_0) \; . \tag{28}$$

Thus

$$\int_0^t dt = \int_0^{P_0} H^{-1}(P_0)dP_0 \; , \tag{29}$$

and integrating

$$t = u(P_0) - u(0) \; , \tag{30}$$

which matches the forward result, equation (16).

Now, rewriting equation (27) using equation (28) yields

$$\frac{dP_1}{dt} = H(P_0)\frac{dP_1}{dP_0} = H'(P_0)P_1 \; . \tag{31}$$

Whence

$$\int_1^{P_1} P_1^{-1} dP_1 = \int_0^{P_0} \frac{H'(P_0)}{H(P_0)} dP_0 \; , \tag{32}$$

and thus

$$P_1 = H(P_0)/H(0) \; . \tag{33}$$

A similar approach yields

$$P_2 = H(P_0)[H'(P_0)-H'(0)]/2H^2(0) \; , \tag{34}$$

and this process may also, in principle, be continued to find any P_n as an *exact, implicit* function of time. These backward results are, of course, identical to the forward ones; the different forms of equations (18), (19) and (33), (34) reflect their different origins.

However, these implicit methods are not restricted to such simple cases.

THE GENERALISED FORWARD EQUATION

The generalised forward equation, equation (1), has so far resisted complete exact solution. In order to allow an approximate solution to be obtained the equation must be simplified by decoupling some of the terms. One possible approach is the 'zero prompt lifetime' assumption of Hurwitz et al. (1964), in which one considers a low power reactor whose reactivity is close to delayed critical, i.e. it is prompt sub-critical. This being the case all prompt neutron chains will be shortlived, typically lasting for 100 generations or 1 millisecond. If we assume that no delayed neutron precursors created by a particular chain decay in the lifetime of that chain, then for a source-free reactor the forward equation becomes

$$\frac{\partial G_{\underline{m}}}{\partial t} = \frac{1}{l}[-x + g_0(x) + \sum_{i=1}^{I} g_i(x)y_i + \sum_{j=1}^{J} \varepsilon_j(z_j-x)]\frac{\partial G_{\underline{m}}}{\partial x} = R(x,y_i,z_j)\frac{\partial G_{\underline{m}}}{\partial x} . \tag{35}$$

So, using Lagrange's method again,

$$dG_{\underline{m}}/0 = dy_i/0 = dz_j/0 = dt/1 = -dx/R(x,y_i,z_j) . \tag{36}$$

Whence

$$G_{\underline{m}} = \Phi \quad \text{a constant} , \tag{37}$$

$$y_i = \theta_i \quad \text{constants} , \tag{38}$$

$$z_j = \zeta_j \quad \text{constants} , \tag{39}$$

and

$$t = \Psi - U(x,\theta_i,\zeta_j) , \tag{40}$$

where

$$U(x,\theta_i,\zeta_j) = \int R^{-1}(x,\theta_i,\zeta_j)dx = \sum_{i=1}^{N} A_i \, ln(x-a_i) , \tag{41}$$

which is integrable because the y_i and z_j are constants.

With the initial condition that the chain starts with a single neutron at t = 0, one obtains

$$U(G_1,0,0) = U(x,y_i,z_j) + t , \tag{42}$$

which is identical in form to the relationship obtained for the simple case, equation (14). Thus G_1 is given as an *exact*, *implicit* function of x, y_i, z_j and t.

THE DELAYED PRECURSOR DISTRIBUTION

The real interest here lies in the distribution of the number of precursors created when the fission chain ends. The probability that m precursors of type i have been created, P_{mi}, is available from

$$\frac{\partial^m G_1}{\partial y_i^m}\bigg|_{\substack{y_i=0\,\underline{y}=1 \\ x=0\,\underline{z}=1}} = m!P_{mi} , \tag{43}$$

where generating function dummy variables x and y_i are set to zero and the remaining variables in $\underline{x}$ (= $(x,\underline{y},\underline{z})$) are set to one. So, for instance,

$$U(P_{0i}) = t + U(x,y_i,z_j)\bigg|_{\substack{y_i=0\,\underline{y}=1 \\ x=0\,\underline{z}=1}} , \tag{44}$$

while other probabilities are obtained from

$$\frac{\partial^m}{\partial y_i^m}\left[U(\underline{x})\right] = \frac{\partial^m}{\partial y_i^m}\left[U(G_1)\right] \quad \text{for } m > 0 . \tag{45}$$

To evaluate the left hand side of equation (45), we remark that

$$U(\underline{x}) = \int R^{-1}(\underline{x})dx , \tag{46}$$

where

$$R(\underline{x}) = \frac{1}{l}[-x+g_0(x)+\sum_{i=1}^{I} g_i(x)y_i+\sum_{j=1}^{J}\varepsilon_j(z_j-x)] , \tag{47}$$

so that, using Leibnitz's theorem to reverse the order of differentiation and integration, we obtain

$$\frac{\partial^m}{\partial y_i^m}\left[U(\underline{x})\right] = \int \frac{\partial^m}{\partial y_i^m}\left[R^{-1}(\underline{x})\right] dx = \int (-1)^m m! g_i^m(x) R^{-(m+1)}(\underline{x}) dx \,. \tag{48}$$

To evaluate the right hand side of equation (45) we note, after Faà de Bruno (1857), that

$$\frac{\partial^m}{\partial y_i^m}\left[U(G_1)\right] = \sum_k \frac{m! \frac{\partial^k U}{\partial G_1^k} \left(\frac{\partial G_1}{\partial y_i}\right)^{k_1} \left(\frac{1}{2!}\frac{\partial^2 G_1}{\partial y_i^2}\right)^{k_2} \cdots \left(\frac{1}{m!}\frac{\partial^m G_1}{\partial y_i^m}\right)^{k_m}}{k_1! k_2! k_3! \ldots k_m!} \,, \tag{49}$$

where the summation in k is performed over all possible integer solutions of

$$m = k_1 + 2k_2 + 3k_3 + \ldots + mk_m \,, \tag{50}$$

and

$$k = k_1 + k_2 + k_3 + \ldots + k_m \,. \tag{51}$$

Given that

$$\frac{\partial^k U}{\partial G_1^k} = (-1)^{k-1}(k-1)! \sum_{i=1}^{N} A_i (G_1 - a_i)^{-k} \,, \tag{52}$$

on setting $x = 0$, $y_i = 0$ and the remainder of $\underline{x} = \underline{1}$, equations (43), (45), (48) and (49) yield

$$\begin{aligned} \frac{\partial^m}{\partial y_i^m}\left[U(G_1)\right] \Big|_{\substack{y_i=0\, \underline{y}=1 \\ x=0\, \underline{z}=1}} &= \sum_k \frac{m!(-1)^{k-1}(k-1)! \sum_{j=1}^{N} A_j (P_{0i} - a_j)^{-k} P_{1i}^{k_1} P_{2i}^{k_2} \ldots P_{mi}^{k_m}}{k_1! k_2! k_3! \ldots k_m!} \\ &= \int (-1)^m m! g_i^m(x) R^{-(m+1)}(\underline{x}) dx \Big|_{\substack{y_i=0\, \underline{y}=1 \\ x=0\, \underline{z}=1}} \,. \end{aligned} \tag{53}$$

But there is only one solution of equation (50) for which $k_m \neq 0$, namely

$$k_1 = k_2 = \ldots = k_{m-1} = 0 \tag{54a}$$

$$k_m = 1 = k \,. \tag{54b}$$

This gives a term $m! \sum_{j=1}^{N} A_j (P_{0i} - a_j)^{-1} P_{mi}$ in the summation - the only term in which P_{mi} occurs. Thus

$$P_{mi} = \frac{\int (-1)^m g_i^m(x) R^{-(m+1)}(\underline{x}) dx \Big|_{\substack{y_i=0\, \underline{y}=1 \\ x=0\, \underline{z}=1}} - \sum_k \frac{(-1)^{k-1}(k-1)! \sum_{j=1}^{N} A_j (P_{0i} - a_j)^{-k} P_{1i}^{k_1} P_{2i}^{k_2} \ldots P_{m-1i}^{k_{m-1}}}{k_1! k_2! k_3! \ldots k_{m-1}!}}{\sum_{j=1}^{N} A_j (P_{0i} - a_j)^{-1}} \,, \tag{55}$$

in which the summation in k is now performed over all possible solutions of

$$m = k_1 + 2k_2 + 3k_3 + \ldots + (m-1)k_{m-1} \tag{56}$$

and

$$k = k_1 + k_2 + k_3 + \ldots + k_{m-1} \,. \tag{57}$$

Hence, starting from P_{0i}, each P_{mi} can be found successively, establishing the required precursor production distribution.

LONGER TERM PRECURSOR VARIATIONS

This distribution can now be used to consider the longer term variations in the precursor population. The relevant forward equation is

$$\frac{\partial V_i}{\partial t} = \lambda_i(v_i(y_i)-y_i)\frac{\partial V_i}{\partial y_i} + s(v_i(y_i)-1)V_i \ , \qquad (58)$$

where $V_i(y_i,t)$ is the generating function for the population of the ith precursor, $v_i(y_i)$ is the associated auxiliary generating function for the production of precursors by a prompt fission chain

$$v_i(y_i) = \sum_{\upsilon=0}^{\infty} P_{\upsilon i} y^{\upsilon} \qquad (59)$$

with $P_{\upsilon i}$ as found above, and s is the rate at which prompt neutron chains occur.

If $v_i(y_i)$ is time invariant, this equation is identical in form to the simple forward equation with a source shown earlier (equation (3)). It will therefore admit an *exact* solution by the same *implicit* method.

Alternatively, if $v_i(y_i)$ has a sufficiently simple dependence on t, it may also be possible to obtain useful analytical solutions, but, in general, a numerical solution will be required.

CONCLUSIONS

We have demonstrated that both the forward and backward Kolmogorov equations allow probability distributions to be found exactly as implicit functions of time for 'simple' problems, even if the generating function cannot be found explicitly. Furthermore, the backward equation allows the same probabilities to be found as explicit functions of time through numerical integration.

Although a completely analytical solution of the generalised forward and backward Kolmogorov equations continues to remain elusive, the exact, implicit solution technique developed here may usefully be extended to more generalised versions of the forward equation, specifically to that produced by using the 'zero prompt lifetime' assumption - an assumption previously employed to facilitate an inexact, numerical solution.

These exact, implicit methods clearly become progressively more impractical and computationally more expensive as higher order probabilities are sought and thus could not realistically form the basis of an efficient general solution technique. They can, however, be used to provide exact results against which the accuracy of faster numerical methods may be tested.

REFERENCES

Faà de Bruno M., 1857, Note Sur Une Nouvelle Formule De Calcul Différentiel, Quart. J. Pure Appl. Math., 1:359.

Hurwitz H., Jr., MacMillan D. B., Smith J. H., and Storm M. L., 1964, Numerical Solution Of The Forward Equation, in: "Naval Reactors Physics Handbook," A. Radkowsky, ed., USAEC, Washington DC.

Lewins J. D., 1988, Solving Low Power Stochastic Equations: Nonlinear Studies, in: "Proceedings of the NATO Advanced Workshop on Noise and Nonlinear Phenomena in Nuclear Systems," J. L. Muñoz-Cobo, ed., Plenum, New York.

Parks G. T., and Lewins J. D., 1985, An Exact Transient Stochastic Solution For Low-Power Neutron Multiplication, Ann. nucl. Energy, 12:65.

Ruby L., and McSwine T. L., 1986, Approximate Solution To The Kolmogorov Equation For A Fission Chain-Reacting System, Nucl. Sci. Eng., 94:271.

CHAPTER III

STOCHASTIC PROCESSES IN LINEAR NUCLEAR SYSTEMS: APPLICATIONS

SOME APPLICATIONS OF STOCHASTIC PROCESSES IN NEUTRON COINCIDENCE: MEASUREMENTS USED IN NUCLEAR SAFEGUARDS

W. Matthes

CCR EURATOM, 21020 Ispra, Italy

ABSTRACT

Pulse-correlation techniques are widely used in non-destructive analysis of the unknown amounts of fissile material in 'Probes'. We present a model of a branching process which we think is flexible enough to cover all neutron multiple correlation counting experiments. Equations describing the statistical properties of pulse trains from detector outputs are given and explicit expression for the moments of the pulse distribution are derived.

By investigating the event sequences of neutron-detector outputs the loss of counts and the modification of the 'pair-correlation' function due to the deadtime of the detection system may become essential. We try to provide an approximate theoretical description of the effect of the deadtime on count rates and correlation functions.

INTRODUCTION

One of the main problems in nuclear safeguards is the following: A piece of material ('probe' or 'sample') is declared to contain a certain amount of fissile material. As this strategically sensitive material has to be safeguarded against diversion, a safeguard inspector is supposed to verify this declaration by independently determining the fissile material content in a non-destructive way.

Due to the presence of the fissile material the sample represents a subcritical system and therefore the standard methods for criticality-measurements might be used.

In actual practice however only the "Noise-Analysis" method is applied, where the sample is irradiated by an external and/or internal neutron source and the statistical properties of the stationary time series of count-events (pulse trains) at neutron-detector outputs are analyzed.

In Fig. 1 we sketch the principle of the method. On the pulse trains Π from the detectors D, we pick an arbitrary set of time intervals I_λ (λ=1,2...L) (non overlapping on the same Π), determine the number of counts N_λ in each of these I_λ

and evaluate finally their moments or cumulants. These measured data are then analyzed by theoretical expressions relating the moments to the material and geometrical sample characteristics.

Expressions for the moments are found in the standard way by a Taylor development of a properly chosen generating function.

As this workshop is mostly interested in non-linear noise, we shall present the main arguments only, all details may be found in the given references.

Theory of Counting Statistics

As basic element for the theory of the count-statistics we take the generating function for the probabilities $P(N_1 N_2 \ldots N_L)$ to find N_λ counts in the time interval I_λ ($\lambda = 1,2,\ldots L$). This 'probability generating function', (PGF in short) defined by:

$$G(X) = \sum_{N_1 N_2 \cdots N_L} P(N)\, X_1^{N_1} X_2^{N_2} \cdots X_L^{N_L} = \left\langle \prod_{\lambda=1}^{L} X_\lambda^{N_\lambda} \right\rangle \qquad (1)$$

where X and N are abbreviations for the sets $\{X_1 X_2 \ldots X_L\}$ and $\{N_1 N_2 \ldots N_L\}$ respectively, has to simulate the physical processes of the

a) source-events emitting neutrons into the system and the

b) generation of neutron populations in the system (by these source neutrons) which eventually lead to counts in the detector.

These physical processes are now to be described in mathematical terms. For this reason we turn first to the characterization of the

a) Distribution of source-events

The neutron-emitting source events are distributed in space and time. The statistical properties of this source-event distribution is characterized by the correlation functions $g_m(\tau_1 \tau_2 \ldots \tau_m)$ $(m=1,2\ldots)$ where τ stands for (x,s) and these variables indicate the coordinates of

$x = (x_1 x_2 x_3)$ space and
s time

$g_1(\tau)d\tau = S(xs)dxds$ is the mean number of source events in the spatial volume element dx in time interval ds.

Further we allow for the possibility that at a source-event more neutrons are emitted (e.g. spontaneous fission source)and counts produced. This feature is characterized by the probability $W(\tau, n_o, n_1 n_2 \ldots n_L)$ that n_o neutrons are emitted in a source event at τ and n_λ counts produced in I_λ (the PGF for $W(\tau, n_o, n_1 n_2 \ldots n_L)$ is denoted by $W(\tau, X_o, X)$ and with $\chi_o(\tau, v)$ we denote the velocity-spectrum of the emitted source neutrons (assumed to be independent of n_o).

b) Generation of neutron populations and their count-distributions by individual (source) neutrons.

A neutron starting at space point x at time s with velocity v generates a neutron population in the multiplying sample and leads with probability

$P(xvs \longrightarrow N)$ (PGF : $f(xvs \longrightarrow X)$)

to the count distribution $(N_1 N_2 \ldots N_L)$ in the time intervals I_λ

A more detailed description of the neutron transport process needs in addition the probability $D_\varrho(xvs)ds$ that a neutron at (x,v,s) makes in ds a reaction of type ϱ and the probability $W_\varrho(xvs, n_o, n_1 n_2 \ldots n_L)$ that n_o neutrons with emission spectrum $\chi_\varrho(xs, v \to w)$ are released at this reaction event and n_λ counts produced in I_λ. ($W_\varrho(xvs, X_o, X)$ = PGF of $W_\varrho(xvs, n_o, n_1 n_2 \ldots n_L)$).

Setting up a balance equation for the probability P(N) in terms of the individual probabilities introduced above leads us finally to equations for G and f (see [1]):

$$\ln G(X) = \sum_{m=1}^{\infty} \frac{1}{m!} \int g_m(\tau_1 \tau_2 \ldots \tau_m) \prod_{i=1}^{m} \{W_o(\tau_i, f_o, X) - 1\} d\tau_i \tag{2}$$

$$\left(\frac{\partial}{\partial s} + v\,\mathrm{grad} + D\right) f = \sum_\varrho D_\varrho(xvs)\, W_\varrho(xvs, f_\varrho, X) \tag{3}$$

where

$f_o(xs \to X) = \int d\chi_o(xs, v)\, f(xvs \to X)$ (average of f over source spectrum)

$f_\varrho(xvs \to X) = \int d\chi_\varrho(xs, v \to w)\, f(xws \to X)$ (average of f over fission spectrum)

These equations are of course too general and have to be tailored to the practical needs.

As a first step towards simplification we note that we hardly have to deal with correlated neutron sources. So we take only the first term of the sum in the equation for G(x). Setting in addition f=1+g (which turns out to be convenient) transforms the equations for G and f into (see [2]):

$$\ln G = \int dS \left\{ \overline{m}_{o1} g_o + \sum_{K \geq 2} \overline{m}_{oK} g_o^K \right\} \tag{4}$$

$$\Omega^+ g = \sum_\varrho D_\varrho \sum_{K \geq 2} \overline{m}_{\varrho K} g_\varrho^K \qquad (\Omega^+ = \text{adjoint Boltzmann operator}) \tag{5}$$

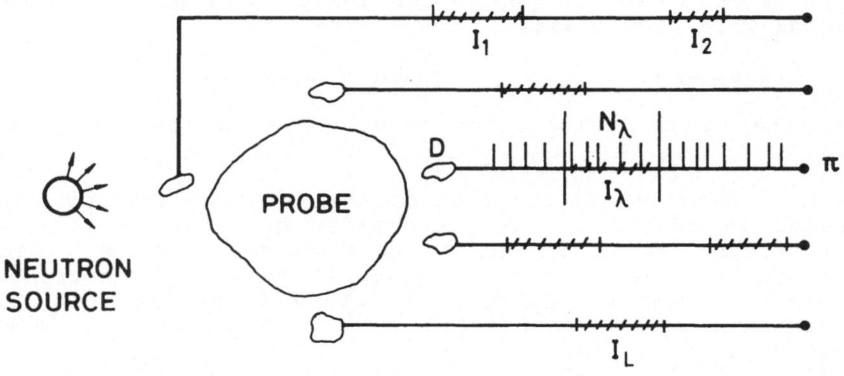

Fig. 1. π is the pulse train at the output of detector D.

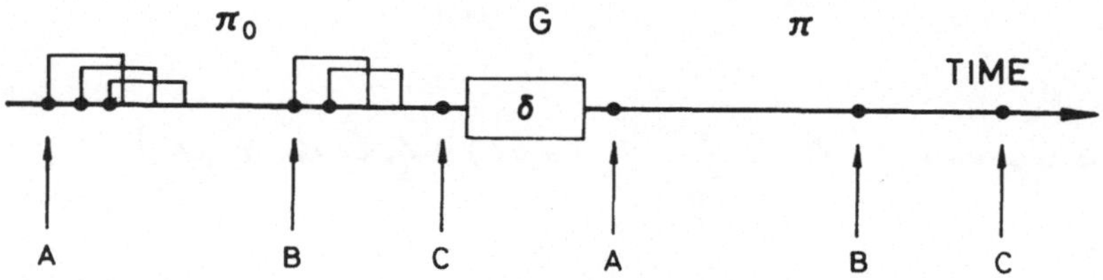

Fig. 2. Action of an updating deadtime gate G in transforming an incoming pulse train π_0 into an outgoing pulse train π.

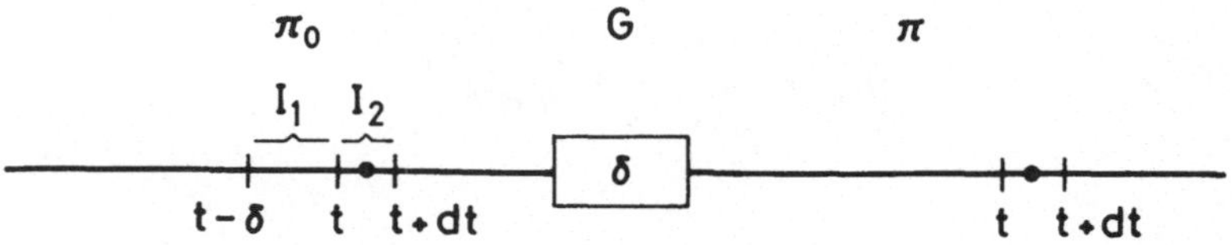

Fig. 3(a) A pulse on π_0 in the interval I_2 $[t, t+dt]$ passes G if there is no pulse in the interval I_1 $[t-\delta, t]$

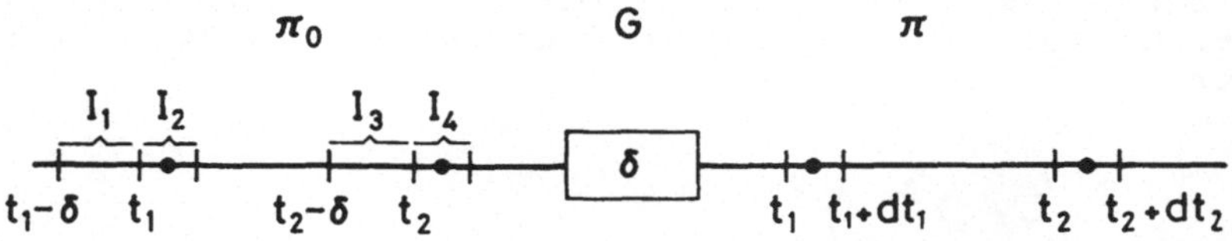

Fig. 3 (b) A pair of pulses on π_0 one pulse in I_2 $[t_1, t_1+dt_1]$ and one pulse in $I_4[t_2, t_2+dt_2]$, passes G if there are no pulses (on π_0) in the intervals $I_1[t_1-\delta, tl]$ and $I_3[t_2-\delta, t_2]$

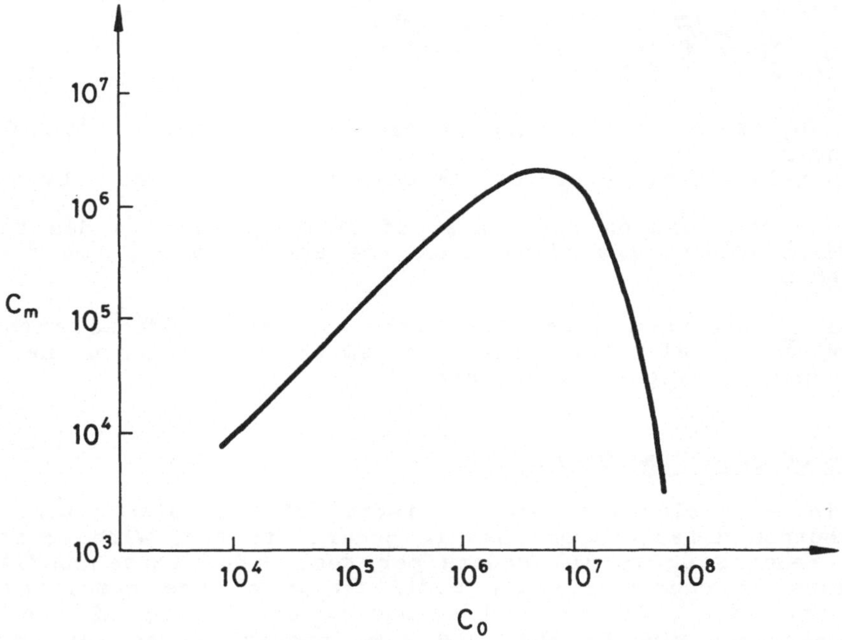

Fig. 4. Measured counting rate C_m as a function of the real C_0. For low real counting rates C_0 we have $C_m \sim C_0$. For high C_0 the measured counting rate C_m drops to zero due to the blocking effect of the updating deadtime.

$$\overline{m}_{xK} = \sum_m W_x (xvs, m, \ldots) \binom{m}{K} \qquad (x = 0, \rho)$$

These equations eventually permit us to construct explicit expressions for the (measurable) cumulants $\Gamma(\nu)$ by applying the operator

$$D(\nu) = \prod_{\lambda=1}^{L} \left(\frac{\partial^{\nu_\lambda}}{\partial x_\lambda^{\nu_\lambda}} \right)_{x=1} \tag{6}$$

on lnG and the equation for g.
The latter equation can then be solved and, using this solution in the equation for $D(\nu)$lnG, leads finally to (see [3]):

$$\Gamma(\nu) = \int dS \sum_{K \geq 2} \overline{m}_{0K} \left[D(\nu) g_0^K \right] +$$
$$\sum_\rho \int dF_\rho \sum_{K \geq 2} \overline{m}_{\rho K} \left[D(\nu) g_\rho^K \right] \tag{7}$$

where dF_ρ (xvs) is the mean number of reactions of type ρ in dxdvds.
The correlation structure of the cumulants $\Gamma(\nu)$ is now obvious.

a) the <u>first</u> term on the r.h.s. of this equation is due to the 'correlations' introduced into the system by all the source-events

b) the <u>second</u> term gives the correlations introduced into the system by all the reactions emitting secondary neutrons (induced fissions, (n,2n) etc.)

Theory of Dead Time Correction

When investigating the time statistics of the pulse trains from the neutron detectors one has in general to deal with counting-rates ranging up to 10^6 counts per sec. Under these conditions the loss of counts and the modification of the cumulants (we shall consider first and second moments only) of the real count-sequence due to the dead time properties of the neutron detector may become significant. We consider two types of dead-time forming counters:

a) Updating dead-time counter

The action is described in the following way:

An updating gate G can be open or closed. If it is open, a pulse arriving at G passes and blocks the gate for a time interval δ.
If G is closed, an arriving pulse will not be counted, the only effect is to block G for another time interval δ.

A pulse arriving at G may therefore pass the gate only if the time interval up to the previous pulse is larger than δ.

An input pulse train Π_o to G is transformed by G into an output pulse train Π (see Fig. 2) with statistical properties differing from the statistical properties of Π_o.

The problem is to obtain information on the parameters describing Π_o by performing measurements on Π. As measurable quantities on Π we choose:

(i) $P(t)dt$ as the probability to find a pulse in dt at t;

and

(ii) $P(t_1,t_2)dt_2$ as the probability to have a pulse in dt_1 at t_1 and a pulse in dt_2 at t_2.

For a stationary pulse train Π the probability $P(t)$ is constant (mean counting rate) and $P(t_1,t_2)$ depends upon the difference t_2-t_1. For $P(t)$ and $P(t_1,t_2)$ in terms of Π_o we have, obviously (see Fig. 3):

(i) $P(t)dt$ = probability that there is:
(a) NO pulse on Π_o in the time interval $I_1[t-\delta,t]$
and
(b) ONE pulse on Π_o in $I_2[t,t+dt]$.

(ii) $P(t_1,t_2)dt_2$ = probability that there is:
(a) NO pulse on Π_o in the time interval $I_1[t_1-\delta,t_1]$,
(b) ONE pulse on on Π_o in $I_2[t_1,t_1+dt_1]$,
(c) NO pulse on Π_o in $I_3[t_2-\delta,t_2]$,
and
(d) ONE pulse on Π_o in $I_4[t_2 t_2+dt_2]$.

What we need therefore are expressions for joint probabilities to have a certain number of pulses in non-overlapping time intervals on Π_o. We consider M non-overlapping time intervals I (=1,2...,M) on Π_o and use the joint probability $P(N_1,N_2,\dots,N_M)$ to have N pulses in I . The general expression for P can then immediately be adapted to two or four intervals and zero or one pulse in these intervals.

The generating function G(X) for the P(N) is given by applying Eq. (1) to this situation.

The $g_m(t_1,t_2\dots t_m)$ are then the correlation functions describing the statistical properties of the pulse train Π_o and are assumed to be known, the $W(t_i,X,f)$ reduce to $W(t_i,X)$.
For the probabilities which we need for our purpose (only ONE or NO pulse in any interval) we find immediately:

a) NO event in all intervals I_λ :

$$P_o = G(X=0)$$

b) ONE event in I_λ , NO event in all I_μ $(\mu\neq\lambda)$

$$P_\lambda = \left.\frac{\partial G}{\partial X_\lambda}\right|_{X=0}$$

c) ONE event in I_λ, ONE event in I_μ, NO event in all I_ν $(\nu \neq (\lambda,\mu))$

$$P_{\lambda\mu} = \left.\frac{\partial^2 G}{\partial x_\lambda \partial x_\mu}\right|_{X=0}$$

To investigate approximately for practical applications the dead-time effect we make the assumption that all $g_m = 0$ for $m>2$.

Equation (2) reduces to

$$\ln G = \sum_{m=1}^{M} (X_m - 1)A_m + \frac{1}{2}\sum_{m,n} (X_m - 1)(X_n - 1)A_{mn} \qquad (8)$$

$$A_m = \int_{I_m} g_1(t)dt \qquad (9)$$

$$A_{mn} = \int_{I_m} dt_1 \int_{I_n} dt_2\, g_2(t_1 t_2) \qquad (10)$$

and the above mentioned calculations leading to P_0, P_λ and $P_{\mu\lambda}$ are easy to perform (when using simple expression for g_1 and g_2.

For more details of the derivation of the formulas and for comparison with experiments see Ref. [4].

In Fig. 4 we show the typical behaviour of the total counting rate C_m measured on Π as a function of the actual counting rate C_o on Π_o.

b) Fixed Dead-time Counter

The action is described in the following way:

A fixed dead-time gate G can be open or closed. If it is open, a pulse at G passes and blocks the gate for a fixed time interval δ. If G is closed, an arriving pulse will not be counted.

An input pulse train Π_o to G is transformed by G into an output pulse train Π (see Fig. 5) with statistical properties differing from the statistical properties of Π_o.

As before, the problem is to obtain information on the parameters describing Π_o by performing measurements on Π.
In this case we may choose as measurable quantities on Π the distribution of the time-intervals between an arbitrarily chosen pulse on Π and its first, second, third....etc. successive pulse (see Fig. 6).

If the probability distribution $P(\tau)$ of the time intervals τ between successive pulses on Π is known, the mean counting rate $C(\text{on } \Pi)$ is then given by

$$\frac{1}{c} = \langle\tau\rangle = \int_{\delta}^{\infty} P(\tau)\,\tau\, d\tau \tag{11}$$

this $P(\tau)$ is obviously given by (refer to Fig. 6).

the probability to have on Π_0

NO pulses in I_2 $[s+\delta, t]$
ONE pulse in I_3^2 $[t+t, dt]$

under the condition of

ONE pulses in I_1 $[s+s, ds]$

This gives us

$$P(\tau)d\tau = \frac{P_0(1 \in I_1, 0 \in I_2, 1 \in I_3)}{P_0(1 \in I_1)} \tag{12}$$

Using again the generating function $G(X_1 X_2 X_3)$ we have

$$P_0(1 \in I_1) = \left.\frac{\partial G}{\partial X_1}\right|_{x_1=0,\ x_2=x_3=1} \tag{13}$$

$$P_0(1 \in I_1, 0 \in I_2, 1 \in I_3) = \left.\frac{\partial^2 G}{\partial X_1 \partial X_3}\right|_{x_1=x_2=x_3=0} \tag{14}$$

We may again use equations (8)-(10) to obtain approximate expressions for $P(\tau)$ and C_0.

Note that for a Poissonian input Π_0 we obtain from (11) the usual result:

$$\langle\tau\rangle = \frac{1+C_0\delta}{C_0} = \frac{1}{C}$$

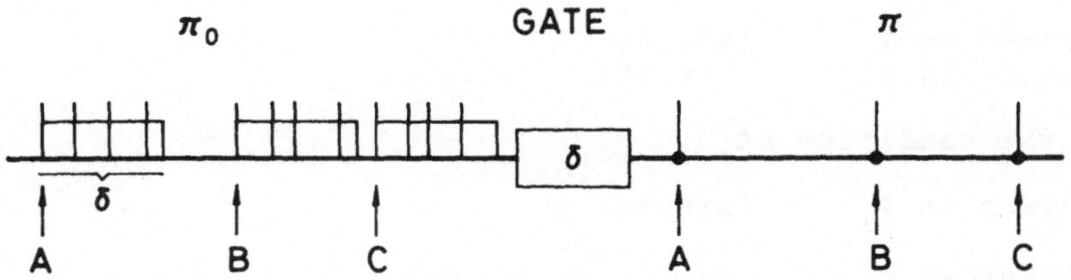

Fig. 5. Action of a "Fixed Dead-time" gate G in transforming an incoming pulse train π_0 into an outgoing pulse train π.

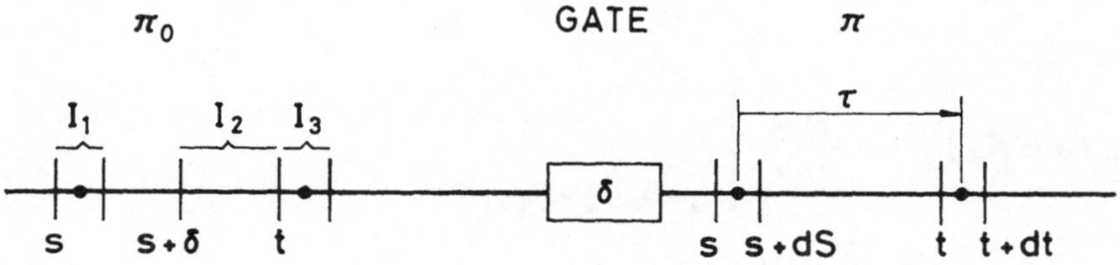

Fig. 6. A pulse in ds on π is immediately followed by a pulse in dt if (on π_0) there was no pulse in I_2 and one pulse in I_1 under the condition of one pulse in ds (which blockes the gate G for a time δ).

References

[1) Van Kampen N.G. (1983) Stochastic Processes in Physics and Chemistry. North-Holland, Amsterdam

[2] Bell I.G. (1965) Nucl. Sci. Eng. 21, 390

[3] Saito K. and Tajo Y. (1967) Nucl. Sci. Eng. 30, 54

[4] Matthes W. and Haas R. (1985) Ann. Nucl. Eng., Vol. 12, No. 12 p. 693

APPLICATION OF STATISTICAL CORRELATION TECHNIQUES FOR MEASUREMENT OF THE Keff OF HIGHLY SUBCRITICAL SYSTEMS

S.B.Degweker*, M. Srinivasan+, K.K. Rasheed+ and C.S.Pasupathy

*Reactor Analysis and Systems Division
+Neutron Physics Division,
Bhabha Atomic Research Centre,
Trombay, Bombay 400 085, India

ABSTRACT

The possibility of measurement of the Keff of highly subcritical (0.4 < Keff < 0.8) systems by statistical correlation techniques is demonstrated. Measurements were carried out using three different techniques on the storage tank of the U-233 uranyl nitrate solution reactor PURNIMA II at Trombay. The first of these is an adaptation of the neutron coincidence counting technique familiar in passive assay of plutonium for safeguards applications. An equation relating the system Keff and the coincidence counts obtained using a shift register has been derived along the lines of Bohnels 1985 paper. The second technique is the so called Dead Time α (first moment) method. Here the loss in count rate measured following introduction of an artificial dead time in the pulse path is related to the kinetic parameters of the subcritical assembly. The complimentarity of these two techniques is pointed out. The third technique adopted is the well known Asymptotic Prompt Variance method of Feynman. In all these techniques it is shown that the subcritical reactivity [(1-Keff)/Keff)] is inversely proportional to a directly measurable quantity characteristic of the technique such as the real coincidence probability or the V/m ratio. The results of the three experiments with the storage tank filled up to maximum solution height as well as those of the shift register coincidence experiment spanning a Keff range of 0.4 to 0.8 are found to be in good agreement with Monte Carlo calculations.

INTRODUCTION

The chain related nature of the neutron multiplicative process in a reactor assembly gives rise to temporal correlations among the neutrons in a time scale corresponding to the inverse of the prompt neutron decay constant α. A number of experimental and analytical techniques had emerged during the formative years of Reactor Noise Analysis[1,2] for the measurement of the degree of subcriticality of reactors. However, since the neutron detection efficiencies (ϵ) with counters located inside a reactor assembly seldom exceeded 10^{-3} earlier attempts were confined to the Keff region of 0.9 < Keff < 1.0. The much higher efficiency requirement at low Keff conditionse follows from the fact that the total number of chain multiplied (correlated) neutrons resulting from a given source neutron is small since multiplication varies as 1/(1-Keff) and the probability of detecting two neutrons belonging to be same chain varies as ϵ^2.

During the last decade, in the context of fissile material safeguards, statistical analysis methods have been developed for the nondestructive assay of Pu in sealed packages. In these, the spontaneous fission of Pu-240 is measured by means of a thermal neutron coincidence counting system having a neutron detection efficiency exceeding 10^{-1}. However, when samples containing large amounts of plutonium (mass < 0.5 Kg) are studied using these, the multiplicity spectrum (MS) of leakage neutrons is found to be considerably altered due to self-multiplication effects within the Pu sample. An increase in neutron multiplication by as small an amount as 10% is found to increase the coincidence counts several times. It is this observation that gave us a hint that it should be possible to measure the Keff in the highly subcritical domain of 0.4 < Keff < 0.8 provided a high efficiency detection system is employed. This paper discusses experiments conducted at Trombay to determine the degree of subcriticality of U-233 uranyl nitrate solution contained in the stainless steel storage tank of the PURNIMA II reactor using three different statistical methods.

SHIFT REGISTER COINCIDENCE METHOD

Plutonium Assay Problem in Safeguards Field

Methods developed during the last decade for Passive Neutron Assay of Pu are based on the measurement of the content of Pu-240 which is a spontaneously fissioning nuclide emitting fast neutrons. These neutron bursts are detected by means of a high efficiency well counter comprising of a large number of BF_3 (or He-3) slow neutron detectors embedded in the form of an annular array inside a hydrogenous medium and connected in parallel to form a bank[4]. The presence of the moderator around the detectors helps to a slow down fast neutrons and increase detection efficiency. Also, the statistical nature of the slowing down process results in temporal separation of detector pulses corresponding to two (or more) simultaneously produced neutrons in a spontaneous fission event. Such time correlated pulses resulting from the same fission event are counted during a gate interval t_g comparable to the neutron die-away time (typically 50 or 100 μs) in the detector-moderator assembly. The shift register[5] (SR) is the most commonly employed instrument for measuring the number of "real" coincidences during such a gate interval.

Using a fixed gate time (say 128 μs) the instrument measures in the (R+A) register, the total number of coincidences comprising of real coincidences caused by truly correlated pulses plus accidental coincidences caused by uncorrelated neutrons. In a second register called the A register, the number of accidental coincidences alone is recorded by delaying the initiating pulse by a time duration several times the die-away time with respect to the initiating pulse. The delay time is typically 1024 μs. The number of real coincidences (R) is thus obtained as the difference of the (R+A) and A register reading. Thus the two quantities obtained from the shift register are the total count rate T and the real coincidence rate R.

These are related[6] to the factorial moments $\widehat{n}$ and $\widehat{n(n-1)}$ of the multiplicity spectrum of leakage neutrons as follows:

$$T = \epsilon S \widehat{n} \tag{1}$$

$$R = \epsilon^2 \frac{FS\widehat{n(n-1)}}{2} \tag{2}$$

where S is the source event rate (henceforth simply referred to as the source) in the assembly, $\hat{n}$ and $\widehat{n(n-1)}$ are the factorial moments of the multiplicity spectrum of neutrons leaking out of the test Pu sample and entering the detection system and ϵ is the efficiency of the detection system defined as the number of neutrons detected per neutron entering it. The factor F is given by

$$F = e^{-t_0/\tau} [1 - e^{-t_g/\tau}] \qquad (3)$$

where t_0 t_g and τ are the predelay, gate width and die-away time of the detection system. Eliminating ϵ, we get

$$\frac{RS}{FT^2} = \frac{\widehat{n(n-1)}}{2\hat{n}} \qquad (4)$$

or,

$$S = [\frac{\widehat{n(n-1)}}{2\hat{n}}][\frac{FT^2}{R}] \qquad (5)$$

Eq. (5) forms the basis for non-destructive assay of Pu in the safegurads field. Here F, T and R are quantities measured in the experiment. For a small Pu metal sample the multiplicity spectrum of leakage neutrons entering the detection system is essentially the Diven spectrum for the spontaneously fissioning isotope, Pu-240. Thus the moments $\hat{n}$ an $\widehat{n(n-1)}$ are known and hence the source S and therefrom the total Pu content in the test sample can be calculated provided the isotopic fraction of Pu-240 is known. In the event that the test sample is in oxide form, the statistical properties of the driving source are modified due to emission of additional neutrons by way of (α, n) reactions. The method of analysis in such cases is discussed in literature.

Relation Between Keff and Shift Register Data for Subcritical Assemblies

Although the above technique has been found to be very useful in the safeguards field for small samples, in the case of large samples of Pu, self-multiplication effects substantially alter the multiplicity distribution of the neutrons entering the detector assembly over what would be expected from spontaneous fissions alone. An elegant method to correct for neutron multiplications within the sample has been proposed by Bohnel[7] recently. Although the method was developed for applying corrections due to self-multiplication, it can also be used to derive the system Keff provided the statistical properties of the driving source are known. For the case when the source is due to (α, n) reactions alone the following expressions for the first and second factorial moments of the multiplicity spectrum of leakage neutrons appearing in Eqs. (1) and (2) have been derived by Bohnel[7] using the probability generating function method.

$$\hat{n} = \frac{1-P_f}{1-\bar{\nu}P_f} \qquad (6)$$

$$\widehat{n(n-1)} = \frac{P_f(1-P_f)^2}{(1-\bar{\nu}P_f)^3} \overline{\nu(\nu-1)} \qquad (7)$$

Starting with these we derive below an expression relating the system Keff to experimentally measureable quantities. Substituting Eqs. (6) and (7) in Eqs. (1) and (2) we get

$$T = \epsilon S \frac{(1-P_f)}{(1-\bar{\nu}P_f)} \tag{8}$$

and

$$R = \frac{FS}{2} \frac{P_f(1-P_f)^2}{(1-\bar{\nu}.P_f)^3} \overline{\nu(\nu-1)} \tag{9}$$

Hence,

$$\frac{R}{T^2} = \frac{F}{2\,S} \frac{P_f\overline{\nu(\nu-1)}}{(1-\bar{\nu}P_f)} \tag{10}$$

or,

$$\frac{RS}{FT^2} = \frac{\overline{\nu(\nu-1)}}{2} \frac{K}{1-K} \tag{11}$$

where we have written Keff as K which is given by

$$K = \bar{\nu}P_f \tag{12}$$

In contrast to Eq. (4) where coincidence counts occur due to one or more neutrons released in single spontaneous fission event, here the coincidence counts occur due to fission chain multiplication of a source neutron. The coincidence count rate would be zero if $K = 0$ i.e. there is no multiplication.

Eq. (11) can also be written in the form

$$\frac{1}{K} = 1 + F\left[\frac{\overline{\nu(\nu-1)}}{2}\right]\left[\frac{T^2}{RS}\right] \tag{13}$$

or,

$$\frac{\Delta K}{K} = F\left[\frac{\overline{\nu(\nu-1)}}{2}\right]\left[\frac{T^2}{RS}\right] \tag{13a}$$

3.0 DEAD TIME α (FIRST MOMENT) METHOD

It is well known from nuclear physics texts that during counting of any truly random (uncorrelated) pulse series such as that met with in radioactivity in the presence of a circuit dead time d (of the nonparalysable type) the decreased count rate (N_d) is related to the original count rate (N_0) as follows.

$$N_d = \frac{N_0}{1+N_0 d} \qquad (14)$$

In the case of fission chain related (correlated) pulses, Srinivasan[8] has earlier shown that if an artifitial dead time d is introduced in the counting apparatus, N_0 is no more given by $N_d/(1-N_d d)$. Instead, the following relation exists between N_0, N_d and the kinetic parameters of the reactor assembly:

$$\frac{N_0}{N_d/(1-N_d d)} = 1 + \frac{\epsilon_f}{2} \frac{K^2}{(1-K)^2} \frac{\overline{\nu(\nu-1)}}{\bar{\nu}^2} (1-e^{-\alpha d}) \qquad (15)$$

where α is the well known Rossi α of reactor noise or the prompt neutron decay constant and ϵ_f is related to ϵ of Eq. (1) through

$$\epsilon(1-P_f) = \epsilon_f P_f \qquad (16)$$

For large d Eq. (15) simplifies to give the asymptotic relation

$$\frac{N_0}{N_d/(1-N_d d)} = 1 + \frac{\epsilon_f}{2} \frac{K}{(1-K)} \frac{\overline{\nu(\nu-1)}}{\bar{\nu}^2} \qquad (17)$$

For the determination of K Eq. (17) requires that the detection efficiency be known. Here the efficiency corresponds to a distributed neutron source which varies with the core dimensions and composition, and hence cannot be regarded as being equivalent to that measured using a calibrated point neutron source. However we can eliminated ϵ between Eqs. (1) and (17) to get

$$\frac{1}{K} = 1 + \frac{\overline{\nu(\nu-1)}}{2\bar{\nu}^2} \left[\frac{T/S}{N_0/[N_d/(1-N_d d)]-1}\right] \qquad (18)$$

or,

$$\frac{\Delta K}{K} = \frac{\overline{\nu(\nu-1)}}{2\bar{\nu}^2} \left[\frac{T/S}{N_0/[N_d/(1-N_d d)]-1}\right] \qquad (18a)$$

The dead time α (first moment) method can be regarded as being complimentary to the shift register coincidence method wherein the correlated pulses which follow each other closely in time are counted. In contrast in the dead time method we filter out the correlated pulses and essentially count only the uncorrelated pulses.

VARIANCE TO MAN RATIO (FEYMAN) METHOD

Asymptotic Prompt Variance Technique

The technique of measuring the subcritical reactivity by the variance to mean ratio method is well known. In the case of highly subcritical systems the method can succeed only if high efficiency detection systems of the kind used in passive neutron assay are employed. However, as in the case of the earlier two techniques, since the efficiency of detection varies with core size and composition and is generally unknown, we proceed as follows: The ratio of the variance and mean of the neutron counts distribution in a time interval t is given by[11]

$$\frac{V}{m} = 1 + \epsilon_f \frac{K^2}{(1-K)^2} \frac{\overline{\nu(\nu-1)}}{\bar{\nu}^2} \left[1 - \frac{1-e^{-\alpha t}}{\alpha t}\right] \tag{19}$$

It is assumed in deriving Eq. (19) that t is such that the delayed neutron contribution to V/m is negligible. For intervals long compared to the prompt neutron decay time, i.e. in the asymptotic prompt variance region, the above equation becomes

$$\frac{V}{m} = 1 + \epsilon_f \frac{K^2}{(1-K)^2} \frac{\overline{\nu(\nu-1)}}{\bar{\nu}^2} \tag{20}$$

Using Eqs. (1) and (16) and rearranging we get

$$\frac{1}{K} = 1 + \frac{\overline{\nu(\nu-1)}}{\bar{\nu}} \left[\frac{T/S}{V/M}\right] \tag{21}$$

or

$$\frac{\Delta K}{K} = \frac{\overline{\nu(\nu-1)}}{\bar{\nu}} \left[\frac{T/S}{V/M}\right] \tag{21a}$$

Eq. (21) or (21a) forms the basis of our third method for measurement of the Keff of highly subcritical systems.

Dead Time α (Second Moment) Method

It was pointed out by Srinivasan and Shani[9] that measurement of the V/m ratio, specially in the asymptotic prompt variance region following the introduction of an artificial dead time gate in the pulse path, can form the basis of determining various reactor kinetic parameters such as α and K. They have derived the following relation in this context

$$\frac{V}{m}(d) = 1 + A(\alpha, d) + \epsilon_f \frac{\overline{\nu(\nu-1)}}{\bar{\nu}} \frac{K^2}{(1-K)^2} f(\alpha, \tau)\, B(\alpha, d)\, e^{-\alpha d} \tag{22}$$

where

$$A(\alpha, d) = C_0 \left[\left(1 - \frac{d}{\tau}\right)^2 (1+P_{loss}) e^{-2P_{loss}} - \frac{1}{1+P_{loss}}\right]$$

$$B(\alpha, d) = \left(1 - \frac{d}{\tau}\right) \frac{f(\alpha, \tau-d)}{f(\alpha, \tau)} e^{-2P_{loss}(1+P_{loss})}$$

The validity of the Dead Time α (Second Moment) method was demonstrated by Edelman[14]. This technique was however not used in the PURNIMA II storage tank experiment reported in this paper.

It may be noted that the three relations presented above between the Keff or the degree of subcriticality ΔK/K and the experimentally measureable statistical properties of the neutron pulse train viz Eqs. (13a), (18a) and (21a) bear a striking similarity to each other. This may be understood on the basis that in each of the three methods it is essentially the desviation from Poisson statistics that is bein measured. In all the three methods apart from the Diven parameters ν and $\nu(\nu-1)$ the only quantity required to be known is the source event rate S for the calculation of the Keff.

EXPERIMENTAL ASPECTS

Description of the Subcritical Assembly and the Neutron Detection System

The experimental set-up used by us is depicted in Fig. 1. The storage tank used for storing the fissile fuel solution whenever the PURNIMA II reactor is in shutdown state is a 168 mm dia, 7 mm thick and 655 mm high vessel welded to the bottom of a glove box. The level of the solution in the storage tank is measured by means of an ultrasonic solution level monitor (USLM) mounted at the bottom of the tank. Surrounding the storage tank are two semi-annular aluminium containers filled with water. 11 numbers of BF3 propotional counters are placed vertically inside the aluminium containers forming a ring around the storage tank. The neutron detection efficiency of this set-up with storage tank empty was determined, with the help of a standard Pu-Be neutron source, to be in the range of 3% to 5% depending on its location inside the tank. This information however is not used in the analysis of the data.

Shift Register Unit and Die-Away Time Determination

The combined output from all the counters is taken through a preamplifier and amplifier to the shift register type coincidence logic unit. Here each pulse is predelayed by 16 μs prior to entry into the SR unit having a coincidence gate width of 128 μs. (The coincidence logic dead time was 2 μs). Each input pulse strobes the accumulator immediately after predelay and again after a long delay of 1024 μs giving the real plus accidental (R+A) and accidental (A) coincidence rates respectively. The neutron die-away time of the detector assembly was determined to be 140 μs by analysing the R values of a Cf-252 neutron source, placed inside the empty storage tank using gates of 16, 32, 64 and 128 μs width. The count rate data obtained with the Cf-252 source are summarised in Table I.

Variation of Coincidence Count with Solution Height

The entire uranyl nitrate solution was first pumped out of the storage tank leaving less than 5 mm of residual solution. The coincidence counts corresponding to this was taken as the background due to extraneous source neutrons. The fissile solution was then transferred in steps into the storage tank, its level being obtained from the USLM reading. At each step the (R+A), A and T register readings were recorded. The experiment was repeated for different solution levels in the tank over the range of 13 cm to 41 cm. The raw experimental data are presented in Table II:

Analysis of SR Data and Keff Results

From the R+A and A readings it can be seen that the real coincidence count rate R is comparable to the accidental coincidence count rate A for all solution heights. This relatively large magnitude of correlated counts has contributed to the success of the experiment. Table III lists the various input quantities that go into the calculation of Keff,

Table I

Coincidence Counts with Cf-252 Source for Determination of Die-away Time

Gate Width t(μs)	R+A cts/s	A cts/s	R cts/s
16	175.2	159.0	16.2
32	347.5	309.1	38.3
64	682.2	611.8	70.4
128	1311.5	1202.4	109.1

Table II

Coincidence Counts for Various Solution Heights

Solution Height(cm)	R+A cts/s	A cts/s	T cts/s	R/T
13.0	13.42	6.76	234.09	1.2154-4
15.0	19.88	9.56	276.67	1.3482-4
17.0	28.06	13.72	326.49	1.3452-4
19.0	37.17	17.69	372.32	1.4053-4
20.8	48.88	21.93	417.16	1.5487-4
22.7	61.99	28.00	468.57	1.5481-4
24.6	76.43	34.11	521.11	1.5684-4
26.4	92.82	40.95	569.17	1.6011-4
28.2	108.91	48.01	616.29	1.6034-4
40.9	207.64	92.66	864.99	1.5368-4

viz, the driving neutron source strength S, $\bar{\nu}$, $\overline{\nu(\nu-1)}$ and F as calculated from eq. (3). $\overline{\nu(\nu-1)}$ was obtained using the relation[10],

$$\sigma = 0.980 + 0.076\,\bar{\nu} \tag{23}$$

for the standard desviation of the Diven spectrum assumed to be approximately Gaussian.

The Keff values for different solution heights determined using Eq. (13) are presented in Table IV. Also given, for comparison, are the corresponding Keff values obtained by the Monte Carlo code KENO using the 16 group Hansen and Roach cross section set.

The results are presented in a different way in Fig. 2 which shows a plot of T^2/RS vs 1/K the reciprocal of the KENO Keff. As

Table III

Data for Keff Calculation

n^0 Source strength per gm of U-223	24.6 n/s
$\bar{\nu}$ for U-223	2.5
$\overline{\nu(\nu-1)}$	5.12
F (for t = 16 μs, t = 128 μs and t = 140 μs)	0.535

Table IV

Comparison of Experimental Keff Using the Shift Register Coincidence Method with KENO Results

Soln. Height (cm)	Keff (Expt)	Keff (KENO)
13.0	0.443	0.440
15.0	0.505	0.470
17.0	0.536	0.510
19.0	0.574	0.545
20.8	0.619	0.605
22.7	0.639	0.617
24.6	0.661	0.630
26.4	0.681	0.644
28.2	0.695	0.659
40.9	0.759	0.715

predicted by Eq. (18), this plot is a straight line passing through (0, 1). The slope of the line is found to be about 0.6 which is close to the value of 0.55 expected from the data given in Table III.

Dead Time α (First Moment) Experiment and Results

The dead time unit is fabricated from a single monostable IC 71423 having two monoshots in a single integrated circuit package. In the dead time unit one monoshot is used in the retriggerable or non-retriggerable modes corresponding to paralysable or non-paralysable dead time units. The desired dead time value determined by the output

pulse width can be varied by adjusting the R-C time constant. The second mono-shot is used to give an output pulse of a fixed width (0.5 μs) at the end of each timing pulse. This is fed to a scalar for counting. The input is from a discriminitator scaled down to suit TTL logic (i.e. 4.5 V).

Measurements by this method were made only for the maximum solution height case viz 40.9 cm. The results are presented in Table V. Fig. 3 shows a plot of N_d/N_0 and $N_d/(N_0(1-N_d d))-1$ against d. By fitting the data in Column 2 of Table V to Eq. (15) we derive a value of 139 μs for the neutron die-away time $1/\alpha$. This is in good agreement with the value of 140 μs obtained from the SR coincidence method. The value of $N_0/(N_d/(1-N_d d))-1$ saturates at 0.21. Using this value in Eq. (18) we get Keff=0.749 at the solution height of 40.9 cm which is good agreement with the Monte Carlo value of 0.715+0.003.

Prompt Variance Results

A microprocessor (Intel 8085) based Probability Distribution Analyser (PDA) was used for collection of data by this method. A histogram of the counts distribution obtained for a given counting time interval was stored in the memory and could be printed for subsequent analysis. As with the dead time α method, measurements were made only for the full solution height case of 40.9 cm. The variance to mean ratio was measured for five counting intervals in the range of 0 to 100 ms. The results are presented in Table VI. Fig. 4 shows a plot of V/m against the counting time t. From the data for large t the asymptotic value of V/m ratio is taken as 1.32. Substituting in Eq. (21) we get Keff=0.694.

Table V

Variation of Count Rate (Nd) with Artificial Dead Time (d)

d(μs)	N_d	$[N_0/ (N_d/(1-N_d d)]-1$
0	779.9	0.0
30	736.8	0.0351
50	711.2	0.0576
92	670.1	0.0921
136	626.9	0.1380
180	600.9	0.1575
225	576.8	0.1766
270	562.6	0.1757
320	537.4	0.2017
360	523.2	0.2099

By fitting the data of Table VI to Eq. (19), we get a value of about 600 μs for the neutron die away time 1/α. This is completely different from the values obtained by the other two methods ands is rather puzzling particularly since the Keff by the three methods agree so well with one another and also with Monte-Carlo calculations.

Table VI

Variation of V/m Ratio with Counting Interval t in Feynman Prompt Variance Method Experiment

V/m	t(ms)
1.185	5
1.235	10
1.295	40
1.315	70
1.320	100

DISCUSSION OF RESULTS

The experimental Keff results obtained by the three different techniques are in reasonably good agreement with KENO computed Keff values for the maximum solution height case of 40.9 cm. The results of the shift register coincidence method, which are available for lower solution heights also, do however show evidence of a small but clearly discernable systematic discrepancy between the experimental and KENO Keff values.

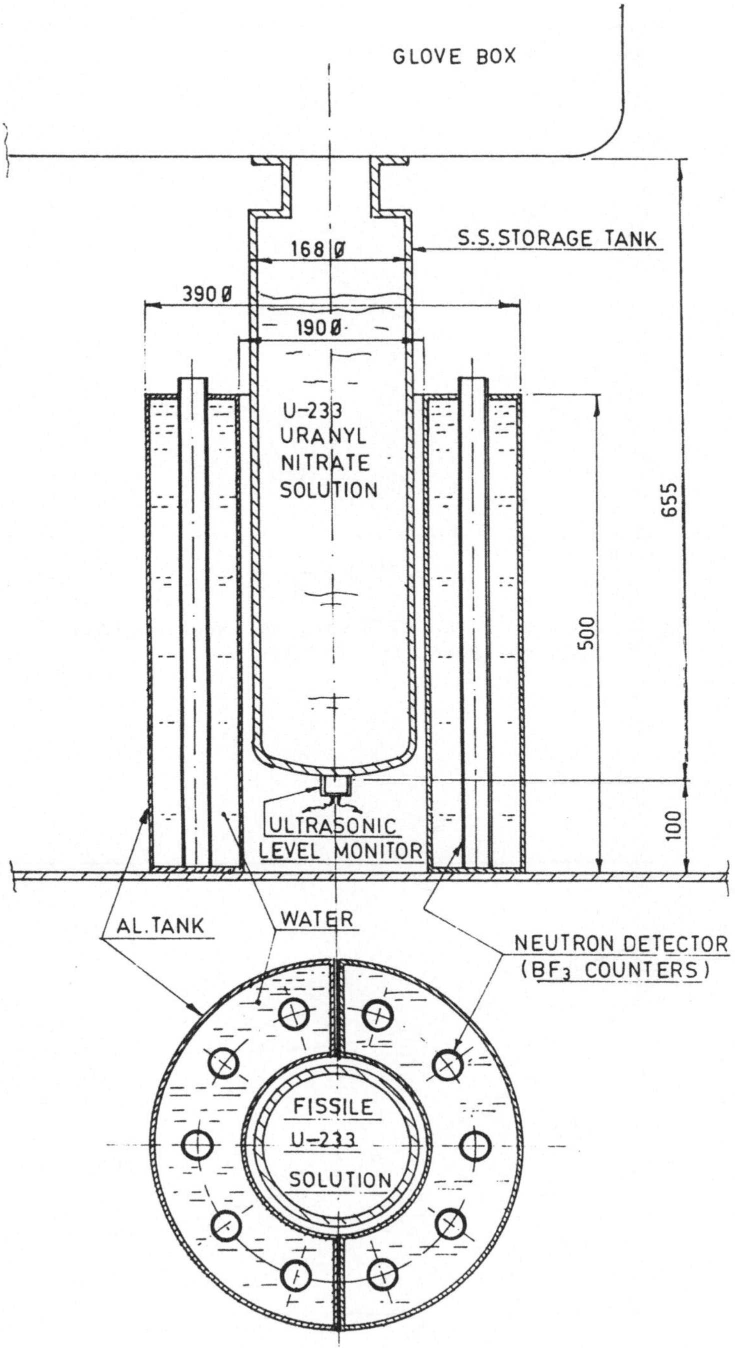

Fig. 1. Experimental set up

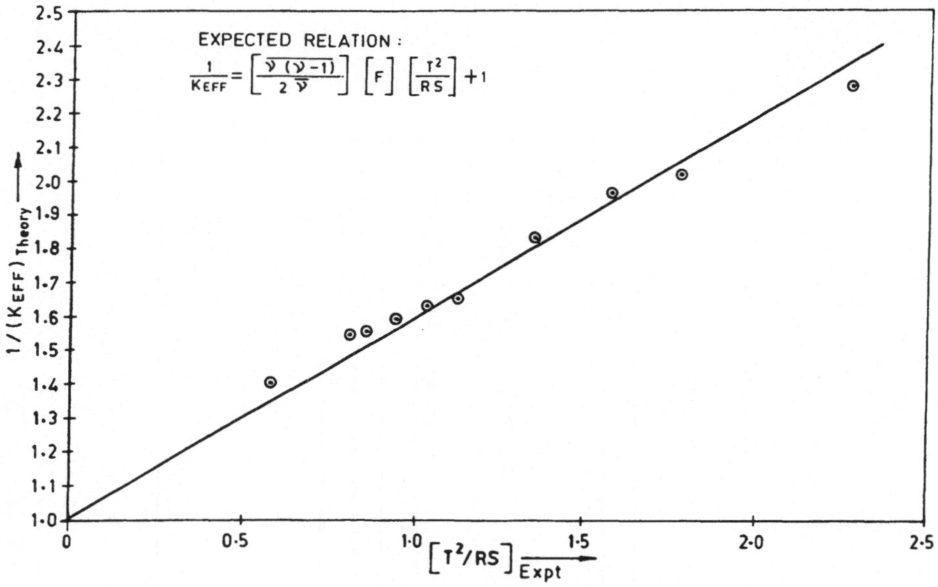

Fig.2. Plot of 1/(Keff) Theory Vs Experimentally Measured Quantity [T^2/RS]

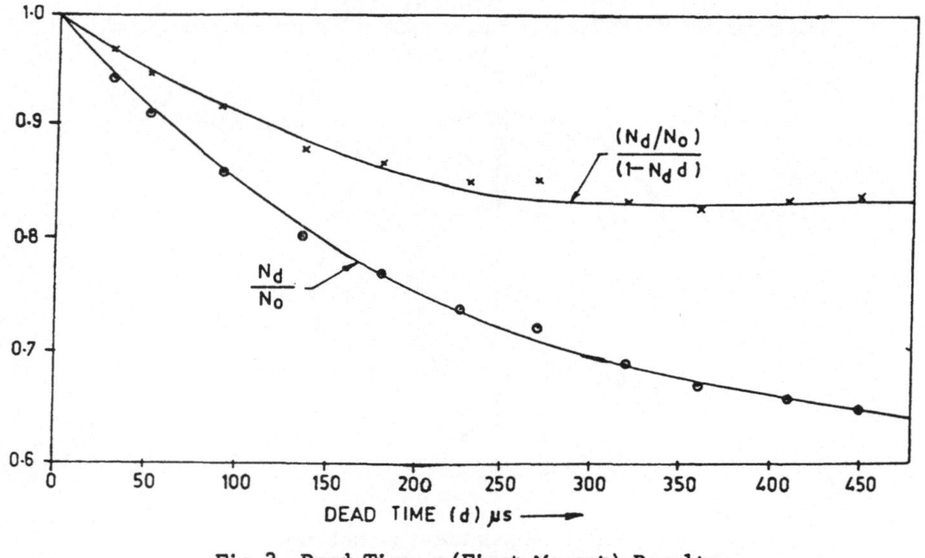

Fig.3. Dead Time α (First Moment) Results

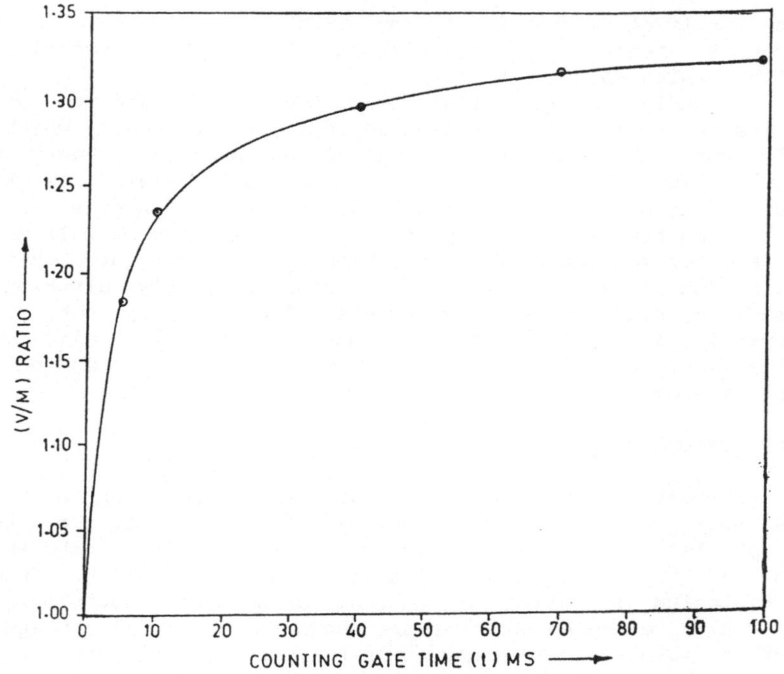

Fig.4. Feynman Prompt Variance Results

PURNIMA II REACTOR: STORAGE TANK

$L\,\hat{l}_c \simeq 15.4\,CM$

$$K_{EFF} \simeq \left[\frac{\hat{l} - \hat{l}_o}{\hat{l}_c - 5.0}\right]$$

$$\hat{l} = \frac{4V}{S} = \left[\frac{h}{0.5 + h/d}\right]$$

$\hat{l}_o \simeq 5.0\ CM$

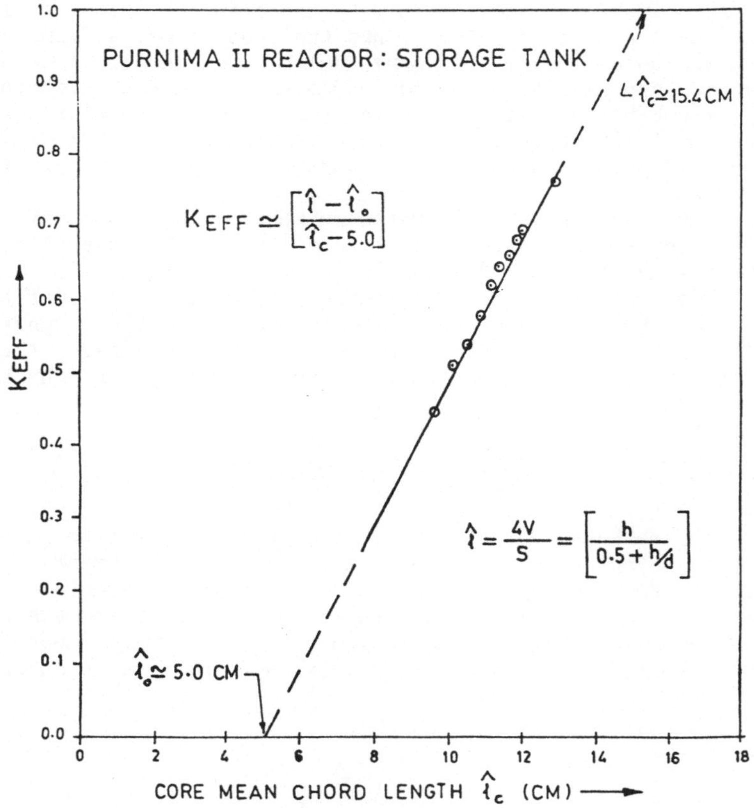

**Fig.5. Experimental Keff Vs Core Size
Expressed as Mean Chord Length**

This can be partly attributed to the uncertainty in the value of F which depends upon the neutron die-away time in the detector assembly and varies slightly with solution height.

Secondly, the analytical model for the SR method assumes that the fission chain multiplication occurs instantaneously which is not exactly true. The rather high value of the neutron die-away time obtained by the prompt variance method is probably related to these effects but is nevertheless difficult to understand in view of the agreement between the $1/\alpha$ values by the first two methods as well as the good agreement between the Keff values. Finally it may be noted that the normal definition of Keff assumes that the neutron density distribution has attained the fundamental mode. However for a highly subcritical system driven by a (flat) source this may not be fully valid. This limitation applies to all the three methods as they are based on the point kinetics model.

SUMMARY AND CONCLUSIONS

The experimental determination of the Keff of a highly subcritical storage tank containing enriched fissile solution by three different statistical techniques is demostrated. The first is an adaptation[12] of the technique developed in recent years for the Passive Assay of Plutonium by means of neutron coincidence counting. The experimental Keff values so obtained are shown plotted in Fig. 5 against the size of the solution core measured in terms of its mean cord length (l_6). As predicted on the basis of the so called Trombay Criticality Formula (TCF)[13], this plot is a straight line confirming the success of the technique. The other two are based on reactor noise theory. The success of the experiment owes mainly to the high detection efficiency employed as well as to the fact that the source being inherent is uniformly distributed inside the core, which for a small system is a reasonable good approximation to the fundamental mode flux distribution. In our experiment the source was due to (α, n) reactions alone. It is, however not difficult to generalise to a spontaneous fission source or mixed (α, n) and spontaneous fission source provided the proportion of each is known.

Since the above experiments measured only two moments of the multiplicity spectrum of leakage neutrons all three have the limitation that the magnitude of the driving source needs to be known. This requirement can possibly be eliminated if additional information contained in the higher moments could be tapped during the experiments. For example if the third moment Q were to be measured, then the following equation, derived by us, can be used for computing the Keff.

$$\frac{1}{K} = 1 + \frac{3}{2} \frac{\overline{\nu(\nu-1)}}{\overline{\nu}\ \overline{\nu(\nu-1)(\nu-2)}} \left[\frac{QT}{R^2} - 2\right] \tag{24}$$

The methods described in this paper appear to have the potential for being developed into a on line instrument for the measurement of Keff of process plant and fissile material storage tanks/bird cages. Such an instrument built around a microprocessor or a personal computer can also be deployed to give advance warning of impending criticality by extrapolating anomalous increase in subcritical multiplication. In this context it is worth noting that most of the criticality alarms systems currently in vogue are essentially dependent on the system becoming accidentally critical first, which is certainly not a very laudable philosophy.

A novel application of these highly subcritical techniques could be for verfying criticality data of special actinide elements such as Neptunium, Americium Curium et. for which it is very difficult to accumulate adequate fissile material inventory to carry out a full scale critical experiment.

REFERENCES

1. N.Pacilio, Reactor Noise Anaylysis in the Time Domain, USAEC Technical Monograph No. TID 24512 (1969)
2. R.E.Uhring. Random Noise Techniques in Nuclear Reactor Systems, Ronald Press, New York (1970)
3. M.S.Krick and J.C.Swansen, Neutron Multiplicity and Multiplication Measurements, Nucl. Instr. Method 219, 384 (1984)
4. M.S.Krick and H.O.Menlove, The High Level Neutron Coincidence Counter (HLNCC): Users Manual, Report LA-7779-M (1979)
5. K.Bohnel, Die Plutoniumbestimmung in Kernbrennstoffen mit der Neutronenkoinzidenzmethode, Report KFK-2203 (1975)
6. R.Dierckx and W.Hage, Neutron Signal Multiplet Analysis for the Mass Determination of Spontaneous Fission Isotopes, Nucl. Sci.Engg. 85, 325, (1983)
7. K.Bohnel, The Effect of Multiplication on the Quantitative Determination of Spontaneously Fissioning Isotopes by Neutron Correlation Analysis, Nucl.Sci. and Engg. 90, 75 (1985)
8. M.Srinivasan, On the Measurement of by a simple dead time method, Nucleonik, 10, 224 (1967)
9. M.Srinivasan and D.C.Sahni, a Modified Statistical Technique for the Measurement of a in Fast Intermediate Reactor Assemblies, Nucleonik, 9, 155 (1967)
10. Huanquiao et al, The average Number of Prompt Neutrons and the Distribution of Prompt Neutrons and the Distribution of Prompt Neutron Number for spontaneous Fission of Pu-240, Cu-242 and Cu-244. Nucl. Sci. Eng. 86, 315 (1984)
11. M.M.R.Williams, Random Processes in Nuclear Reactors, Pergamon Press, 1974
12. M.Srinivasan, K.K.Rasheed, C.S.Psapathy and S.B.Degweker, Experimental Determination of the Keff of Highly Subcritical Enriched Fissile Units by a Statistical Correlation Technique, Int. Sem. Criticality Safety, Tokyo, 1987
13. Anil Kumar, M.Srinivasan and K.Subba Rao, Characterisation of Neutron Leakage Probability, Keff and Critical Core Surface Mass Density of Small Reactor Assemblies through the Trombay Criticality Formulae, Nucl. Sci. Engg. 84, 155 (1983)
14. M. Edelmann, et. al., Investigations of the Two Detector Covariance Method for the Measurement of Coupled Reactor Kinetics Parameters, Ann. of Nuclear Energy. 2, 207 (1975)

STOCHASTIC ASPECTS OF NEUTRON MULTIPLICATION IN FUSION BLANKET MATERIALS AND SPALLATION TARGETS

S.B.Degweker*, Amar Sinha+ and M.Srinivasan+

*Reactor Analysis and Systems Division
+Neutron Physics Division
Bhabha Atomic Research Centre,
Trombay, Bombay 400 085, India

ABSTRACT

A Stochastic description of neutron multiplication in fusion blanket materials and spallation targets is presented. An asymptotic solution of the multigroup Kolmogorov equation for a point model is obtained. The solution gives the multiplicity spectrum (MS) of leakage neutrons and its factorial moments as well as the neutron multiplicity equation of Kumar and Srinivasan. The MS (or its factorial moments) are measurable and can yield valuable information regarding the individual nuclear processes such as (n, 2n) reactions. The paper also discusses the use of stochastic transport theory for calculation of the MS. Results of calculations using the two methods for Be-9, a promising candidate for fusion blankets are discussed.

INTRODUCTION

Statistical methods for studying the multiplication of neutrons in a nuclear reactor have been in use for well over three decades. The techniques developed in the field of reactor noise analysis[1] have been very successful in measuring a number of parameters of interest in reactor physics such as the delayed neutron fraction β, the prompt neutron lifetime l, and the reactivity ρ.

During the last decade techniques akin to those of reactor noise analysis have been developed for the passive non-destructive assay of plutonium[2-4] in sealed containers for safeguards applications. In a recent paper[5] Srinivasan has suggested the use of similar techniques for the study of neutron multiplication in fusion blanket materials and in spallation targets. While the mean number of neutrons produced per incident neutron (or proton) is of direct relevance to the design of these assemblies, much more information about the individual nuclear reactions can can be extracted from the multiplicity spectrum (MS) of leakage neutrons. The sensitivity of the MS to basic nuclear data has been studied[6-8] in the case of Be-9 a promising candidate for fusion blankets and Pb-207 a common spallation target.

The present paper discusses a point model multigroup Kolmogorov equation approach for calculating the MS. An asymptotic solution is obtained by introducing the probability generating function (pgf). The solution yields the neutron multiplicity equation[6] of Kumar and Srinivasan as well as simple recursion relations for the factorial moments of the MS. The paper also discusses the Stochastic Transport Theory method[9] of Bell for computing the multiplicity spectrum. Results of calculations by the two methods are discussed for Be-9.

THE POINT MODEL

The Kolmogorov Equation for Fusion Blanket Materials

We consider the problem of a high energy neutron injected inside a finite multiplying medium. we divide the energy range below this into G groups such that the secondary neutrons in (n, 2n) reactions occuring in any group appear only in lower energy groups. Fission neutrons (in case the medium is fissionable) of course may be produced in higher energy groups also. Let $P(N_1, N_2, \ldots\ldots, N_G, R, T)$ denote the probability of finding the system in a state characterised by N_i neutrons in the i-th group and R neutrons having leaked out of the system at time t. The probability at time t+dt can be related to the probability at time t as follows:

$$
\begin{aligned}
P(N_1,N_2,\ldots\ldots,N_G,R;t+dt) &= \sum_i \lambda_{ci}(N_i+1)P(N_1,N_2,\ldots,N_i+1,\ \ldots,N_G,R;t)dt \\
&+ \sum_i \lambda_{li}(N_i+1)P(N_1,N_2,\ldots,N_i+1,\ldots,N_G,R-1;t)dt \\
&+ \sum_{j,i} \lambda_{ij}(N_i+1)P(N_1,N_2,\ldots,N_i+1,\ldots,N_j-1,\ldots,N_G,R;;t)dt \\
&+ \sum_{j,k,i}\lambda_{ijk}(N_i+1)P(N_1,N_2,\ldots,N_i+1,\ldots,N_j-1,\ldots,N_k-1,\ldots,N_G,R;t)dt \\
&+ \sum_i \lambda_{fi}(N_i+1) \sum_\nu P_{\nu i} \sum_{\nu_1+\ldots\nu_G=\nu} \frac{\nu!}{\nu_1!\ldots\nu_G!} X_1^{\nu_1}\ldots X_G^{\nu_G} \\
&\qquad P(N_j-\nu_j+\delta_{ij},R;t)dt \\
&+ (1-\sum_i \lambda_i N_i dt)P(N_1,N_2,\ldots,N_G,R;t) \qquad (1)
\end{aligned}
$$

where,

λ_{li} is the probability of leakage of a neutron per unit time

λ_{ci} is the probability of capture per neutron per unit time

λ_{fi} is the probability of fission in group i per neutron per unit time

λ_{ij} is the probability per unit time of a neutron being scattered from group i to j

λ_{ijk} is the probability per unit time of a neutron in group i producing a neutron in group j and another in group k

$P_{\nu i}$ is the probability that ν neutrons are produced in a fission in group i.

λ_i is the sum $\lambda_{ci}+\lambda_{li}+\lambda_{fi}+\sum_j \lambda_{ij}+\sum_{j,k}\lambda_{ijk}$

Writing $P(N_1,\ldots,N_G,R,t+dt)$ as

$$P(N_1,N_2,\ldots,N_G,R;t+dt) = P(N_1,N_2,\ldots,N_G,R;t) + \frac{dP}{dt}$$

the above equation becomes

$$
\begin{aligned}
\frac{dP}{dt}(N_1,N_2,\ldots,N_G,R;t) &= \sum_i \lambda_{ci}(N_i+1)P(N_1,N_2,\ldots,N_i+1,\ldots,N_G,R;t) \\
&+ \sum_i \lambda_{li}(N_i+1)P(N_1,N_2,\ldots N_i+1,\ldots,N_G,R-1;t) \\
&+ \sum_{j,i} \lambda_{ij}(N_i+1)P(N_1,N_2,\ldots,N_i+1,\ N_j-1,\ldots,N_G,R;t)
\end{aligned}
$$

$$+\sum_{j,k,i} \lambda_{ijk}(N_i+1)P(N_1,N_2,\ldots,N_i+1,\ldots,N_j-1,\ldots,N_k-1,\ldots,N_G,R;t)$$

$$+\sum_i \lambda_{fi}(N_i+1)\sum_\nu P_{\nu i}\sum_{\nu_1+\ldots\nu_G=\nu}\frac{\nu!}{\nu_1!\ldots\nu_G!}X_1^{\nu_1}\ldots X_G^{\nu_G}$$

$$P(N_j-\nu_j+\delta_{ij},R;t)$$

$$+\sum_i \lambda_i N_i P(N_1,N_2,\ldots,N_G,R;t) \tag{2}$$

This is the Kolmogorov equation for the process. The initial condition for the problem is

$$P(N_1,N_2,\ldots,N_G,R;t=0) = \delta_{N1},\delta_{N20}\ldots\delta_{NG0}\delta_{R0} \tag{3}$$

Solution of the Kolmogorov Equation

We introduce the probability generating function defined by

$$F(x_1,x_2,\ldots,x_G,z;t) = \sum_{N_1}\sum_{N_G}\sum_R P(N_1,N_2,\ldots,N_G,R;t)x_1^{N_1}\,x_2^{N_2}\ldots x_G^{N_G}.z^R \tag{4}$$

Multipliying (2) by $x_1^{N_1}.x_2^{N_2}\ldots x_G^{N_G}.z^R$ and summing over all Ns an R we get

$$\frac{\partial F}{\partial t} = \sum_i [\lambda_{ci}+z\lambda_{1i}+\sum_j \lambda_{ij}x_j+\sum_{j,k}\lambda_{ijk}x_jx_k+\sum_\nu P_{\nu i}(X_1x_1+X_2x_2+\ldots+X_Gx_G)^\nu\lambda_{fi}$$

$$-\lambda_i x_i]\frac{\partial F}{\partial t} \tag{5}$$

The initial conditions (3) become

$$F(x_1,x_2,\ldots,x_G\ ,z;t=0) = x_1 \tag{6}$$

Apart from this we also have the condition

$$F(1,1,\ldots,1,1;t) = 1 \tag{7}$$

since the sum of all the probabilities is unity.

The partial differential equation (5) is completely equivalent to the following set of ordinary differential equations defining the characteristic curves.

$$\frac{dx_i}{dt} = \lambda_{ci}+z\lambda_{di}+\sum_j \lambda_{ij}x_j+\sum_{j,k}\lambda_{ijk}x_jx_k+\lambda_{fi}\sum_\nu P_{\nu i}(X_1x_1+X_2x_2+\ldots+X_Gx_G)^\nu$$
$$-\lambda_i x_i \tag{8a}$$

$$\frac{dz}{dt} = 0 \tag{8b}$$

$$\frac{dF}{dt} = 0 \tag{8c}$$

To obtain the solution of (5) with the initial condition (6) we must solve the characteristic equations starting from all possible points in the hyperplane t=0. However, for obtaining the multiplicity spectrum it is enough to solve Eq. (5) in the $t\to\infty$ limit. In this limit, F is a function of z alone. This is because at large times no neutrons of any energy whatsoever are expected to be present inside the system. Hence any one parameter (z) curve in the t=0 hyperplane is sufficient for the purpose. The characteristics are simplest if we begin from the curve defined by

$$\lambda_{ci}+z\lambda_{ci}+\sum_j \lambda_{ij}x_j+\sum_{j,k}\lambda_{ijk}x_jx_k+\lambda_{fi}\sum_\nu P_{\nu i}(X_1x_1+X_2x_2+\ldots+X_Gx_G)^\nu-\lambda_i x_i=0 \tag{9}$$

At these points $dx_i/dt=0$ and therefore the characteristics originating from these points are simply straight lines parallel to the t axis. The assymptotic solution is therefore obtained by eliminating $x_2, x_3, \ldots, x_G$ from equations (9) and setting $F(z) = x_1$. It is clear from this argument that $x_1(z)$ is the pgf of the multiplicity spectrum when the source neutron is injected in the first group. Likewise $x_i(z)$ is the pgf of the MS with the source neutron in the ith group. Dividng (9) by λ_i we can write

$$x_i(z) = P_{ci}+P_{ci}z+\sum_j P_{ij}x_j+\sum_{j,k} P_{ijk}x_jx_k-P_{f_i}\sum_\nu P_{\nu i}(\sum_j X_jx_j)^\nu \tag{10}$$

where $P_{xi} = \lambda_{xi}/\lambda_i$ is the probability of a neutron in the ith energy group undergoing the reaction x.

Generalisation to Other Processes

The equation (2) derived above and its asymptotic solution can be generalised to processes involving several species of particles such as spallation. However the solution to the problem requires that a large number of transition probabilities (the λ-s) be known. In principle these could be obtained from Monte Carlo codes such as HETC, but the method is rather clumsy. A better method perhaps would be to include intra-nuclear cascades and other nuclear models directly in the analytical approach for obtaining the multiplicity spectrum for spallation.

Deductions from the Kolmogorov Equation

Eq. (10) forms the basis of the point model for the calculation of the multiplicity spectrum. If the material is not fissionable, the above equations can be solved recursively to give the exact expression for the pgf of the multiplicity spectrum. By differentiating Eq. (10) r times with respect to z an setting z=0 we get an equation similar to the neutron multiplicity equation of Kumar for the multiplicity spectrum $Q_i(\nu)$, the probability of ν neutrons emerging when a neutron is injected in the ith group.

$$Q_i(\nu) = \sum_j p_{ij}Q_j(\nu) + \sum_{j,k} p_{ijk}\sum_{l=0}^{\nu} Q_j(l)Q_k(\nu-1) \tag{11}$$

On the other hand by differentiating Eq.(10) with respect to z and setting z=1 we get the following relations for the factorial moments of the MS.

$$M^{(r)}{}_i = \sum_j p_{ij}M^{(r)}{}_j+\sum_{j,k} p_{ijk}(2M^{(r)}{}_j+\sum_{s-1}^{r-1}\binom{r}{s}M^{(s)}{}_jM^{(r-s)}{}_k) \tag{12}$$

When fission is present, the equations for the MS are non-linear instead of the simple recursion relations (11). The equations for the factorial moments are nevertheless linear even in the presence of fission.

STOCHASTIC TRANSPORT THEORY

In this section we derive the space-energy dependent equation for the pgf of the multiplicity spectrum of leakage neutrons from a system. We assume that neutron multiplication is through (n, 2n)reactions alone. We need the following properties of the pgf.

(1) If f1 (z) and f2(z) are the pgfs of two integer random variables n1

and n2, the pgf of the variable n1+n2 is f1(z)f2(z).

(2) If $g_i(z)$ be the pgf formed with the conditional probability P(n/i) which describes the probability that the variable takes value n under the condition i and if p_i is the probability of condition i occurring, then the pgf of the random variable n without conditions is

$$\sum_i p_i g_i(z)$$

Let $P_{ni}(\vec{r}, \vec{\Omega})$ be the probability that n neutrons escape from the system when 1 neutron is initially present at the point $\vec{r}$ in the direction $\vec{\Omega}$ in group i. The pgf is defined as before by

$$G_i(z, \vec{r}, \vec{\Omega}) = \sum_n P_{ni}(\vec{r}, \vec{\Omega}) z^n \qquad (13)$$

A neutron at $\vec{r}$ in group 1 indirection $\vec{\Omega}$ may either escape without a collision or it may collide at a distance between s and s+ds from $\vec{r}$. The probabilities for these events are $\exp(-\Sigma b)$ and $ds\exp(-\Sigma s)\Sigma$ respectively where b is the distance of the boundary from the point $\vec{r}$ in the direction $\vec{\Omega}$. In the latter case the neutron may get captured, scattered in group j or undergo multiplication by (n, 2n) reaction leading to a neutron, in group j in direccion $\vec{\Omega}1$ and another ingroup k in direction $\vec{\Omega}_2$. The probabilities for these events are, respectively $\Sigma c/\Sigma_i, \Sigma_{ij}(\vec{\Omega}\to\vec{\Omega}')/\Sigma_i, \Sigma_{ijk}(\vec{\Omega}\to\vec{\Omega}_1, \vec{\Omega}_2)/\Sigma_i$. Using the properties 1 and 2 given above we can write

$$G_i(z,\vec{r},\vec{\Omega}) = ze^{-s\Sigma_i} + \int_0^b ds\, e^{-b\Sigma_i} [\Sigma_{ci} + \sum_j \int d\Omega\, \Sigma_{ij}(\vec{\Omega}\to\vec{\Omega}') G_j(z,\vec{r}+s\vec{\Omega},\vec{\Omega}')$$

$$+ \sum_{jk} \int d\Omega_1 d\Omega_2 \Sigma_{ijk}(\vec{\Omega}\to\vec{\Omega}_1\ \vec{\Omega}_2) G_j(z,\vec{r}+s\vec{\Omega},\vec{\Omega}_1) G_k(z,\vec{r}+s\vec{\Omega},\vec{\Omega}_2) \qquad (14)$$

Rewriting the above equation for the point $\vec{r}+\delta\vec{\Omega}$ we have

$$G_i(z,\vec{r}+\vec{\Omega}\delta,\vec{\Omega}) = ze^{-(b-\delta)\Sigma_i} + \int_\delta^b ds\, e^{-(s-\delta)\Sigma_i} [\Sigma_{ci} + \sum_j \int d\Omega'\, \Sigma_{ij}(\vec{\Omega}\to\vec{\Omega}') G_j(z, \vec{r}+s\vec{\Omega},\vec{\Omega}')$$

$$+ \sum_{jk} \int d\Omega_1 d\Omega_2\, \Sigma_{ijk}(\vec{\Omega}\to\vec{\Omega}_1\vec{\Omega}_2) G_j(z, \vec{r}+s\vec{\Omega},\vec{\Omega}_1) G_k(z,\vec{r}+s\vec{\Omega},\vec{\Omega}_2)] \qquad (15)$$

Subtracting (14) from (15) dividing by δ and taking the limit as $\delta\to 0$ we get

$$\vec{\Omega}.\vec{\nabla}G_i(z,\vec{r},\vec{\Omega}) = \Sigma_i z e^{-b\Sigma_i} + \Sigma_i \int_0^b ds\, e^{-s\Sigma_i}[\Sigma_{ci} + \sum_j \int d\Omega' \Sigma_{ij}(\vec{\Omega}\to\vec{\Omega}') G_j(z,\vec{r}+s\vec{\Omega}, \vec{\Omega}') + \sum_{j,k} \int d\Omega_1 d\Omega_2\, \Sigma_{ijk}(\vec{\Omega}\to\vec{\Omega}_1\vec{\Omega}_2)\, G_j(z,\vec{r}+s\vec{\Omega},\vec{\Omega}_1)\, G_k(z,\vec{r}+s\vec{\Omega},\vec{\Omega}_2)]$$

$$- \{\Sigma_{ci} + \sum_j \int d\Omega' \Sigma_{ij}(\vec{\Omega}\to\vec{\Omega}') G_j(z,\vec{r},\vec{\Omega}') + \sum_{j,k} \int d\Omega_1 d\Omega_2\, \Sigma_{ijk}(\vec{\Omega}\to\vec{\Omega}_1\vec{\Omega}_2) G_j(z,\vec{r},\vec{\Omega}_1) G_k(z,\vec{r},\vec{\Omega}_2)\} \qquad (16)$$

Combining (14) and (16) we get

$$-\vec{\Omega}.\vec{\nabla}G_i(z,\vec{r},\vec{\Omega}) + \Sigma_i G_i(z,\vec{r},\vec{\Omega}) = \Sigma_{ci} + \sum_j \int d\Omega' \Sigma_{ij}(\vec{\Omega}\to\vec{\Omega}') G_j(z,\vec{r},\vec{\Omega}') +$$

$$+\sum_{j,k} \int d\Omega_1 \Omega_2 \Sigma_{ijk}(\vec{\Omega}\to\vec{\Omega}_1\vec{\Omega}_2) G_j(z,\vec{r},\vec{\Omega}_1) G_k(z,\vec{r},\vec{\Omega}_2) \qquad (17)$$

The equations for the factorial moments are readily obtained by differentiating (17) with respect to z and settingz=1. Thus we have for the mean

$$-\vec{\Omega}.\vec{\nabla}M^{(1)}_i + \Sigma_i M^{(1)}_i = \sum_k \int d\Omega' \, [\Sigma_{ij}(\vec{\Omega}\to\vec{\Omega}') + 2\Sigma^{(n,2n)}_{ij}(\vec{\Omega}\to\vec{\Omega}')] M^{(1)}_j(\vec{r},\vec{\Omega}') \qquad (18)$$

and for the rth factorial moment

$$-\vec{\Omega}.\vec{\nabla} M^{(r)}_i + \Sigma_i M^{(r)}_i = \sum_j \int d\Omega' [\Sigma_{ij}(\vec{\Omega}\to\vec{\Omega}') + 2\,\Sigma^{(n,2n)}_{ij}(\vec{\Omega}\to\vec{\Omega}')] M^{(r)}_j(\vec{r},\vec{\Omega}')$$

$$+\sum_{s=1}^{r-1} \binom{r}{s} \sum_{j,k} \int d\Omega_1 d\Omega_2 \; \Sigma_{ijk}\,(\vec{\Omega}\to\vec{\Omega}_1\vec{\Omega}_2)\; M^{(r-s)}_j(\vec{r},\vec{\Omega}_1) M^{(s)}_k(\vec{r},\vec{\Omega}_2) \qquad (19)$$

Equation (17) is simply the time independent version of Bell's[9] equation. The equations for the factorial moments have the same form as the adjoint of the linear Bolzmann transport equation and can be solved by the methods employed for solving the transport equation. The factorial moments of any order can be obtained from knowledge of all the factorial moments of lower order. The boundary conditions for these equations can be directly obtained from physical considerations. Thus for outgoing neutrons at the boundaries, the mean is clearly 1 while all other moments are 0. It is interesting to note that even for calculating the mean multiplicity Eq. (18) offers an advantage over the direct transport equation since a single run gives the mean multiplicity for all positions and velocities of the initiating neutron.

RESULTS AND DISCUSSION

Nuclear Data for Computing MS in Fusion Blanket Materials

For calculating the MS using either Eq. (10) or Eq. (17) a knowledge of all the cross sections including detailed energy and angle dependence of the type implied in say $\Sigma_{ijk}(\vec{\Omega}\to\vec{\Omega}_1,\vec{\Omega}_2)$ is necessary. For practical applications in cross section data adjustment or testing of nuclear models this data should be derived from the model under consideration. The results presented here are only meant to illustrate the methods outlined above and to study the sensitivity of the MS to nuclear data and the point approximation. For this purpose we have used the data and the point approximation. For this purpose we have used the data available in a 35 group data library of Garg and Sinha[10]. Correlations between secondary neutrons in (n, 2n) reactions have been obtained in an approximate way from simple energy conservation requirements. The leakage probability per unit time for the point model is obtained by an Sn calculation using the code ANISN λ_1 is taken as the leakage rate in group i divided by the volume integrated flux. The multiplicity distribution of fission neutrons is not available in the cross section data library. Only the mean number of neutrons per fission is available. Huanquiao et al[11] have shown that this distribution is approximately Gaussian with a standard deviation which varies linearly with the mean. Their formula has been fitted for spontaneous fission. For the purpose of our calculations we assume that the same is applicable to the case of neutron induced fission.

Results of the Point Model

Table I shows the MS computed Eq.(10) for Be-9 spheres for

TABLE I
MS of Leakage Neutrons from Be-9 Spheres with a centrally located 14 Mev source

	5 CM RADIUS TYPE OF CORR.		10 CM RADIUS TYPE OF CORR	
n	(i)	(ii)	(i)	(ii)
0	0.222-1	0.222-1	0.390-1	0.385-1
1	0.660+0	0.660+0	0.407+0	0.408+0
2	0.238+0	0.244+0	0.344+0	0.347+0
3	0.626-1	0.610-1	0.150+0	0.154+0
4	0.133-1	0.107-1	0.504-1	0.440-1
5	0.230-2	0.118-2	0.140-1	0.740-2
6	0.350-3	0.760-4	0.320-2	0.720-3
7	0.460-4	0.259-5	0.650-3	0.370-4
8	0.560-5	0.384-7	0.120-3	0.860-6

TABLE II
MS of Leakage Neutrons form Th-232 Spheres with a Centrally Located 14 Mev Source
P(n)

n	5 cm radius	10 cm radius
0	0.897023-2	0.394516-1
1	0.535857	0.397470+0
2	0.309727+0	0.407322+0
3	0.119996+0	0.128072+0
4	0.169550-1	0.195386-1
5	0.660145-2	0.641469-2
6	0.156752-2	0.141157-2
7	0.267229-3	0.256695-3
8	0.474107-4	0.514157-4
9	0.646302-5	0.747752-5

for correlated and uncorrelated secondary neutrons. The effect of correlation is clearly discernible at high multiplicities. Table II shows the MS for Th-232 spheres. Th-232 being a fissionable material, Eq. (10) was solved numerically at equally spaced z values between 0 and 1. The derivatives at z=0 were obtained using Newton's interpolation formula.

Results Using Stochastic Transport Theory

A program called MOMENT using the DSn method has been written to solve for the equations for the factorial moments in spherical geometry. This was necessary because the code ANISN does not admit an inhomogeneous boundary condition or a distributed angle dependent source. Preliminary results obtained show a slight (about 3%) discrepancy between the mean leakage calculated form ANISN and MOMENT even though the same data library is used. The difference is however larger for the higher moments calculated using the point model described above with leakage probabilities obtained form ANISN. It is not yet clear whether this is an amplification of the differences in the mean or is due to the use of the point model. A definite conclusion is possible only after the difference in the mean (which leads to different leakage probabilities) is resolved.

CONCLUSION

The point model multigroup Kolmogorov equation and the stochastic transport theory methods have been presented as alternatives to the Monte Carlos method for studying the stochastic behaviour of neutron multiplication in fusion blanket materials. The methods can be used for studying the spallation process after some modification. The point model leads to simple expressions for the multiplicity spectrum and its factorial moments. The stochastic transport equation method is expected to give fast and accurate results. It has the added advantage of giving the mean multiplication and the higher moments for all possible positions and velocities of the initiating neutron.

REFERENCES

1. Williams M.M.R., Random Processes in Nuclear Reactors, Pergammon Press (1974)
2. Bondar L., Proc. Int. Mtg. Monitoring Pu Contaminated Wastes, Ispra, Italy 25-28 Sept(1979)
3. Bondar L., Proc. Int. Conf. Nucl. Safeguards, Vienna, IAEA-SM-260/54 p137(1980)
4. Bohnel K. Nucl. Sic Engng. 90, 75(1985)
5. Srinivasan M. (1985), Feasibility of the Application of Statistical Correlation Technique for Experimental Study of Multiplication of 14 MeV Neutrons in Be Assemblies, Ann.Nucl.Energy, 12, no.3, pp125-135
6. Kumar A. and Srinivasan M.(1984), Neutron Multiplicity Equation and its Application for (n, 2n) Multiplication Measurements by Statistical Correlation Technique; Submitted for publication to Nucl.Sci.Eng.
7. Sinha A. Degweker S.B. and Srinivasan M.Neutron multiplication in Be-9; A Stochastic Approach Ann.Nucl. Energy 14, 447(1987)
8. Sinha A. Garg S.B. and Srinivasan M. Sensitivity Studies of the Neutron Multiplicity Spectrum in the Spallation of Lead Targets Ann.Nucl.Energy 13, 579 (1986)
9. Bell G.I. Stochastic Neutron Transport Theory Nucl. Sci. Engng. 21,390 (1965)
10. Garg S.B. and Sinha A., BARC-35 A 35 Group Cross-Section Library With P3 Anisotropic Scattering Matrices and Resonance Self-Shielding Factors, Report BARC-1222(1984)
11. Huanquiao Z. et al, The Average Number of Prompt Neutrons and the Distributions of Prompt Neutron Emission Number for Spontaneous Fission of Plutonium-240 Curium-242 and Curium-244, Nucl.Sci. and Engg.86,315(1984)

STOCHASTIC CALCULATIONS OF THE EFFECTIVE DENSITY FACTOR

M. Enosh, U. Salmi, D. Shalitin, J.J. Wagschal, and Y. Yeivin

Racah Institute of Physics
The Hebrew University of Jerusalem
91904 Jerusalem, Israel

KEYWORDS

delta eigenvalue / effective density factor / Monte Carlo / Monte Carlo schemes / collision scheme / modified collision scheme / production / leakage / transport equation / Jezebel

ABSTRACT

Methods for calculations of the effective density factor, the delta eigenvalue, by Monte Carlo schemes are presented. One procedure is to use a modified collision scheme, and to calculate iteratively delta as the ratio of the net production to the total leakage. It comes out that in each iteration we need to have a previous ,calculated or guessed, approximate value of delta in order to get a better one. Another alternative is to use a collision scheme, and calculate delta as an effective density factor. The delta calculations of some simple problems by both methods are demonstrated, and compared to Sn calculations.

DELTA EIGENVALUE

Several eigenvalue-type representations of the neutron transport equation have been considered in previous works[1-4]. Comparisons of various physical parameters based on the different eigenvalue equations, were usually performed by solving the appropriate discrete ordinate equations. For complex geometries one has to resort to Monte Carlo calculations. In this paper we present some of the problems and solutions, encountered in calculating the effective density factor, delta, by Monte Carlo methods.

The transport equation in the delta-formulatin is

$$\underline{\Omega}\cdot\nabla\Psi_\delta(\underline{r},E,\underline{\Omega}) = \{S[\Psi_\delta(\underline{r},E,\underline{\Omega})] - \Sigma_t(\underline{r},E)\Psi_\delta(\underline{r},E,\underline{\Omega})\}/\delta \qquad (1)$$

where $S(\underline{r},E,\underline{\Omega})$ is the total fission and scattering source.

Note that both terms on the rhs of Eq. (1) depend linearly on the nuclear density ρ, while the lhs of Eq. (1) does not depened on ρ. Thus, delta may be interpreted as the factor by which one has to divide all the nuclear densities in the assembly, in order to get an exactly critical assembly. This representation of delta justifies the name "effective density factor".

Another meaning for delta eigenvalue is secured by integrating Eq. (1) over all the variables: space, energy and direction. On the lhs we get the total leakage from the assembly. On the rhs we get the total source less the total removal. Thus we have another interpretation of delta as the ratio of the net production (source minus removal) to the leakage in the whole assembly, where the flux is the eigenfunction of the transport equation (1).

MONTE CARLO CALCULATION OF DELTA

An obvious suggestion for Monte Carlo calculations of the δ eigenvalue is to use this meaning of delta. The problem is what sampling scheme to use in the delta calculations. Should one try to follow a generation of neutrons up to their first collision as in the calculation of the effective collision multiplication factor gamma defined by:

$$(\underline{\Omega}\cdot\nabla+\Sigma_t)\Psi_\gamma = [S(\Psi_\gamma)]/\gamma \quad (2)$$

or to follow a generation of neutrons up to their first fission as in calculatios of the effective fission multiplication factor k? It is clear that following a generation of neutrons up to their first collision or first fission (gamma or k calculations), one ends up with a flux Ψ which is not adequate for delta calculations, and therefore with wrong results.

In order to solve this problem we re-write eq. (1) as:

$$(\underline{\Omega}\cdot\nabla+\Sigma_t)\Psi_\delta = [S(\Psi_\delta)+(\delta-1)\Sigma_t\Psi_\delta]/\delta \quad (3)$$

Thus, the procedure for the Monte Carlo calculations of delta should be as follows:

Follow a generation of neutrons up to their first collision as in γ calculations. In addition to the source S, δ-1 neutrons emerge from the collision in the *same direction* Ω as that of the incident neutron. An approximate value for δ is calculated as the ratio of the net production to the total leakage. Then follow the next generation until the δ values converge.

This procedure has some obvious drawbacks. First, the calculated value delta is needed in order to calculate the distribution of the neutrons in each generation. We can overcome this difficulty by using delta from the previous calculation in order to calculate next delta. A more serious difficulty is that in the case of subcritical or about critical assemblies, the δ-1 source term yields negative "particles", which considerably complicates the calculations.

Another point to consider is that for assemblies with small leakage, the calculated values of δ may fluctuate violently.

In order to overcome these difficulties, we use the definition of δ as the effective density factor. We follow the n-th generation of neutrons up to their first collision. Then we calculate the ratio:

$$R_n = (\text{net production})/(\text{leakage}) \qquad \text{for the n-th generation} \quad (4)$$

and calculate the instantaneous density ρ_n as:

$$\rho_n = \rho_{n-1}/R_n \quad (5)$$

Table I. Effective density multiplication factor δ calculations by S_n and Monte Carlo methods

No.	Inner zone		Outer zone		delta	
	radius (mean free paths)	c (neutrons per collision)	radius (mean free paths)	c (neutrons per collision)	S_n	Monte Carlo
1.	2	1.5	1	1.0	1.398	1.405±0.001
2.	2	1.5	4	1.0	1.581	1.587±0.007

and the n-th generation eigenvalue as:

$$\delta_n = \rho_o/\rho_n \qquad (6)$$

where ρ_o is the initial physical density. In the (n+1)-th generation we use the most recent density ρ_n, as the density of the assembly.

We can improve the statistics by using the average of ρ's obtained in all previous generations instead of the instantaneous density ρ_n, and by using the average of R_n over all the generations, or by re-defining R_n as:

$$R'_n=\text{(net production)/(leakage)} \quad \text{total of all generations} \qquad (7)$$

NUMERICAL EXAMPLES

In order to check the procedure, we calculate δ of two simple systems. Two spherical assemblies, each consisting of 2 concentric spheres, were calculated in an isotropic one energy group approximation. The description of the assemblies and the results of the calculations are given in Table I.

We see from Table I, that Monte Carlo calculations of the effective density factor delta, yield results very close to those obtained by S_n calculations.

To check the procedure in case of an only slightly super-critical system, a Jezebel like assembly was calculated. It is a spherical assembly, consisting of pure Pu_{239}, its radius is 6.3849cm (1.379 mean free paths), and its density is 15.61gr/cm^3. It was calculated in an isotropic one energy group approximation, with c=1.684 (neutrons per collision). The results of the calculations by different schemes are given in Table II.

DISCUSSION

We see in Table II (and the remark), that for an only slightly super-critical assembly , the Monte Carlo calculation of delta is tricky. The calculation of δ as the ratio of net production to total leakage, with a directed source proportional to δ-1, involves fluctuations which lead to a negative source when δ<1. We can overcome these difficulties by increasing the number of neutrons per generation, or better, by using ratio of net production to total leakage summed over al the generations. But all these methods would not help in the case of negative δ (sub-critical assembly). Only the

Table II. Effective density multiplication factor δ calculations of a Jezebel like assembly by S_n and Monte Carlo methods

method of calculation \ neutrons in a generation	100	1000
Scheme of Eq.(3),ratio of net production to total leakage in each generation.	1.115±0.002[a]	1.051±0.002[a]
Scheme of Eq.(3),ratio of net production to total leakage, summed over all generations.	1.031±0.001	1.035±0.001
Scheme of Eqs. (5)-(7) (density modifications)	$1.0338 \pm 5*10^{-5}$	$1.0349 \pm 9*10^{-5}$
S_n method.	1.0318	

[a]In many generations, delta fluctuations were big enough to get $\delta<1$. In order to avoid calculations with negative "particles", $\delta=0$ was assumed in such cases. This explains the too high value of delta.

method of density modifications can avoid calculation with negative particles.

In conclusion, we have seen some encouraging preliminary results of δ calculations. More effort is needed to find the best way for calculating the δ eigenvalue by Monte Carlo methods.

REFERENCES

1. Y. Ronen, D. Shalitin and J.J. Wagschal , Trans. Am. Nucl. Soc., 24:474 (1976).
2. D. Shalitin , J.J. Wagschal and Y. Yeivin, Trans. Am. Nucl. Soc., 27:845 (1977).
3. G. Velarde, C. Ahnert and J.M. Aragones , Nucl. Sci. Eng., 66:284 (1978).
4. D.G. Cacuci, Y. Ronen, Z. Shayer, J.J. Wagschal and Y. Yeivin, Nucl. Sci. Eng., 81:432 (1982).

STATISTICAL CORRELATION CALCULATION OF THE TIME CONSTANT ALPHA

IN MONTE CARLO SIMULATION OF SUBCRITICAL ASSEMBLIES

U. Salmi, and Y. Yeivin

Racah Institute of Physics
The Hebrew University of Jerusalem
91904 Jerusalem, Israel

KEYWORDS

inverse period/time constant/Rossi Alpha/Monte Carlo/subcritical assembly/auto-correletion/correlation/chain reactions/neutron source/mean life time/effective multiplication factor/probability/

ABSTRACT

The asymptotic inverse reactor period, time constant, alpha, is calculated by a Monte Carlo program, for some subcritical assemblies. For each assembly the neutrons population is sampled successively in short time intervals, and auto-correlation function is computed. Alpha is calculated from the auto-correlation function, and compared with the results of direct calculations.

GENERAL

In this Monte Carlo simulation,we discuss sub-ctitical idealised spherical assemblies. In such an assembly, the behaviour is determined only by the neutron source and the growth or decay of the neutron chain reactions.

We shall use the lumped model (one energy group, independence of space coordinates). We shall discuss only fast assemblies, and therefore the delayed neutrons will be neglected.

Let l be the mean life time of a neutron in the assembly, and let k be the prompt effective multiplication factor, then (neglecting the delayed neutrons), the neutron balance equation is:

$$dN/dt=N(k-1)/l \qquad (1)$$

If we define:

$$\alpha=(1-k)/l \qquad (2)$$

then the neutron balance equation becomes:

$$dN/dt = -\alpha N \quad (3)$$

and its solution is:

$$N = N_o e^{-\alpha t} \quad (4)$$

So α is the time constant of the assembly.

The experimental measurement of α by correlations was suggested by Rossi[1]. The many developments that followed are referred to in some classical works[2-5]. It has been used for example, in the area of sub criticality safety[6,7]. It consists of is counting neutrons by various neutron detectors in different times. Co-incident counts of the detectors may result by pure chance due to neutrons produced at separate fissions in different times, but also by correlated counts of neutrons from the same fission event. If the latter term is greater than the first, it is possible to get α by correlations.

ROSSY-α OF MONTE CARLO SIMULATION

The Monte Carlo simulation is different from the experimental method by some important details. First, the computer program can count any neutron anywere any time, so that the question of detector efficiency does not exist at all. Second, a neutron detector removes from the assembly each neutron it detects, while the program can count a neutron as many times as we wish, without affecting it at all. The last point is that a detector needs some time interval to count, while the program can count neutrons instantaneously

We discuss the probability of the event: ν neutrons emerge out a fission at time t_0, They decay until we count them at time t_1, and they further decay until the second count at time t_2.

If the fission rate in the assembly is F_0 fissions per time unit, then the number of fissions during time interval dt_0 at time t_0 is $N_0 = F_0 dt_0$. From each fission ν neutrons emerge, and they decay exponentially. If we want to discuss independent probabilities, we have to take care of all the ν neutrons simultaneously, because the neutrons born from the same fission are not independant.

The number of neutrons that survived at time t_1 from the ν neutrons that emerged from one fission at time t_0 is:

$$N_1 = \nu \exp[-\alpha(t_1 - t_0)] \quad (5)$$

We can discuss number of neutrons and not the probability of counting neutrons during dt, because the Monte Carlo program does not need time interval dt in order to count neutrons, and it counts every neutron.

Similarly, the number of neutrons that survived at time t_2 from the ν neutrons that emerged from the same fission at time t_0 is:

$$N_2 = \nu \exp[-\alpha(t_2 - t_0)] \quad (6)$$

Therefore, the joint (correlated) probability for neutron count at time t_1 and neutron count at time t_2, is the product of the three quantities N_0, N_1, N_2, summed over all the times t_0 from $t_0 = -\infty$ to $t_0 = t_1$:

$$P_{cor}(t_1,t_2) = \int_{-\infty}^{t_1} F_0 N_1 N_2 dt_0 =$$

$$= \int_{-\infty}^{t_1} F_0 \nu \exp[-\alpha(t_1-t_0)] \nu \exp[-\alpha(t_2-t_0)] dt_0$$

$$= F_0 \overline{\nu}^2 \exp[-\alpha(t_2-t_1)]/2\alpha \qquad (7)$$

There is also uncorrelated probability to count at t_1 a neutron from some fission and at t_2 another neutron from another fission:

$$P_{uncor}(t_1,t_2) = [F_0\overline{\nu}l/(1-k)][F_0\overline{\nu}l/(1-k)] = (F_0\overline{\nu}/\alpha)^2 \qquad (8)$$

Therefore, the total probability to count a neutron at time t and a neutron at time t+T is:

$$P(T) = F_0\overline{\nu}^2(F_0/\alpha + e^{-\alpha T}/2)/\alpha \qquad (9)$$

which is of course proportional to the auto-correlation function of counting n neutrons in the assembly:

$$P(T) \sim C(T) = [\sum_{i=1}^{N} n(t_i)n(t_i+T)]/N \qquad (10)$$

Insofar as the left term in the brackets in Eq. (9) is negligible relatively to the right term, we can get alpha as the slope of the graph of lnP(T) as the function of T. In that case we can get alpha also as the slope of lnC(T) of Eq. (10) as the function of T. In order to have C(T) in Monte Carlo simulation all we need is to count n, the number of neutrons in the assembly, with very short time intervals between the counts, and calculate C(T) by Eq. (10).

From Eq. (9) we see that a necessary condition for the successful determination of α by correlations is fission rate F_0 as small as possible. In Monte Carlo simulation (as in experiment), it amounts to maintain a very weak source for a long period of time, in order to gain enough statistics.

We can check the calculation of α by correlations, through comparison to direct calculations by the Monte Carlo program. From the Monte Carlo program we get the effective multiplication factor k as:

$$k = P/(L+A+F) \qquad (11)$$

where P= neutrons produced by fissions, L= leakage of neutrons out of the assembly, A= number of absorptions (captures), and F= number of fissions.

The mean life time l is calculated as:

$$l = (\sum_{i=1}^{N} l_i)/N \qquad (12)$$

where l_i is the actual life time of the i-th neutron.

The time constant Rossi-α is calculated by Eq. (2).

NUMERICAL EXAMPLES

In order to check numerically the Monte Carlo calculation of α, we calculate α of three sub-critical fast assemblies, both by correlations and by direct counting. All three assemblies consist of Pu_{239} with density of 15.872 gr/cm^3. Assembly No. 1 has radius of 4.0 cm, and mass of 4.255 kg.

Table I. Time constant α of fast sub-critical assemblies calculated by Monte Carlo and by correlations

No	No. of neutrons in source	No. of neutrons born by fission	k	l (nsec)	$\alpha = (1-k)/l$	α by correlations
1	182 460 1969	360 1064 4068	0.672 0.698 0.674	2.311 2.222 2.230	142.1 135.7 141.6	147.3 71.3 38.1
2	93	1780	0.950	3.389	14.65	14.41
3	22 59	792 1813	0.972 0.968	3.801 3.331	7.39 9.46	15.83 18.00

Assembly No. 2 has radius of 5.5 cm, and mass of 11.061 kg. Assembly No. 3 has radius of 6.0 cm, and mass of 14.361 kg.

The three assemblies were simulated by Monte Carlo program for a period of 1 μsec. For each one of them, one or several netron sources of different strength were distributed with equal probability in time. The program counted the number of neutrons that leaked, absorbed, caused fission, and born by fission, in order to calculate the effective multiplication factor k by Eq. (11).

The life time of the neutrons was registered in order to calculate the mean life time l by Eq (12). From k and l, the time constant α was calculated directly by Eq. (2).

The Monte Carlo program also counted n, the neutron population in the assembly, every 0.8 nsec, and stored n(t). After all the source neutrons

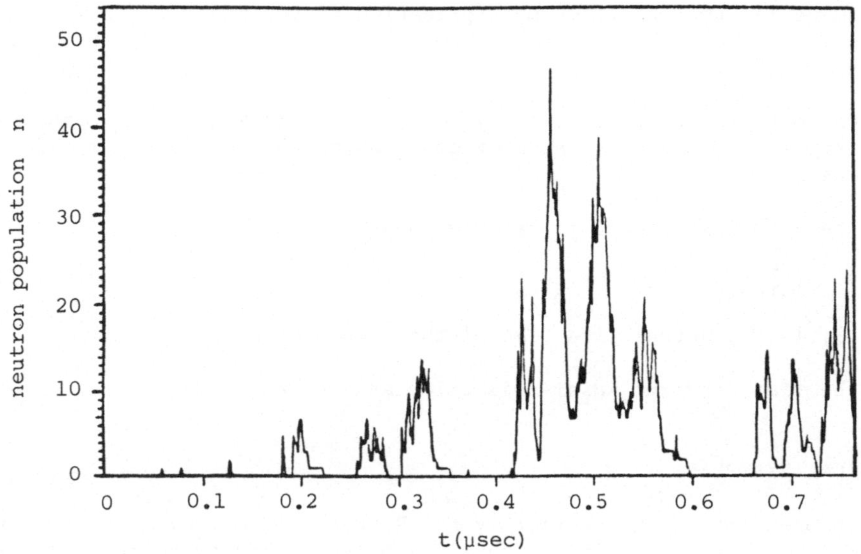

Fig. I. Neutron population of assembly 2 as function of time

and their progenitors were processed, the program calculated the auto correlation function C(T) of n(t), and its natural logarithm. A straight line was fitted by least squares to the graph of the natural logarithm of C(T) as function of T, and its slope was considered as α calculated by correlations.

As an example, the neutron number n as function of time is given for assembly No. 2 in figure I. For the same assembly, lnC(T) as function of correlation time T is given in figure II, together with its slope.

The results of the calculations for the three assemblies by both methods are given in Table I.

DISCUSSION

We see in Table I, that the time constant α can be calculated by the correlation method. As was predicted by the theory, the weaker is the neutron source - the better is the agreement between direct calculation of α and the calculation by correlations. We also see that for assembly No. 3, which is very close to critical, the agreement is quite poor, as it is well known.

These calculations of the time constant α by correlations for Monte Carlo simulated assemblies, demonstrate the possibilities of this method of calculation for more practical use.

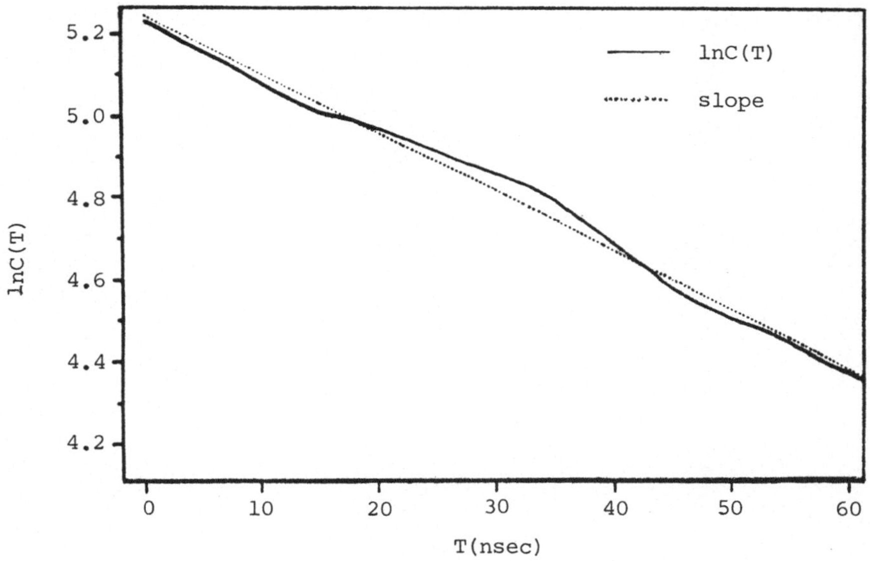

Fig. II. Natural logarithm of the auto-correlation function C(T), for assembly 2

REFERENCES

1. J.D. Orndoff, Nucl. Sci Eng., 2:450 (1957).
2. J.A. Thie, Reactor Noise, Rowman and Littlefields, (1963).
3. N. Pacilio, Reactor Analysis in the Time Domain, Atomic Energy Commission Critical Review Series,TID-24512 (1969).

4. R.E. Uhrig, Random Noise Techniques in Nuclear Reactor Systems, Ronald Press, New York (1970).
5. M.M.R. Williams, Random processes in Nuclear Reactors, Pergamon Press (1974).
6. J.T. Mihalczo, Nucl. Sci. Eng., 53:393 (1974)
7. J.T. Mihalczo, W.T. king, and J.P. Renier, Trans. Am. Nucl. Soc., 41:619 (1982).

CHAPTER IV

STOCHASTIC PROCESSES IN NONLINEAR NUCLEAR SYSTEMS : THEORY

NONLINEAR STOCHASTIC THEORY AND IDENTIFICATION OF NONLINEARITY IN NUCLEAR REACTORS

H. Konno

Institute of Materials Science
University of Tsukuba
Tsukuba, Ibaraki, 305 Japan

ABSTRACT

Presented are a nonlinear stochastic theory and a method for the identification of non-linearity using empirical models in nuclear systems. First I describe a stochastic dynamical power reactor model. Here I proposed are a method (i) of identifying BWR stability with the use of the covariance and the irreversible circulation of fluctuation, and (ii) a method of inferring reactor parameters such as feedback coefficients. Second, the identification theory of 2D random vibration using of a generalized stochastic process and a fluctuation-dissipation theorem is described. Third, the feasibility for the identification of fluid is also demonstrated.

1. INTRODUCTION

The linear aspects of nuclear stochastic phenomena have been widely observed and utilized [1,2,3]. In zero-power reactors (which are characterized by the terms; calm, clean and cold), the neutron and high-energy photon statistics are formulated successfully [3] by the fluctuations belonging to the linear, Markovian stochastic processes.

In at-power reactors which are extremely noisy, dirty and hot, there are diverse noise sources due to the thermal hydraulics and structural effects. The intrinsic nonlinear nature of the at-power reactor noise is derived from the fact that:

(a) the reactivity feedbacks affect parametrically the neutron field in the bi-linear form;

(b) the thermal hydraulics and the motions of structural components are generally subjected to nonlinear equations;

(c) there are many nonlinear transfer elements with limiter and/or dead band in the mechanical and electrical equipment.

At the steady (normal) state, the nonlinear nature does not make its existence evident. But in anomalous situations (when accidents and/or malfunctions take place), the nonlinearity is abruptly actualized.

After the accidents at the Chernobyl and the TMI nuclear power plant, the sequence of events which took place during these accidents have been attentions of many researchers [4].

This paper presents a non-linear stochastic theory and an identification method for the random processes in nuclear systems. The nonlinear modelling of phenomena becomes important not only to better understand the practical reactor noises but also to extract information concerning the recognition of their patterns for anomaly detection.

We explain first the main idea of nonlinear stochastic theory in section 2. Stochastic modelling of nonlinear phenomena in nuclear systems is described in the later sections. In section 3, I show the stochastic dynamical power reactor model. Section 4 shows a 2D nonlinear random vibration of a reactor's internals. Section 5 sketches briefly the analysis of noises in one-phase flows of coolant. The final section is devoted to summary and future problems.

2. NONLINEAR STOCHASTIC THEORY

If we can perform the first principle calculation of neutron transport, the fluid dynamical and heat conduction equation in 3D space with the boundary condition, stochastic theory is not needed. However, the calculation is difficult from a practical point of view. Therefore, the need for the stochastic theory arises.

Spatio-temporal "coarse-graining" is the central idea in non-equilibrium statistical mechanics. The methods to demonstrate this idea up to now are
(a) the contraction of state variables (Mori et al.[5], Haken [6]);
(b) system size expansion (Van Kampen [7], Kubo et al.[8]);
and (c) the reductive perturbation method (Taniuchi & Washimi [9]).

With the use of the above methods, the dynamical equations of practical nuclear systems can be simplified greatly. In the reduced equations, the effects of rapidly varying state variables upon the global state variables are taken into account as external noises.

There are (i) space-independent and (ii) space-dependent models. The space-independent models are inevitably empirical and qualitative. To better incorporate the qualitative features, we must improve the models by accounting for space-dependence. A review of the nonlinear nuclear stochastic theory was presented by Saito (1983) [10]. The nonlinear aspects of nuclear reactor kinetics have also been described by Ash (1979) [11].

Nonlinear Stochastic Equation

Consider the global behaviour of the state variables X in nuclear systems. By virtue of coarse-graining, the global dynamics of X is described by the nonlinear stochastic differential equation (SDE) with Gaussian-white external noises. When one needs to consider a non-white, non-Gaussian process for the external noise, the introduction of new state variables may lead to an equation with Gaussian white noise.

Note here the fact that:
According to measurements at the steady state in BWRs, almost all kinds of reactor noises are subjected to the logarithmic-normal probability distribution function [12].

From the physical point of view this means that:
(i) reactor noises are composed of the state variables with infinite degrees of freedom; (ii) there exists a class of the energy cascading process and the mixing process among an infinite number of modes similar to turbulence. These facts guarantee an empirical analysis of reactor noises with the use of space-independent stochastic models.

Identification Theory of Nonlinearity

The nonlinear stochastic systems give rise to: (1) Non-Gaussianity in the probability density function (pdf); (2) Non-Markovianity (large intensity at low frequencies) in the power spectral density (PSD) [Long-time-tail in the correlation function]; (3) Pulsation and/or bursting; (4) Anomalously large fluctuation; (5) Successive bifurcations and/or phase transitions. These features are ubiquitous in non-equilibrium open systems.

The detection and quantification of the above properties are the basic goal of identification theory. In the anomalous situations, the nonlinear properties are enhanced and observed as noble spatio-temporal phenomena. Non-Gaussianity and non-Markovianity can be quantified by the standard signal processing method with the use of the pdf and the PSD. However, there are many microscopic processes which give a similar profile of the pdf and the PSD. The next problem is how to quantify (or identify) the differences of the microscopic processes with various nonlinearities. We will show briefly the three identification methods for system parameters using (i) the stochastic process of the jump event in a time-domain, (ii) a generalized fluctuation-dissipation theorem and (iii) the higher order correlation functions.

(a) Stochastic process of the jump event

We frequently come across the jump phenomena in nonlinear systems. Some typical examples of jump events in nuclear systems are: (i) hump events when an anomalous core barrel motion takes place [13]; (ii) pulsation of pressure in two-phase flow [14]; (iii) random bursts of pressure fluctuation in a turbulent boundary layer [15] and acoustic emission from bubbles [16] in two-phase flow.

The stochastic processes of jump events in many cases can be identified by the generalized Polya process [17,18] which represents various non-stationary processes. The non-linear processes generally have non-Makovian properties. This fact explains why one can identify a stationary non-Markovian process such as the hump phenomenon [17] with the nonstationary process by counting the pulse number distribution during a short duration of time T.

The existence of the long-time-tail

$$< X(t) > \propto t^{-1/2} \quad (t \to \infty),$$

has been proven by Suzuki [19] for the parametric SDE. This fact indicates the nonstationary aspects of the non-Markovian process.

(b) A generalized fluctuation-dissipation theorem (GFDT)

Consider a nonlinear diffusion process

$$P_t(x,t) = - [A(x)P(x,t)]_x + [B(x)P(x,t)]_{xx}. \qquad (2.1)$$

Okabe [20] has proposed a GFDT for eq.(2.1). The GFDT relates the generalized diffusion constant D_F, the variance R(0) and the generalized friction coefficient $C_{\beta,\gamma}$ in the following form:

$$D_F = R(0)\, C_{\beta,\gamma} \quad . \qquad (2.2a)$$

The quantities D_F, R(0) and $C_{\beta,\gamma}$ are expressed in terms of the pdf $P_s(x)$, as

$$D_F = \int xA(x)P_s(x)dx \Big/ \int P_s(x)dx,$$

$$R(0) = \int x^2 P_s(x)dx \Big/ \int P_s(x)dx \qquad (2.2b)$$

and

$$C_{\beta,\gamma} = \int xA(x)P_s(x)dx \Big/ \int x^2 P_s(x)\, dx.$$

Since $C_{\beta,\gamma}$ is constructed to satisfy automatically the relation (2.2), one needs another expression of $C_{\beta,\gamma}$ which does not include the pdf for the use of the system identification. There is a relation between $C_{\beta,\gamma}$ and the coefficients α,β, and γ of the corresponding linear non-Markovian Langevin equation to the Fokker-Planck equation (2.1);

$$\frac{d}{dt} x = -\beta x - \int_{-\infty}^{0} \gamma(t+s)\, x(s) ds + \alpha dW(t) \quad (2.3)$$

Specifically, the generalized friction $C_{\beta,\gamma}$ is expressed in terms of α,β, and the memory function $\gamma(t)$ as

$$C_{\beta,\gamma} = \pi \int | \beta + \gamma(\omega) - i\omega |^{-2} d\omega . \quad (2.4)$$

If α, β and $\gamma(\omega)$ are estimated from the correlation function R(t) or the PSD (Ref.[20] and [21]), the FDT can be used to infer the system parameters A(x) and B(x). The extension of the GFDT to a class of nonlinear systems is examined [21].

(c) Higher Order Correlation Functions

There are a number of stochastic processes which give rise to the same profile of the pdf and/or the PSD. To distinguish the differences among them, one may utilize the higher order correlation functions.

For example, the Fourier transform of the third order correlation function is the bispectrum $B(\omega_1,\omega_2)$. Note that the third order moment

$$\mu_3 \propto \iint B(\omega_1,\omega_2)\, d\omega_1\, d\omega_2.$$

The quantity is, therefore, sensitive to the growth of non-linear terms which break the symmetry of the effective nonlinear potential of relevant systems. The $B(\omega_1,\omega_2)$ also represents the interactions among the nonlinear modes.

An availability of the bispectrum for anomaly detection has been demonstrated [18] for several types of bursting phenomena. This quantity is used to demonstrate that a relevant fluctuation observed possess a nonlinear nature. The interpretation of the bispectrum from physical view point becomes complex when the interaction among a large number of modes takes place. Improvement of analytical model becomes important in this case.

3. NONLINEAR STOCHASTIC REACTOR DYNAMICS

3.1 Stochastic Dynamical Power Reactor Model

Consider the stochastic dynamical power reactor model:

$$\frac{d}{dt} N = \frac{\rho-\beta}{\ell} N + \lambda C + F_N(t),$$

$$\frac{d}{dt} C = \frac{\beta}{\ell} N - \lambda C + F_C(t),$$

$$\frac{d}{dt} T = \frac{1}{c_R} (qN - hT) + F_T(t) \qquad (3.1)$$

$$\frac{d^2}{dt^2} X_\alpha + \zeta(X_\alpha) \frac{d}{dt} X_\alpha + \omega_0^2 X_\alpha = kT + p(N-N_s) + F_\alpha(t),$$

and $$\rho(t) = \gamma_\alpha X_\alpha(t) - \gamma_T T(t), \qquad (3.2)$$

where N = neutron density,
C = precursor density,
T = reactor temperature
and X_α = reactivity feedback from the void and/or automatic control of the control rod system.

The external noises are assumed to be

$$\langle F_i(t)\rangle = 0 \text{ and } \langle F_i(t)\, F_j(t')\rangle = D_{ij}\, \delta(t-t').$$

The model (3.1) has a non-linear term in the field X_α;

$$\zeta(X) = (g_0 + g_1 X^2 + \Delta(t)),$$

where $\Delta(t)$= a parametric pink noise. The analysis neglecting the temperature and the power feedback (viz., k=0 and p=0) has been reported [22] by the present author. When one neglects the external noises, the power feedback (viz., $F_N = F_C = F_T = F_\alpha = 0$ and p=0) and the non- linearity in $\zeta(X_\alpha)$ and the parametric noise $\Delta(t)$ (viz., $\zeta(X_\alpha) = g_0$), the model reduces to the nonlinear BWR model which has been proposed by J.M. Leuba et al.[23].

The steady-state values are expressed in terms of T_s:

$$N_s=(h/q)T_s,\ T_s=(\rho_0+X_{\alpha s})/\gamma_T,\ C_s=(\beta/\ell\lambda)(h/q)T_s \text{ and } X_{\alpha s}=(k/\omega_0^2)T_s. \qquad (3.3)$$

When N_s, T_s and ρ_0 are given, the other reactor parameters such as the feedback coefficient γ_T are inferred.

The transformation of the variables with

$$n= (N-N_s)/N_s,\ c=(C-C_s)/N_s,\ \theta=(T-T_s)/N_s,\ \rho_\alpha=(X_\alpha-X_{\alpha s})/N_s, \qquad (3.4)$$

leads to the dynamical power reactor model around the mean values:

$$\dot{n} = (\rho-\beta)n/\ell + \rho/\ell + \lambda c + f_n(t),$$

$$\dot{c} = (\beta/\ell)\, n - \lambda c + f_c(t),$$

$$\dot{\theta} = (qn - h\,\theta)/c_R + f_\theta(t), \qquad (3.5a)$$

$$\ddot{\rho}_\alpha + \zeta(\rho_\alpha,\Delta(t))\,\dot{\rho}_\alpha + \omega_0^2\,\rho_\alpha = k\theta + pn + f_\alpha(t),$$

where $\rho = \Gamma_\alpha \rho_\alpha - \Gamma_T\theta$, $\Gamma_\alpha = N_s\gamma_\alpha$ and $\Gamma_T = N_s\,\gamma_T$. (3.5b)

Equation (3.5b) shows that the magnitudes of the void and the temperature feedback become large as the mean power N_s increases. The introduction of non-linearity $\zeta(X_\alpha)$ and the parametric noise $\Delta(t)$ in the field X_α is important to explain the violent intermittent behaviour of power oscillations which were observed in the BORAX-II [22] (cf.[1]).

3.2 Linear Stability and the irreversible circulation of fluctuation

The fluctuation $\delta Y = \mathrm{col}(\delta N, \delta C, \delta T, \delta X_\alpha, \delta X_\alpha)$ around the mean values (3.4) become

$$\delta\dot{Y} = L\,\delta Y + R(t), \qquad (3.6)$$

and R(t) is the random force in the linearized Langevin equation. When R=F, the PSD matrix $P(\omega)$ can be calculated with use of the Green's function $G[s]=(s+L)^{-1}$ for eq.(3.6a) as $P(\omega) = G[i\omega]\; D\; G[-i\omega]$.

The characteristic equation is given by $\det(G[s]^{-1}) = 0$. For brevity, I shall neglect the precursor from now on. The characteristic equation reduces to

$$s^4 + C_1 s^3 + C_2 s^2 + C_3 s + C_4 = 0, \qquad (3.7a)$$

where

$$C_1 = \mathrm{Tr}(L'), \quad C_2 = (C_1^{\,2} - \mathrm{Tr}(L'^2))/2,$$

$$C_3 = \mathrm{Tr}(L'^{-1})\mathrm{Det}(L') \text{ and } C_4 = \mathrm{Det}(L'), \qquad (3.7b)$$

and L' is the residual matrix of L after neglecting the precursor. Routh-Hurwitz stability criterion [6] leads to the inequalities:

$$C_1 > 0,\quad C_2 > 0,\quad C_3 > 0,\ C_4 > 0,\ C_1C_2C_3 > C_3^{\,2} + C_1^{\,2}C_4 \quad . \qquad (3.8)$$

$C_4=0$ corresponds to the onset point of the soft mode instability. On the other hand, $C_1C_2C_3 = C_3^{\,2} + C_1^{\,2}C_4$ corresponds to that of the hard mode instability (the limit cycle oscillation). A classification of the reactor state for p=0 and $\zeta(X)=g_0$ is examined in the next subsection.

The covariance and the irreversible circulation of fluctuations [24,25,26] in the steady state, viz.

$$\sigma_{ij} = (1/2\pi) \int \mathrm{Re.}(P_{ij}(\omega))\, d\omega,$$

and

$$\alpha_{ij} = (1/2\pi) \int \omega\ \ \mathrm{Im.}(P_{ij}(\omega))\, d\omega, \qquad (3.9)$$

are, therefore, given exactly after the analytical integration by

$$\sigma_{ij} = (L_1^{(ij)}C_2C_3 + L_2^{(ij)}\,C_3 + L_3^{(ij)} + L_4^{(ij)}/C_4)/2J_4,$$

$$\alpha_{ij} = (K_1^{(ij)}C_2C_3 + K_2^{(ij)}C_3 + K_3^{(ij)})/2J_4, \qquad (3.10)$$

where $J_4=C_1C_2C_3-C_3^2-C_1^2C_4$, $K_\ell^{(ij)}$ and $L_\ell^{(ij)}$ (ℓ=1-4) are the coefficients which are written in terms of the matrix elements of L'. When the system approaches the unstable thresholds, viz. $C_4 \to 0$ and/or $J_4 \to 0$, the values of the covariance σ and the irreversible circulation α becomes large. In the stable region (3.10), the expressions (3.13) σ and α provide analytical measures of the topological distance in the phase space from the unstable region.

3.3 Discussion

(a) BWR stability

Now let us examine a classification of the stability of the BWR model by adopting the parameters ($q/c_R=a_1$, $h/c_R=a_2$, $\zeta(X)=g_0=a_3$, $\omega_0^2=a_4$ and p=0) which have been used by J.M. Leuba et al. [23]. I obtain:

$$C_1=a_2+a_3 \ (>0), \quad C_2=a_2a_3+(\gamma_T N_s/\ell)a_1-a_4,$$

$$C_3=a_1a_3(\gamma_T N_s/\ell)-a_2a_4 \text{ and } C_4=a_1(N_s/\ell)(\gamma_\alpha k-\gamma_T a_4).$$

The conditions of the hard ($C_1C_2C_3 < C_3^2+C_1^2C_4$) and the soft mode instability ($C_4 <0$) lead to inequalities in the following forms:

$$A_1Z_T^2+A_2Z_T+A_3 < kZ_\alpha, \quad \text{(Hard Mode)}$$

$$kZ_\alpha < a_4 Z_T, \quad \text{(Soft Mode)}$$

$$\text{where } Z_T=\gamma_T N_s/\ell, \; Z_\alpha=\gamma_\alpha N_s/\ell,$$

$$A_1=a_1a_2a_3/(a_2+a_3)^2,$$

$$A_2=a_2a_3(a_2a_3+a_3^2+2a_4)/(a_2+a_3)^2,$$

$$A_3=-a_2a_3a_4(a_2^2+a_2a_3-a_4)/a_1/(a_2+a_3)^2.$$

The instability regions are classified in the phase space (Z_T,kZ_α) in Fig.1. There exists a cross-over region of instabilities wherein the hard and soft mode instabilities take place simultaneously.

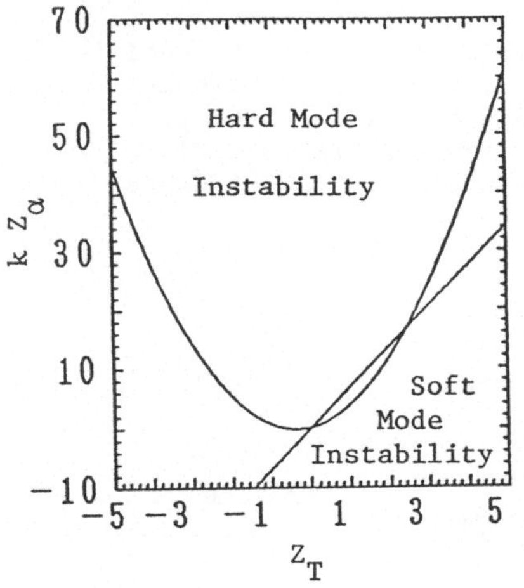

Figure 1. The classification of reactor state in parameter space (kZ_α,Z_T) based on the Routh-Hurwitz criterion. Here I adopted the BWR parameter as used in Leuba et al. [23] (a_1=25.04, a_2=0.23, a_3=2.25 and a_4=6.82; The other parameters also are taken as the same values).

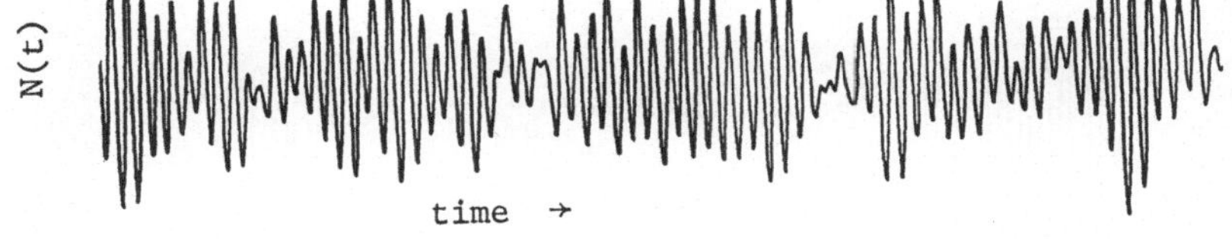

Figure 2. An intermittent power oscillation obtained from the stochastic power reactor model near the point of onset of instability.

(b) Intermittency of the power oscillations

An intermittent oscillation near the unstable point is also shown in Figure 2. The effect of the feedback term kT in the field X_α for the small values of k does not create a significant difference on the temporal development of the intermittent property of the neutrons.

Figure 3 shows the typical power oscillation which is obtained [22] from the power reactor model (3.1) which accounts for the non-linear term and the parametric noise in the field X_α. According to the model, the strong intermittency is ascribed mainly to the slow and small variation of the parametric noise $\Delta(t)$ in the field X_α.

Power oscillations have been observed [27] in the NSRR, Japan (JAERI) and in the FORSMAK1 & 2, Sweden (STUDSVIK) [28,29] (cf. BORAX-II [1]). The oscillations have transient and intermittent properties in any case as shown with a schematic illustration in Fig.4.

The NSRR is a water pooled reactor wherein the reactor core is cooled by a natural convection mechanism. In high power operations (300kW), power oscillations sometime arise. The amplitude of oscillation is relatively small (±2%). The origin of the oscillation was considered [27] due to the reactivity effect from the automatic control system which has the transfer function of the damped harmonic oscillator. The types of oscillation were classified into the three ((i) sawtooth, (ii) rectangular and (iii) sinusoidal) types.

In the FORSMAK1 & 2 [28,29], a BWR, the amplitude of the oscillations are rather large (±16%). According to the reports, the analysis with the use of the multiple-component ARMA model [29] showed that there is a strong spatial coherence in the APRM and LPRM signals although the flow rates in the different channels are out of phase.

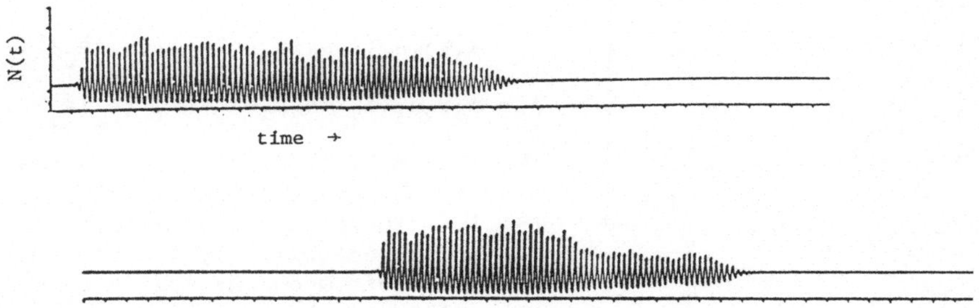

Figure 3. An intermittent violent power oscillations obtained from the the stochastic dynamical power reactor model (3.1) (cf.[22]).

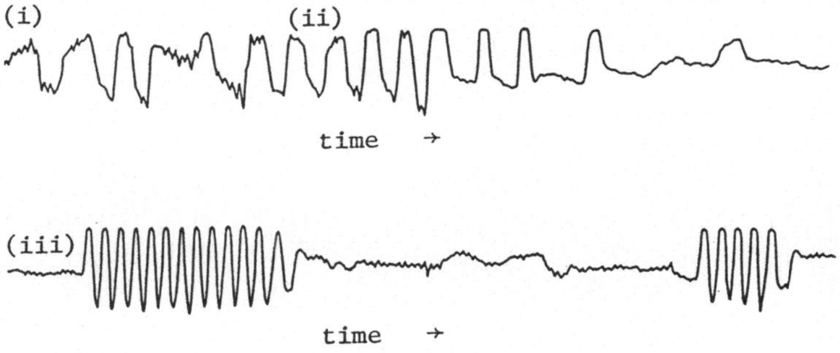

Figure 4. A schematic illustration of power oscillations which have been observed in the NSRR (JAERI) by Hayashi et al.[27].

(c) the single component model and the FDT When the effect of the void-reactivity feedback is small, the power reactor model may be approximated by (cf. Konno [30,31])

$$\dot{N} = (\rho_0 + \rho(t) - \gamma T)/ \ell + F_N(t)$$

and

$$\dot{T} = (q/C_R)N - (h/C_R)T + F_T(t). \qquad (3.14)$$

When the heat transfer from the reactor is relatively fast, the reactor temperature can be eliminated with the use of the contraction method [5] in the following form [30]:

$$\dot{N} = (\alpha + \rho(t))N - \beta N^2 - \gamma N^3 + f_N(t) . \qquad (3.15)$$

By assuming $<\rho(t)\rho(0)>=D_{\rho\rho}\delta(t)$, $<\rho(t)f_N(t')>=0$ and $<f_N(t)f_N(0)>=D_{NN}$ $\delta(t)$, the probability density function at the steady state $P_s(N)$ [30] becomes

$$P_s(N)= C_f(D_{\rho\rho}N^2 + D_{NN})^{u_0-1} \exp(- u_1 N - u_2 N^2 + u_3 \tan^{-1} u_4 N), \qquad (3.16)$$

where $u_0 = (\alpha D_{\rho\rho} + \gamma D_{NN})/2D_{\rho\rho}$,

$u_1= \beta/D_{\rho\rho}$, $u_2=\gamma/2D_{\rho\rho}$,

$u_3= \beta\, D_{NN}/D_{\rho\rho}{}^{3/2}$ and $u_4 = (D_{\rho\rho}/D_{NN})^{1/2}$.

Note here that (i) $P_s(N)$ is not the Gaussian distribution, (ii) the GFDT (2.8) holds for the reduced model (3.15).

The model (3.15) reduces to the various single component models which are exactly solvable:

$$\dot{n} = an - fn^2 + f(t) \quad \text{(cf. Suzuki [32],1987)}$$

and $$\dot{n} = (a_0 + \rho(t))n - b_0 n^2 \quad \text{(cf. Karmeshu [33], 1981).}$$

The stochastic equation which has a new parametric noise b(t) in the coefficient of the 2nd order nonlinear term, viz.

$$\dot{n} = (a_0 + \rho(t))n - (b_0 + b(t))n^2,$$

has been also analyzed by Williams [34] and Sako [35].

Although the above models are simple ones, they reveal the following qualitative characteristics of nonlinear nuclear dynamics:

(a) anomalous transient fluctuation may take place when the state goes from an initial unstable state to the final stable state (cf. [22,30,31] and references cited therein);
(b) due to the existence of feedback loops, oscillations with multiple-periodicity arise (cf.[30]);
(c) the interaction of nonlinear term with the external noise gives rise to higher order nonlinear terms [30];
(d) the existence of correlations among the noise sources works to suppress the fluctuations [30,31];
(e) the intermittency of the void fluctuation and the flow rate may lead to the intermittent behaviour of power oscillation [22].

4. NONLINEAR RANDOM VIBRATION OF REACTOR'S INTERNALS

4.1 2D Nonlinear Random Vibration

Let us now consider the nonlinear vibration of the reactor's internals. The 3D rigid-body beam element which accounts for the interaction with fluid flow [36] leads to the following nonlinear lumped model with the effective potential V(X,Y) [21,34]:

$$\ddot{X} + k\dot{X} + \frac{\partial}{\partial X}V(X,Y) = F_X(t)$$

and

$$\ddot{Y} + k\dot{Y} + \frac{\partial}{\partial Y}V(X,Y) = F_Y(t), \qquad (4.1)$$

where $V(X,Y) = A_1(X+Y) + A_2(X^2+Y^2)/2 + A_4(X^4+Y^4+2X^2Y^2)/4.$ (4.2)

The fluctuating forces F_X and F_Y due to the turbulent pressure is assumed to be Gaussian white with null mean;

$$\langle F_i(t)\rangle=0 \text{ and } \langle F_i(t)F_j(0)\rangle = D\,\delta(t)\,\delta_{ij}\,.$$

In the normal situation, the term of linear bias $A_1(X+Y)$ takes null value, and $A_2 > 0$ and $A_4 > 0$. On the other hand, in an anomalous situation such as a case in which the cramping is lost, the coefficients become: $A_1 \neq 0$ and $A_2 < 0$. The physical situation with a non-zero A_1 may be due to the existence of a limiter (cf. [38,39]). A negative A_2 comes from the interaction of fluid motion and the loss of cramping (cf. Konno and Tanaka [36], 1988). When the vibration becomes one-dimensional, the effective potential is reduced to the double-well type [17,37].

Figure 5 shows the numerically simulated sequences. The contour map of the 2D probability density function is shown in Fig. 6(A). The corresponding 1D probability density function

$$P_s(X) = \int P_s(x,y)\,dy$$

is shown in Fig.6(B).

The simulated PSD does not have strong intensity at low frequencies. On the other hand, the observed PSDs [39] of the neutron signals show a growth intensity at low frequencies. Further, there exists no coherence below 1 Hz between neutron signals. This suggests that the low-frequency components of the PSD in the WWER-440 type reactor may come across during the process of converting information into neutron signals.

4.2 Generalized Fluctuation-Dissipation Theorem

The Fokker-Planck equation for the Langevin equation (4.1) with Gaussian white noise (4.2) gives the pdf [21]:

$$P_s(X,Y) = N_0 \exp(-\,2kV(X,Y)/D), \qquad (4.3)$$

where N_0=the normalization constant. The generalized FDT (2.8) is available to identify the system parameters.

For 1D motion, the study of the stochastic process becomes a powerful means of determining the system parameters [17]. The FDT is a supplementary means for identification in the 1D case. For 2D motion, however, the non-Markovianity becomes stronger than that of the 1D motion. The study of the stochastic process of the hump event and the use of the generalized FDT provides a feasible identification method to determine V(X,Y), k and D.

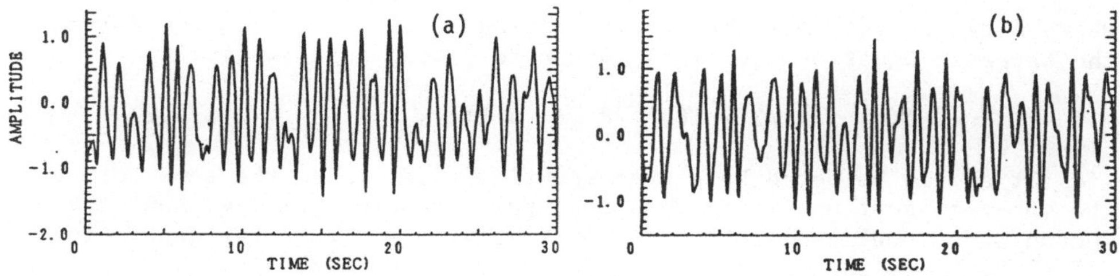

Figure 5. The numerically simulated sequences x (= Xcosθ - Ysinθ) of 2D nonlinear random vibration (the parameters: k=1,D=20, A_1=10, A_2=-50 and A_4=100): (a) θ=0 and (b) θ=-π/4.

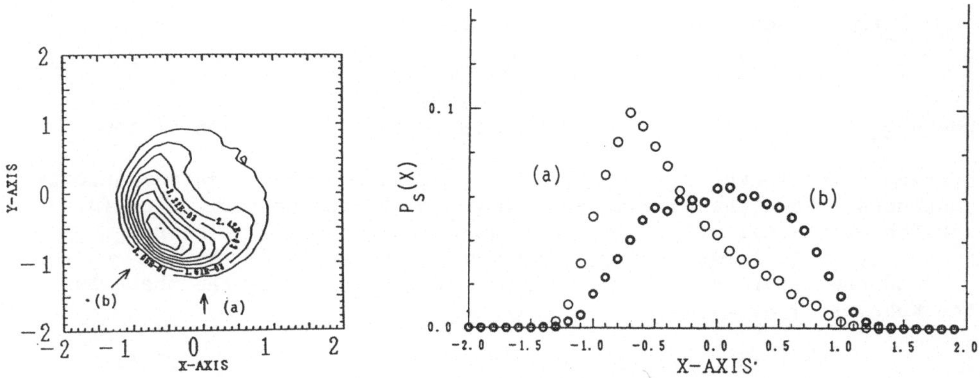

Figure 6. (A) The contour map of the 2D probability density function P_s(X,Y); (B) The integrated 1D probability density function P_s(X) in the two different directions: (a) θ=0 and (b) θ=-π/4.

5. TURBULENT SEQUENCE IN ONE-PHASE FLOW

5.1 Weakly Turbulent Flow

In at-power reactors, there are many noise signals which are induced and/or governed by fluid motion. We frequently encounters noble phenomena which cannot be understood by linear nature. I have solved partially the one-phase flow problem in a nuclear reactor: The bursting phenomena has been observed in the pressure fluctuation signals [15,40,41]; (b) There is a similarity between the time derivative of the velocity fluctuation in the turbulent boundary layer and the pressure fluctuation [42,43]; (c) Many peaks can be seen in the power spectral density [15,40,41]; (d) The origin of the multiplicity of the peaks is not explained; (e) Although a theory of intermittency [44], which is based on the piece wise linear map, has attempted to explain the multiplicity of the peaks in the PSD, the pdf in general takes unphysical distributions; (f) To construct the stochastic model which explains consistently both the PSD and the pdf profiles.

5.2 A Generalized Narrow Band Process

To construct an empirical model, let us consider the generalized narrow band process in the following form:

$$X(t)=A(t)\ \sin(\ \omega_0 t + \phi(t)\ ), \tag{5.1}$$

where $A(t)$ and $\phi(t)$ are the slowly varying functions. When we assume the Langevin equations for $A(t)$ in the following form [45]:

$$\dot{A}(t) = \alpha A/2 - \beta A^3/8 + A\,\xi(t), \qquad (5.2)$$

It should be noted here that equation (5.2) takes the same form as the one-component nonlinear dynamical power reactor model without the second order nonlinear term. One can easily see that the generalized FDT (2.8) holds for the equation of the amplitude $A(t)$.

5.2 PSD and Flatness

Let us study first the PSD based on the model (5.2). The exact solution of the amplitude $A(t)$ is given by the confluent hypergeometric function [46]. The PSD for the amplitude is expressed as

$$P_A(\omega) = \int dt \exp(-i\omega t) < A(t)\, A(0) >,$$

$$= \sum_n g_n^2 / (\omega^2 + \lambda_n^2) + \int g^2(\lambda)/(\omega^2 + \lambda^2)\, d\lambda\ , \qquad (5.3)$$

where $g_n = \int xP_n(x)\, dx$ and $P_n(x)$ is the n-th eigenfunction of the corresponding Fokker-Planck eq. Since the eigenvalue spectrum becomes continuous above the threshold as shown in the second term in eq.(5.3), the long-time-tail arises in the correlation function. The PSD intensity at low frequencies, therefore, increases.

On the other hand, if we approximate the motion of the phase by the Kubo oscillator [47];

$$\dot{\phi} = i(\omega_1 + \xi(t))\, \phi, \qquad (5.4)$$

the exact solution can be also obtained. The PSD for the phase becomes

$$P_\phi(\omega) = \sum_n nD / [\ (\omega - n\omega_1 - \omega_0)^2 + (nD/2)^2\ ]. \qquad (5.5)$$

The quasi-periodic oscillation of the phase provides equally spaced Lorentzian lines.

Consider next the effect of amplitude modulation upon the the growth of the flatness (kurtosis) of the sequence. To simplify discussion, we will consider the case of simple phase diffusion. In this case, the flatness of the sequence X becomes

$$F = F(A)\ F(\phi) = (1 + D/2\alpha)\ (3/2). \qquad (5.6)$$

It is easily seen that the growth of the flatness is determined only by the ratio (D/α).

To demonstrate the nonlinear nature of the sequence in detail, one must analyze higher order correlation functions such as the bispectrum. An qualitative analysis of the pressure fluctuations which was measured at the Borssele nuclear reactor has been analyzed along this line [45]. For a detailed analysis, the purified turbulent data without contamination is necessary.

5. SUMMARY AND FUTURE PROBLEMS

5.1 Summary and Remarks

We have shown (1) the stochastic modelling of the nonlinear phenomena in nuclear systems and (2) identification theories.

In regard to the dynamical power reactor model, we have shown (a) an identification method of reactor parameters such as the feedback coefficients and (b) the concept of the irreversible circulation as a measure of stability. The concept of the irreversible circulation has been introduced [24] to explain the limit cycle behaviour in non-equilibrium open systems. Intensive studies with the use of this concept may become useful to measure and understand power oscillations.

Concerning the nonlinear random vibrations, we have shown that the 3D nature of a structural component and the interaction of fluids are important to explain the anomalous motions of the core barrel in the Palisades and the WWER-440 reactors. The study of the stochastic process of humps and the fluctuation-dissipation relation can be available to have detailed information for identification.

On the one and the two-phase flow problems, we have shown the necessity for the construction of empirical models and precise analyses with the use of higher order correlation functions.

5.2 Future Problems

(a) Power Reactor Dynamics

An analysis of space-dependent stochastic models is not presented in this paper. Studying the stochastic two-group theory with the two-region model of heat transfer with the fuel and coolant temperatures fluctuations may be the next problem. In this case, one must adopt a reliable model to account for the void-reactivity feedback with various flow types (bubbly, slug or froth).

This problem may be related to the transient nonlinear wave propagation which is described by the generalized stochastic KdV-Burgers type equation (cf. [48,49]). This type of equation can be derived by applying the reductive perturbation method [9] to nonlinear dynamical coupled equations for plasma and two-phase flow systems. This model exhibits soliton-like wave propagation and chaotic behaviours depending on the parameters and the initial conditions [49].

The spatio-temporal behaviours are essentially that of an infinite degrees of freedom. The model can not be reduced to a set of nonlinear equations with a few degrees of freedom.

If one tries to use this type of equation to analyze BWR stability problems which are related to the occurrence of violent power oscillation, one of the main difficulties is: how to take into account the correlation among the different coolant-flow channels in a reactor.

(b) Anomalous Motion of Tube

In section 4, I have considered the non-linear random vibration model in a situation in which the deformation of the rigid body is not so large. However, in the case of the flow-induced vibration of instrument tubes, the situation is as simple as that described in this paper. The governing equation becomes the 3D nonlinear stochastic integro-differential equation.

ACKNOWLEDGEMENT

This study was partially supported by the Japanese Ministry of Education, Science and Culture (No. 62780251).

REFERENCES

[1] M.M.R. Williams, "Random Processes in Nuclear Reactors", Pergamon, Oxford (1974).
[2] J. Thie, "Power Reactor Noise", American Nuclear Society, La Grange Park (1981).

[3] K. Saito, Ann. Nucl. Energy 1: p31-48,107-128,209-221 (1974).
[4] USSR State Committee on the Utilization of Atomic Energy; The Accident at Chernobyl Nuclear Power Plant and Its Consequences (1986).
[5] H. Mori, T. Morita and K. T. Mashiyama, Progr.Theor.Phys. 63 ,1865 and 64: 500 (1980).
[6] H. Haken,"Synergetics", Springer, Berlin (1977).
[7] Van Kampen, Can J. Phys. 39: 551 (1961).
[8] R. Kubo, K. Kitahara and K. Matsuo, J. Stat. Phys. 9: 51 (1973).
[9] T. Taniuch and H. Washimi, Phys. Rev. Lett. 21: 209 (1968).
[10] K. Saito, "Non-linear Nuclear Stochastic Theory" in: Advances in Nuclear Science and Technology, J. Lewins and M. Becker ed., Plenum Press, New York (1983).
[11] M. Ash, "Nuclear Reactor Kinetics", 2nd Ed., McGraw-Hill, New York (1979).
[12] e.g., K. Fukunishi, Nucl. Sci. Engng 67: 298 (1978); Y. Andoh et al., J. Nucl. Sci. Technol. 20: 769 (1983).
[13] D.N. Fry, R.C. Kryter and J.C. Robinson, ORNL-TM-4570 (1974) .
[14] T. Fukano and S. Osaka, J. Atomic. Energy Society of Japan 27: 1147 (1985) (in Japanese).
[15] E. Turkan (1984) Summary of SMORN-IV Benchmark Test Problem.
[16] Dentico G., Pacilio V. Paparia B. and Tosi V., Progr. Nucl. Energy 9: 255 (1982).
[17] H. Konno, Ann. Nucl. Energy 13: 185 (1986).
[18] H. Konno, Progr. Nucl. Energy 21 (1988).
[19] M. Suzuki, Progr. Theor. Phys. 67: 1756 (1982); 68: 98 (1982).
[20] Y. Okabe, Commn. Math. Phys. 98: 449 (1985).
[21] H. Konno, J. Phys. Soc. Japn. 57: 777 (1988).
[22] H. Konno, Progr. Nucl. Energy 15: 209 (1985).
[23] J. M. Leuba et al., Nucl. Sci. Engng. 93: 111 and 124 (1986).
[24] K. Tomita and H. Tomita, Progr. Theor. Phys. 51: 1731 (1974).
[25] K. Kishida et. al., J. Nucl. Sci. Technol. 13: 161 (1976).
[26] K. Kishida et al., Progr. Nucl. Energy 1: 247 (1977).
[27] K. Hayashi et al., JAERI-M 84-056; 84-137 (1986) and private communications.
[28] B. G. Bergdhal et al., BWR-stablilty Investigation at FORSMAK 1, STUDSVIK/NI-88/12.
[29] R. Oguma et al., Investigation of BWR stability in FORSMAK 2 Based on Multivariable Noise Analysis, STUDSVIK/NI-88/3.
[30] H. Konno, Ann. Nucl. Energy 10: 451 (1983).
[31] H. Konno, Ann. Nucl. Energy 11: 405 (1984).
[32] M. Suzuki, J. Stat. Phys. 49: 977 (1988).
[33] Karmeshu , Ann. Nucl. Energy 8: 141 (1981).
[34] M.M. R. Williams, Ann. Nucl. Energy 10: 51 (1983).
[35] O. Sako, J. Nucl. Sci. & Technol. 17: 323 (1980).
[36] H. Konno and T. Tanaka (1988) (unpublished).
[37] H. Konno and K. Saito, Ann. Nucl. Energy 11: 1 (1984).
[38] J. A. Thie, Nucl. Technol. 45: 4 (1979).
[39] P. Liewers, W. Schmitt, P. Schumann and F.P. Weiss, Progr. Nucl. Energy 21: (1988).
[40] E. Turkcan, Porgr. Nucl. Energy 9: 437 (1981).
[41] R. Oguma and E. Turkcan, Progr. Nucl. Energy 15: 863 (1985).
[42] U. Frisch and R. Morf, Phys. Rev. A23: 2673 (1981).
[43] U. Frisch, Physica Scripta T9: 137 (1985).
[44] H. Mori et al., Physica Scripta T9: 27 (1985).
[45] H. Konno, H. Soneda, T. Tanaka and K. Hayashi, Ann. Nucl. Energy 18: (1988).
[46] A. Schenzle and H. Brand, Phys. Rev. A20: 1628 (1979).
[47] R. Kubo, J. Math. Phys. 4: 174 (1963).
[48] L. van Wijngaarden, Ann. Rev. Fluid. Mech. 4: 369 (1972).
[49] H. Konno, J. Phys. Soc. Japn 57: 1163 (1988).

STOCHASTIC NONLINEAR DYNAMICS OF NUCLEAR POWER PLANTS: NOISE EFFECTS ON NONLINEAR STABILITY

J.L. Muñoz-Cobo
and J. Sancho García

Universidad Politécnica de Valencia
Departamento Ingeniería Química y Nuclear P.O. Box 22012
46071 Valencia

INTRODUCTION

It is well known from previous work of Bell [1] and Williams [2] that in a linear reactor model, fluctuations become very important during start-up operation with weak sources. At this point, it is important to distinguish two different types of random process which are at the origin of neutron density fluctuations: firstly we have the inherent randomness of the emission and absortion events, and secondly we have the parametric noise sources, i.e. external noise due to fluctuations in the reactor parameters.

Recently Rodriguez, San Miguel and Sancho [3] have developed point reactor models, with a joint description of internal fluctuations and parametric noise. They then computed the thermodynamic limit of their models which is equivalent to considering large reactors where the internal noise is neglected.

Particular examples of their equations are similar to those proposed by Saito [4] and Quabili [5], with some provisos.

We have assumed in this paper a large reactor where the termodynamic limit is valid. This paper considers a point reactor model, with one group of delayed neutrons, Newtonian cooling, and temperature feedback, under the influence of parametric Gaussian white noise in the reactivity. The aim of this paper is to study the nonstationary behaviour of this nonlinear power reactor model, and to try to gain an insight into the stability of this model. Previous work in this field of research has recently been done by Konno [6] using the standard Langevin technique and the fluctuation dissipation theorem.

DERIVATION OF THE FOKKER-PLANCK EQUATION FOR THE POINT REACTOR KINETIC EQUATIONS WITH FUEL-TEMPERATURE FEEDBACK AND PARAMETRIC WHITE GAUSSIAN NOISE IN THE REACTIVITY

The nonlinear power reactor model with a single group of delayed neutrons is given by the following set of equations:

$$\frac{dn}{dt} = \frac{\rho + \xi - \beta}{\Lambda} n + \lambda c \qquad (1)$$

$$\frac{dc}{dt} = \beta \frac{n}{\Lambda} - \lambda c \qquad (2)$$

$$\frac{dT}{dt} = Kn - \gamma (T - T_c) \qquad (3)$$

$$\rho = \rho_0 + \alpha T , \qquad (4)$$

where Λ is the neutron mean reproduction time, β is the delayed neutron fraction, ρ is the reactivity, λ is the decay constant of delayed neutron precursors, α is the temperature feedback coefficient ρ_0 is the reactivity evaluated at T_0, T is the increase of fuel temperature above T_0, and T_c the coolant temperature above or below T_0. Finally γ and K are given by:

$$\gamma = \frac{h_f}{MC_p} \quad \text{and} \quad K = \frac{30\times 10^{-12}}{\Lambda MC_p} \text{ (C/s neutron)} ,$$

where h_f is the heat transfer coefficient, M is the fuel mass, and C_p the specific heat at constant pressure. We have assumed in this model that the coolant can be considered as a heat sink with constant temperature. Finally, ξ is a white gaussian noise in the reactivity. Under these assumptions we can write:

$$<\xi> = 0 \tag{5}$$

$$<\xi(t)\,\xi(t')> = 2\,D\,\delta(t-t') . \tag{6}$$

The stochastic differential equation system (1) to (4) is of the general form [7]

$$du/dt = f(u) + g(u)\,\xi , \tag{7}$$

where u is the column vector:

$$u = col\,[n,\ c,\ T] , \tag{8}$$

f(u) is the vector field:

$$f(u) = col\left[\frac{\rho_0+\alpha T-\beta}{\Lambda}.n + \lambda c,\ \beta\frac{n}{\Lambda} - \lambda c,\ Kn - \gamma\,(T-T_c)\right] , \tag{9}$$

and $g(u)\xi$ is the stochastic vector field:

$$g(u)\,\xi = (n/\Lambda\xi,\ 0\ ,\ 0\ ) . \tag{10}$$

Now it has been shown elsewhere (see the works of Sancho [7] and Fox [9]) that the stochastic differential equation system (7) is equivalent to the following Fokker Plank equation for the probability density P(u):

$$\frac{\partial P(u,t)}{\partial t} = \hat{f}(u)\,P(u,t) + \exp[t\hat{f}(u)]\sum_{n=1}^{\infty} G^n(u,t)\,\exp[-t\hat{f}(u)]\,P(u,t) , \tag{11}$$

where $\hat{f}(u)$ is the operator:

$$\hat{f}(u) = -\,\nabla\,.\,f(u) , \tag{12}$$

and $G^n\,(u,t)$ is the n-ordered cumulant of $\hat{F}(u,t)$, given by:

$$\hat{F}(u,t) = \exp\,(-\hat{f}(u)t)\,\hat{g}(u,t)\,\exp\,(\hat{f}(u)t) , \tag{13}$$

with

$$\hat{g}(u,t) = -\nabla \cdot g(u,t)\,\xi(t) \,. \qquad (14)$$

For white gaussian noise all the cumulants are zero except for the second order one which, according to Fox [9], is equal to:

$$G^2(u,t) = \int_0^t \langle \hat{F}(u,t)\,\hat{F}(u,S)\rangle\, dS = D\exp[-\hat{f}(u)t]\,[\nabla . g(u)]^2 \exp[\hat{f}(u)t] \,. \qquad (15)$$

on account of equation (15), the Fokker-Planck equation (11) simplifies to:

$$\frac{\partial P(u,t)}{\partial t} = -\nabla . f(u)\, P(u,t) + D(\nabla . g(u,t))^2\, P(u,t) \,. \qquad (16)$$

Now, on account of the expression for the vector fields f(u) and g(u) that can be obtained from (9) and (10), it is found that P(n,c,T,t) obeys the following master equation:

$$\frac{\partial}{\partial t} P(n,c,T,t) = -\frac{\partial}{\partial n}\left[\left(\frac{\rho_0+\alpha T-\beta}{\Lambda} . n + \lambda c\right) P(n,c,T,t)\right] -$$

$$-\frac{\partial}{\partial c}\left[\left(\frac{\beta}{\Lambda} - \lambda c\right) P(n,c,T,t)\right] - \frac{\partial}{\partial T}\left[(Kn - \gamma(T-T_c))\, P(n,c,T,t)\right] +$$

$$+ D\frac{\partial}{\partial n}\frac{n}{\Lambda}\frac{\partial}{\partial n}\frac{n}{\Lambda} P(n,c,T,t) \qquad (17)$$

Operating on equation (17), we obtain after, some calculus:

$$\frac{\partial}{\partial t} P(n,c,T,t) = -\left[\frac{\rho_0+\alpha T-\beta}{\Lambda} n + \lambda c\right]\frac{\partial P(n,c,T,t)}{\partial n} - \left[\frac{\beta}{\Lambda} n - \lambda c\right]\frac{\partial P(n,c,T,t)}{\partial c} -$$

$$-[Kn - \gamma.(T-T_c)].\frac{\partial}{\partial T} P(n,c,T,t) - \frac{\rho_0+\alpha T-\beta}{\Lambda} P(n,c,T,t) + \gamma\, P(n,c,T,t) +$$

$$+\lambda P(n,c,T,t) + D\left[\frac{1}{\Lambda^2} + \frac{3n}{\Lambda^2}\frac{\partial}{\partial n} + \left(\frac{n^2}{\Lambda^2}\right)\frac{\partial^2}{\partial n^2}\right] P(n,c,T,t) \,. \qquad (18)$$

DYNAMICAL BEHAVIOUR OF THE MEAN VALUES

To get the evolution equations for the average values (i.e.

$\langle n\rangle$, $\langle c\rangle$, $\langle T\rangle$) we multiply equation (18) by n, c and T, respectively, and then we integrate over the whole phase space $U \subset R^3$.

After some calculus the following set of differential equations is obtained for the average values:

$$\frac{d}{dt}\langle n\rangle = \frac{(\rho_0+D/\Lambda)-\beta}{\Lambda}\langle n\rangle + \frac{\alpha}{\Lambda}\langle Tn\rangle + \lambda\langle c\rangle \tag{19a}$$

$$\frac{d}{dt}\langle c\rangle = \frac{\beta}{\Lambda}\langle n\rangle - \lambda\langle c\rangle \tag{19b}$$

$$\frac{d}{dt}\langle T\rangle = K\langle n\rangle - \gamma.[\langle T\rangle - T_c]\ . \tag{19c}$$

Equations (19a) to (19c) deserve some comment: first we note that the white gaussian reactivity noise gives rise in equation (19a) to an effective reactivity $\rho_{ef} > \rho_0$ equal to:

$$\rho_{ef} = \rho_0 + D/\Lambda\ . \tag{20}$$

To solve equations (19a), (19b) and (19c), we need to know the time evolution of $\langle Tn\rangle$. This is a typical problem that appears in parametric noise studies, where the time evolution equations for the lower moments are expressed in terms of higher ones.

To find the evolution equations for the second moments, $\langle n^2\rangle$, $\langle c^2\rangle$, $\langle T^2\rangle$, $\langle nc\rangle$, $\langle nT\rangle$ and $\langle cT\rangle$, we multiply equation (18) by n^2, c^2, T^2, nc, nT and cT, respectively, and then we integrate these equations over the whole phase space. After some calculus the following set of coupled differential equations is obtained:

$$\frac{d\langle n^2\rangle}{dt} = 2\left[\frac{\rho_0+D/\Lambda-\beta}{\Lambda}.\ \langle n^2\rangle + \frac{\alpha}{\Lambda}\langle n^2T\rangle + \lambda\langle cn\rangle\right] \tag{21}$$

$$\frac{d\langle c^2\rangle}{dt} = 2\left[\frac{\beta}{\Lambda}\langle nc\rangle - \lambda\langle c^2\rangle\right] \tag{22}$$

$$\frac{d\langle T^2\rangle}{dt} = 2\left[K\langle nT\rangle - \gamma\,(\langle T^2\rangle - T_c\langle T\rangle)\right] \tag{23}$$

$$\frac{d\langle nc\rangle}{dt} = \frac{\rho_0 + D/\Lambda - \beta}{\Lambda}\langle nc\rangle + \frac{\alpha}{\Lambda}\langle Tnc\rangle + \lambda\langle c^2\rangle + \frac{\beta}{\Lambda}\langle n^2\rangle - \lambda\langle nc\rangle \tag{24a}$$

$$\frac{d\langle nT\rangle}{dt} = \frac{\rho_0 + D/\Lambda - \beta}{\Lambda}.\langle nT\rangle + \frac{\alpha}{\Lambda}\langle nT^2\rangle + K\langle n^2\rangle - (\langle nT\rangle - \langle n\rangle T_c) \tag{24b}$$

$$\frac{d\langle cT\rangle}{dt} = \frac{\beta}{\Lambda}\langle nT\rangle - \lambda\langle cT\rangle + K\langle nc\rangle - \gamma\,(\langle cT\rangle - \langle c\rangle T_c)\ . \tag{24c}$$

To try to solve the infinite hierarchy of equations which arise in the nonlinear stochastic examples it is necessary to use some kind of truncation. We have chosen to perform as the truncation the cumulant neglect closure technique [10].

At this point we remind that the n-th cumulant of the random variables X_1, X_2, ... X_n is a measure of the correlation among these random variables . For this reason the n-th cumulant is called the n-th correlation in the well-known book by Stratonovich [11].

This physical meaning is revealed in the following relationship between moments and cumulants:

$$\langle X_j\rangle = K_1\,(X_j)\ ,$$

$$\langle X_j\,X_k\rangle = K_2\,(X_j,\ X_k) + K_1(X_j)\,K_1(X_k)\ ,$$

$$\langle X_j\,X_k\,X_l\rangle = K_3(X_j,\ X_k,\ X_l) +$$

$$+\ 3[K_1(X_j)K_2(X_k,\ X_l)]_S + K_1(X_j).K_1(X_k)K_1(X_l)\ , \tag{25}$$

where $K_n(\quad)$ is the n-order joint cumulant of the bracketed random variables and $[\quad]_S$ indicates the symmetrizing operation with respect to all arguments of the cumulants:

$$[K_1\,(X_j)\,K_2\,(X_k,\ X_l)]_S = 1/3\,(K_1(X_l)K_2(X_j,\ X_k) +$$

$$+\ K(X_j)K_2(X_k,\ X_l) + K_1(X_k)\,K_2(X_j,\ X_l))\ . \tag{26}$$

(i) <u>K_2=0 Approximation</u>

From these previous relations it is possible in the K2= 0

approximation to set:

$$\langle nT\rangle \approx \langle n\rangle \langle T\rangle \,. \tag{27}$$

ii) $K_3 = 0$ Approximation

From equation (25), and setting $K_3 = 0$, the following relation can be deduced:

$$\langle X_j X_k X_l\rangle \approx \langle X_j\rangle \langle X_k X_l\rangle + \langle X_k\rangle \langle X_j X_l\rangle + \langle X_l\rangle\langle X_j X_k\rangle$$

$$- 2 \langle X_j\rangle \langle X_k\rangle \langle K_l\rangle \,, \tag{28}$$

$$\langle X_j X^2_k\rangle \approx \langle X_j\rangle \langle X^2_k\rangle + 2\langle X_k\rangle \langle X_j X_k\rangle - 2\langle X_j\rangle \langle X_k\rangle^2 \,, \tag{29}$$

therefore in the $K_3 = 0$ approximation we can write:

$$\langle Tnc\rangle \approx \langle T\rangle \langle nc\rangle + \langle n\rangle\langle cT\rangle + \langle c\rangle \langle Tn\rangle - 2 \langle n\rangle \langle c\rangle \langle T\rangle \tag{30}$$

$$\langle nT^2\rangle \approx \langle n\rangle \langle T^2\rangle + 2\langle T\rangle \langle nT\rangle - 2 \langle n\rangle \langle T\rangle^2 \tag{31}$$

$$\langle n^2T\rangle \approx \langle T\rangle \langle n^2\rangle + 2\langle n\rangle \langle nT\rangle - 2 \langle T\rangle \langle n\rangle^2 \,; \tag{32}$$

therefore the evolution of $\langle n\rangle$, $\langle c\rangle$ and $\langle T\rangle$ in the $K_2 = 0$ approximation is governed by equations (19a), (19b) and (19c), with $\langle nT\rangle$ given by (27). In the $K_3 = 0$ approximation the evolution of $\langle n\rangle$, $\langle c\rangle$ and $\langle T\rangle$ with time is governed by the differential equation system formed by (19a), (19b), (19c), (22), (23), (24c) and the following three differential equations obtained from (21), (24a), (24b), on account of equations (30), (31) and (32):

$$\frac{d}{dt} \langle n^2\rangle = 2\left[\frac{\rho_0 + D/\Lambda - \beta}{\Lambda} \langle n^2\rangle + \frac{\alpha}{\Lambda} (\langle T\rangle\langle n^2\rangle + 2\langle n\rangle\langle nT\rangle -\right.$$

$$\left. - 2\langle T\rangle\langle n\rangle^2) + \lambda\langle cn\rangle\right] \,, \tag{33}$$

$$\frac{d}{dt} \langle nc\rangle = \frac{\rho_0 + D/\Lambda - \beta}{\Lambda} \langle nc\rangle + \frac{\alpha}{\Lambda} (\langle T\rangle\langle nc\rangle + \langle n\rangle\langle cT\rangle + \langle c\rangle\langle Tn\rangle -$$

$$- 2\langle n\rangle\langle c\rangle\langle T\rangle) + \lambda\langle c^2\rangle + \frac{\beta}{\Lambda} \langle n^2\rangle - \lambda\langle nc\rangle \,, \tag{34}$$

$$\frac{d}{dt}\langle nT\rangle = \frac{\rho_0 + D/\Lambda - \beta}{\Lambda}\langle nT\rangle + \frac{\alpha}{\Lambda}(\langle n\rangle\langle T^2\rangle + 2\langle T\rangle\langle nT\rangle -$$

$$- 2\langle n\rangle\langle T\rangle^2) + K\langle n^2\rangle - \gamma(\langle nT\rangle - \langle n\rangle T_c) . \qquad (35)$$

NOISE INFLUENCE ON THE STABILITY IN THE $K_2 = 0$ APPROXIMATION

To try to get an insight into the noise influence on the stability and trajectories we have first calculated the $K_2 = 0$ approximation. In this approximation, equations (19a) to (19c) can be written in the form:

$$\frac{dx_1}{dt} = \frac{\rho_{ef} - \beta}{\Lambda} x_1 + \frac{\alpha}{\Lambda} x_1 x_3 + \frac{\alpha}{\Lambda}\langle n\rangle_0 x_3 + \frac{\alpha}{\Lambda}\langle T\rangle_0 x_1 + \lambda x_2 \qquad (36a)$$

$$\frac{dx_2}{dt} = \frac{\beta}{\Lambda} x_1 - \lambda x_2 \qquad (36b)$$

$$\frac{dx_3}{dt} = K x_1 - \gamma x_3 , \qquad (36c)$$

where the new state variables x_1, x_2 and x_3 are defined as follows:

$$x_1 = \langle n\rangle - \langle n\rangle_0 \qquad (37)$$

$$x_2 = \langle c\rangle - \langle c\rangle_0 \qquad (38)$$

$$x_3 = \langle T\rangle - \langle T\rangle_0 , \qquad (39)$$

and $\langle n\rangle_0$, $\langle c\rangle_0$ and $\langle T_0\rangle$ define the fixed or equilibrium non trivial point of equations (19.a) to 19.c) in the $K_2 = 0$ approximation, given by:

$$\langle n\rangle_0 = \frac{\left(\frac{\rho_{ef} - \beta}{\Lambda} + \frac{\alpha}{\Lambda} . T_c + \frac{\beta}{\Lambda}\right)}{\left(-\frac{\alpha}{\Lambda}\right)\frac{K}{\gamma}} \qquad (40)$$

$$\langle c\rangle_0 = \frac{\beta}{\Lambda\lambda} . \langle n\rangle_0 \qquad (41)$$

$$\langle T\rangle_0 = \frac{K\langle n\rangle_0+\gamma T_c}{\gamma} . \qquad (42)$$

To construct a Lyapunov function $V(x_1, x_2, x_3)$ we follow the variable gradient method [12]. In order to make $\nabla V(x_1, x_2, x_3)$ integrable, it has be shown [12] that:

$$\nabla V(x_1, x_2, x_3)= \begin{bmatrix} \sum_{j=1}^{3} a_{1j}\, x_j \\ \sum_{j=1}^{3} a_{2j}\, x_j \\ \sum_{j=1}^{3} a_{3j}\, x_j \end{bmatrix} , \qquad (43)$$

where $a_{ij}= 0$ for $i \neq j$,

and a_{11}, a_{22} and a_{33} are arbitrary functions of the state variables. In this way the total derivative of the Lyapunov function with time will be:

$$\frac{dV}{dt} = f.\nabla V = a_{11}\, x^2_1 \left(\frac{\rho_{ef}-\beta}{\Lambda} + \frac{\alpha}{\Lambda} \langle T\rangle_0 \right) + x_1\, x_3 \left[a_{11} \frac{\alpha}{\Lambda} (x_1 +\langle n\rangle_0) + \right.$$

$$\left. + a_{33}\, K\right] + x_1 . x_2 . \left[a_{11}\ \lambda + a_{22} \frac{\beta}{\Lambda}\right] - a_{22}\, \lambda\, x^2_2 - a_{33}\, \gamma\, x^2_3 . \qquad (44)$$

We now seek values for a_{11}, a_{22} and a_{33} that will make dV/dt, semidefinite or definite negative. To this end we proceed as follows, first we observe that the term in $x_1\, x_3$ vanishes if we select

$$a_{11}= \frac{-a_{33} . K}{\alpha/\Lambda . (x_1+\langle n\rangle_0)} . \qquad (45)$$

Now we note that $\alpha<0$, $x_1 + \langle n\rangle_0= \langle n\rangle > 0$, and $K>0$. Therefore it is deduced that the sign of a_{11} is the same as that of a_{33}. Next we select a_{33} equal to

$$a_{33}= (-\alpha/\Lambda) / K . \qquad (46)$$

In this way the coefficient a_{11} reduces to:

$$a_{11} = \frac{1}{(x_1 + \langle n \rangle_0)} \, . \tag{47}$$

It is convenient at this point to remind ourselves of the essential ideas of this method. The procedure is to compute dV/dt in terms of the variable coefficients, select coefficients such that dV/dt is negative definite or semidefinite negative in some region $W_S \subset R^3$, and then obtain V as a line integral in state space such that V = 0 at the origin. If the resulting V is positive definite in W_S then W_S is the region of stability (see Hetrick [12] for more details). With these coefficient values we can write equation (44) as follows:

$$\frac{dV}{dt} = x_1^2 \left[\frac{1}{(x_1 + \langle n \rangle_0)} \cdot \left(\frac{\rho_{ef} - \beta}{\Lambda} + \frac{\alpha}{\Lambda} \langle T \rangle_0 \right) \right] +$$

$$+ x_1 \, x_2 \cdot \left(\frac{\lambda}{x_1 + \langle n \rangle_0} + a_{22} \frac{\beta}{\Lambda} \right) - a_{22} \, \lambda \, x_2^2 + \frac{\alpha \gamma}{K \Lambda} x_3^2 \, . \tag{48}$$

Next, we apply the Sylvester theorem to the quadratic form of equation (48). To be definite negative, the following three conditions must hold simultaneously:

$$A_1 = \frac{1}{x_1 + \langle n \rangle_0} \left(\frac{\rho_{ef} - \beta}{\Lambda} + \frac{\alpha}{\Lambda} \langle T \rangle_0 \right) < 0 \tag{49}$$

$$A_2 = - \frac{a_{22} \, \lambda_{ef}}{x_1 + \langle n \rangle_0} \left(\frac{\rho_{ef} - \beta}{\Lambda} + \frac{\alpha}{\Lambda} \langle T \rangle_0 \right) - \frac{1}{4} \left[\frac{\lambda}{x_1 + \langle n \rangle_0} + a_{22} \frac{\beta}{\Lambda} \right]^2 > 0 \, , \tag{50}$$

$$\det [A] < 0 \, , \tag{51}$$

where A is the matrix of the quadratic form.

Now, turning to a_{22}, in order that conditions (50) and (51) hold, the method we have used is to make:

$$a_{22} = \epsilon / \lambda \tag{52}$$

where ϵ is an arbitrary number that is selected in order to ensure that inequality (50) remains valid in the stability region. We denote by ϵ_+ and ϵ_- the ϵ values that make $A_2 = 0$, and these values are given by:

$$\epsilon_{\pm} = \frac{-(2b_1+b_2.b_3)\pm 2(b^2{}_1+b_1.b_2.b_3)^{1/2}}{b^2{}_3}, \tag{53}$$

where:

$$b_1 = \frac{\rho_{ef}-\beta}{\Lambda} + \frac{\alpha}{\Lambda}\langle T\rangle_0 < 0 \tag{54}$$

$$b_2 = \frac{\lambda}{x_1+\langle n\rangle_0} > 0 \tag{55}$$

$$b_3 = \frac{\beta}{\Lambda.\lambda} > 0, \tag{56}$$

for large reactors b_2 is very small in a large range of x_1 values, and $A_2(\epsilon)$ has a máximum between ϵ_- and ϵ_+

$$\frac{d^2A_2(\epsilon)}{d\epsilon^2} = -2\, b_3^2 < 0 \tag{57}$$

$$\epsilon_{max} = -\frac{b_1+2b_2b_3}{2b_3^2} > 0. \tag{58}$$

In general for large reactors ϵ_- will be located very close to zero and positive, so that we have a interval $[\epsilon_-, \epsilon_+]$ of positive ϵ values where $A(\epsilon)$ remains positive. Obviously if the neutron population changes then $[\epsilon_-, \epsilon_+]$ changes, but for practical puposes the lower limit does not change and the upper limit changes only slightly and so we can pick up an ϵ value such that $A_2(\epsilon) > 0$ for a large range of x_1 values. Therefore for ϵ to exist we must have the following condition deduced from (53):

$$\langle n\rangle > \langle n\rangle_{lim}, \tag{59}$$

where

$$\langle n\rangle_{lim} = \frac{\beta}{[-(\rho_{ef}-\beta + \alpha\langle T\rangle_0)]}. \tag{60}$$

From equation (60) it is deduced that $\langle n\rangle_{lim}$ is very small. For large reactors we have found that $\langle n\rangle_0 \sim 10^{15}$, so we can always find an ϵ value such that $A_2(\epsilon) > 0$. This ϵ value exist for a large interval of neutron population values.

Once we have chosen the appropriate ϵ-value, we perform an integration and get:

$$V(x_1, x_2, x_3) = \int_0^{x_1} \frac{1}{(x_1+\langle n\rangle_0)} x_1\, dx_1 + \int_0^{x_2} \frac{\epsilon}{\lambda_{ef}} x_2\, dx_2 + \int_0^{x_3} \frac{(-\alpha/\Lambda)}{K} x_3\, dx_3 =$$

$$= x_1 - \langle n\rangle_0 \log\left(1 + \frac{x_1}{\langle n\rangle_0}\right) + \frac{\epsilon}{\lambda}\frac{x_2^2}{2} + \frac{(-\alpha/\Lambda)}{K}\frac{x_3^2}{2} . \qquad (61)$$

This Lyapunov function $V(x_1, x_2, x_3)$ vanishes at the origin $x(0, 0, 0)$ and is definite positive, provided that $\epsilon>0$, $\lambda > 0$ and $\alpha <0$.

Otherwise dV/dt is definite negative if $A_1<0$, $A_2>0$ and $A_3<0$, the conditions to make $dV/dt < 0$, which has been explained previously except for condition (51). Now on account of the A matrix given by:

$$A= \begin{bmatrix} A_1 & \frac{1}{2}\left(\frac{\lambda}{x_1+\langle n\rangle_0} + \frac{\epsilon}{\lambda} \frac{\beta}{\Lambda}\right) & 0 \\ \frac{1}{2}\left(\frac{\lambda}{x_1+\langle n\rangle_0} + \frac{\epsilon}{\lambda} \frac{\beta}{\Lambda}\right) & -\epsilon & 0 \\ 0 & 0 & \frac{\alpha\gamma}{K\Lambda} \end{bmatrix}, \qquad (62)$$

we have that condition (51) reduces to:

$$\det [A]= -A_1\, \epsilon \frac{\alpha\gamma}{K\Lambda} - \frac{1}{4}\left[\frac{\lambda}{x_1+\langle n\rangle_0} + \frac{\epsilon}{\lambda} . \frac{\beta}{\Lambda}\right]^2 . \frac{\alpha\gamma}{K\Lambda} < 0 . \qquad (63)$$

Therefore if we assume $A_1<0$, $\alpha<0$ and $\epsilon>0$, it is deduced from (63) that det [A] is smaller than zero, provided condition (63) holds.

NUMERICAL CALCULATIONS

A series of numerical calculations have been performed for which we have considered the following PWR reactor parameter values: $\rho= 0.67\times10^{-2} - 0.21\times10^{-4}\, T$ T_0= (reference temperature for ρ_0) = 350 C, β= 0.007, with β/λ= 0.089 s, and λ= 0.0786 s^{-1}. A value for Λ between 10^{-4} and 10^{-5} s is quite usual, being the specific calculations carried out with $\Lambda = 10^{-5}$ s. The coolant temperature T_c has been taken as being constant and equal to 308.65 C, so we have: T_c= 308.65-350= -41.35 C. Other thermal parameters have been set as: K= $1.636.10^{-18}/\Lambda$ (C/s. neutron) and γ= 0.333 s^{-1}.

Without noise it is found that the above parameter values give the following equilibrium point: $n_e= 0.7333x10^{15}$, $c_e= 0.6555x10^{19}$ and $T_e= T_f - T_0 = 319.03$ C, which yield an average fuel temperature of 669.03 C.

In Figure 1 we have represented trajectories and curves of constant V, the ϵ value selected when D= 0 was 10^{-30}; these curves are the intersection of the surface V with the plane x_2=3.275 x 10^{17}. On the y-axis we have represented $y= n/n_e$ (i.e., the ratio of the neutron population and the neutron equilibrium population), while on the x-axis we have represented the excess of temperature over the equilibrium one. We observe that when D= 0, the fixed point is asymptotically stable and all the trajectories in the stability region approach the equilibrium poinht when $t \longrightarrow \infty$ (See Figure 1)

Now if the noise parameter is not zero we have observed, in the K_2= 0 approximation, that the equilibrium point changes with D (table 1) toward higher values of $\langle n \rangle_0$, $\langle T \rangle_0$ and $\langle c \rangle_0$, and this approximation can still be applied to the Lyapunov theorem. However, in order to study the stability it is convenient to apply the Hartman-Grobman theorem [13] and the stable manifold theorem for a fixed point [13].

In Table 1 we give the change in the equilibrium point for different values of the noise parameter D. In table 2, we show how the eigenvalue changes with D.

Finally, in Figure 2 we have represented curves of constant V in the K_2= 0 approximation for various D values, and we have observed without plotting it that the fixed point attracts all the trajectories in the stability region.

Table 1. Fixed points in the K_2= 0 Aproximation

D	$\langle n \rangle_0$	$\langle T \rangle_0$	$\langle c \rangle_0$
0	0.7332 x 10^{15}	669.04	0.655 x 10^{19}
0.50 x 10^{-10}	0.7347 x 10^{15}	669.28	0.655 x 10^{19}
0.50 x 10^{-9}	0.7390 x 10^{15}	671.42	0.658 x 10^{19}
0.50 x 10^{-8}	0.7827 x 10^{15}	692.85	0.697 x 10^{19}
0.25 x 10^{-7}	0.9767 x 10^{15}	788.09	0.870 x 10^{19}
0.50 x 10^{-7}	1.2192 x 10^{15}	907.14	1.086 x 10^{19}

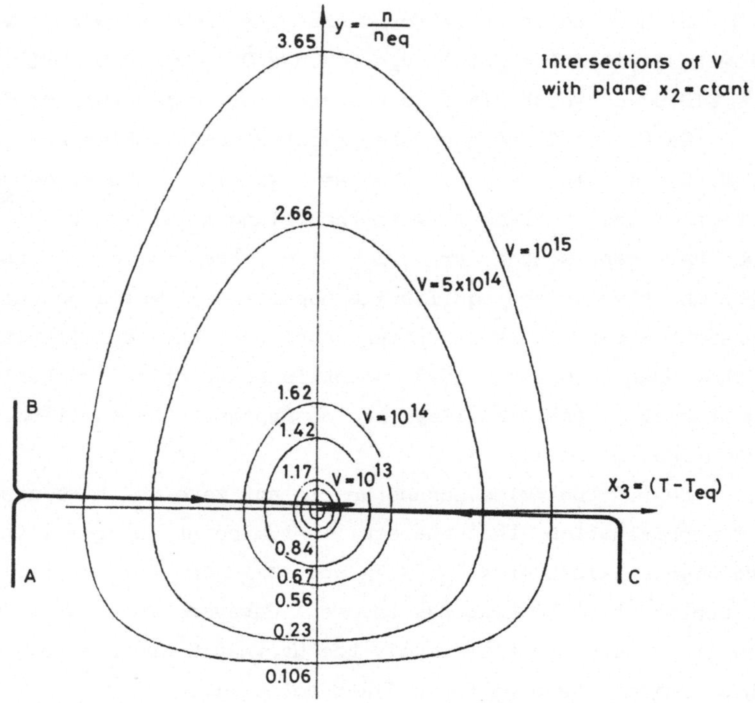

Figure 1. Trajectories and curves of constant V, for a case with D = 0. Parameter values: $\alpha = -0.21 \times 10^{-4}\ {}^{o}C^{-1}$, $\beta = 0.7 \times 10^{-2}$, $\gamma = 0.3333\ s^{-1}$, $K = 0.16366 \times 10^{-12}\ {}^{o}C/s$, $\Lambda = 10^{-5}$ s, $\rho_0 = 0.67 \times 10^{-2}$, and $T_c = 308.65^{o}C$.

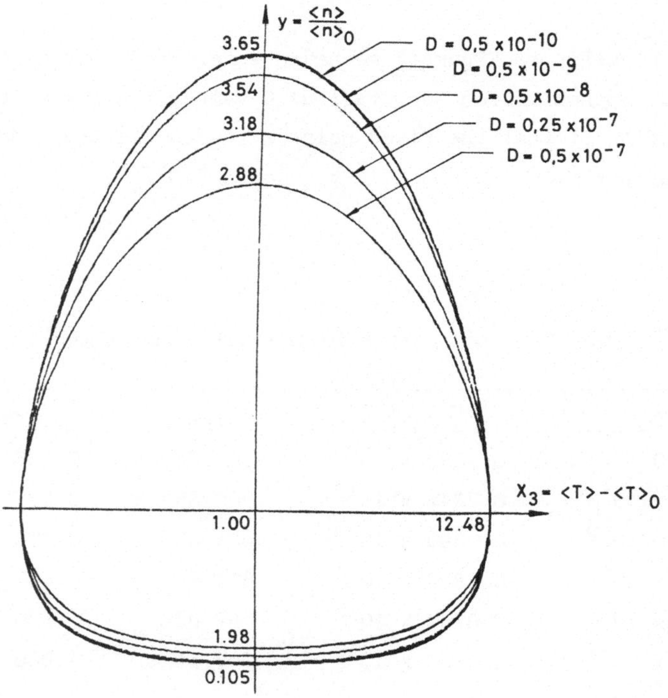

Figure 2. In this figure we give a set of curves with $V=10^{15}$ for different noise intensities $D = 0.5 \times 10^{-13}$, 0.25×10^{-12}, 0.5×10^{-12}, 0.5×10^{-11}.

Table 2. Eigenvalues of the Jacobian Matrix at the equilibrium point

D	λ_1	λ_2	λ_3
0.50×10^{-10}	- 699.71	- 0.3607	- 0.3333
0.50×10^{-9}	- 699.71	- 0.3628	- 0.3333
0.50×10^{-8}	- 699.69	- 0.3843	- 0.3333
0.25×10^{-7}	- 699.59	- 0.4796	- 0.3333
0.50×10^{-7}	- 699.47	- 0.3607	- 0.3333

We conclude with some remarks about the application of the Hartman-Grobman theorem which states: "If the Jacobian matrix of the vector field, evaluated at the equilibrium point, has no zero or purely imaginary eigenvalues, then there is a homeomorphism defined on some neighborhood U of the equilibrium point, locally taking orbits on the non linear flow, to those of the linear flow. The homeomorphism preserves the sense of the orbits and can also be chosen to preserve parametrization by time."

In order to apply this theorem, we note that equations (36.a) to (36c) can be expressed in the form

$$dx/dt = f(x), \tag{64}$$

where the components of the vector field f(x,t) are given by the right hand sides of equations (36a) to (36c) and the Jacobian matrix is:

$$Df(x)=\begin{bmatrix} \frac{\rho_{ef}-\beta}{\Lambda} + \frac{\alpha}{\Lambda} x_3 + \frac{\alpha}{\Lambda} \langle T\rangle_0 & \lambda & \frac{\alpha}{\Lambda} x_1 + \frac{\alpha}{\Lambda} \langle n\rangle_0 \\ \frac{\beta}{\Lambda} & -\lambda & 0 \\ K & 0 & -\gamma \end{bmatrix}, \tag{65}$$

The Jacobian matrix, evaluated at the equilibrium point, is obtained by setting $x_1 = x_2 = x_3 = 0$ in (65). The next step is to apply the stable manifold theorem if the fixed point is hyperbolic; in this case there exist stable and unstable manifolds of the same dimension as those of the eigenspaces of the linearized system and tangent to them at the equilibrium point.

The eigenvalues of $Df(x_e)$, evaluated at the fixed point, show that for $D < 0.5 \times 10^{-7}$ the fixed point x_e (0, 0, 0) is hyperbolic and the eigenvalues are all real and negative.

Therefore there exists a local stable manifold $W^S{}_{loc}(x_e) \subset R^3$ of dimension n= 3.

CONCLUSIONS

In the first stage of this paper we have developed the Fokker-Planck equation for the point reactor kinetic equations with one group of delayed neutrons, fiel temperature feedback, Newtonian cooling and white Gaussian noise in the reactivity. Then we have obtained the evolution equations for the average values and we have developed the $K_2 = 0$ and $K_3 = 0$ cumulant neglect closure approximations.

The evolution equation for the average values shows that we have an effective reactivity, given by equation (20), that is larger than the deterministic one. To try to get an insight into the noise influence on the stability, we have obtained in the $K_2 = 0$ approximation the conditions that must hold in order to have a Liapunov function $V(x_1, x_2, x_3)$ such that $dV/dt < 0$. Then we have developed some numerical examples and observed that when the noise parameter D increases, the fixed point in the $K_2 = 0$ approximation moves toward higher values of $\langle n \rangle_0$, $\langle c \rangle_0$ and $\langle T \rangle_0$.

Finally we have applied the Hartman-Grobman and the stable manifold theorem to the vector field in the $K_2 = 0$ approximation, and we have found that the fixed point is hyperbolic and that the three eigenvalues are real and negative for all physical D values investigated. This means that we have a local stable manifold of dimension 3.

REFERENCES

1.- G.I. Bell, Ann. Phys. (N.Y.) 21, 243 (1963).

2.- M.M.R. Williams, Random Processes in Nuclear Reactors. Pergamon Press, Oxford (1974).

3.- M.A. Rodriguez, M. San Miguel, J.M. Sancho, Ann. nucl. Energy, Vol 10, No 5., pp 263-269 (1983), and Ann. nucl. Energy, Vol 11, No7, pp 321-336 (1984).

4.- K. Saito, Ann. nucl. Energy 1, 31 (1974), Prog. nucl. Energy 3, 157 (1979). Ann. nucl. Energy 6, 591 (1979).

5.- E.R. Quabili, and M. Karasulu. Ann. nucl. Energy 6, 133 (1979).

6.- H. Konno, Ann. nucl. Energy. Vol 11, No 8, pp 405-417 (1984).

7.- J.M. Sancho, in Stochastic Processes Applied to Physics. Ed. L. Pesquera and M. Rodriguez, pp 96-109. World Scientific (1985).

8.- P. Hangi, in Stochastic Processes Applied to Physics. Ed. L. Pesquera and M. Rodriguez, pp 69-95. World Scientific (1985).

9.- R.F. Fox, Journal of Mathematical Physics, Vol 16, No 2 (1975).

10.- W.F. Wu, Y. K. Lin, Int. Journal Non Linear Mechanics. Vol 19. No 4 (1984).

11.- R.L. Stratonovich, Topics in the Theory of Random Noise. Gordon and Breach (1963).

12.- D.L. Hetrick, Dynamics of Nuclear Reactors. Univ. of Chicago Press (1971).

13.- J. Gluckenheimer, P. Holmes, Non Linear Oscillations, Dynamical Systems and Bifurcations of Vector Fields. Springer Verlag, 2nd Ed. (1986).

THE ROLE OF FLUCTUATIONS IN NUCLEAR REACTOR TRANSIENTS

M.A. Rodríguez+ and A. Díaz-Guilera*

+ Dpto. de Física Moderna, Univ. de Cantabria, E-39005, Santander, Spain
* Dpto. de Física, Univ. Autónoma de Barcelona, 08193 Bellaterra, Barcelona, Spain

1. INTRODUCTION

Stationary fluctuations in stable physical systems scale as the inverse of volume, hence they are so small that in most cases can be neglected. In nonstationary stuations there are different behaviours depending on the nature of the transient and also on the initial conditions. In the following we will refer as normal the cases in which fluctuations scale as the inverse of volume and anomalous all the other cases.

The study of anomalous fluctuations have a great interest from a statistical point of view since fluctuations dominate the evolution of the system.[1] This point has been experimentally observed in a variety of physical systems and it is usually asociated with the decay from unstable states. Anomalous fluctuations appear when starting up a system with a small number of initial elements (particles, individuals...). There is an initial time in which fluctuations dominate the evolution of the system followed by other time interval where the initial fluctuations propagate. As a result there is an anomalous enhancement of fluctuations.[1,2] This anomalous behaviour has been refered by many authors from different fields and with distinct words. For instance biologists deal with the spread of an epidemic calculating incubation times[3] and physicists are concerned with the initial fluctuation when starting up electronical devices,[4] chemical reactors[5] and lasers.[6]

Focusing on nuclear systems there is an experiment performed by Wimett[7] (1960) that shows the anomalous behaviour of fluctuations in a nuclear reactor. This experiment was made in the Godiva reactor. This reactor is essentially a small sphere of U^{235} composed by several parts that control the reactivity. No artificial source was present so initially the number of neutrons was very small and the conditions for the appearance of anomalous fluctuations were favourable. The reader interested on details is refered to the original paper of Wimett and to chapter 4 of William's book[8] where an interesting review of weak source fluctuations is presented. The reactor was made supercritical by assembling its components and the neutron density was measured as a function of time. After an initial time in which fluctuations were dominant the behaviour becomes normal. Then a temperature feedback mechanism took place and the system shut itself down.

As concern our study on transient fluctuations we can deduce interesting consecuences from this experiment. For instance, the variance of the time needed to reach the maximum neutron density was extremely large. It denotes a very anomalous behaviour, more than the one found in the decay from unstable states. We will treat this interesting point in Section 4.

On the other hand there are other situations which do not present anomalous fluctuations in the sense above defined but they present a large enhancement of fluctuations due to an amplification in the propagation of fluctuations. As we will show from a point of view of control these situations are almost so interesting as the case of anomalous fluctuations.

Our aim in this paper is to analyse the stochastic behaviour of a nuclear reactor during a transient with special interest in the implication of this analysis on the control of the system. We believe that this analysis could be of practical interest in a control framework. For instance we can define undesirable values of the control parameters, predict anomalous behaviours governed by random events and situations with amplificated fluctuations. In Section 2 we introduce the basic definitions and methods needed for doing an analysis of transient fluctuations. Also we present from a phenomenological point of view the mecanism of generation and propagation of fluctuations. In section 3 we introduce the stochastic models of nuclear reactor used in the analysis of fluctuations. With these models we take into account the effect of delayed neutrons and delayed temperature response. In section 4 we analyse the possibility of anomalous fluctuations and in general we classify the kind of transient depending on initial and final values of the control parameter. In this section we use a simple model of nuclear reactor. Finally in section 5 we present a numerical study of fluctuations with more complicated models of nuclear reactors taking into account delayed neutrons and temperature feedback effects. In this case only normal fluctuations are analysed.

2. DEFINITIONS AND SCOPE OF THE ANALYSIS

In this paper we restrict our attention to internal fluctuations since we are more interested on the initial evolution of the system. A recent study[9] with both external and internal fluctuations applied to lasers shows that internal fluctuations are relevant at initial times and external fluctuations dominate for long times. This is just what one expects form physical arguments and it must also be valid in other systems as a nuclear reactor.

As it is usual in standard theories of reactor kinetics we assume time dependent lumped models. Fluctuations are taken into account by means of a balance probability method giving rise to a master equation for the relevant variables of the system.[8] We remark that a Langevin treatment of fluctuations can be erroneous when considering weak sources at initial times since this approach does not keep the natural absorbing boundary which appears in the master equation without neutron source.[10]

Let us assume that in this frame of internal fluctuations and using balance probability methods we have a master equation for the probabilities $P(\bar{x}, \bar{\lambda}, t)$ of having the reactor in state $\bar{x}$ with control parameters $\bar{\lambda}$ at time t. Details of this equation will be given in the following section. Now we focus on the mathematical description of a transient.

Following deterministic theories we can consider a time dependent variation of control parameters $\lambda(t)$ as the imput of the system. From stochastic considerations we take $P(x, \lambda(t), t)$ as the response. The solution of a master equation with arbitrary time dependent parameters seems to be a impossible task. However we are able to deal with particular cases that represent two extreme situations and can give an idea of the phenomenology of general transients. Such cases are called adiabatic $\dot{\lambda}(t)=0$, and instantaneous $\dot{\lambda}(t) = \Delta\,\delta(t)$. The adiabatic transient represent a limit case in which the variation of control parameters are much slower than the evolution of the system. In this limit fluctuations are the corresponding to a stationary state with parameter $\lambda(t)$. Fluctuations in steady states are rather known. The instantaneous transient represents the limit case in which the evolution time of the control parameter are much smaller than the characteristic evolution times of the system. In such a case the problem reduces to one without time dependent parameters and initial conditions given by the stationary probability of the initial state. As in the previous case this problem is rather tractable.

2.-1 Analysis of transients

Normal fluctuations can be taken as a perturbation of a deterministic value originated by the discrete character of the matter and the finitess of realistic systems. From the more practical point of view of the control theory we can see them as an inevitable error around a deterministic and consequently controlable value of system variables. Bearing this in mind and in order to connect with control problems we will analyse normal fluctuations by defining fluctuations $\sigma^2(t)$ around a deterministic trajectory $x_d(t)$ as

$$\sigma^2(t)= \langle (x(t) - x_d(t))^2 \rangle$$

The deterministic trayectory $x_d(t)$ appears with probability one when doing a thermodynamic limit on the master equation. We identify the error due to inherent process (internal fluctuations) with $\sigma(t)$, the standard deviations around the deterministic trayectory. A very suitable method for dealing with normal fluctuations taking a deterministic trayectory as reference is the system size expansion method due to Van Kampen.[11] We will use this method in section 5.

Anomalous fluctuations require a more detailed analysis since as we will show later one can distinguish several time intervals with different behaviours. Anomalous fluctuations are properly generated at initial times, when the internal fluctuations drive to the system far from its unstable initial condition. The best manner for analizing the behaviour of the system at this initial time is by using passage time statistics for reaching a determinated level. After crossing this level, fluctuations can be considered as normal, so one can define deterministic trayectories with an initial probability given by the statistics of the initial fluctuations. An interesting and unknown question is about the possibility of controlling a system exhibiting anomalous fluctuations. We will refer to this question along the paper.

2.-2 Phenomenology of fluctuations in transients

In order to avoid complications let us consider in this section an one dimensional model. It is convenient to visualize the dynamics of the system and a good method is to employ a mechanical picture. Supose that by means of a given transformation we have a variable **x**, related with the original variable of the system, whose probability density P(x,t) obeys a Fokker-Planck equation

$$\frac{\partial P}{\partial t} = \frac{\partial}{\partial x}\left[\frac{\partial U}{\partial x} P\right] + \frac{\varepsilon}{2}\frac{\partial^2 P}{\partial x^2}$$

or equivalently a Langevin equation

$$\dot{x}(t) = f(x) + \varepsilon^{\frac{1}{2}}\xi(t)$$

with $f(x) = -\frac{\partial U}{\partial x}$ and being $\xi(t)$ a Gaussian white noise with intensity one. With this transformation we obtain a good visualization since the Langevin equation represents the dynamics of a overdamped particle in a potential $U(x)$ and driven by a small random force $\varepsilon^{\frac{1}{2}}\xi(t)$ The system control parameters are included in the potential, so we can write $U(x, \lambda)$ and visualize a general transient by considering the time dependent shape of the potential $U(x, \lambda(t))$ and its effect over the particle. In such a manner the phenomenology of transients can be easily explained.

First we consider an adiabatic transient. If the system is globally stable, so there is one well defined potential well slowly changing in time, the fluctuations are always normal. Bistable or critical situations are difficult to analyze since strictly an adiabatic limit does not exist when passing throught a critical point. In this paper only stable situations are analysed.

The instantaneous transient can be analysed suposing that just at the origin of times an instantaneous change of the potential shape occurs. Before this time one assumes that the system is in a stationary state so it is possible to know the probability of presence in a determined interval or point by calculating the stationary probability with the old potential. After this time, the system evolves with the new potential shape. Normal fluctuations are generated when the particle is in a nonzero potential anywhere. In such a case the systematic force f(x) dominates over the fluctuating one $\varepsilon^{\frac{1}{2}}\xi(t)$ since ε is a small parameter giving the strengh of internal fluctuations.

Anomalous fluctuations are generated when the particle is in a flat region of the potential. Indeed, in this case the systematic force f(x) is zero and the fluctuating force dominates the evolution of the system. Since the intensity of internal fluctuations ε is usually very small, the system will spend many time in the neighboring of the unstable point. In the limit case $\varepsilon \rightarrow 0$ the system would remain indefinitely in this point.

The shape of the potential generates different degrees of anomalous fluctuations. In table I we present the most known a y b and others with relevance in nuclear reactors (c y d). We classify transients by studying the assymptotic form of the mean and variance of the time needed for crossing a significant level (Table I). Case a is the decay from an unstable state. It is the most studied and best known case of generation of anomalous fluctuations. It is different to the other cases refered since the variance is not dependent on the noise intensity ε. Roughly speaking, one can say that this case

Table 1. Generation of anomalous fluctuations. Mean and variances of the decay time as a way of classificating anomalous transients.

	a: Unstable	b: Marginal	c: Diffusive	d: Logaritmic

$\langle t \rangle$	$\ln\left(\frac{2a^2}{\varepsilon}\right)$	$\frac{1}{(a^2\varepsilon)^{1/3}}$	$\sim \frac{1}{\varepsilon}$	$\frac{\ell^2}{\varepsilon(1-\alpha)}$
σ_t^2	$\frac{\ln 2}{a}$	$\frac{1}{(a^2\varepsilon)^{2/3}}$	$\sim \frac{2}{3\varepsilon^2}$	$\frac{\ell^4(8-2\alpha)}{\varepsilon^2(1-\alpha)^2(3-\alpha)}$
$U(x)$	$-ax^2$	$-ax^3$	0	$\alpha \ln x$ $(\alpha < 1)$

will generate the least anomalous fluctuations. From the point of view of the control we can see that the waiting time needed to start up the system have a small error ($(\ln 2/a)^2$. However the knowledge of the side by which the ball will descent (see figure in table I) has associated a great uncertainty. Case b is a marginal unstability. Sparce results are known about the anomalous fluctuations generated by this potential. We can see that both, the meantime and variance go to infinite when the intensity of the noise ε tends to zero. This seems to be an uncontrolable situation. Case c and d are very similar. They represent a diffusive potential. They are the least known, most anomalous and worst controlable of all cases. As we will see in section 4 all these situations are possible in a instantaneous transient of a nuclear reactor.

To finish this section we visualize in fig. 1 the propagation of fluctuations depending on whether the potential is convex or concave. The image is very expresive. In the first case small initial deviations amplificate during the evolution while in the second the initial deviations reduce itself. Even in a normal case, fluctuations could become very intense as a consequence of a persistent amplification in the propagation. Summarizing we note that an enhancement of fluctuations can be due to several causes. One is the generation of anomalous fluctuations followed by a relaxation to a new stable configuration, crossing consequently a region of concavity in the potential. Other could be an amplification of normal fluctuations when the particle cross a convex potential region followed by a relaxation to a stable configuration. The remarkable difference between the above situations is that in the first case fluctuations are of orden one, that is independent of the system size while in the second case fluctuations are normal (order V^{-1}) and could be reduced taking a larger system.

3. STOCHASTIC NUCLEAR REACTOR MODELS

Our aim in this work is to analize the role of most relevant phenomena in transient fluctuations, so we consider the stochasticity associated to the fission, absorption and source emission of neutrons as the basic neutronic events. Processes associated with the heat production and transfer are taken into account at the same description level as the neutronic processes. In table II we summarize the element needed to stablish a probability balance equation. Relevant variables are N, the number of neutrons, C, the number of precursors, and U, the number of thermal energy units in the fuel region. U is related to the difference of temperatures between fuel and coolant, assuming that the coolant is an infinite heat reservoir of fixed temperature:[12,13]

$$U = \frac{c_v \rho V}{q} (\theta_f - \theta_c)$$

The unit is taken to be q, the recovered energy per fission.

We assume that with these variables our system can be modelized as a discret Markovian process composed by several independent events. An event is characterized by the change induced in the relevant variables of the system. For example a fission event implies the emergence of (m_o-1) instantaneous neutrons, m_i precursors and one thermal energy unity in the fuel. The conditional probability for the occurrence of an event is given in table II where $\mathcal{L}$ is the probability per unit of time that a neutron induces a fission, $P_F(m_o,m_i)$ is the probability of producing m_o prompt neutrons and m_i precursors in a fission event. The strength of the source is s, λ_c is the capture rate, the decay constant of precursors is λ and the heat transfer from fuel to coolant coefficient is h_t.

Following a probability balance method we obtain the master equation governing the stochastic evolution of the above model. Taking $P(N,C,U,t)$ as the probability of having our reactor in the state $[N,C,U]$ this master equation reads:

$$\begin{aligned}\frac{\partial P(N,C,U,t)}{\partial t} = \sum_{m_o,m_i} & \mathcal{L} P_F(m_o,m_i)(N-m_o+1)\, P(N-m_o+1, C-m_i, U-1, t) \\ & + sV\, P(N-1, C, U, t) + \lambda (C+1)\, P(N-1, C+1, U, t) \\ & + h_t (U+1)\, P(N, C, U+1, t) + \lambda_c (N+1)\, P(N+1, C, U, t) \\ & - \{sV + (\lambda_c + \mathcal{L})N + \lambda C + h_t U\}\, P(N, C, U, t) \qquad (1)\end{aligned}$$

Fig. 1. Propagation of fluctuations. A convex potential amplificates initial deviations. A concave potential reduces initial deviations.

Table II. Elementary events

EVENT	NET CHANGE			PROBABILITY
	N	C	U	
Fission	m_0-1	m_1	+1	$\Lambda P_F(m_0,m_1) N$
External source of neutrons	+1	0	0	sV
Decay of precursors	+1	-1	0	λC
Heat transfer from fuel to coolant	0	0	-1	$h_t U$
Neutron capture	-1	0	0	$\lambda_c N$

where we have followed the same order for the description of events that in the table II.

This equation modelizes a linear evolution. Nonlinearities are taken into account when considering the explicit dependence of parameters with the value of variables. In this paper we consider a linear dependence of the absorption coefficient λ_c with the fuel temperature: $\lambda_c = \lambda_{c0} + \alpha\theta = \lambda_{c0} + (\gamma/V)\,U$. The model so defined provides a good description of the basic phenomena taking place in a general nuclear reactor and concerning to the generation of internal fluctuations. These phenomena induce several time scales on the evolution of the system. Mainly we have a time scale related with the instantaneous generation of neutrons, other with the existence of delayed neutrons and a last related with the delay in the heat transference. In order to analyze the effect of each phenomena separately we will consider three models A, B and C, as special cases of the more general model previously introduced.

Model A: It represents a limit situation in which the delayed neutrons and heat transference are considered as instantaneous - ($\lambda \to \infty$, $h_t \to \infty$). Defining the instantaneous feedback coefficient Γ as

$$\Gamma = \frac{\Lambda \gamma}{h_t}$$

and performing an adiabatic elimination of variables C and U we obtain a master equation for the single variable N. Taking the dual aproximation (two neutrons emerge per fission) for $P(m_0)$ this equation reads:

$$\begin{aligned}\frac{\partial P(N,t)}{\partial t} = {} & (N+1)\left(\lambda_c + \frac{\Gamma}{V} N\right) P(N+1,t) \\ & + sV\, P(N-1,t) + \Lambda (N-1)\, P(N-1,t) \\ & - \left[sV + (\Lambda + \lambda_c) N + \frac{\Gamma}{V} N(N-1) \right] P(N,t) \qquad (2)\end{aligned}$$

This is the master equation version of the Verhulst model. With this model we get two objectives, one is to isolate a time scale related with instantaneous processes, other is to obtain exact results for stationary moments.

Model B: It takes into account the effect of delayed neutrons. It comes from the general model with an adiabatic elimination of the variable u. The master equation for this model is:

$$\begin{aligned}\frac{\partial P(N,C,t)}{\partial t} &= (N+1)(\lambda_{c0} + \Gamma/v\, N)\, P(N+1,C,t) \\ &+ \lambda(C+1)\, P(N-1,C+1,t) + sv\, P(N-1,C,t) \\ &+ \Lambda \sum_{m_0,m_1} P_F(m_0,m_1)\,(N+1-m_0)\, P(N+1-m_0, C-m_1, t) \\ &- \left\{ sv + (\lambda_{c0}+\Lambda)N + \Gamma/v\, N(N-1) + \lambda C \right\} P(N,C,t)\end{aligned} \quad (3)$$

Model C: It takes as instantaneous the decay of precursors keeping the effect of thermal transference delay. It comes from the general model with an adiabatic elimination of the variable C. The master equation in this case is

$$\begin{aligned}\frac{\partial P(N,u,t)}{\partial t} &= (N+1)(\lambda_{c0} + \frac{\gamma}{v} u)\, P(N+1,u,t) \\ &+ h_t(u+1)\, P(N,u+1,t) + sv\, P(N-1,u,t) \\ &+ \Lambda \sum_{m_0=0}^{\infty} P_F(m_0)\,(N+1-m_0)\, P(N+1-m_0, u-1, t) \\ &- \left\{ sv + (\lambda_{c0} + \Lambda + \gamma/v\, u)N + h_t u \right\} P(N,u,t)\end{aligned} \quad (4)$$

Comparing the results obtained with models A and B,C we can extract information of the effect due to delayed neutrons and the delay in a thermal transference.

Before concluding with this section it is convenient to note that a discret formulation for the heat removal seems to be a rather artificial method. We are conscious of this fact and on spite of this we believe that this formulation is adequate. In fact, a more natural formulation keeping discreteness for neutrons and precursors and continuity for the energy is easy to derive. Internal fluctuations for the heat transfer would be taken into account by a fluctuation disipation theorem.[14] The reason for using the discret formulation is that at the phenomenological description level used along the paper it is equivalent to the above mentioned discret-continous formulation. Roughly speaking we are using an inverse Langevin approach.

4. ANALITICAL RESULTS FOR MODEL A

As we have mentioned in the above section model A is the well known Verhulst model, used in population dynamics and chemical reaction contexts. In most studies in these fields the model is used jointly with a Langevin description for the internal fluctuations. Indeed, the master equation formulation is equivalent to the Langevin equation when the system fluctuates far from unstable points. We must keep the master equation formulation since anomalous fluctuations are originated just in these unstable points.

Stationary exact results for the probability and moments are easily derived. Let us define a generating function as usually

$$G(x,t) = \sum_{N=0}^{\infty} x^N P(N,t) \tag{5}$$

The equation obeyed by G(x,t) is from (3):

$$\frac{\partial G}{\partial t} = (1-x)\left[\frac{\Gamma x}{V}\frac{\partial^2 G}{\partial x^2} + (\lambda_c - \lambda x)\frac{\partial G}{\partial x} - \delta V G\right] \tag{6}$$

A stationary solution is easily found. It is

$$G_s(x) = \frac{M\left(\frac{\delta V}{\lambda}, \frac{\lambda_c V}{\Gamma}, \frac{\lambda V}{\Gamma} x\right)}{M\left(\frac{\delta V}{\lambda}, \frac{\lambda_c V}{\Gamma}, \frac{\lambda V}{\Gamma}\right)} \tag{7}$$

with M(a,b,x) being the Hypergeometric Confluent function. From (7) the first moments can be immediately calculated taken into account the relation of factorial moments $C_K = \langle N(N-1)\dots(N-K+1)\rangle$

$$C_K = \left.\frac{\partial^K G(x,t)}{\partial x^K}\right|_{x=1}$$

These results can be used to calculate exactly fluctuations in adiabatic transients.

Let us focus our attention in some of the key questions of this paper namely, when anomalous fluctuations occur, what kind of fluctuations are and what is its possible implication in a control theory. In order to answer these questions we introduce a transformation due to De Pasquale and Tartaglia.[15] In terms of the generating function this transformation reads:

$$G(z,t) = \int_0^{\pi/2} dx\, P(x,t)\, \exp\left[\frac{\lambda V}{\Gamma}(z-1)\,\mathrm{sen}^2 x\right] \tag{8}$$

where P(x,t) obeys a Fokker-Planck equation which can be obtained substituting (8) in (6). This equation is

$$\frac{\partial P}{\partial t} = \frac{\partial}{\partial x}\left[\frac{du}{dx} P(x,t)\right] + \frac{1}{2}\frac{\Gamma}{\lambda V}\left[\frac{\partial^2}{\partial x^2} P(x,t)\right] \tag{9}$$

where u(x), the potential in the mechanical frame of section II, is

$$u(x) = -\mu_1 \ln \mathrm{sen}^2 x - \mu_2 \ln \cos^2 x + \frac{1}{2}\cos^2 x \quad ; \quad x \in \left(0, \frac{\pi}{2}\right) \tag{10}$$

with

$$\mu_1 = \frac{\Gamma}{2\lambda V}\left(\frac{\delta V}{\lambda} - \frac{1}{2}\right) > -\frac{\Gamma}{4\lambda V}$$

$$\mu_2 = \frac{1}{2}\left[\frac{\lambda_c}{\lambda} - \frac{\Gamma}{\lambda V}\left(\frac{\Delta V}{\lambda} + \frac{1}{2}\right)\right] > 0 \qquad (11)$$

Equation (9) indicates that the new process x(t) can be visualized as the position of a overdamped particle moving in the potential u(x) with a stochastic Gaussian white force with a small intensity given by $\Gamma/\lambda V$.

The relation between our new stochastic process x(t) and the original one N(t) can be stablished from (8). In terms of moments it writes

$$C_K(t) = \left(\frac{\lambda V}{\Gamma}\right)^K < sen^{2K} x(t) > \qquad (12)$$

Exact relations between passage time statistics of the original and new system are difficult to obtain, however it is easy to see from (8) that in the limit $V \to \infty$ the position of the particle x and the neutron density n=N/V are related as $n = (\lambda/\Gamma)\, sen^2 x$ and then we can use the fact that passage times for crossing thresholds n_o, x_o obeying the relation $n_o = (\lambda/\Gamma)\, sen\, x_o$ are aproximately equal, at least in the lowest order.

The relation between fluctuations is easy to do taking into account the expression (12). From this expression it is easy to see that a moderate amplification (reduction) of fluctuation in x implies an amplification (reduction) of fluctuations in N (taking as strength of fluctuations the reduced variance $N_{\bar{N}N} = C_2 - C_1^2$). Bearing in mind these relations we can easily translate the behaviour of the process x(t), position of the particle, to the variable N(t) neutron number. The behaviour of x(t) can be analyzed taking into account the shape of the potential. In figure 2 we have distinguished four regions in the parameter space, (μ_1, μ_2) depending on the shape of the potential u(x). In region II the potential has only one minimum, so it is anywhere concave. In region I the potential has one inflection point and one minimum, so below the inflection point the potential is convex and it will be concave above. Both regions are separated by the curve[15]

$$108(\mu_1 + \mu_2)^2 + (4\mu_1 + 4\mu_2 - 1)^3 = 0 \qquad (13)$$

In region IV there are neither inflection point nor minimum and in region III there are one maximum, minimum and inflection point. The frontier between these regions is given approximately by the curve $\mu_2 = 0.5$ when $|\mu_1|$ is small. A more detailed image of the potential with determined values of parameters μ_1 and μ_2 are given in figures 2.- a,b,c.

It is important to know physical implications of this analysis. Regions I and II correspond to states with significant independent sources $\mu_1 = 0 \Rightarrow \Delta V > (\lambda/2)$. In this parameter region the generation of anomalous fluctuations is not possible. Only in region I and when the particle is below the inflection point the amplification of normal fluctuations is possible. Regions II and IV correspond to situations with weak independent sources $\Delta V < \lambda/2$. On the other hand, region IV corresponds to parameters of a subcritical reactor and III to a supercritical one. A reactor with parameters in region IV must be in a extinction state.

As concern transient states we note that variation of the

parameter λ_c (for example variation of a control rod position) imply variations of parameter μ_2 with μ_1 fixed. To introduce a control rod means in our figure 2 to vary parallely to μ_2 axis from the right to the left (for example the path a, b, c, d). Variations of the source imply variations of parameter μ_1 (for example the path $\alpha, \beta, \gamma, \Omega, \xi$ when retiring a source from the reactor).

Generation of anomalous fluctuations in region III and IV can be analyzed by studying the asymptotic form of u'(x) for x going to zero:

$$u'(x) \underset{x \to 0}{\sim} -\frac{2\mu_1}{x} + (2\mu_2 - 1)x + (2\mu_2 + 1)x^2 \tag{14}$$

This expression indicates that depending on the value of parameters we can have an unstable behaviour ($\mu_1 \sim 0$, $\mu_2 \neq 1/2$) or marginal ($\mu_1 \sim 0$, $\mu_2 = 1/2$) or diffusive ($\mu_1 < 0$)

A more detailed analysis of anomalous fluctuations in this model will be the subject of a further publication. Now let us summarize the most important facts:

i) Using the Fokker-Plank equation (9) is possible to extract a complete information about passage time statistics of variable x. More difficult is to relate it with the statistics of the original process N.

ii) Previous results (14) indicate a diversified behaviour of the generation of anomalous fluctuations. This point seems to be very interesting since possibly the experimental situation found in the Godiva reactor (see Section I) is included in one of these cases.

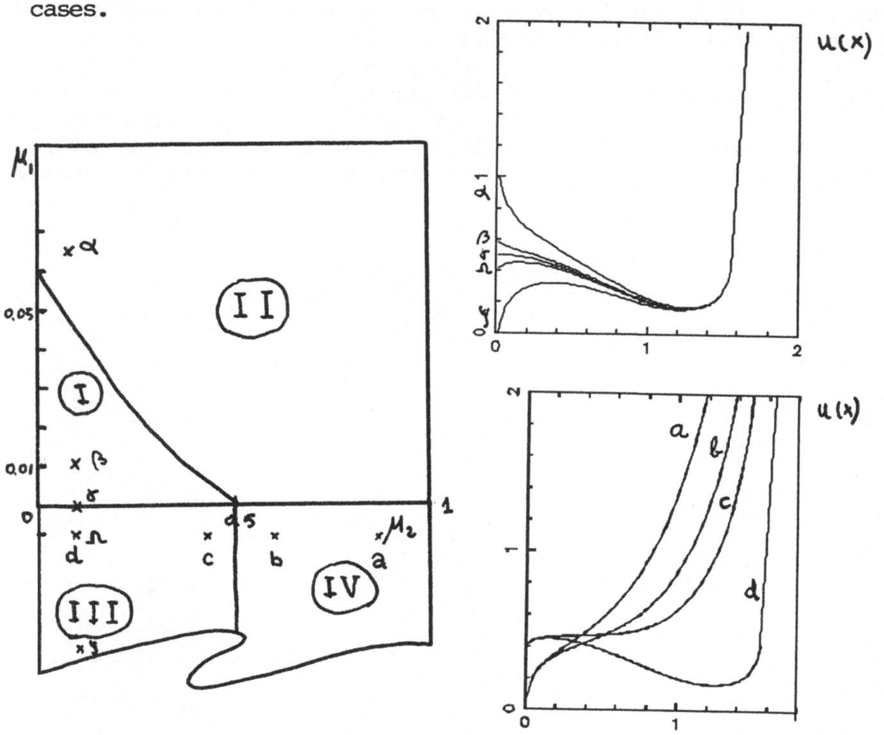

Fig. 2. Parameter space (μ_1, μ_2) with definitions of different regions. Some potentials are plotted.

iii) In a instantaneous transient the initial stationary state gives the probability of starting with a determined value. The potential in this initial situation also gives information about this probability. The potential of the final state gives the behaviour of the fluctuations in the transient. With the help of figure 2(a,b,c) we can qualitatively analyze the possible behaviours of fluctuations. For instance, transients from region IV have a high probability of exhibiting anomalous fluctuations when going to region III and great amplification of fluctuations when going to region I. Transients from region II have a moderate probability of exhibiting anomalous fluctuations when going to region III and high probability of moderate amplification of fluctuations when going to region II. We note that quantitative results are easy to obtain numerically.

5. NUMERICAL ANALYSIS OF MODELS B AND C

In deterministic kinetic theories the effect of delay in the neutron production or heat removal plays an important role defining new time scales or even damped oscilatory behaviours. We have been interested on investigating this effect over fluctuations in transient states.[16,17] In these papers we studied fluctuations of power reactors with source. With words of section 4 we dealt with parameters in region I and II. In this case it is possible to find normal amplification of fluctuations and it is just what we investigated.

One of the problems requiring our attention was the strength of relative error reached in these fluctuations. This is an interesting problem since gives information about the controlability of the transient. Our conclussions were that contrarily to a previous study[18] (giving an error of 758%) the greatest relative error stimated was about 10%. Indeed, as we have seen along the paper normal fluctuations scale as the inverse of volume, and generally nuclear reactors have a great volume. But this result only states that a deterministic evaluation is reliable. Possibly, it has a greater interest to know the ratio between fluctuations in transient states and stationary

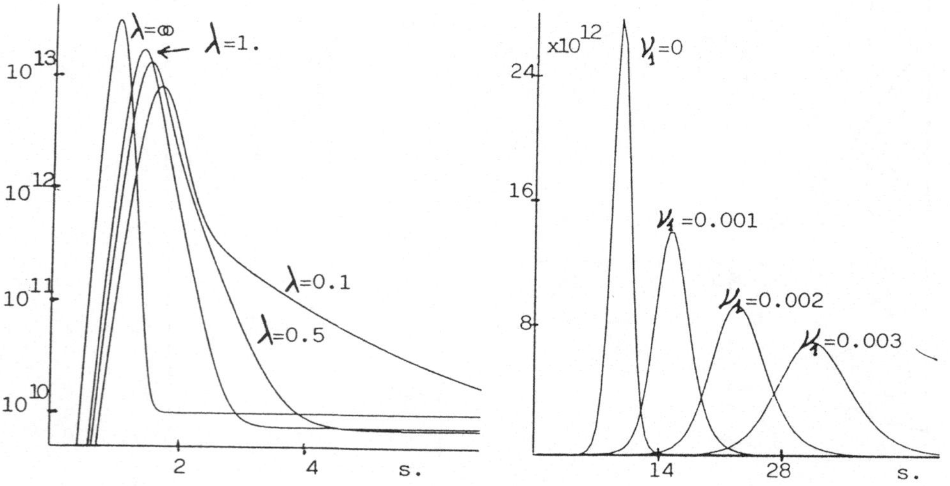

Fig. 3. Relative neutron fluctuations versus the time for some different values of the parameters. Left : Δ =2; ν_1 =0,001. Right: Δ =1; λ =1.

states with the same value of the deterministic trajectory. This would permit to use a more adjusted and best known confidence levels in a programmed transient. This ratio is just the relation between the variance of an instantaneous and adiabatic transient as function of the same value of the deterministic trajectory. To study this point together with a brief discussion about the effect of delay in the neutron production and heat removal of fluctuations in transients will be the objectives of this section.

Normal fluctuations can be studied by a method of linearization around the deterministic trajectory. Starting with a master equation it is convenient to use the system size expansion. The deterministic trajectory comes as a solution of deterministic equation for densities $(n = N/V,\ c = C/V,\ u = U/V)$. In the thermodynamic limit $V \to \infty$ these equation follow from (2), (3) and (4) and are respectively for model A, B and C:

$$\frac{dn}{dt} = -\Gamma n^2 + (\Delta - \nu_1 \Lambda) n + \lambda c + s \qquad (15)$$

$$\begin{cases} \dfrac{dn}{dt} = -\Gamma n^2 + \Delta n + s \\[2ex] \dfrac{dc}{dt} = \Lambda \nu_1 n - \lambda c \end{cases} \qquad (16)$$

$$\begin{cases} \dfrac{dn}{dt} = -\gamma n u + \Delta n + s \\[2ex] \dfrac{du}{dt} = \Lambda n - h_t u \end{cases} \qquad (17)$$

with $\bar{\nu} = \langle m_0 \rangle$, $\nu_1 = \langle m_1 \rangle$, $\Delta = (\bar{\nu} - 1)\Lambda - \lambda_{c0}$

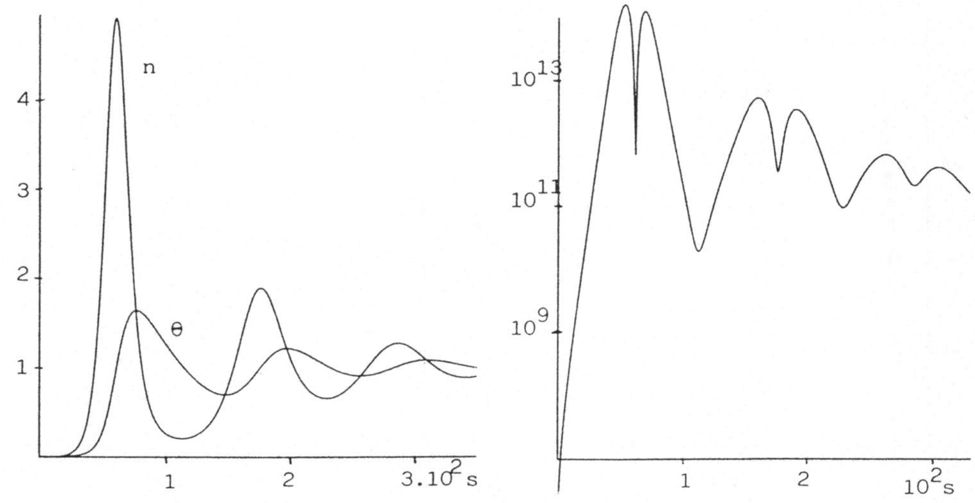

Fig.4. Neutron density (n) and temperature (θ) normalized to their steady values, and relative neutron fluctuations respectively versus the time. The values of the parameters are: Δ=0,17; h_t=0,02.

Fluctuations are calculated with a perturbative method around the deterministic trajectory. For example for the neutron number N(t) we use the following expansion

$$N(t) = V n(t) + V^{\frac{1}{2}} \xi_n(t) + O(V^{-1}) \tag{18}$$

A detailed description of this method is given in refs. 16 ,17 . Here we only show the main results obtained in a numerical analysis of models B and C with the following fixed values of parameters $\Lambda = 1000\ s^{-1}$, $s = 100\ n \cdot s^{-1} cm^{-3}$, $\Pi = 6.14 \times 10^{-8}$. Initial state from which the transient starts has for Δ a value Δ =-0.1. We have considered instantaneous and adiabatic changes. The instantaneous transient is choosed in such a manner that it appears enhancement of fluctuations. The stationary value of n_{ST} is the same for the three models, so differences found in the analysis are only due to delay effects. In figure 3 we show the effect of delayed neutrons over fluctuations, defined as $1/V \langle (N(t) - V n(t))^2 \rangle$ in a instantaneous transient. The effect of delay can be increased by taking a smaller λ or a greater ν_1 . The first case which corresponds to take a longer time of appearance of delayed neutrons affects mainly to the relaxation of fluctuations. The second case corresponds to changes in the delayed neutron fraction. This change mainly affects to the time of maxima fluctuations keeping the shape of the curve. In both cases the effect of delay is two fold, on one hand the intensity of fluctuations is reduced on the other the time in which fluctuations are amplified is longer.

In figure 4 we show an important effect due to a delay in the heat removal. From a stability analysis of deterministic equations (17) is possible to find the parameter region in which a damped oscillatory behaviour occurs:

$$\Delta > 4 h_t$$

In such a situation we show in figure 4 that fluctuations follow also

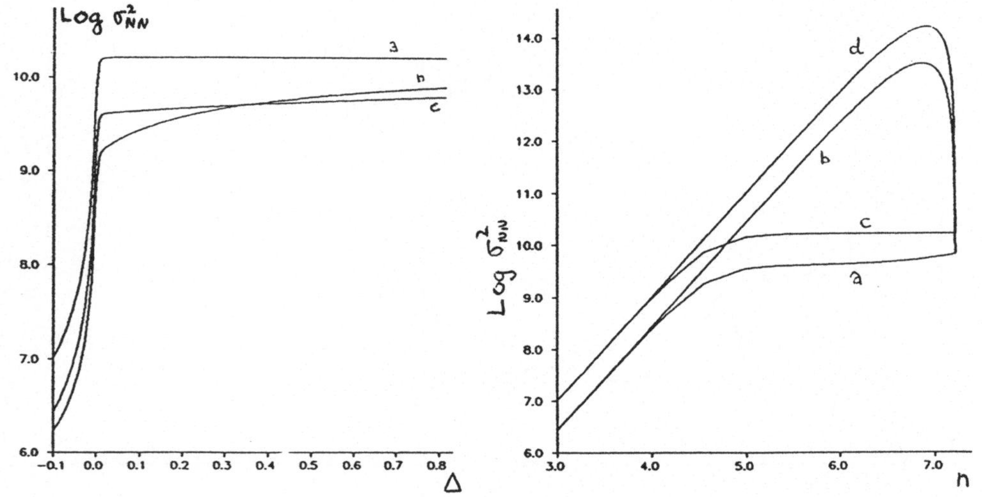

Fig. 5a.- Adiabatic fluctuations for a ($\nu_1 = 0$), b ($\nu_1 = 0.001$, $\lambda = 0.1$) and c ($\nu_1 = 0.003$, $\lambda = 0.1$)

Fig. 5b.- A comparison between instantaneous [d, b] and adiabatic [a, c] transients. (a, b; $\nu_1 = 0.001$, $\lambda = 0.1$) (c, d; $\nu_1 = 0$).

an oscilatory behaviour. It is easily explained taken into account our phenomenological ideas of propagation of fluctuations, depending on the curvature of the potential.

Finally in figure 5 we have a plot of adiabatic fluctuations. In 5a we present the effect of delayed neutrons over the adiabatic fluctuations. As in the instantaneous case a reduction of about 10% is appreciated. The effect of delay in the neutron generation λ can be more efective than the effect of a change in the delayed neutron fraction depending on the value of Δ. In 5b we present a comparison between fluctuations in a instantaneous and adiabatic transient as function of the value of n(t), the deterministic trajectory. Differences can have a factor 10^4.

ACKNOWLEDGEMENT

Financial support of the C.A.I.C.Y.T. is acknowledged.

REFERENCES

1. M. Suzuki, "Order and Fluctuations in Equilibrium and Non-equilibrium Statistical Mechanics", Edited by Nicolis G., Dewel G. and Turner J.W., Wiley, New York (1981).
2. K. Kawasaki, M.C. Yalabik and J.D. Gunton, Phys. Rev. A 17: 455 (1978).
3. J. Gani, "Stochastic Model for Bacteriophage", Methuen Rev. Series, 1965.
4. S. Kabashima, M. Itsumi, T. Kawakubo, T. Nagashima, J. Phys. Soc. Jap. 39: 1183 (1975).
5. F. de Pasquale, P. Tartaglia, P. Tombesi, Il Nuovo Cimento, 69B: 228 (1982).
6. F.T. Arecchi, U. Degiorgio and B. Querzola, Phys. Rev. Lett. 19: 1168 (1967).
7. T.F. Wimett et al., Nucl. Sci. Engng., 8:691 (1960).
8. M.M.R. Williams, "Random Processes in Nuclear Reactors". Pergamon Press, Oxford (1974).
9. F. de Pasquale, J.M. Sancho et al., Phys. Rev. Lett 56:2473 (1986).
10. M.A. Rodriguez, A. Díaz-Guilera and J.M. Sancho, Ann. nucl. Energy 13:49 (1986).
11. N.G. Van Kampen, "Stochastic Processes in Physics and Chemistry", North-Holland, Amsterdam (1983).
12. N. Morishima, J. nucl. Sci. Technol., 10:29 (1973).
13. M. Matthey, Ann. nucl. Energy 6: 345 (1979).
14. A.M.S. Tremblay, M. Arai and E.D. Siggia, Phys. Rev. A23:1451 (1981).
15. F. de Pasquale, P. Tartaglia and P. Tombesi, Phys. Lett. 78A:129 (1980).
16. A. Díaz-Guilera, J.M. Sancho and M.A. Rodríguez, Ann. nucl. Energy 12:501 (1985).
17. A. Díaz-Guilera, M.A. Rodríguez and J.M. Sancho, Ann. nucl. Energy 12:441 (1985).
18. E.R. Quabili, Ann. nucl. Energy 7:185 (1980).

THE EFFECT OF FUEL DEPLETION ON STOCHASTIC BEHAVIOUR IN SUPERCRITICAL SYSTEMS

M. M. R. Williams

Nuclear Engineering Department
The University of Michigan
Ann Arbor, MI 48109, U.S.A.

INTRODUCTION

Extensive studies have been carried out of the probability distribution of neutron number in multiplying systems and a comprehensive list of references is available[1]. Essentially, the procedure involves the construction of a probability balance equation for $P(n,t)$, the probability that exactly n neutrons are present in the system at time t. Various extensions to this distribution may easily be developed which include delayed neutron precursors, number of counts recorded, etc. An aspect of this general problem that has by and large been neglected is the inclusion of burn-up of the fissile material. That is, the destruction of nuclei that occurs when fission takes place. The justification for such neglect is that during the time of measurement the number of nuclei destroyed is negligible compared with those remaining. This is indeed true except in one special case; namely that of a nuclear explosion. In this case, the time scale is such (10^{-7}s) that a large fraction of the fissile material is consumed before the period of interest is over. Thus it is necessary to consider the two parameter probability distribution function $P(n_1,n_2,t)$; the probability of finding n_1 neutrons and n_2 fissile nuclei at time t.

This problem is not unique to neutron multiplication. An analogous although somewhat simpler situation is to be found in epidemiology, i.e. the theory of infectious diseases.[2] Consider, for example, a community which at a given time has a certain number of susceptibles (i.e., persons who can catch a disease, exluding those who are immune) and a certain number of infectives (i.e., persons who have the disease and can communicate it). Then, clearly, each infective, on interacting with a susceptible, will destroy a susceptible and produce two infectives. This is analogous to fission in which two neutrons are emitted. The actual destruction of infectives by interacting with a susceptible does not occur so that the analogy with non-fission capture is absent. Infectives are removed by death or recovery and this is analogous to neutron leakage which also occurs without destroying a fissile nucleus.

In the work to be described here, we shall set up a probability balance equation for the above problem, discuss methods of solving this equation and deduce what the results can tell us about the physical nature of the system.

THEORY

We define the probability distribution $P(n_1,n_2;t)$ as the probability that, at a given time t, the system contains n_1 neutrons and n_2 fissile nuclei. We assume that the system contains only fissile material although non-fission capture is allowed. We will also account for leakage out of the system and the presence of an artificial source of neutrons. We shall assume that all neutrons have the same speed and ignore the spatial variation of the neutron field and the delayed neutrons.

The other probability distribution needed is that of the neutron yield in a fission. We let p(k) be the probability that k neutrons are emitted in a fission.

It is now necessary to formulate a probability balance and we do this by considering the various mutually exclusive events that contribute to

$$P(n_1, n_2 \quad ; t + \Delta t) \tag{1}$$

in the time Δt.

<u>Fission</u>:

$$\alpha\Delta t(n_2 + 1) \sum_{k=0}^{\infty} p(k)(n_1 + 1 - k)P(n_1 + 1 - k, n_2 + 1; t) \tag{2}$$

where α is the probability of fission per unit time per fissile nucleus.

<u>Capture</u>:

$$\beta\Delta t(n_1 + 1)(n_2 + 1)P(n_1 + 1, n_2 + 1; t) \tag{3}$$

where β is the probability of non-fission capture per unit time per fissile nucleus.

<u>Leakage</u>:

$$\mu\Delta t(n_1 + 1)P(n_1 + 1, n_{2;t}) \tag{4}$$

where μ is the probability of neutron leakage per unit time.

$$S\Delta t\, P(n_1 - 1, n_2; t) \tag{5}$$

where S is the probability that the source emits a neutron per unit time.

The probability that none of these events occur in Δt is equal to

$$[1 - (\alpha + \beta)\Delta t\, n_1 n_2 - \mu\Delta t\, n_1 - S\Delta t]\, P(n_1, n_2; t) \tag{6}$$

Rearranging the terms, dividing by Δt and allowing $\Delta t \to 0$, leads to

$$\frac{\partial P(n_1, n_2; t)}{\partial t} = \alpha\,(n_2 + 1) \sum_{k=0}^{\infty} p(k)(n_1 + 1 - k) P(n_1 + 1 - k, n_2 + 1; t)$$

$$+ \beta(n_1 + 1)(n_2 + 1) P(n_1 + 1, n_2 + 1; t) + \mu(n_1 + 1) P(n_1 + 1, n_2; t) + SP(n_1 - 1, n_2; t)$$

$$- [(\alpha + \beta) n_1 n_2 + \mu n_1 + S] P(n_1, n_2; t) \tag{7}$$

Now following the usual procedure, we introduce the probability generating function

$$G(z_1, z_2; t) = \sum_{n_1=0}^{\infty} \sum_{n_2=0}^{\infty} z_1^{n_1} z_2^{n_2} P(n_1, n_2; t) \tag{8}$$

Multiplying eq. (7) by

$$z_1^{n_1} z_2^{n_2}$$

and summing over all n_1 and n_2, we obtain the following partial differential equation.

$$\frac{\partial G}{\partial t} = [\alpha\, f(z_1) + \beta - (\alpha + \beta) z_1 z_2] \frac{\partial^2 G}{\partial z_1 \partial z_2} + \mu(1 - z_1)\frac{\partial G}{\partial z_1} + (z_1 - 1) SG \tag{9}$$

where the auxilliary generating function

$$f(z) = \sum_{k=0}^{\infty} z^k p(k) \tag{10}$$

α, β, μ and S can be time-dependent if the physical situation warrants it.

In the special case of 2 neutrons per fission where $f(z)=z^2$ and zero non-fission capture (i.e., $\beta=0$), we find

$$\frac{\partial G}{\partial t} = \alpha\left[z_1^2 - z_1 z_2\right]\frac{\partial^2 G}{\partial z_1 \partial z_2} + \mu(1 - z_1)\frac{\partial G}{\partial z_1} + S(z_1 - 1)G \tag{11}$$

which, is the equation for the infectious disease problem mentioned above with immigration of infectives. The reason for β being zero in this case is that there are generally no 'Kamikazi' events in epidemiology. If there is immigration of susceptibles or, in neutronic language, replenishment of fuel, the immigration term becomes $S_2(z_2-1)G$ where S_2 is the immigration rate.

Although the equations for P and G are linear, they actually describe a non-linear system as we shall demonstrate below. Before that, however, let us note that the conventional probability equation for neutron fluctuations assumes that the number of fissile nuclei, n_2, remains constant with time. If this is the case, we can write

$$P(n_1, n_2; t) = P(n_1; t)\delta_{n_2,\overline{n}_2} \tag{12}$$

where $\overline{n_2}$ is the number of fissile nuclei. The corresponding generating function

$$G(z_1, z_2, t) = G(z_1, t)z_2^{\overline{n}_2} \tag{13}$$

and hence the equation for $G_1(z_1,t)$ becomes

$$\frac{\partial G_1}{\partial t} = \left\{\overline{n}_2\left[\alpha f(z_1) + \beta - (\alpha + \beta)z_1\right] + \mu(1 - z_1)\right\}\frac{\partial G_1}{\partial z_1} + (z_1 - 1)SG_1 \tag{14}$$

Equation (14) is the standard probability generating function equation from which the conventional reactor kinetics equation may be derived (without delayed neutrons) as well as the variance and higher moments. Such equations are linear and can be solved recursively. The same remarks cannot be made of eqn (9) as we shall discuss in the next section.

THE MOMENT EQUATIONS

Although it is possible to obtain explicit solutions to eqn (14) and hence obtain the corresponding probabilities and moments, eqn (9) is not so amenable. In such cases, methods of appoximation must be introduced and it is the purpose of this section to examine some possible methods of approach.We shall assume throughout that there is no steady source but that a burst of neutrons is introduced into the system at t=o. Thus the initial condition is

$$G(z_1, z_2; 0) = z_1^{n_{10}} z_2^{n_{20}}$$

Mean value approximation

In this technique we use the approximation given by eqn (13) but allow $\bar{n}_2$, the fissile nuclei concentration, to be unknown and time dependent.

Differentiating eqn (14) with respect to z_1 in the usual fashion we obtain the following set of equations:

$$\frac{d\bar{n}_1}{dt} = (\alpha - \beta)\bar{n}_1\bar{n}_2 - \mu\bar{n}_1 \tag{15}$$

$$\frac{d\bar{n}_2}{dt} = -(\alpha + \beta)\bar{n}_1\bar{n}_2 \tag{16}$$

$$\frac{dC_{11}}{dt} = 2(\alpha - \beta)\bar{n}_2 C_{11} - 2\mu C_{11} + (\alpha + \beta)\bar{n}_1\bar{n}_2 + \mu\bar{n}_1 \tag{17}$$

where the autocovariance

$$C_{11} = \overline{n_1^2} - \bar{n}_1^2 \tag{18}$$

and for simplicity we have assumed a simple birth and death process such that $f(z)=z^2$, i.e., 2 neutrons per fission.The initial conditions are

$$\overline{n_1}(0) = n_{10} \quad \overline{n_2}(0) = n_{20} \quad C_{11}(0) = 0$$

The set of equations (15)-(17) can be solved analytically by a recursive method.Eqns(15) and (16) are the conventional burn-up equations for a single isotope system.

The Gaussian Approximation

In obtaining eqns (15)-(17) we employed an approximation that neglected the correlation between n_1 and n_2. Let us obtain the moment equations directly for eqn (9) by differentiation. We shall omit the steady source term. The result, for $f(z)=z^2$, is

$$\frac{d\bar{n}_1}{dt} = (\alpha - \beta)\overline{n_1 n_2} - \mu\bar{n}_1 \tag{19}$$

$$\frac{d\bar{n}_2}{dt} = -(\alpha - \beta)\overline{n_1 n_2} \tag{20}$$

$$\frac{d\overline{n_1^2}}{dt} = 2(\alpha-\beta)\overline{n_1^2 n_2} - 2\mu\overline{n_1^2} + (\alpha+\beta)\overline{n_1 n_2} + \mu\overline{n}_1 \tag{21}$$

$$\frac{d\overline{n_2^2}}{dt} = (\alpha+\beta)\overline{n_1 n_2} - 2(\alpha+\beta)\overline{n_1 n_2^2} \tag{22}$$

$$\frac{d\overline{n_1 n_2}}{dt} = -(\alpha-\beta+\mu)\overline{n_1 n_2} - (\alpha+\beta)\overline{n_1^2 n_2} + (\alpha-\beta)\overline{n_2^2 n_1} \tag{23}$$

These five equations involve 7 unknowns. The triple correlations

$$\overline{n_1^2 n_2} \quad \text{and} \quad \overline{n_2^2 n_1}$$

can only be related to higher order moments and thus our system is not closed. We effect a closure by using the cumulant expansion and retain only second order terms. This leads to the following 'Gaussian' approximation.

$$\overline{n_1 n_2 n_3} \cong \overline{n}_1 \overline{n}_2 \overline{n}_3 + \overline{n}_3\left(\overline{n_1 n_2} - \overline{n}_1 \overline{n}_2\right) + \overline{n}_1\left(\overline{n_2 n_3} - \overline{n}_2 \overline{n}_3\right) + \overline{n}_2\left(\overline{n_1 n_3} - \overline{n}_1 \overline{n}_3\right) \tag{24}$$

which for the two special cases $n_2=n_3$ and $n_1=n_3$, gives

$$\overline{n_1 n_2^2} \cong \overline{n_2^2}\,\overline{n}_1 + 2\overline{n}_2\left(\overline{n_1 n_2} - \overline{n}_1 \overline{n}_2\right) \tag{25}$$

$$\overline{n_2 n_1^2} \cong \overline{n_1^2}\,\overline{n}_2 + 2\overline{n}_1\left(\overline{n_1 n_2} - \overline{n}_1 \overline{n}_2\right) \tag{26}$$

Using these relations in eqns (19)-(23) and rearranging, we find

$$\frac{d\overline{n}_1}{dt} = (\alpha-\beta)\overline{n}_1\overline{n}_2 - \mu\overline{n}_1 + (\alpha-\beta)c_{12} \tag{27}$$

$$\frac{d\overline{n}_2}{dt} = -(\alpha+\beta)\overline{n}_1\overline{n}_2 - (\alpha+\beta)c_{12} \tag{28}$$

$$\frac{dC_{11}}{dt} = 2(\alpha-\beta)\bar{n}_2 C_{11} - 2\mu C_{11} + [\alpha+\beta+2(\alpha-\beta)\bar{n}_1]C_{12} + (\alpha+\beta)\bar{n}_1\bar{n}_2 + \mu\bar{n}_1 \tag{29}$$

$$\frac{dC_{22}}{dt} = -2(\alpha+\beta)\bar{n}_1 C_{22} + (\alpha+\beta)(1-2\bar{n}_2)C_{12} + (\alpha+\beta)\bar{n}_1\bar{n}_2 \tag{30}$$

$$\frac{dC_{12}}{dt} = [(\alpha-\beta)\bar{n}_2 - (\alpha+\beta)\bar{n}_1 - \alpha+\beta-\mu]C_{12}$$

$$-(\alpha+\beta)\bar{n}_2 C_{11} + (\alpha-\beta)\bar{n}_1 C_{22} - (\alpha-\beta)\bar{n}_1\bar{n}_2 \tag{31}$$

Thus we have a closed set which can be solved numerically.The initial conditions are

$$\bar{n}_1(0) = n_{10} \quad \bar{n}_2(0) = n_{20} \quad C_{ij}(0) = 0$$

NUMERICAL SOLUTION OF THE MOMENT EQUATIONS

In order to assess the importance of correlations between neutron number and nuclei number, we have solved the two sets of equations (15)-(17) and (27)-(31). Time is measured in units of neutron lifetime $l=1/n_{20}(\alpha+\beta)$, multiplication in term of $k=2\alpha/(\alpha+\beta)$ and leakage in term of μl. For actual values we have set $n_{10}=5\times10^7$, $n_{20}=2.5\times10^{25}$, $\mu l=0.1$ and $k=1.8$. These values correspond to fissile data for U^{235} at 1 MeV in a sphere of radius 5 cms. However, the value of k assumes 2 neutrons per fission and not the correct physical value of 2.5. Figure 1 shows the values of the relative variance of the neutrons, defined by $RV1=(C_{11})^{1/2}/\bar{n}_1$, for the uncorrelated (u) and Gaussian (G) approximations. Whilst the uncorrelated results are very close to those for the Gaussian approximation up to about 50 neutron lifetimes, they fail in the physically interacting region when burn-up is appreciable. This indicates that there is significant correlation between nuclei and neutron number in the burn-up region. In fact, calculations show that during the burn-up phase the number of neutrons present exceeds the number of fissile nuclei and therefore one might expect the populations to be statistically dependent. Figure 2 shows the relative variances of neutrons and nuclei using the Gaussian approximatin from which we note that for up to about 50 neutron lifetimes the behaviour is virtually the same as for no burn-up. However, in the burn-up region where the neutron number increases significiantly and the fissile nuclei number decreases correspondingly, the variances respectively decrease and increase. The reason for the maximum and minimum can be understood from Figure 3 which illustrates the relative variances of neutrons and the neutron number. During the initial phases of the burn-up region, the neutron number increases due to multiplication and the relative

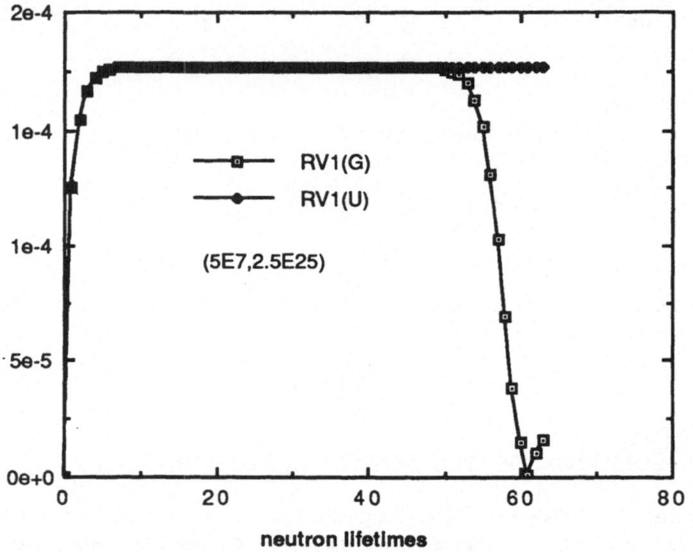

Fig.1 Relative variance for uncorrelated and Gaussian approximations

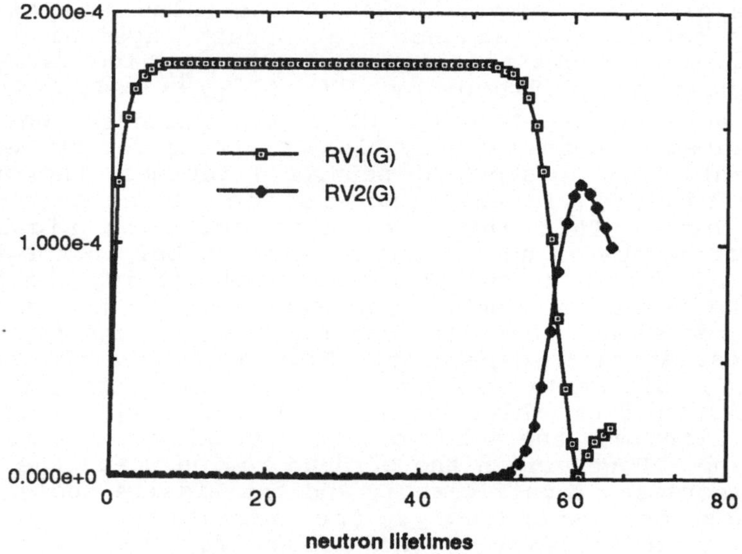

Fig.2 Relative variance of neutrons (1) and nuclei (2)

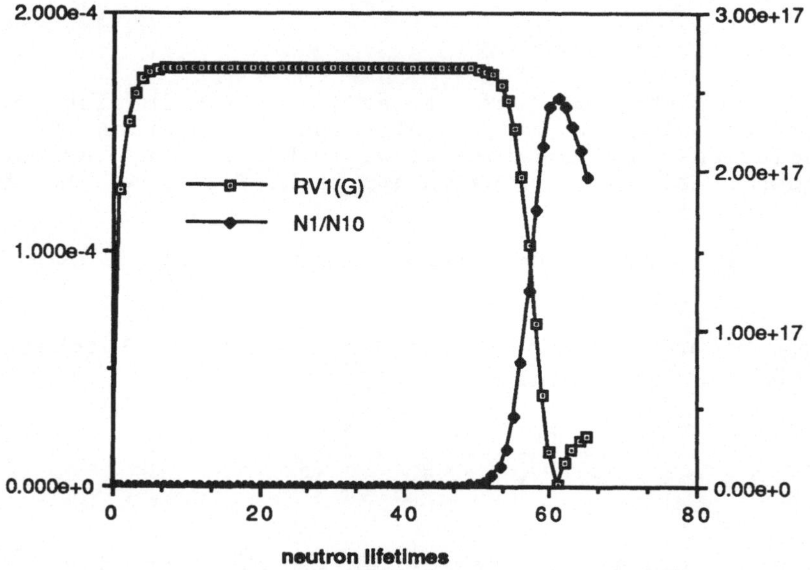

Fig. 3 Neutron density and relative variance

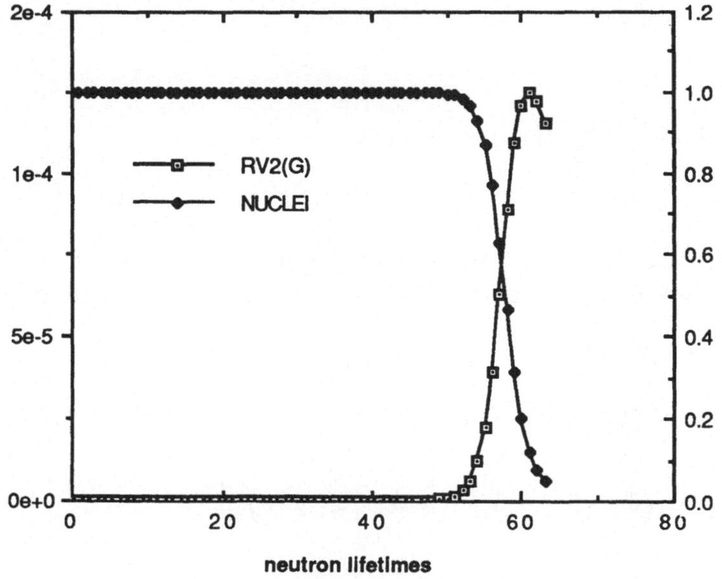

Fig. 4 Nuclei density and variance.

variance reduces accordingly. However, at later times, leakage reduces the neutron density and hence its variance increases. Figure 4 shows the analogous results for the fissile nuclei.

THE PROBABILITY DISTRIBUTION

To obtain the complete probability distribution it is necessary to solve the partial differential equation (9). No exact solution seems possible so we look for an approximation based upon a Gaussian-diffusion approach. First, we change to the new variables

$$z_1 = \exp\left(i\theta_1 / \sqrt{n_o}\right) \text{ and } z_2 = \exp\left(i\theta_2 / \sqrt{n_o}\right) \tag{32}$$

where n_o is an unspecified but large number. Further, we define two new random variables

$$u = \frac{n_1 - n_o\eta(t)}{\sqrt{n_o}}, \quad w = \frac{n_2 - n_o\xi(t)}{\sqrt{n_o}} \tag{33}$$

where η and ξ are as yet unspecified functions of time. Then we find that

$$G\left(i\theta_1 / \sqrt{n_o}, i\theta_2 / \sqrt{n_o}; t\right) \equiv M(\theta_1, \theta_2, t) \tag{34}$$

$$= e^{i\theta_1\sqrt{n_o}\,\eta + i\theta_2\sqrt{n_o}\,\xi}\, C(\theta_1, \theta_2, t) \tag{35}$$

where

$$C(\theta_1, \theta_2, t) = \sum_{u,w} f(u, w, t)\exp\left(i\theta_1 u + \theta_2 w\right) \tag{36}$$

Converting the source free form of eqn (9) to the new variables and using (35), we obtain after setting $\alpha=\alpha_o/n_o$ and $\beta=\beta_o/n_o$,

$$\begin{aligned}\frac{\partial c}{\partial t} &= i\sqrt{n_o}\left\{(\theta_1 - \theta_2)\alpha_o\xi\eta - \theta_1\dot{\eta} - \theta_2\dot{\xi} - \mu\eta\theta_1 - \beta_o(\theta_1 + \theta_2)\xi\eta\right\}c \\ &+ \left\{[\alpha_o(\theta_1 - \theta_2) - \beta_o(\theta_1 + \theta_2)]\xi - \mu\theta_1\right\}\frac{\partial c}{\partial\theta_1} \\ &+ \left\{\alpha_o(\theta_1 - \theta_2) - \beta_o(\theta_1 + \theta_2)\right\}\eta\,\frac{\partial c}{\partial\theta_2} \\ &- \tfrac{1}{2}\,\xi\eta\left\{\alpha_o(\theta_1 - \theta_2)^2 + \beta_o(\theta_1 + \theta_2)^2\right\}c - \tfrac{1}{2}\,\mu\theta_1^2\eta c\end{aligned} \tag{37}$$

The coefficients of θ_1 and θ_2 in the term in $i\sqrt{n_o}\{\cdots\}$ are now set equal to zero leading to

$$\dot{\eta} = (\alpha_o - \beta_o)\xi\eta - \mu\eta \tag{38}$$

$$\dot{\xi} = -(\alpha_o + \beta_o)\xi\eta \tag{39}$$

which we note are the equations for the mean values of the random variables n_1/n_o and n_2/n_o in the deterministic approximation. However, as eqns (27) and (28) indicate, η and ξ are not precisely the mean values since there are correlations between n_1 and n_2 which are neglected here.

We now apply the inverse Fourier transform to eqn (37) to obtain for the probability distribution function f(u,w,t):

$$\frac{\partial f}{\sigma t} = -(\alpha_o - \beta_o)\frac{\partial}{\partial u}[\xi u + \eta w]f + (\alpha_o + \beta_o)\frac{\partial}{\partial w}[\xi u + \eta w]f$$

$$+ \mu\frac{\partial}{\partial u}(uf) + \tfrac{1}{2}\mu\eta\frac{\partial^2 f}{\partial u^2}$$

$$+ \tfrac{1}{2}\xi\eta\left\{(\alpha_o + \beta_o)\frac{\partial^2 f}{\partial u^2} + (\alpha_o + \beta_o)\frac{\partial^2 f}{\partial w^2} - 2(\alpha_o - \beta_o)\frac{\partial^2 f}{\partial u \partial w}\right\} \tag{40}$$

Equation (40) is a Forward Kolmogoroff or Fokker-Planck equation of the type discussed by Chandrasekhar.[3] The procedure adopted in arriving at eqn (40) is very similar to a two-dimensional analogue of a technique developed by Van Kampen.[4]

The complete probability distribution may be regained by assuming a solution of the form

$$C(\theta_1, \theta_2, t) = \exp\left\{-\tfrac{1}{2}\nu_1\theta_1^2 - \tfrac{1}{2}\nu_2\theta_2^2 - \nu_3\theta_1\theta_2\right\} \tag{41}$$

and inserting it into eqn (37). Equating coefficients of θ_1^2, θ_2^2 and $\theta_1\theta_2$, we readily find the following differential equations

$$\frac{dC_{11}}{dt} = 2(\alpha - \beta)\bar{n}_2 C_{11} + 2(\alpha - \beta)\bar{n}_1 C_{12} + (\alpha + \beta)\bar{n}_1\bar{n}_2 + \mu\bar{n}_1 - 2\mu C_{11} \tag{42}$$

$$\frac{dC_{22}}{dt} = -2(\alpha + \beta)\bar{n}_1 C_{22} - 2(\alpha + \beta)\bar{n}_2 C_{12} + (\alpha + \beta)\bar{n}_1\bar{n}_2 \tag{43}$$

$$\frac{dC_{12}}{dt} = \left[(\alpha-\beta)\bar{n}_2 - (\alpha+\beta)\bar{n}_1\right]C_{12} - (\alpha+\beta)\bar{n}_2 C_{11} + (\alpha-\beta)\bar{n}_1 C_{22} - (\alpha-\beta)\bar{n}_1\bar{n}_2 - \mu C_{12} \tag{44}$$

where we have set $n_1 = n_o\eta$, $n_2 = n_o\xi$, $C_{11} = n_o v_1$, $C_{22} = n_o v_2$ and $C_{12} = n_o v_3$. Apart from terms of order $1/n_1$, these are precisely the equations obtained using the Gaussian approximation of an earlier section. The complete probability distribution is obtained from the Fourier inverse of eqn (41) leading to the following Gaussian distribution:

$$P(n_1, n_2, t) = \frac{1}{2\pi}\frac{1}{\sqrt{C_{11}C_{22} - C_{12}^2}}\exp\left\{-\frac{C_{22}(n_1-\bar{n}_1)^2 - 2C_{12}(n_1-\bar{n}_1)(n_2-\bar{n}_2) + C_{11}(n_2-\bar{n}_2)^2}{2(C_{11}C_{22} - C_{12}^2)}\right\} \tag{45}$$

Integrating over n_2, we find

$$P(n, t) = \frac{1}{\sqrt{2\pi C_{11}}}\exp\left\{-\frac{(n_1-\bar{n}_1)^2}{2C_{11}}\right\} \tag{46}$$

While this Gaussian distribution undoubtedly describes the behaviour of the neutrons and nuclei adequately in practice, it does have some limitations and inconsistencies. For example, we know that n_1 and n_2 are positive integers, whereas the Gaussian assumes that n_1 and n_2 are continuous variables in the range $(-\infty, \infty)$. Negative values are, of course, unphysical but for large populations these contribute insignificantly and can be neglected. However, such behaviour prevents eqn (46) from being used to calculate the extinction probability $P(o, n_2; t)$; i.e., the probability that all neutron chains have become extinct. Certainly for large n_1, this probability is very small, but early in the burst lifetime this may not be the case. In such a situation, it is probably bettter to use eqn (14) neglecting burn-up. This follows from the fact that little burn-up takes place until around 30 or 40 neutron lifetimes and by that time the chain will either have become extinct or developed to such an extent that its likelihood of extinction is negligible. Whilst such arguments are true for neutronic systems, for which the random variable is large, the same argument cannot always be used for epidemiological problems where typical numbers of infectives and susceptibles may be less than 50 in some cases. It remains, therefore, to obtain a more accurate solution of eqn (9) which is valid for small numbers. One such approach would be to solve directly by numerical means the finite set of probability balance equations (7), which are similar to typical finite difference equations.

ACKNOWLEDGEMENT

The author is grateful to Dr. W.Matthes for a helpful comment.

REFERENCES

1. M. M. R. Williams, "Random Processes in Nuclear Reactors," Pergamon Press, Oxford (1974).

2. N. T. J. Bailey, "The Elements of Stochastic Processes," John Wiley, NY (1964)

3. S. Chandrasekhar, "Stochastic Problems in Physics and Astronomy," Rev. Mod. Phys. 115:1(1943).

4. N. G. Van Kampen, "Birth and Death Processes in Large Populations," Biometrika, 60:419 (1973).

BACKWARD KOLMOGOROV EQUATIONS WITH FEEDBACK : CONTRIBUTION TO THE DISCUSSION

Jeffery D. Lewins

University of Cambridge
Engineering Department
Trumpington Street
Cambridge CB2 1PZ
England

Two challenges were issued during this stimulating paper, one by the author and one by me. Both challenges turn out to be linked and their resolution casts some interesting light on the model. I use Professor Williams' notation as far as possible.

Williams models a system in which there is fuel burn-up as well as neutron behaviour. In ordinary deterministic parlance, the system with such feedback would show nonlinearities although, interestingly, this is not necessarily the result of a stochastic representation. He was able for this model to derive the forward Kolmogorov equation in the form

$$\frac{\partial G}{\partial t} = -\alpha[z_1 z_2 - g(z_1)]\frac{\partial^2 G}{\partial z_1 \partial z_2} + \beta[1 - z_1 z_2]\frac{\partial^2 G}{\partial z_1 \partial z_2} + \mu[1 - z_1]\frac{\partial G}{\partial z_1} + S[z_1 - 1]G . \qquad (1)$$

This equation displays the properties of a linear partial differential equation, linear in its arguments, but of second order. The additional order, compared to a low power equation modelling neutron behaviour alone, is a severe disadvantage since the customary solution methods of Cauchy-Lagrange for first-order equations are not available. Solution seems a formidable task.

The first challenge, issued by Professor Williams, was to derive the backward equations for this model. The difficulty to be anticipated is that the conventional backward equations employ an independence of neutron behaviour which as a convolution of the z-transforms of the generating function would lead to a single, albeit nonlinear equation. That device will not be open to us in the feedback model.

Tackling the forward equations, the author considered the equations for the moments and demonstrated a problem of closure. With two equations for the rates of change of the first moments (means) of both neutron and fissile atom populations, terms in higher moments are introduced by the second-order differential term. Repeated derivation of equations to obtain these higher moments invokes further higher moments and so on. Some method of closure seems essential and Williams used a conventional relation for the higher moments, neglecting covariance terms. At the end of the paper, I posed my challenge: how is that the neutrons know what to do in such a system, when we cannot even write down the (exact) moment equations describing their behaviour.

It turns out that the answer to the first challenge leads to an unexpected answer to the second: there is strictly no problem of closure and we could in principle write down the equations the neutron and fissile atom moments satisfy.

To derive the backward Kolmogorov equations, recollect that the model encompasses n_1 neutrons and n_2 fissile atoms which can suffer interactions with mean rates as follows:

$\alpha n_1 n_2$ fission rate, gain two neutrons, lose one fissile atom,

$\beta n_1 n_2$ capture rate, lose one neutron and one fissile atom,

μn_1 leakage rate, lose one neutron,

and independent source of mean rate

S gain one neutron.

For simplicity we assume that every fission produces exactly two neutrons; this simplification was employed by Williams in his numerical work.

With Williams, let $P(n_1,n_2,m_1,m_2;t|s)$ be the probability that m_1 neutrons and m_2 fissile atoms at an initial time s will lead to n_1 neutrons and n_2 fissile atoms at a later time t. Write vector states $\underline{m} = (m_1,m_2)$ and $\underline{n} = (n_1,n_2)$. The backward equation is obtained by considering a conservation of probability for the initial state $\underline{m}$ after a short interval δs where either nothing has happened to change the state or something has happened, changing the state to a new state which nevertheless will lead to the final state $\underline{n}$ at t. That is

$$\begin{aligned} P(\underline{n},\underline{m};t|s) = {} & [1-\alpha m_1 m_2\delta s-\beta m_1 m_2\delta s-\mu m_1\delta s-S\delta s]P(\underline{n},\underline{m};t|s+\delta s) && \text{no change} \\ & + \alpha m_1 m_2\delta s P(\underline{n},m_1+1,m_2-1;t|s+\delta s) && \text{fission} \\ & + \beta m_1 m_2\delta s P(\underline{n},m_1-1,m_2-1;t|s+\delta s) && \text{capture} \\ & + \mu m_1\delta s P(\underline{n},m_1-1,m_2;t|s+\delta s) && \text{leakage} \\ & + S\delta s P(\underline{n},m_1+1,m_2;t|s+\delta s)\ . && \text{source event} \end{aligned} \tag{2}$$

Rearrange and divide by δs. Employ vector Kronecker deltas, $\underline{\delta}_1$ and $\underline{\delta}_2$, to show the addition of one neutron or one fissile atom. This yields the backward equation

$$\begin{aligned} -\frac{\partial P(\underline{m})}{\partial s} = {} & -[m_1 m_2(\alpha+\beta)+\mu m_1+S]P(\underline{m}) + \alpha m_1 m_2 P(\underline{m}+\underline{\delta}_1-\underline{\delta}_2) \\ & + \beta m_1 m_2 P(\underline{m}-\underline{\delta}_1-\underline{\delta}_2) + \mu m_1 P(\underline{m}-\underline{\delta}_1) + SP(\underline{m}+\underline{\delta}_1)\ . \end{aligned} \tag{3}$$

The appropriate boundary condition is that

$$P(\underline{n},\underline{m};t|t) = \underline{\delta}_{nm}\ . \tag{4}$$

Define the generating function $F(\underline{m},t|s;\underline{z})$ where $\underline{z}$ is a vector of (z_1,z_2), dummy variables between 0 and 1. Employ the Lewins dot symbol so that

$$F(\underline{m};\underline{z}) = \underline{z}^{\cdot\underline{n}}P(\underline{n},\underline{m}) \tag{5}$$

with the convention of summing over repeated $\underline{n}$-indices. Operate on the backward equation with $\underline{z}^{\cdot\underline{n}}$. The equation can then be written

$$\begin{aligned} -\frac{\partial F(\underline{m})}{\partial s} = {} & -[m_1 m_2(\alpha+\beta)+\mu m_1+S]F(\underline{m}) + \alpha m_1 m_2 F(\underline{m}+\underline{\delta}_1-\underline{\delta}_2) + \beta m_1 m_2 F(\underline{m}-\underline{\delta}_1-\underline{\delta}_2) \\ & + \mu m_1 F(\underline{m}-\underline{\delta}_1) + SF(\underline{m}+\underline{\delta}_1)\ ; \quad F(\underline{m};t|t) = \underline{z}^{\cdot\underline{m}}\ . \end{aligned} \tag{6}$$

It is at this stage that we would have wished to introduce the assumption of neutron independence appropriate to a low power system. Nevertheless, the equation as such is valid in the more general model.

If we seek moment equations from this representation, we have the same problem of lack of closure or self-consistency. But consider the equation in the absence of a source, where, for clarity, we write the generating function as G (no source) as opposed to F (source). For the particular initial values $m_1 = m_2 = 1$ it will be found that the equation for the rate of change of G_{11} involves G_{10} and also G_{20}. But if we examine the equations for G_{10} and G_{20} in turn, no new terms are invoked and the equations are a closed set:

$$-\frac{\partial G_{10}}{\partial s} = -\mu[G_{10}-G_{00}] \ ; \quad G_{00} = 1 \ ; \quad G_{10}(t|t) = z_1 \ . \qquad (7)$$

$$-\frac{\partial G_{20}}{\partial s} = -2\mu[G_{20}-G_{10}] \ ; \quad G_{20}(t|t) = z_1^2 \ . \qquad (8)$$

$$-\frac{\partial G_{11}}{\partial s} = -\alpha[G_{11}-G_{20}] - \beta[G_{11}-G_{00}] - \mu[G_{11}-G_{01}] \ ; \quad G_{01} = z_2 \ ; \quad G_{11}(t|t) = z_1 z_2 \ . \qquad (9)$$

Indeed we give the complete solution for this special case later. Alternatively, we could set $\underline{z} = \underline{0}$ to obtain three simultaneous equations for extinction probabilities following the introduction of one neutron and one fissile atom, one neutron or two neutrons and no fissile atom. Higher probabilities could be found recursively in the manner discussed in other papers at this meeting (Lewins and Parks).

If the conventional source term is reintroduced, however, it is found that the equations are no longer a closed set; the source adding one neutron always invokes one higher term. So why does the system behave this way?

The origin of the behaviour is the nature of the feedback model: there are finite resources of fissile atoms to produce the neutrons. Lacking a source, but with a finite number of initial neutrons, there is evidently a finite number of fissile atoms (the original loading). But the maximum number of neutrons is also finite. It cannot be higher than that given by every neutron fissioning every fissile atom and releasing the maximum number of neutrons per fission (two in our model of course). And the spirit of the model ensures that this maximum of neutrons per fission is itself finite. Correspondingly $<\upsilon>$ is bounded finite (otherwise the forward and backward equations would not be identical).

It follows that all higher probabilities are identically zero and the equations for the probability generating functions naturally close. Equally, it must be the case that the number of independent moments is finite. All higher moments could in principle be expressed as combinations of the lower moments.

Thus, although it would not be practicable perhaps to identify where the independence of moments ceases and what the higher moments are in terms of lower moments, the set of moment equations is in principle self-closing.

Why then do we lose closure on reintroducing the conventional source? Because the conventional source, if it is constant, describes the possibility of continuing to introduce source neutrons indefinitely, without limit. And this is not in the spirit of the feedback model with its finite resources. The solution to this problem is straightforward: represent the source as a finite resource. Thus we could add to the state description of the system the number of nuclides, n_3, whose decay (at mean rate λ say) introduces a source neutron. With this extension, the maximum state of the system is still finite and higher probabilities identically zero. The equations are now as before with G_{10} going to G_{100}, G_{20} to G_{200} and G_{11} to G_{110} and with

$$-\frac{\partial F_{001}}{\partial s} = -\lambda[F_{001}-G_{100}] \ ; \quad F_{001}(t|t) = z_3 \ , \qquad (10)$$

and we see one addition to the closed set of equations.

It is also to be remarked that the definition of criticality, super-criticality etc. in this model is not clear; ultimately either neutrons or fissile atoms must be extinguished.

Some exact solutions

We solve the elementary case with one neutron and one fissile atom (no source) with a view to displaying some properties of the exact solution vis à vis its approximate counterparts.

It is readily found that the following solve the source-free problem:

$$G_{10} = [1 - e^{-\mu t}] + z_1 e^{-\mu t} . \quad (11)$$

$$G_{20} = [1 - e^{-\mu t}]^2 + 2z_1 e^{-\mu t}[1 - e^{-\mu t}] + z_1^2 e^{-2\mu t} = G_{10}^2 . \quad (12)$$

$$G_{11} = \frac{\alpha+\beta}{\alpha+\beta+\mu}[1 - e^{-(\alpha+\beta+\mu)t}] - \frac{2\alpha e^{-\mu t}}{\alpha+\beta}[1 - e^{-(\alpha+\beta)t}] + \frac{\alpha e^{-\mu t}}{\alpha+\beta-\mu}[e^{-\mu t} - e^{-(\alpha+\beta)t}]$$
$$+ 2\alpha z_1 e^{-\mu t}[\frac{1 - e^{-(\alpha+\beta)t}}{\alpha+\beta} - \frac{e^{-\mu t} - e^{-(\alpha+\beta)t}}{\alpha+\beta-\mu}] + \frac{\mu z_2}{\alpha+\beta+\mu}[1 - e^{-(\alpha+\beta+\mu)t}] \quad (13)$$
$$+ \frac{\alpha z_1^2 e^{-\mu t}}{\alpha+\beta-\mu}[e^{-\mu t} - e^{-(\alpha+\beta)t}] + z_1 z_2 e^{-(\alpha+\beta+\mu)t} .$$

The various probabilities are coefficients of the terms in z_1 and z_2.

First moments, the expected number of neutrons and fissile atoms, are found by differentiating with respect to z_1 or z_2 respectively and putting $\underline{z} = \underline{1}$ to find, for $\underline{m} = \underline{1}$, that

$$\langle n_1 \rangle = \frac{2\alpha e^{-\mu t}}{\alpha+\beta}[1 - e^{-(\alpha+\beta)t}] + e^{-(\alpha+\beta+\mu)t} , \quad (14)$$

$$\langle n_2 \rangle = \frac{\mu}{\alpha+\beta+\mu} + \frac{\alpha+\beta}{\alpha+\beta+\mu} e^{-(\alpha+\beta+\mu)t} . \quad (15)$$

We may also consider the deterministic model for the numbers of neutrons and precursors $\bar{n}_1$ and $\bar{n}_2$ starting with one of each:

$$\frac{d\bar{n}_1}{dt} = + \alpha\bar{n}_1\bar{n}_2 - \beta\bar{n}_1\bar{n}_2 - \mu\bar{n}_1 \ ; \ \bar{n}_1 = 1 \text{ at } t = 0 , \quad (16)$$

$$\frac{d\bar{n}_2}{dt} = - \alpha\bar{n}_1\bar{n}_2 - \beta\bar{n}_1\bar{n}_2 \ ; \ \bar{n}_2 = 1 \text{ at } t = 0 . \quad (17)$$

A first integral is readily found:

$$(\alpha+\beta)\bar{n}_1 = 2\alpha - (\alpha-\beta)\bar{n}_2 + \mu\, ln(\bar{n}_2) . \quad (18)$$

A second integral is more difficult but for short times, when $\bar{n}_2$ is close to unity, an expansion of the logarithm allows us to integrate to find

$$\bar{n}_2 = \frac{2\alpha+\beta}{\alpha+\beta-\mu+(\alpha-\beta)e^{(2\alpha+\mu)t}} . \quad (19)$$

In general it is seen that $\bar{n}_2$ is not $\langle n_2 \rangle$ (except indeed for very short times).

Returning to the exact first integral, consider the situation after long time when

$$\langle n_1 \rangle \to 0 , \quad (20)$$

and

$$\langle n_2 \rangle \to \mu/(\alpha+\beta+\mu) . \quad (21)$$

But, for $\bar{n}_1 = 0$, and specialising to the case where $\alpha = \beta$, we find

$$\bar{n}_2 = e^{-2\alpha/\mu} \ ; \ \langle n_2 \rangle = \mu/(\mu+2\alpha) . \quad (22)$$

And these results are only compatible in the limit of $2\alpha/\mu$ very small, when the system would be essentially one of neutron leakage only and hence lose its feedback interest.

Even at this trivial level then, the weakness of the deterministic model shows up as a failure of the behaviour in the mean to account for correlation between neutrons and fissile atoms.

It can also be seen from the model that the variance and covariance of the system is of significance but that higher order factorial moments involving more than one differentiation of z_2 and two of z_1 are inherently zero, providing for closure of the moments equations as anticipated.

Conclusion

A feedback model implies some finiteness of resources. When this implication is represented faithfully throughout the model, it seems that the equations for the moments are in principle closed. The forward equations have the disadvantage of introducing a second-order PDE; some form of approximation, as elegantly explored by Williams, seems essential. Yet the backward equations demonstrate the structure of a set of linear first-order PDEs (strictly linear in the generating functions) even in the presence of a (finite) source. It is not suggested that it would be feasible to solve these exactly in any but the elementary cases given here, but there is no doubt that the backward equations display the structure of the problem more openly than the forward Kolmogorov equations.

STEADY STATE DYNAMICS OF NONLINEAR NUCLEAR REACTOR SYSTEMS

A. Hernández-Machado and J. Casademunt

Dept. E.C.M., Facultat de Física
Universitat de Barcelona
Avda. Diagonal, 647. E-08028 Barcelona- Spain

1. INTRODUCTION

The effects of the nonlinearities in the steady state dynamics of nuclear reactor models have not been considered until recently[1,2]. The nonlinear terms appear, for example, in the Langevin equation for the number of neutrons, due to an adiabatic elimination of the fast variables (delayed neutrons, fuel temperature, refrigerator temperature, etc.) The reduction in the number of variables gives a more tractable problem in the sense that the validity of the approximations that one uses are more easily known, but the price that is paid is the nonlinearity of the equations. Then, the usual procedure is the linearization of the resulting equations around the deterministic steady state. The validity of this approximation has been considered by many authors[3-6]. All of these studies agree that the linearization is a good approximation far from the instability points, but it breaks down near them.

A good characterization of the dynamics of the steady state is given by the steady state correlation function. This quantity gives an idea of how fast the fluctuations decay in the steady state and what are the characteristic time scales in the system. From the point of view of the nuclear reactors, the correlation function provides interesting information. For example, one can compare the experimental results with the theoretical predictions in order to determine some parameters, like the reactivity. This allows one to know how far the reactor is from the instability points. To have this information we need to handle with theoretical methods capable to take into account the nonlinear terms of the equations. The aim of this paper is to convince the audience of the field of nuclear reactors that there are methods simple enough to calculate correlation functions very near the instability points, and which give good qualitative and quantitative results for all time regimes.

In the case of a one-dimensional linear Langevin equation with additive noise the steady state correlation function is exactly known. It corresponds to an exponential. The constant with dimensions of time in the exponent gives a time scale which characterizes the short time behaviour (it is the inverse of the so-called effective eigenvalue which corresponds to the first derivative of the correlation function at time equal zero) and also the long time or global behaviour (it is the so-called relaxation time, the area under the curve of the correlation function).

In the case of a one-dimensional nonlinear Langevin equation the correlation function is, in general, not known. An infinite hierarchy of coupled differential equations has to be solved. Many scales will be present in the system. The existence of noise of external origin appearing in a multiplicative way will produce a continuous spectrum. For a particular class of models the exact steady state correlation function is known[1]. In particular, it corresponds to a nonlinear point nuclear reactor model, with instantaneous temperature feedback, and reactivity fluctuations. In this case the evidence of the complexity of the time scales can be checked. Only far from the instability points all these time scales are frozen and the simple exponential behaviour of the linearized model is right. So a practical criterion to check the validity of the linear approximation is the comparison between the inverse of the effective eigenvalue and the relaxation time. The linear approximation will be valid when these two time scales are essentially equal.

Recently a systematic method that has no linearization hypothesis inside was applied to the nonlinear point nuclear reactor model[2]. It is based in a Mori-Zwanzig projection operator technique[7]. The method gives the Laplace transform of the correlation function, C(w), as a continued fraction expansion (CFE). It can be seen that the CFE is related to a high frequency expansion of C(w)[2]. This is an accordance [2] with the fact that the method gives good results for short times in regions near the instabilities, where the linearization fails.

In this paper we present a new method[8] which gives good qualitative and quantitative results for the correlation function in all time regimes and can be applied very near instability points. It has been proposed in an other context[9] but it has not been fully explored and it has never been applied in the context of nuclear reactors. The basic idea of the method is that in the same way that the linear problems only require a characteristic time scale, implicit in the result for the correlation function, a nonlinear problem needs more than one effective time scale. Depending on the particular interest one uses effective time scales for short times (related to higher order derivatives of the correlation function at time equal zero) or for long time or global behaviour (related to higher order of the so-called relaxation moments). In section 2 we present the nuclear reactor model that we will consider. In sec.3 we present the method and we do some remarks in connexion with other methods. In sec.4 we do the application of the method to the model presented in sec.2. Finally, in sec. 5 we sumarize the main the results. In the Appendix we give more detalis about the calculation of the relaxation moments.

2. THE MODEL

The method that we will present in sec. 3 is applicable to any Langevin equation of the form

$$\dot{q} = v(q) + g(q)\sigma(t) \tag{2.1}$$

with v(q), g(q) any function of q. As the simplest situation we take σ(t) as a Gaussian white noise with zero mean and correlation function given by

$$\langle \sigma(t)\sigma(t') \rangle = 2D\delta(t-t') \tag{2.2}$$

where D is the intensity of the noise. The method is also applicable to a colored noise case. (In this last case, which will not be treated here, the results are given to first order in the relaxation time of the noise).

As an application of the method to the field of nuclear reactors we will consider as a prototype model a nonlinear point nuclear reactor model with instantaneous temperature feedback, that introduces the nonlinear term in the equation, and an independent source of neutrons. We consider reactivity fluctuations which model the external noise of structural and hydrodynamical origin. The Langevin equation for the variable number of neutrons N is

$$\frac{dN}{ds} = -\Gamma N^2 + (a + \bar{\sigma}(s))N + S \tag{2.3}$$

Γ is the temperature coefficient, 'a' the reactivity, 'S' the independent neutron source and D the intensity of the noise. After the reparametrization

$$\sigma(t) = \frac{\bar{\sigma}(s)}{D} \; ; \quad t = Ds \; ; \quad n = \frac{\Gamma}{D}N \; ; \quad \alpha = \frac{a}{D} \; ; \quad \beta = \frac{S\Gamma}{D^2} \tag{2.4}$$

the equation (2.3) takes the form

$$\dot{n} = \alpha n - n^2 + \beta + n\sigma(t) \tag{2.5}$$

with

$$\langle \sigma(t)\sigma(t') \rangle = 2\delta(t-t') \tag{2.6}$$

For β=0 Dutré and Debosscher[1] obtained the exact normalized steady state correlation function which for α<1 (only contributions from the continuous spectrum) reads

$$C(t) = \frac{1}{2\Gamma(\alpha+1)} \int_0^\infty dx \, e^{-(x+\frac{\alpha^2}{4})t} \left|\Gamma\left(\frac{\alpha}{2}+1+i\sqrt{x}\right)\right|^2 \frac{\sinh 2\pi\sqrt{x}}{\cosh^2\pi\sqrt{x} - \cos^2\pi\frac{\alpha}{2}} \tag{2.7}$$

We will use this result to check the validity of our approximative results. For $\beta \neq 0$ the exact result is not known. Here we will present the results of our method for different values of α and β.

3. THE METHOD

We are interested in the calculation of the normalized steady state correlation function C(t)

$$C(t) = \frac{\langle q(t)q(0)\rangle - \langle q\rangle^2}{\langle q^2\rangle - \langle q\rangle^2} \tag{3.1}$$

where $\langle q\rangle$ and $\langle q^2\rangle$ are steady state moments. We consider the Laplace transform of C(t), C(w) given by

$$C(\omega) = \int_0^\infty dt\, e^{-\omega t} C(t) \tag{3.2}$$

The starting point of the method is the low and high frequency formal expansions of C(w)

$$C(\omega) = \sum_{K=0}^{\infty} C_K^{\infty}\, \omega^{-K} \tag{3.3}$$

with $$C_0^{\infty} = 0\ ;\quad C_0^{\infty} = 1\ ;\quad \underset{(K\geq 2)}{C_K^{\infty} =} \frac{d^{K-1}}{dt^{K-1}} C(t)\Big|_{t=0} \equiv (-1)^K \mu_{K-2} \tag{3.4}$$

$$C(\omega) = \sum_{K=0}^{\infty} C_K^{0}\, \omega^{K} \tag{3.5}$$

with $$C_K^0 = \frac{(-1)^K}{K!} \int_0^\infty dt\, t^K C(t) \equiv \frac{(-1)^K}{K!} T_K \tag{3.6}$$

The equations (3.3-4) could be obtained from eq.(3.2) derivating with respect to 1/w and using one representation of the Dirac delta function. The result of eqs. (3.5-6) is obtained by a simple expansion of the exponential in eq.(3.2). The expansion (3.3) will be useful if one is interested in the short time behaviour of C(t). This is the limit that the conventional CFE considers when one cuts at some order of approximation. In fact, the zero and first order of the CFE are equivalent to an expanion to w^{-2} and w^{-4} respectively. The low frequency limit (3.5-6) will be the appropriate one in the study of the long time behaviour. More precisely, it provides a global characterization. This fact is easily understood from eqs. (3.5-6). We see that for instance, T_0 is the usual relaxation time. T_0 does not give information about any precise region of time, but a global one; it is the total area under the curve of C(t). The interpretation of the relaxation moments of higher order is clear from (3.6). T_K is proportional to the k-th moment of the correlation function normalized by T_0, if we think of $C(t)/T_0$ as a probability distribution. Then, T_K contains information about the distribution of area under the curve C(t). Eqs. (3.4) and (3.6) define the new quantities μ_K and T_K. μ_o corresponds to the effective eigenvalue and T_o is the relaxation time. For the linear case $T_o=\mu_o^{-1}$ and this only time scale characterizes in a complete way the correlation function. On the other hand, in the nonlinear case this is no longer true and, as we will see later, a good description of C(t) near the instability points needs to take into account more than one of this "effective time scales", going to higher orders in μ_K and T_K.

We can write the two expansions in a compact way as

$$C(w) = \sum_{K=0}^{\infty} c_K^L w_L^K \tag{3.7}$$

with L=0 or ∞, $w_o=w$ and $w_\infty= w^{-1}$. In practice we will know a finite number of coefficients c_K^L. An appropriate way to incorporate this information in order to determine an approximate form for C(t) is to consider a Padé approximant of C(w), that is, a quocient of two polinomials, that can be decomposed as

$$C_M(w) = \sum_{m=1}^{M} \frac{A_m}{w+\delta_m} \tag{3.8}$$

The conditions to determine the constants A_m and δ_m are obtained imposing that $C_M(w)$, given by (3.8), has to have the correct expansion up to a corresponding order in the appropriate expansion L=0 or L=∞

The inverse Laplace transform of (3.8) takes the form

$$C_M(t) = \sum_{m=1}^{M} A_m e^{-\delta_m t} \tag{3.9}$$

Eq. (3.9) is the final approximate expression for C(t). The coefficients A_m and δ_m contain information about the short or global behaviour given by the high or low frequency expansions eq. (3.7) and extrapolate the other regimes.

In practice the way to calculate $C_M(t)$ is to assume as the starting point an expression of the form (3.9). To determine the constants A_m and δ_m one asks first for the normalization condition C(0)=1. Then, depending on wheter we are interested in the short time or global behaviour of C(t) we impose that $C_M(t)$ has to give correctly 2M-1 derivatives at the origin, $(\mu_o,\ldots,\mu_{2M-2})$ or 2M-1 relaxation moments, $(T_o,\ldots,T_{2M-2})$. This implies 2M algebraic equations to solve.

Nevertheless, the two points of view can be combined through the so-called two-point Padé approximation of C(t). In this case one imposes that $C_M(t)$ has to give correctly d derivatives at the origen and 2M-1-d relaxation moments. We will see later that this is the most appropriate method to get a good qualitative and quantitative behaviour of C(t) in all the temporal regions.

The last question to answer is how to calculate the quantities μ_K and T_K. The calculation of the derivatives μ_K is not a problem. We only need to know the steady state moments $\langle q^n \rangle$. These are obtained from the steady state distribution $P_{st}(q)$.

To calculate the relaxation moments T_K some algebra is needed. It appears in the Appendix. The result for T_K is given as a quadrature

$$T_K = \frac{1}{\langle q^2\rangle - \langle q\rangle^2} \int_a^b \frac{G_0(q)\, G_K(q)}{D g^2(q)\, P_{st}(q)}\, dq \qquad (3.10)$$

with

$$G_0(q) = -\int_a^q (q' - \langle q\rangle)\, P_{st}(q')\, dq' \qquad (3.11a)$$

$$G_K(q) = \int_a^q P_{st}(q') \left[\int_a^{q'} \frac{G_{K-1}(q'')}{D g^2(q'')\, P_{st}(q'')}\, dq'' - \left\langle \int_a^{q} \frac{G_{K-1}(q')}{D g^2(q')\, P_{st}(q')}\, dq' \right\rangle \right] dq' \qquad (3.11b)$$

Before considering the application of the method to the nuclear reactor model, we want to do a remark about the relation between the method we present here and the conventional CFE. When one cuts the CFE to some order, one gets precisely a Padé approximant of the 1/w-expansion of C(w) truncated at the corresponding order.

It is very simple to see this fact. For example, the first order of the CFE corresponds to the result (3.9) with M=2 imposing the condition on the first three derivatives μ_K. In this sense the CFE is equivalent to the method for $L=\infty$ and is the complementary of the L=0 that we present for the long time behaviour of C(t). The advantage of our method is that the low frequency regime is treated at the same level than the high frequency regime. It is an open question what is the equivalent of the coventional CFE formalism in the low frequency regime. However, to give the complete behaviour of C(t) in all regimes of time it is very important to have information of this low frequency regime.

4. APPLICATION

We will apply the method to the example presented in sec.2 defined by eqs. (2.5), (2.6).

$\underline{\beta=0.}$ As a first result for this case we have obtained from eq. (3.10) that the relaxation moment of zeroth order (the relaxation time) is given by

$$T_0 = \frac{1}{\alpha} \qquad (4.1)$$

This is an exact result. We have no notice that this result has been given before in the literature. For this particular example the linearization gives the same result (4.1), but this is clearly a casuality. For example, the linearization does not give the exact result for T_o in the case $\beta \neq 0$ or for other types of nonlinearities.

Some recent exact results concerning dynamical moments $\langle q^n(t) \rangle$ [10] and correlation functions[11] predicted the occurrence of critical slowing down at this type of instability points we are dealing with here. However, from the explicit knowledge of the long time tails it was not possible to conclude about the behaviour of the relaxation time in the limit of α->0. Here, instead, we have obtained a divergence of the relaxation time in this limit. It is remarkable that this occurs for a finite noise intensity (D=1), so it does not involve the deterministic limit D->0. It corresponds to a deterministic instability point (α=0). Critical slowing down does not appear, instead, in the so-called noise-induced transition, defined as the point where the steady-state distribution changes its maximum (α=1).

In the other hand it is also remarkable that the CFE gives to any order, a finite value of the relaxation time in the limit α->0 [12], so it is clearly apparent how the CFE cannot predict global properties of C(t), involving long time behaviour.

The lowest order approximations of both expansions are given in this case by

$$(\omega \gg 1) \quad ; \quad C_1(t) = e^{-\mu_0 t} \quad ; \quad \mu_0 = -\frac{d}{dt}C(t)\Big|_{t=0} = \alpha + 1 \qquad (4.2)$$

$$(\omega \ll 1) \quad ; \quad C_1(t) = e^{-t/T_0} \quad ; \quad T_0 = \int_0^\infty dt\, C(t) = \frac{1}{\alpha} \qquad (4.3)$$

We can compare both results with the exact one[1] given by eq.(2.7). The three curves appear in fig.(1). Neither of both approximative results (4.2) and (4.3) give good results near the instability point (α<<1). Eq. (4.3) corresponds to the result of the linearization method that for casuality is in this case the lowest approximation of the low frequency expansion. This has been the only order of approximation considered in ref.9b. Only one effective time scale is taken into account in this case. Here, we go beyond this order.

One of the aspects we are interested in is the sensitivity of the results of $C_M(t)$ to the number of conditions on derivatives and relaxation moments that we impose in the expression (3.9). We can see from Fig.1 that we obtain excellent results for a very large range of times if me compare with the exact one [1]. By considering $C_M(t)$ given by eq.(3.9) with M=3 and the conditions of three derivatives μ_0, μ_1, μ_2 and two relaxation moments T_0, T_1. The result is very good in a regime of times very much larger than the one which corresponds to only three derivatives (M=2) in eq.(3.9). The last one is the first order of approximation of the conventional CFE. We can see from Fig.2 that to the lowest orders the result is very sensitive to the number of conditions in the relaxation moments that we impose. From Fig.2 we also conclude that this is not the case for the number of conditions on the derivatives.

Hence in this case it seems to be necessary to impose at least two relaxation moments in order to get a qualitatively good approximation of the global form of the curve, and then one can proceed by adding derivative conditions (which are easier to determine from a practical point of view).

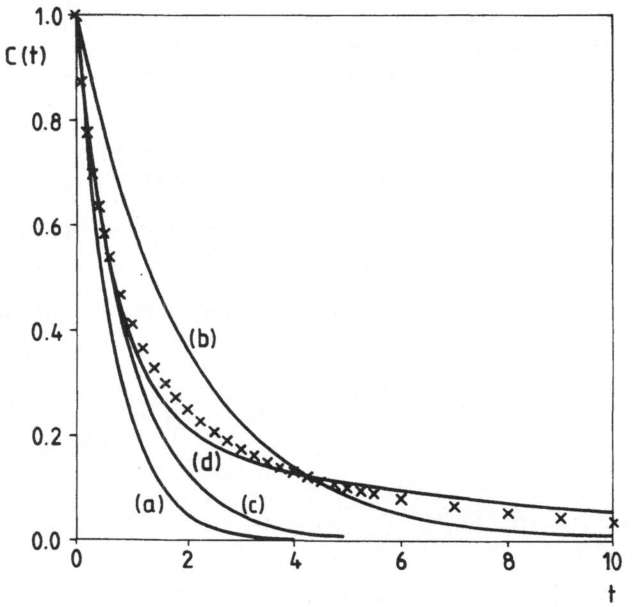

Fig. 1. (a) Eq.(4.2) (zeroth order of CFE); b) Eq. (4.3), (linearization); (c) 1st order of CFE (3 derivatives); (d) M=3 with 3 derivatives and 2 relaxation moments; X, exact solution.

In any case it is not necessary to go very far in the number of conditions to have very good results. In general, this practical method will be useful for models in which there are long time tails in the correlation function. The convergence of the long frequency expansion will in this case be very slow. The reason is because it is very difficult to fit the long time tail imposing conditions at the origin of times. In these cases some few global conditions can drastically improve the approximation simultaneously for all time regimes without having to go to very high orders of approximation.

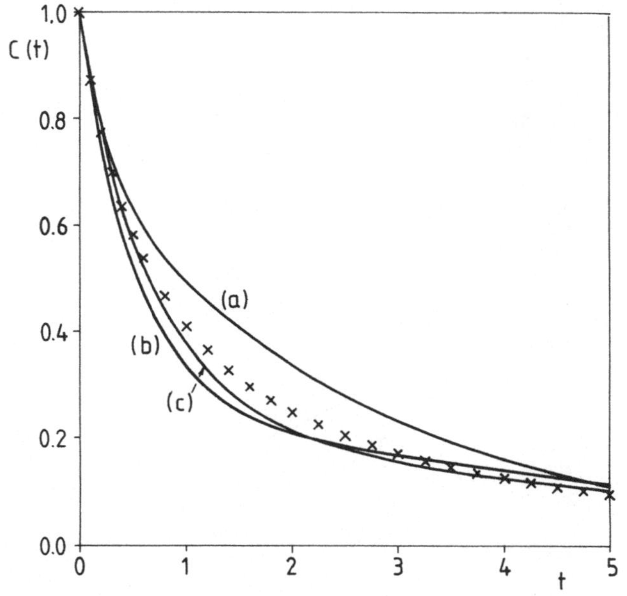

Fig. 2. (a) M=2 with 2 derivatives and 1 relaxation moment ; (b) M=2 with 1 derivative and 2 relaxation moments; (c) M=3 with 3 derivatives and 2 relaxation moments; X, exact solution.

<u>$\beta \neq 0$.</u> In this case no exact results are known for C(t). It is only possible to compare with the approximation given by the CFE in Ref.2. This approximation is valid only for short times. The interest of our result consists in the extension of the range of validity to practically all times. Then we obtain, in our knowledge for the first time, conclusions about the long time behaviour of C(t). From fig.(4) we can conclude that in the presence of a finite source of neutrons the long time tail of C(t) dissappears. The decay of C(t) for all times is very fast and increasing β, the decay is faster.

From Fig.4 it is also clear that the correlation function is more sensitive to β than to α, when $\beta \neq 0$. The unsensitivity to α seems to be even stronger for $\alpha < 0$, as is clearly shown in Fig.5, where the curves for distinct α are undistinguishable.

At this point two remarks are appropiated. One in the direction to justify the approximation that we use for $\beta \neq 0$. An other, an intuitive explanation of the disapearance of the long time tail for $\beta \neq 0$.

Respect to the first one, we consider conditions on the two first derivatives of C(t). This tells us that we can be confidents about the validity of the procedure for short times. The decay is so fast that only taking into account as a new condition the relaxation time, it is reasonable to

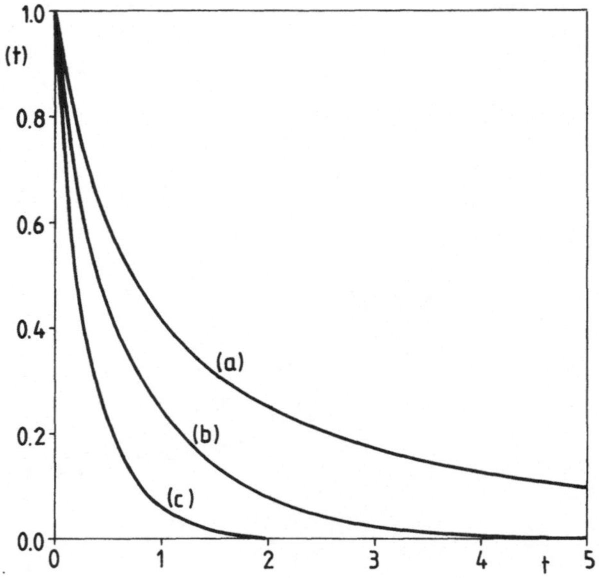

Fig. 3. (a) α=.5, β=0 (exact solution); (b) α=.5, β=.1 (M=2 with 2 derivatives and 1 relaxation moment); (c) α=.5, β=1 (same approximation).

think that the result is good for all times. The area under the curve is then correct and there is no much place for correction of higher order.

The second remark will help us to reinforce this argument. For β=0 the distribution probability P(n,t) is very peacked around the value n=0 for α->0. The critical slowing down appears because the deterministic force vanishes at the same point where the effect of the noise also disappears (n=0). This is translated to a long time tail in the correlation function. For $\beta \neq 0$ this is not the case. The maximum of P(n,t) corresponds to a finite value of n. Then for $\beta \neq 0$ there is no reason for a long time tail. In fact it can be seen that the relaxation time for $\beta \neq 0$ is finite in the limit of α->0.

Therefore, the practical rule to combine the low and high frequency expansions is the following. A small number of relaxation moments is necessary in order to obtain global properties of the curve. If a long time tail is not expected, the lowest order can be sufficient. In case that a long time tail occurs, it can be necessary to go beyond (but not very much) in order to get a better knowledge of the long time behaviour (that is, not only the total area but some information about its distribution). Finally, if one wants to improve the approximation, one can proceed by adding derivative (local) conditions, which will automatically improve not only the initial regime but the global one provided the relaxation moments conditions are maintained.

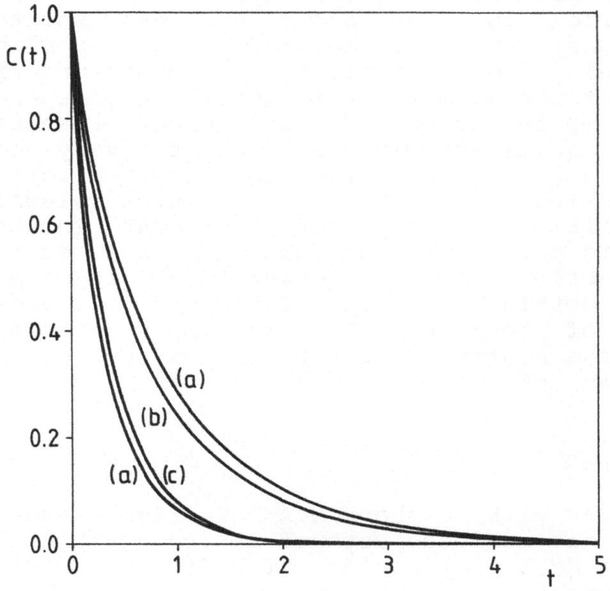

Fig. 4. (a) α=.1, β=.1; (b) α=.5, β=.1; (c) α=.1, β=1; (d) α=.5, β=1; all the curves have M=2 with 2 derivatives and 1 relaxation moment.

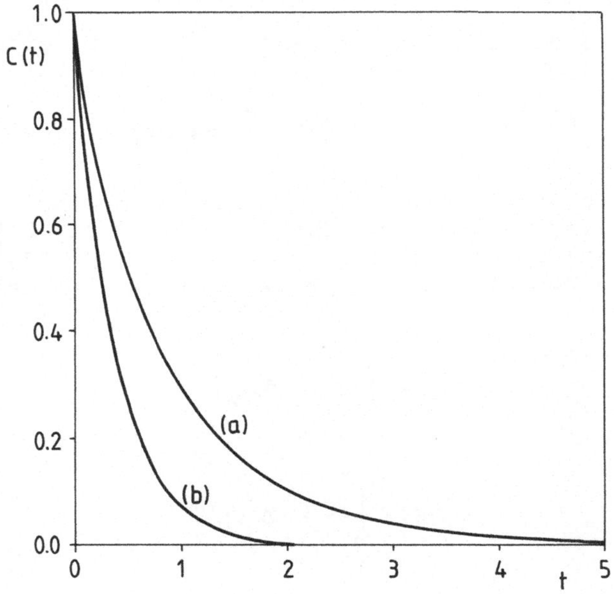

Fig. 5. (a) β=.1 for α=-.5 and α=-.1 (superposed); (b) β=1 for α=-.5 and α=-.1 (superposed)

SUMMARY

We have presented an approximate method to calculate steady state correlation functions of nonlinear stochastic Langevin equations. We have applied it to a nonlinear nuclear reactor model with instantaneous temperature feedback and reactivity fluctuations. For vanishing independent neutron source we compare the method with exact results[1]. The agreement is qualitatively and quantitatively very good. The method enlarges very much the region of validity of a previous approximative method[2]. The exact relaxation time is obtained. Critical slowing down is present in the stationary deterministic point. A long time tail is present in the correlation function. For a nonvanishing independent neutron source we obtain that the long time tail disappears. The sensitivity of the result is very important respect to the values of the neutron source β but not to the linear coefficient α, when $\beta \neq 0$.

ACKNOWLEDGEMENT

The authors wish to thank M.A. Rodríguez and L. Pesquera for helpful discussions. Both of us are grateful to the Comisión Asesora de Investigación Científica y Técnica, project nº361/84 (Spain) for support.

APPENDIX

In this appendix we will calculate the expression of the relaxation moments T_K. We will obtain T_K as a recursive quadrature.

Taking into account the expression for C(t) given by

$$C(t) = \frac{1}{\langle \delta q^2 \rangle} \int_a^b dq\, \delta q\, e^{L(q)t} \delta q\, P_{st}(q) \tag{A.1}$$

where a and b are the natural boundaries of the process q(t) defined by (2.1), L(q) is the corresponding Fokker-Planck operator, $\delta q = q - \langle q \rangle$ and $P_{st}(q)$ is the steady state distribution and interchanging the integrals in the expression for T_K we have

$$T_K = \int_a^b dq\, \delta q\, R_K(q) \tag{A.2}$$

with

$$R_K(q) = \int_0^\infty dt\, t^K\, W(q,t) \tag{A.3}$$

and

$$W(q,t) = \frac{1}{\langle \delta q^2 \rangle} e^{L(q)t} \delta q\, P_{st}(q) \tag{A.4}$$

Hence, to calculate T_K we only need to obtain $R_K(q)$. Now, we will get the equation for $R_K(q)$. Defining the Laplace transform of W(q,t) as

$$R(q,\omega) = \int_0^\infty dt\; e^{-\omega t}\, W(q,t) \tag{A.5}$$

and expanding the exponential in (A.5) we obtain

$$R(q,\omega) = R_0(q) - \omega R_1(q) + \frac{\omega^2}{2!} R_2(q) - \ldots \tag{A.6}$$

Now from the equation for W(q,t)

$$\frac{\partial}{\partial t} W(q,t) = L(q) W(q,t) \tag{A.7}$$

by Laplace transforming we get the corresponding equations for $R_K(q)$

$$L(q)\, R_K(q) = R_{K-1}(q) \tag{A.8}$$

$$L(q) R_0(q) = -W(q,0) = -\frac{1}{\langle \delta q^2 \rangle}\, \delta q\, P_{st}(q) \tag{A.9}$$

The solution of this equations[8] gives for T_K the expressions (3.10) and (3.11).

REFERENCES

1. Dutré, W.L. and Debosscher, A.F.; Nucl. Sci. Engn. 62, 355 (1977).
2. Hernández-Machado, A., Rodríguez, M.A. and San Miguel, M.; Ann. Nucl. Energy. 12, 471 (1985).
3. Williams, M.M.R.,"Random Processes in Nuclear Reactors", Pergamon Press, Oxford (1974).
4. Kishida, K., Kanemoto, S. and Sekiya, T.; J. Nucl. Sci. Tech. 13(1), 19 (1976).
5. Sako, O., Taniguchi, A. and Kuroda, Y.; Ann. Nucl. Energy 9, 325 (1982).
6. Rodríguez, M.,"Análisis de fluctuaciones en reactores nucleares: modelos no lineales y no markovianos". Tesis Doctoral. Universidad de Santander (1983).
7. Zwanzig, R.,"Lectures in Theoretical Physics" Eds. W. Brittin and L. Dunham, vol.3 (Wiley and Sons, New York, 1961).
8. Casademunt, J. and Hernández-Machado, A., preprint 1988.
9. a) Nadler, W. and Schulten, K.; J. Chem. Phys. 82, 151 (1985).
   b) Nadler, W. and Schulten, K.; Z. Physik B59, 53 (1985).
10. Brenig, L. and Banai, N.; Physica 5D, 208 (1982).
11. Graham, R. and Schenzle, A.; Phys. Rev. A25, 1731 (1982).
12. Fujisaka, H. and Grossman, S.; Z. Physik B43, 69 (1981).

NON-LINEAR EFFECTS IN STRUCTURAL VIBRATION INDUCED FLUCTUATIONS

L. Pesquera and M.A. Rodríguez

Departamento de Física Moderna, Universidad de Cantabria
Avda. de los Castros, E-39005 Santander, Spain

1.- INTRODUCTION

In a nuclear reactor an important source of noise is that induced by the random motion of structural components and internal disturbances. Two significant examples are the random motion, formation and collapse of steam bubbles in boiling water reactors and the random vibration of control rods and fuel elements in pressurized water reactor (Williams, 1974).

In recent years the possibility of locating and predicting anomalous rod vibrations has been one of the interests of reactor noise analysis (Pázsit and Glöckler, 1983, 1984; Glöckler and Pázsit, 1987). It is clear that localization is possible only if the spatial dependence of neutron noise is taken into account. The usual method followed in the investigation of space-dependent effects is the study of a single vibrating absorber in the Feinberg Galanin approximation (Williams, 1970; Pázsit, 1977, 1984; Analytis, 1980; Antonopoulos-Domis, 1976; Gotoh and Yasuda, 1983; I. Martínez and M.A. Rodriguez, 1985).

In this paper we analyze, using the Feinberg-Galanin model, the non-linear transference from the random vibration of control rods to neutron noise. This non-linear effect consists of the appearance of peaks at ω_o and $2\,\omega_o$ frequencies in the power spectral density (PSD) of neutron noise, while the vibration PSD has only a peak at ω_o (see Fig. 1). This non-linear transference has been observed by Lucia et al. (1973). The bi-frequency peak in the neutron PSD has been attributed to space-dependent effects (Antonopoulos Domis, 1976; Pázsit 1977). An attempt to explain this peak has been made by Gotoh (1982) using a point reactor model. In this case the non-linear transference is due to the multiplicative character of the point reactivity fluctuations. However, the correlation time of the reactivity noise required to get the peak at $2\,\omega_o$ is too large. This result was derived using diagrammatic techniques.

The aim of the present work is to analyze non-linear effects in structural vibration induced fluctuations avoiding linearization procedures. We consider simple models in order to get exact results and/or well justified approximations. In this way recent exact results for the spacedependent model (I. Martínez and M.A. Rodríguez, 1985) are used to test the closure approximation (Williams, 1970, 1974). A well justified approximation for the spatial dependence of neutron noise comes out from this analysis. A relation between the amplitudes of peaks at frequencies ω_0 and $2\,\omega_0$ in the neutron PSD is derived.

As concerns the point model used by Gotoh (1982), we obtain an exact result for the neutron PSD. We show that, unlike the result derived by Gotoh (1982), the bi-frequency peak can be explained with realistic values for the correlation time of the reactivity fluctuations.

The outline of the paper is as follows. In Section 2 we analyze in detail the experimental results (Lucia et al., 1973). In Section 3 we introduce the Feinberg-Galanin model. The non-linear transference due to the spatial dependence of neutron noise is investigated. Section 4 is devoted to the study of a point model with reactivity fluctuations due to the vibration of a control rod.

2. EXPERIMENTAL RESULTS

Non-linear effects associated with the random motion of structural components can be classified in two groups. First, the nonlinearity is due to the noise source, but its propagation to the neutron field is treated as linear (H. Konno and K. Saito, 1984; R. Sanchis, 1986; R. Sanchis et al., 1988). Second, the vibration is linear and its effect upon the neutron field is nonlinear.

In this work we consider the non-linear transference observed by Lucia et al. (1973). In Fig. 1 the noise measurements at the ECO-reactor are presented. It can be seen that the vibration PSD has a peak at $\omega_0 = 0.25\,Hz$ while the neutron noise has peaks at $\omega_0 = 0.25\,Hz$ and $2\,\omega_0 = 0.5\,Hz$.

Using these experimental results we estimate the parameters that characterize the absorber vibration. Denoting by $\varepsilon(t)$ the random position of the absorber with mean position at x=0, we have

$$\langle \varepsilon(t) \rangle = 0 \quad , \quad \langle \varepsilon(t)\, \varepsilon(t') \rangle = \Delta^2\, C(t-t') \, , \tag{1}$$

where $C(\theta)$ is the correlation function (C(0) =1) and Δ is the displacement magnitude. If we describe the flow-induced vibration of the absorber by a linear damped oscillator (Païdoussis, 1982), we get

$$C(\theta) = e^{-\lambda \omega_0 |\theta|} \left[\cos \omega_0 \theta + \frac{\lambda}{1-\lambda^2} \sin \omega_0 |\theta| \right] \, , \tag{2}$$

where ω_0 is the first eigenfrequency of the absorber and λ is the corresponding damping factor. The correlation time is $t_c = (\lambda \omega_0)^{-1}$. An estimation of t_c is given by the inverse of the peak width at ω_0 in the displacement PSD. From Fig. 1 we see that $t_c \sim 10$s. and, then, $\lambda \sim 0.4$.

If we assume that $\varepsilon(t)$ is Gaussian, the noise source is completely specified by its correlation function. The problem is to obtain the effect of the absorber vibration, $\varepsilon(t)$, upon the neutron field. It is clear from Fig. 1 that is an example of non-linear transference. In the following section we analyze non-linear effects due to the spatial dependence of neutron noise.

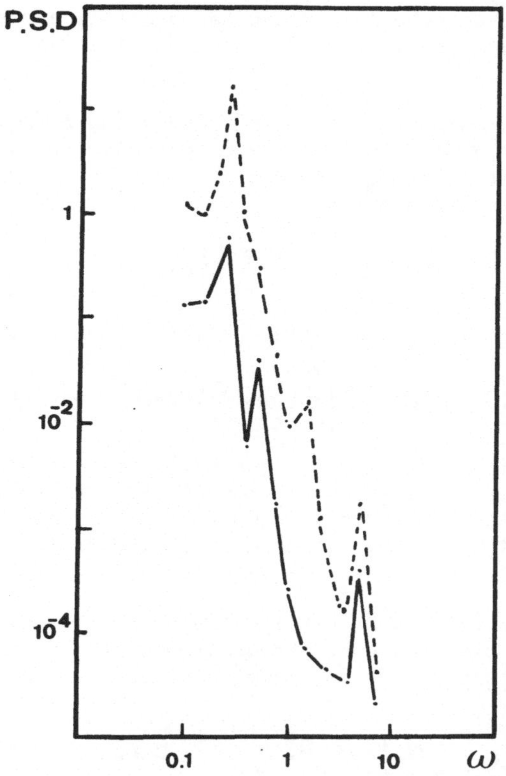

Fig. 1. Neutron PSD (——) and displacement PSD (---) taken from Lucia et al. (1973)

3. SPATIAL DEPENDENCE OF NEUTRON NOISE

Let us consider a bare subcritical reactor. Embedded in the medium is a thin absorbing plate which executes random vibrations in the direction normal to its surface. We assume the one-speed diffusion theory and the Feinberg-Galanin model for the absorber. Then, we write the equation for neutron density, neglecting delayed neutrons, as

$$\frac{\partial N(x,t)}{\partial t} = D_o \frac{\partial^2 N}{\partial x^2} - \alpha N - \gamma\, \delta(x-\varepsilon(t)) N(x,t) + S_o , \tag{3}$$

where D_o is the diffusion coefficient, $1/\alpha$ is the mean lifetime of a thermal neutron, S_o is a uniformly distributed source and γ is the Galanin constant for the plate, whose random position is given by $\varepsilon(t)$.

In the steady state the differential equation (3) can be written in the following integral form

$$N(x,t) = \frac{S_o}{\alpha} - \gamma \int_0^\infty du \int_V dx' \, G(x,u/x',o)\, \delta(x'-\varepsilon(t-u)) N(x',t-u) , \tag{4}$$

where G(x,u/x',o) is the Green's function associated with the operator $D_o \partial^2/\partial x^2 - \alpha$. It is clear from equation (4) that the equations for mean values are unclosed because of the flux-dependency of noise effect (multiplicative noise). The absorber vibration has a non-linear effect due to the coupling between the position, $\varepsilon(t)$, and the flux depression induced by the absorber.

An exact method to study the mean value $\langle N(x,t)\rangle$ has been derived when the vibration, $\varepsilon(t)$, is a Markovian process (I. Martínez and M.A. Rodríguez, 1985). To get $\langle N(x)\rangle_s$, the stationary average density, we need to calculate the following quantity

$$A(x) = \langle \delta(x-\varepsilon(t)) \; N(x,t)\rangle_s \quad . \tag{5}$$

When $\varepsilon(t)$ is a Markovian proces, A(x) satisfies the following integral equation

$$A(x) = N_o(x,\gamma=0)\, P_s(x) - \gamma\int_0^\infty du \int_V dx'\, G(x,u/x',0)\, w(x,u/x',0)\, A(x') \, , \tag{6}$$

where $N_o(x,\gamma=0)$ is the stationary density in the absence of the absorber and $P_s(x)\,(w(x,u/x',0))$ is the stationary probability density (conditional probability density) of the vibration. The equations obtained are closed, so they are useful to test approximation methods as the closure approximation (Williams, 1970, 1974).

Equation (6) can be easily solved when the evolution time of the system is shorter than the correlation time of the noise. This is the adiabatic case that corresponds to approximate w(x,u/x',o) by δ(x-x'). The resulting A^{ad} is

$$A^{ad}(x) = \frac{N_o(x,\gamma=0)}{1+\gamma\int_0^\infty du\, G(x,u/x,0)} \cdot P_s(x) \quad . \tag{7}$$

To study the validity range of this approximation we neglect size effects. The Green's function of a 1-D infinite reactor is

$$G(x,u/x',0) = (4\pi D_o u)^{-1/2} \exp[-\alpha u - (x-x')^2/4D_o u] \; . \tag{8}$$

The adiabatic approximation is valid when G(x,u/x',o) does not change in equation (6) due to the effect of the noise, that is when $G(\varepsilon(u),u/\varepsilon(0),0) \simeq G(\varepsilon(0),u/\varepsilon(0),0)$. If we evalutate this condition using average values, we get the condition $\Delta^2/t_c \ll D_o$, which is equivalent to $\Delta^2/t_c \ll L^2\alpha$, where $L=(D_o/\alpha)^{1/2}$ is the diffusion lenght. Therefore the diffusion due to the absorber vibration, Δ^2/t_c , must be much smaller than the diffusion coefficient D_o.

If we choose the following numerical values for the parameters: $\alpha = 100\,s^{-1}$, L= 2cm. and $t_c \gtrsim 1\,s.$ (see Sect. 2), we see that the adiabatic approximation is valid for noise such that $\Delta \sim L$. However if we consider delayed neutrons, $\alpha^{del.} t_c \sim 1$ and the displacement must be small, $\Delta \ll L$, for the adiabatic approximation to be valid.

Now, we analyze the connection between the closure approximation and the adiabatic limit. For a 1-D infinite reactor we have from eq. (7), $A^{ad.}(x) = N_o(0,\gamma)\, P_s(x)$, where $N_o(0,\gamma)$ is the flux depression due to a static absorber. In the closure approximation the following assumption is made, $A(x) \propto \langle N(\varepsilon(t),t)\rangle_s P_s(x)$. Then a close relationship exists between both results. In fact, Williams (1970) has shown that in the adiabatic limit $\langle N(\varepsilon(t),t)\rangle_s = N_o(0,\gamma)$. Therefore the closure approximation is justified in the adiabatic case. The reason for this is that the vibration is so slow that the flux rises and falls in phase with it and, hence, the depression is unaffected by the vibration. As a consequence $N(\varepsilon(t),t)$ is given by the deterministic depression, $N_o(0,\gamma)$, and the closure approximation follows.

The closure approximation has been tested using a three-level Markovian vibration (I. Martínez and M.A. Rodríguez, 1985). In this case an analytic solution of equation (6) can be obtained. In Fig. 2 we represent the relative error of the global absorption, $\gamma\int_V dx\, A(x)$, as a function of the inverse correlation time for several values of the noise magnitude, Δ , and the parameter $r = N_0(0,\gamma)/N_0(0,\gamma=0)$, that measures the strength of the absorber. We see that, even for strong absorbers (r= 0.1) and large displacements $(\Delta = L)$ the closure approximation fails only when the correlation time is short.

To obtain the spatial dependence of neutron noise we consider the fluctuations $\delta N(x,t) = N(x,t) - \langle N(x)\rangle_s$. From equation (4) we have

$$\delta N(x,t) = -\gamma\int_0^\infty du\left[G(x,u/\varepsilon(t-u),0)\,N(\varepsilon(t-u),t-u) - \langle GN\rangle_s\right]. \qquad (9)$$

Setting $x = \varepsilon(t)$ gives a stochastic integral equation for $N(\varepsilon(t),t)$

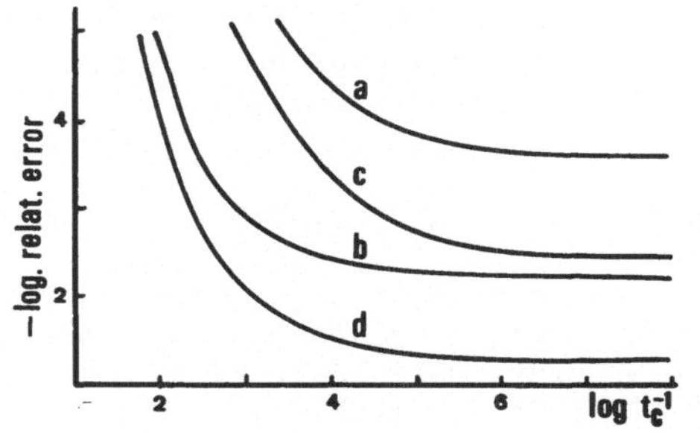

Fig. 2. Relative error of the closure approximation (a)r= 0.5, Δ = 0.1L; (b)r= 0.5, Δ = L; (c)r= 0.1, Δ = 0.1L; (d)r= 0.1, Δ = L (I. Martínez and M.A. Rodríguez, 1985)

$$\delta N(\varepsilon(t),t) = -\gamma\int_0^\infty du\left[G(\varepsilon(t),u/\varepsilon(t-u),0)\,N(\varepsilon(t-u),t-u) - \langle GN\rangle_s\right]. \qquad (10)$$

As we have shown before, in the adiabatic case, $\Delta^2 \ll L^2\alpha t_c$ we have $G(\varepsilon(t), u/\varepsilon(t-u),0) \simeq G(\varepsilon(t),u/\varepsilon(t),0)$, which is a non random function if we neglect size effects. Then we obtain

$$\delta N(\varepsilon(t),t) = -\gamma\int_0^\infty du\; G(0,u/0,0)\,\delta N(\varepsilon(t-u),t-u)\quad, \qquad (11)$$

whose solution is $\delta N(\varepsilon(t),t) = 0$. We find again the result that the depression does not change when the vibration is slow. Therefore we get from eq.(9)

$$\delta N(x,t) = -\gamma\int_0^\infty du\left[G(x,u/\varepsilon(t-u),0) - \langle G\rangle_s\right]N_0(0,\gamma)\;. \qquad (12)$$

This result was also obtained by Pázsit (1984) for small vibrations. However, we have shown that it is valid even for large displacements, $\Delta \sim L$, when the vibration is slow, $\alpha t_c \gg 1$.

It is clear form eqs. (9) and (12) that in the adiabatic limit the substitution of $\delta(x-\varepsilon(t))N(x,t)$ by $\delta(x-\varepsilon(t))N_0(0,\gamma)$ in eq. (3) is justified when we neglect transients. Note that this is different from the linearization procedure.

Equation (12) shows that non-linear effects are brought about by the dependence of the Green's function on the absorber position.This relation can be used to calculate the noise at points close to the absorber. However, in order to estimate the relation between the amplitudes of peaks at frequencies ω_0 and $2\omega_0$ in the neutron PSD, we develop the Green's function in powers of the noise. For the space dependent part of G we have

$$\exp[-(x-\varepsilon)^2/4D_0u] = e^{-\frac{x^2}{4D_0u}}\left[1+\frac{x\varepsilon}{2D_0u}+\frac{\varepsilon^2}{4D_0u}\left(\frac{x^2}{2D_0u}-1\right)+\dots\right] \quad . \tag{13}$$

This development is valid when $\Delta \ll x \ll L^2/\Delta$, that is, $x \gg \Delta$ and $\Delta \ll L$ (Williams, 1970). The ε and ε^2 terms are responsible for the appearance of the ω_0 and $2\omega_0$ frequencies, respectively. Using the value of the noise at L,we get that the relation between the amplitudes of both peaks is of order L/Δ . From the experimental results (Fig. 1) we obtain $L/\Delta \sim 15$. Therefore small displacements can explain the bi-frequency peak even in an infinite reactor owing to the space dependence of the Green's function.

4.- POINT MODEL WITH VIBRATION INDUCED REACTIVITY FLUCTUATIONS

In this Section we study a point model with reactivity fluctuations which are assumed to be proportional to the amplitude of rod vibrations. Similar models have been previously analyzed using approximation methods (Kosály and Williams, 1971; Gotoh, 1982). We consider a simple model without delayed neutrons and we present an exact result for the neutron PSD. This PSD exhibits all the harmonics of the frequency vibration.

The basic neutronic equation is

$$\frac{dn}{dt} = -\varrho n + \xi(t) n + s \quad , \tag{14}$$

where ϱ is proportional to the mean value of the subcritical reactivity and s is the neutron source. The reactivity fluctuations are assumed to be Gaussian and its correlation function proportional to that of rod vibration. In a first approximation, when the damping is small, we have from eq. (2)

$$\langle \xi(t)\xi(t+\theta)\rangle = \sigma^2 \cos\omega_0\theta \exp[-|\theta|/t_c] = \mathrm{Re}\left\{\sigma^2 \exp[-\mu|\theta|]\right\} \quad , \tag{15}$$

where we have introduced the parameter $\mu = 1/t_c - i\omega_0$. Following the same method than in M.A. Rodríguez and L. Pesquera (1983), exact expressions for mean values and the correlation function can be obtained using the following exact stationary solution

$$n(t) = s\int_0^\infty du \; \exp\left[-\varrho u + \int_0^u du' \, \xi(t-u')\right] \quad , \tag{16}$$

and the Gaussian property of the noise, $\xi(t)$.

The mean value is given by (M.A. Rodríguez, 1983)

$$\langle n\rangle_s \simeq \frac{s}{\varrho - D^*} \quad , \quad D^* = \frac{\sigma^2 t_c}{1+\omega_0^2 t_c^2} \quad , \tag{17}$$

where we have neglected terms of order Dt_c, being D the intensity of the noise, $D=\sigma^2 t_c$. The stability of $\langle n\rangle_s$ is governed by an effective intensity D^*. The mean value is finite when $\varrho_0 = \varrho - D^* > 0$ and, hence, the stability increases with ω_0. As concerns the variance, the stability region, $\varrho - 2D^* > 0$, is smaller (M.A. Rodríguez, 1983). This is a non-linear effect due to the multiplicative character of the noise.

An exact expression for the correlation function can be given in the following form

$$\begin{aligned}\langle n(t)n(t+\theta)\rangle_s &= s^2\int_0^\theta du\int_0^\infty du'\, e^{-\varrho_0(\theta-u+u')} \exp\{\beta(e^{-\mu(\theta-u')}-1)+\beta(e^{-\mu u'}-1) \\ &-\beta(e^{-\mu\theta}-e^{-\mu u})(1-e^{-\mu u'})+c.c.\} + 2s^2\int_{-\infty}^0 du\int_{-\infty}^u du'\, e^{-\varrho_0(\theta-u-u')-2D^*u} \\ &\cdot \exp\{\beta(e^{-\mu(\theta-u')}-1)+2\beta(e^{-\mu u}-1)+\beta(e^{\mu u'}-1) \\ &-\beta(e^{-\mu\theta}-1)(1-e^{-\mu u'})-\beta(1-e^{-\mu u})(e^{\mu u}-e^{\mu u'})+c.c.\}\ ,\end{aligned} \tag{18}$$

where $\beta = \sigma^2/2\mu^2$. Making a development in powers of $\exp[-\mu\theta]$ and $\exp[-\mu(\theta-u)]$, it is possible to get the neutron PSD as a summ of Lorentzians centered at $n\omega_0$. Here we only give the result. Details may be found elsewhere (M.A. Carpintero, 1986). The development in harmonics is governed by the parameter $|\beta| = D^* t_c/2$. Then the analysis that follows is valid whenever $D^* t_c \ll 1$.

The PSD has peaks at $n\omega_0$ with a width of order t_c^{-1}. Its intensity $S(n\omega_0)$ is of order $s^2\varrho_0^{-2} t_c (D^* t_c/2)^n$. The background, $S_b(\omega)$, considered as the contribution of the Lorentzian centered at $\omega = 0$, is of order $s^2\varrho_0^{-2} D^* (\varrho_0^2+\omega_0^2)^{-1}$. If we take frequencies $\omega_0 \sim 1\,Hz$, we have $\varrho_0 < \varrho \ll \omega_0$. As an example we can consider a reactivity $1-k \sim 10^{-5}$ and a neutron lifetime $\ell \sim 10^{-3} s.$, that lead to $\varrho \sim 10^{-2} s.^{-1}$. The relation between the peak and the background intensity can then be roughly estimated by

$$S(n\omega_0)/S_b \sim (D^* t_c)^{n-1}(\omega_0 t_c)^2 . \tag{19}$$

Since we have assumed that $D^* t_c \ll 1$, it is necessary that $\omega_0 t_c \gg 1$ for the peak at $2\omega_0$ to be observable. As concerns the relation between the intensity of the peaks at ω_0 and $2\omega_0$, we get the estimate $S(2\omega_0)/S(\omega_0) \sim D^* t_c$. According to the experimental results (Fig. 1), $S(\omega_0)/S(2\omega_0) \sim 15$ and $\omega_0 t_c \gtrsim 1$, (see Sect. 2), that are consistent with our model.

We have investigated the behaviour of $S(n\omega_0)/S_b\,(n=1,2)$ with $\gamma = D^*/\varrho$ and $\omega_0 t_c$. The parameter γ measures the effect of the noise on the system. The instability for the neutron variance appears for $\gamma = 0.5$. In Fig.3 we show in logarithmic scale the first peak intensity for several values of ω_0 and t_c when $\varrho = 10^{-2} s.^{-1}$ and D changes. We see that, except for very large t_c ($t_c = 10^2 s.$)

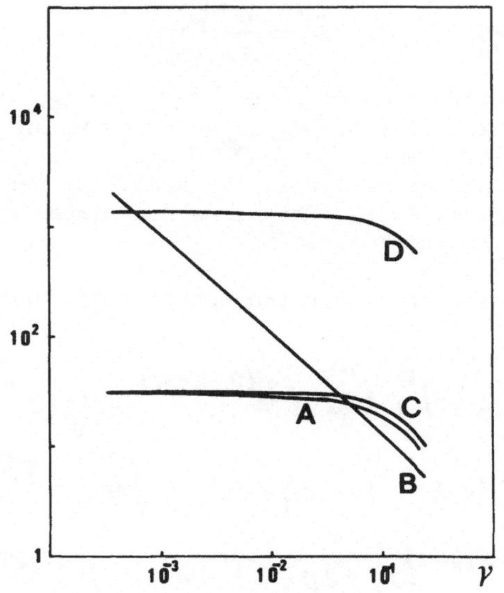

Fig. 3. Plot of the first peak intensity as a function of $\gamma = D^{*}/\varrho$ with $\varrho = 10^{-2}s^{-1}$ kept fixed. Values of the parameters: $A(t_c = 10s., \omega_o = 1H_z)$; $B(t_c = 10^2 s., \omega_o = 1Hz)$; $C(t_c = 1s., \omega_o = 10Hz)$; $D(t_c = 1s, \omega_o = 10^2 Hz)$.

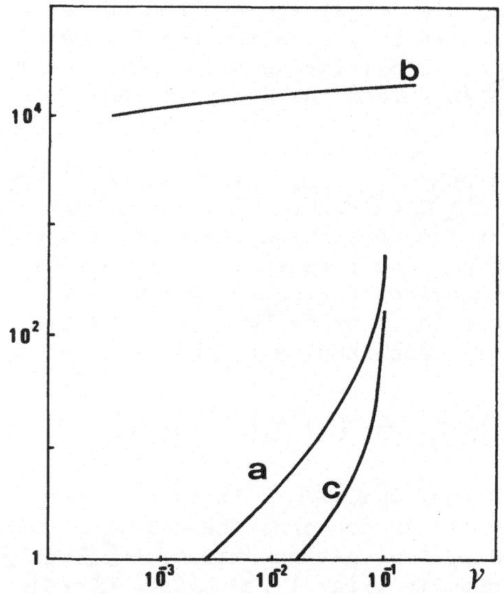

Fig. 4. Plot of the second peak intensity as a function of $\gamma = D^{*}/\varrho$ with $\varrho = 10^{-2}s^{-1}$ kept fixed. Values of the parameters: $a(t_c = 10s., \omega_o = 1Hz)$; $b(t_c = 10^2 s., \omega_o = 1Hz)$; $c(t_c = 1s., \omega_o = 10Hz)$.

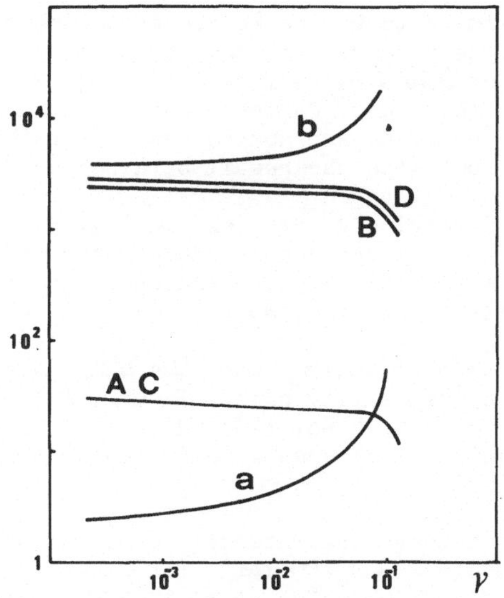

Fig. 5. Plot of first (A,B,C,D) and second (a,b) peak intensities as a function of γ with D= $10^{-2}s^{-1}$ kept fixed. Same values of parameters than in Fig. 3 and 4.

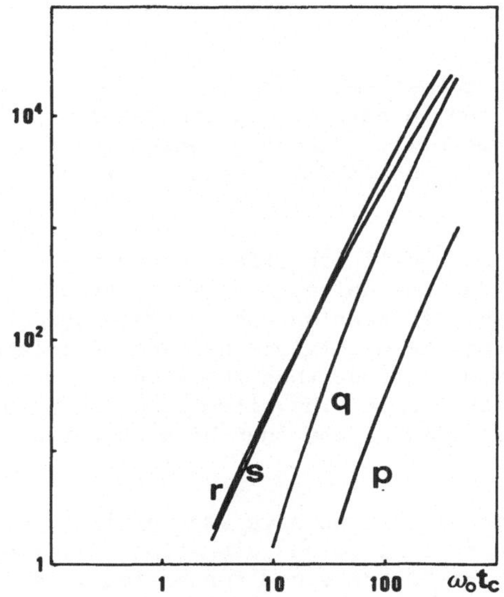

Fig. 6. Plot of first (r,s) and second (p,q) peak intensities as a function of $\omega_o t_c$. Values of the parameters:
r(D= $10^{-4}s.^{-1}$,ω_o=1Hz); s(D=$10^{-4}s.^{-1}$,t_c=1s.);
p(D= $10^{-4}s.^{-1}$;ω_o=1Hz); q(D=$4.10^{-3}s.^{-1}$,ω_o=1Hz).

, $S(\omega_0)/S_b$ is not sensitive to D when we are not close to the instability point, $\gamma = 0.5$. Under these conditions the rough estimate (19) can then be used. A similar plot for $S(2\omega_0)/S_b$ is made in Fig. 4. We do not show curve d because $S(2\omega_0)/S_b < 1$ in this case. Curves a and c are in agreement with (19), showing that large values of t_c are not necessary for the peak at $2\omega_0$ to be observable. This disagrees with the result obtained by Gotoh (1982).

The effect of varying ρ is shown in Fig. 5. The value of D is kept fixed, $D = 10^{-2} s^{-1}$. The results are nearly independent of the value of ρ in qualitative agreement with (19). As a consequence the inclusion of delayed neutrons will not modify these results.

In Fig. 6 we show $S(\omega_0)/S_b$ and $S(2\omega_0)/S_b$ as a function of $\omega_0 t_c$ in situations far from the instability point, $\gamma = 0.5$. The results for the first peak, curves r and s, agree with eq. (19). However, the intensity of the second peak, curves p and q, increases with t_c faster than in the estimate given by eq. (19).

The relation between the intensity of peaks at ω_0 and $2\omega_0$ is analyzed in Fig. 7. We plot $S(\omega_0)/S(2\omega_0)$ for several values of t_c as a function of γ keeping fixed the value of $\rho = 10^{-2} s^{-1}$. This relation is proportional to γ^{-1} in agreement with eq. (19). These results can be used to fit experimental data (Fig. 1). We see that this is possible if we take $t_c \sim 15 s.$ and $\gamma \sim 10^{-2}$, which corresponds to a reactor operating in normal conditions. In this case we get the experimental result $S(\omega_0)/S(2\omega_0) \sim 15$ .For these values of the parameters the stability is due to the large value of $\omega_0 t_c$. The variation in the reactivity owing to the noise, σ/ρ , is of order 1. However, the effective intensity $D^* = \sigma^2 t_c/(1+\omega_0^2 t_c^2)$, is much smaller than $\rho(\gamma = 10^{-2})$. The rapid oscillations in the noise correlation function stabilize then the system.

5. CONCLUSIONS

The non-linear transference from the random vibration of control rods to neutron noise has been investigated using one-dimensional models. This analysis is useful to identify sources of non-linearity, but not to locate anomalous rod vibrations. Localization requires to consider at least two-dimensional models.

Two sources of non-linear effects have been analyzed. First, the spatial dependence of neutron noise is shown to produce nonlinear propagation of rod vibration because of Green's function dependence on the absorber position. This result has been obtained neglecting delayed neutrons, but it still holds with delayed neutrons when the random displacement is small with respect to the diffusion length. The bifrequency peak in the neutron PSD observed by Lucia et al. (1973) can then be attributed to space-dependent effects.

The second model that we have studied is a point model with reactivity fluctuations proportional to rod vibrations amplitude. We have analyzed an exact solution for the PSD obtained for a simple model without delayed neutrons. The intensity of peaks at ω_0 and $2\omega_0$ frequencies is shown to be independent of the value of the reactivity in regions far from the instability point. Therefore the inclusion of delayed neutron will not modify our conclusions. In this point model the appearance of the bifrequency peak in the neutron PSD is due to the multiplicative character of the reactivity fluctuations. Experimental data are recovered with reasonable values of the parameters.

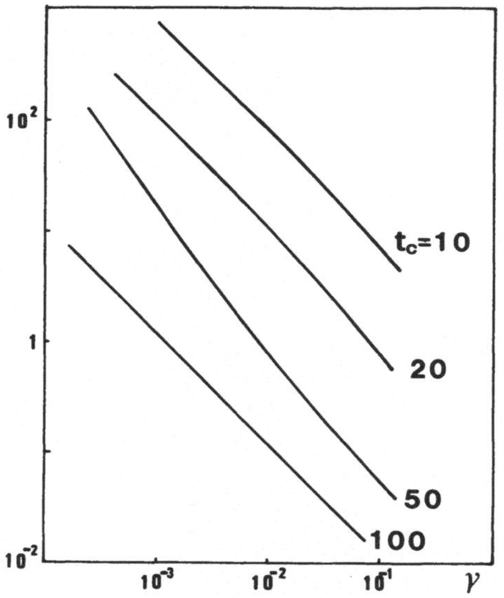

Fig. 7. Plot of the relation between peak intensities as a function of γ with $\rho=10^{-2}s^{-1}$ and $\omega_0=1$ Hz kept fixed.

ACKNOWLEDGEMENT

Part of this work has been made in collaboration with M.A. Carpintero. Financial support from C.A.I.C.Y.T., Spain (361/84) is acknowledged.

REFERENCES

Analytis, G.Th., 1980, Ann. Nucl. Energy, 7: 351.
Antonopoulos-Domis, M., 1976, Ann. Nucl. Energy, 3: 451.
Antonopoulos-Domis, M., 1981, Ann. Nucl. Energy, 8: 91.
Carpintero, M.A., 1986, Master Degree work, Univ. de Cantabria, Spain.
Glöckler, O., and Pázsit, I., 1987, Ann. Nucl. Energy, 14: 63.
Gotoh, Y., 1982, Prog. Nucl. Energy, 9: 303.
Gotoh, Y. and Yasuda, N., 1983, Ann. Nucl. Energy, 10: 589.
Konno, H. and Saito, K., 1984, Ann. Nucl. Energy, 11: 1.
Kosály, G. and Williams, M.M.R., 1971, Atomkernenergie, 18: 203.
Lucia, A., Ohlmer, E. and Schwalm, D., 1973, Atomkernenergie, 22: 4.
Martínez, I. and Rodríguez, M.A., 1985, Ann. Nucl. Energy, 12: 113.
Païdoussis, M.P., 1982, Nucl. Eng. Des., 74: 31.
Pázsit, I., 1977, Atomkernenergie, 30: 29.
Pázsit, I., 1984, Ann. Nucl. Energy, 11: 441.
Pázsit, I. and Analytis, G.Th., 1980, Ann. Nucl. Energy, 7: 171.
Pázsit, I. and Glöckler, O., 1983, Nucl. Sci. Eng., 85: 167.
Pázsit, I. and Glöckler, O., 1984, Nucl. Sci. Eng., 88: 77.
Rodríguez, M.A., 1983, Ph.D. Thesis, Univ. de Cantabria, Spain.
Rodríguez, M.A. and Pesquera, L., 1983, J. of Nucl. Sci. Tech., 2: 174.
Sanchis, R., 1986, Ph.D. Thesis, Univ. Politécnica de Valencia, Spain.
Sanchis, R., Verdú, G. and Muñoz-Cobo, J.L., 1988, Ann. Nucl. Energy, in press
Williams, M.M.R., 1970, Nucl. Sci. Energy, 39: 144.
Williams, M.M.R., 1974, "Random Processes in Nuclear Reactors", Pergamon Press, Oxford.

CHAPTER V

STOCHASTIC PROCESSES IN NONLINEAR NUCLEAR SYSTEMS: APPLICATIONS

BISPECTRUM ESTIMATION USING OVERLAPPED SEGMENTS

J. Piñeyro+ and K. Behringer

Paul Scherrer Institute

CH-5303 Würenlingen, Switzerland

I. INTRODUCTION

The power spectral density (PSD) function cannot detect nonlinear random processes which effect non-Gaussian contributions to the noise signal. For the characterization of non-Gaussian noise contributions, higher-order cumulant functions or their corresponding Fourier transforms must be considered (e.g. Hasselmann et al., 1963; Brillinger, 1965; Bendat and Piersol, 1982). The triple correlation function or the bispectrum, respectively, are the next higher approaches, provided that a skewness exists. There are also interesting applications to deterministic signals, which are often obscured by background noise (Sato and Sasaki, 1977; Lohmann and Wirnitzer, 1984).

While the conventional PSD estimation via fast Fourier transform (FFT) techniques has reached a highly developed level, it seems that the bispectrum estimation still requires some basic considerations (Nikias and Raghuveer, 1987). The indirect method, comparable with the Blackman-Tukey procedure in the PSD estimation, first estimates the triple correlation function from measured data, windows or tapers the function estimate in an appropiate manner, and then transforms the windowed data by double FFT to obtain the bispectrum estimate (Brillinger, 1965; Rosenblatt and Van Ness, 1965; Sasaki et al., 1975; Rao and Gabr, 1984). A two-dimensional correlation function window for minimum bispectrum bias has been derived by Sasaki et al. (1975), and for minimum mean squares error by Rao and Gabr (1984). An analytical expression for the Rao-Gabr window has been given by Saito and Tanaka (1985).

The direct method, comparable with the Cooley-Tukey procedure in the PSD estimation, approximates an equivalent definition of the bispectrum. The signal record is divided into segments, the data in each segment are Fourier transformed, and an averaged triple product is used as estimate of the bispectrum (e.g. Brillinger and Rosenblatt, 1967; Huber et al., 1971; Helland et al., 1985). The direct PSD estimation procedure has been modified by Welch (1967), who introduced appropiate signal windows for leakage reduction and side lobe surpression and applied overlapped signal segmentation techniques in order to retrieve the loss in the degrees of freedom due to windowing. There are remarks where overlapped signal segmentation has been used in bispectrum estimation or its use is suggested (Lii and Helland, 1981), but we could not find criteria for suitable window functions and the required degree of segment overlap.

The aim of this work is to quantify the Welch method for the estimation of the bispectrum and to attempt combining the PSD estimation and the bispectrum estimation within

+On leave of absence from CNEA, Buenos Aires, Argentina.

one code by using the same signal window and the same degree of segment overlap in the estimation of both spectra. Since the bispectrum can be structured, as can the PSD, high frequency resolution and good side lobe surpression with economic computational efforts are aimed for.

Section II recalls briefly the definitions of the statistical functions. In section III the direct spectra estimation procedure by the Welch method is explained. Emphasis is given to the development of a combined method for the simultaneous estimation of the PSD and the bispectrum. Section IV deals with the problem of minimizing the variances of the estimates by proper segment overlap. Suitable windows and their characteristics are considered in section V. In section VI an application example, mainly for checking the method, is shown. In section VII final remarks are given.

II. DEFINITIONS

We assume that the random noise signal $x(t)$ belongs to a process which is at least weakly ergodic up to the 6th order. The ordinary autocorrelation function (ACF) and its Fourier transform, the power spectral density (PSD), are defined by

$$R(\tau) = \langle x(t)x(t+\tau)\rangle = \lim_{T\to\infty} \frac{1}{2T} \times \int_{-T}^{+T} x(t)x(t+\tau)\,dt \qquad (1)$$

$$S(\omega) = \int_{-\infty}^{+\infty} R(\tau)e^{-i\omega\tau}\,d\tau \qquad (2)$$

The brackets in equation (1) denote ensemble averaging. ω in the equation (2) is the frequency in radians per time unit. Among higher-order correlations we are essentially interested here in the triple correlation function (TCF) and its Fourier transform, which is called the bispectral density (BSD) or briefly the bispectrum, introduced by Tukey (1959) :

$$R(\tau_1,\tau_2) = \langle x(t)x(t+\tau_1)x(t+\tau_2)\rangle = \lim_{T\to\infty} \frac{1}{2T} \times \int_{-T}^{+T} x(t)x(t+\tau_1)x(t+\tau_2)\,dt \qquad (3)$$

$$B(\omega_1,\omega_2) = \int\int_{-\infty}^{+\infty} R(\tau_1,\tau_2)e^{-i\omega_1\tau_1-i\omega_2\tau_2}\,d\tau_1 d\tau_2 \qquad (4)$$

While the PSD is a real-valued function of ω, the BSD is in general a complex function of the two frequencies ω_1 and ω_2. There are symmetry properties. The ACF and the PSD are both symmetric functions. The TCF and the BSD have the following main symmetry properties (e.g. Rosenblatt and Van Ness, 1965; Lii and Helland, 1981), from which further symmetry properties can be derived :

$$R(\tau_1,\tau_2) = R(\tau_2,\tau_1) = R(\tau_1-\tau_2,-\tau_2) \qquad (5)$$

$$B(\omega_1,\omega_2) = B(\omega_2,\omega_1) = B^*(-\omega_1,-\omega_2) = B(\omega_1,-\omega_1-\omega_2) \qquad (6)$$

In particular, when the BSD is known in the sector $\omega_1 \geq 0, 0 \leq \omega_2 \leq \omega_1$, it is determined over the whole frequency plane.

If the signal x(t) has zero-mean, the ACF and TCF represent covariance functions. According to the theorem of Wang and Uhlenbeck (1945) all odd-order correlations of Gaussian noise with zero-mean average to zero. Hence, the triple covariance function or the BSD may only exist if x(t) contains non-Gaussian and skewed noise contributions. If the mean of x(t)is different from zero, or if an apparent DC component is involved in the estimation procedure, the spectra are biased at zero frequency (the BSD on the frequency axes). We exclude the zero-frequency region in our considerations.

III. DIRECT SPECTRA ESTIMATION USING THE WELCH METHOD

For simplicity and brevity we prefer the mathematical presentation by continuous functions in time and frequency and wish to avoid presentation by the usual digital approach.

For the estimation of the PSD the Welch method is now the most popular one. It involves segmenting the record, weighting appropriately the data in each segment, computing instantaneous PSDs by fast Fourier transform (FFT) techniques in these segments, and averaging the instantaneous PSDs. It is a straightforward extension to apply this procedure also to the BSD estimation and to try to combine the estimation of both spectral functions in a common algorithm. It not only lowers the computational costs but also offers more consistent results when comparing the information contained in both spectra.

We assume that the noise record x(t) is available in the time interval $0 \leq t \leq T_0$ and is to be analyzed over this time span. We segment this time span into the intervals :

$$(0,T),(s,s+T),(2s,2s+T),\cdots,((P-1)s,(P-1)s+T) \tag{7}$$

where T is the segment length, P is the number of segments, and $s(>0)$ is the amount of shift which each successive segment undergoes. It must be

$$(P-1)s+T \leq T_0 \tag{8}$$

If $s < T$, the second and each successive segment overlaps by the amount $T-s$ with the preceding one.

The direct estimates of the PSD and the BSD, denoted by the superscript (E), are given by

$$S^{(E)}(\omega) = \frac{1}{PT}\sum_{p=1}^{P}|X_p^{(S)}(\omega)|^2 \tag{9}$$

$$B^{(E)}(\omega_1,\omega_2) = \frac{1}{PT}\sum_{p=1}^{P}X_p^{*(B)}(\omega_1+\omega_2)X_p^{(B)}(\omega_1)X_p^{(B)}(\omega_2) \tag{10}$$

where $X_p^{(S)}(\omega)$ and $X_p^{(B)}(\omega)$ are the Fourier transforms given by

$$X_p^{(S),(B)}(\omega) = \int_{segment\ p} w^{(S),(B)}(t-(p-1)s)x(t)e^{-i\omega t}\,dt \tag{11}$$

$w^{(S)}(t)$ and $w^{(B)}(t)$ are signal window (SW) functions (temporal windows, data windows) introduced to reduce leakage and side lobe production in the estimates due to the finite Fourier transform. In a combined estimation procedure they are essentially of the same type and differ only by a factor. If the signal data are non-uniformly weighted with weigths ≤ 1, there is an amplitude loss in the statistical averaging which is different for the different spectral moments. Hence, we write

$$w^{(S)}(t) = \xi_S \cdot w(t)$$
$$w^{(B)}(t) = \xi_B \cdot w(t). \tag{12}$$

where ξ_S and ξ_B are constants which must be recovered from statistical considerations, and $w(t)$ is a common SW defined by

$$w(t)\begin{cases} >0 & \text{in } 0<t<T \\ =0 & t<0; t>T \end{cases} \tag{13}$$

$w(t)$ is assumed to be symmetric around T/2, i.e.

$$w(T-t) = w(t) \tag{14}$$

Usually, $w(t)$ is normalized $w(T/2) = 1$. There are many known SWs which are zero at the boundaries, i.e. $w(0) = w(T) = 0$. But this is not a necessary condition. The expected values of the spectral estimates, denoted by the superscript ^ can be expressed by

$$\hat{S}(\omega) = \int_{-T}^{+T} W(\tau)R(\tau)e^{-i\omega\tau}\, d\tau \tag{15}$$

$$\hat{B}(\omega_1,\omega_2) = \iint_{-T}^{+T} W(\tau_1,\tau_2)R(\tau_1,\tau_2)e^{-i\omega_1\tau_1 - i\omega_2\tau_2}\, d\tau_1 d\tau_2 \tag{16}$$

$W(\tau)$ and $W(\tau_1,\tau_2)$ are correlation function window (CFW) functions (lag windows). In the limit as T tends to infinity, these functions must go to 1, and equations (9) and (10) exhibit asymptotically unbiased spectra estimates. In the indirect spectra estimation procedure, when the ACF and the TCF are estimated at first, the CFWs are functions to be specified appropriately. Here, they follow from a given SW. $W(\tau)$ results from the convolution

$$W(\tau) = \frac{1}{T}\int_0^{T-|\tau|} w^{(S)}(t)w^{(S)}(t+|\tau|)\, dt \tag{17}$$

It is symmetric around $\tau = 0$ and extends over the range $|\tau| \leq$T. $W(\tau_1,\tau_2)$ is given by

$$W(\tau_1,\tau_2) = \frac{1}{T}\int_0^{T-\tau_{max}} w^{(B)}(t)w^{(B)}(t+\tau_{min})w^{(B)}(t+\tau_{max}) \tag{18}$$

where

$$\tau_{min} = |max(0,\tau_1,\tau_2) - max(\tau_1,\tau_2) + min(\tau_1,\tau_2)| \tag{19}$$

$$\tau_{max} = max(|\tau_1|,|\tau_2|,|\tau_1-\tau_2|) \tag{20}$$

$W(\tau_1,\tau_2)$ fulfils the same symmetry conditions as given for the TCF by equation (5) and extends over the hexagonal area $\tau_{max} \leq T$. In particular,equation (18) defines a class of two-dimensional CFWs which is different from that considered by other authors (e.g. Sasaki et al. 1975; Rao and Gabr, 1984), where a product presentation by one-dimensional CFWs, $W(\tau_1,\tau_2) = W(\tau_1)W(\tau_2)W(\tau_1-\tau_2)$, is assumed. Such a decomposition is here not possible.

Equations (15) and (16) can be expressed by convolutions in the frequency domain. If $W(\omega), W(\omega_1,\omega_2)$ are the one and two dimensional spectral CFWs, and $w^{(S)}(\omega), w^{(B)}(\omega)$ are the spectral SWs, one can write

$$\hat{S}(\omega) = \frac{1}{2\pi}\int_{-\infty}^{+\infty} W(\omega')S(\omega-\omega')\, d\omega' \tag{21}$$

$$\hat{B}(\omega_1,\omega_2) = \frac{1}{(2\pi)^2}\iint_{-\infty}^{+\infty} W(\omega_1',\omega_2')B(\omega_1-\omega_1',\omega_2-\omega_2')\, d\omega_1' d\omega_2' \tag{22}$$

and identify

$$W(\omega) = \frac{1}{T}|w^{(S)}(\omega)|^2 \tag{23}$$

$$W(\omega_1,\omega_2) = \frac{1}{T}w^{*(B)}(\omega_1+\omega_2)w^{(B)}(\omega_1)w^{(B)}(\omega_2) \tag{24}$$

Good SWs are, in the frequency domain, sharply peaked functions at $\omega = 0$. If the theoretical spectra $S(\omega)$ and $B(\omega_1,\omega_2)$ do not change significantly within the bandwidth of the windows, equations (21) and (22) can be approximated by

$$\hat{S}(\omega) \simeq S(\omega)\frac{1}{2\pi}\int_{-\infty}^{+\infty} W(\omega')\, d\omega' = S(\omega)W(\tau=0) \tag{25}$$

$$\begin{aligned}\hat{B}(\omega_1,\omega_2) &\simeq B(\omega_1,\omega_2)\frac{1}{(2\pi)^2}\iint_{-\infty}^{+\infty} W(\omega_1',\omega_2')\, d\omega_1' d\omega_2' \\ &= B(\omega_1,\omega_2)W(\tau_1=0,\tau_2=0)\end{aligned} \tag{26}$$

from which one obtains the normalization conditions for the factors ξ_S and ξ_B in equation (12) in the time domain :

$$W(\tau=0) = \frac{\xi_S^2}{T}\int_0^T w^2(t)\,dt = 1 \tag{27}$$

$$W(\tau_1=0,\tau_2=0) = \frac{\xi_B^3}{T}\int_0^T w^3(t)dt = 1 \tag{28}$$

Suitable SWs usually have simple forms in the time domain and the numerical determination of the ξ factors can be treated more easily there than in the frequency domain.

IV. VARIANCE OF THE SPECTRA ESTIMATES

The expected values of the spectral estimates do not depend upon the degree of segment overlap or on a gap separating successive segments. Windowing, however, affects the statistical scattering of the spectra data points by a loss in the degrees of freedom. Welch's concept consists in retrieving this loss by the choice of a proper segment overlap. In a combined method for the simultaneous estimation of the PSD as well as of the BSD, a compromise must be made, since in general the optimum segment overlap conditions are different for each spectrum.

The variance of the PSD estimate is given by (Nuttall, 1971)

$$Var(S^{(E)}(\omega)) = \frac{1}{(2\pi)^2 P}\sum_{k=-(P-1)}^{P-1}(1-\frac{|k|}{P})\left|\int_{-\infty}^{+\infty} S(\omega')W(\omega-\omega')e^{iks\omega}\,d\omega'\right|^2 \tag{29}$$

Equation (29) is exact for frequencies different from the zero frequency region ($|\omega| > b$, b= bandwidth of $w(\omega)$, see definitions in section V) and under the assumption of Gaussian noise. When the noise is non-Gaussian, equation (29) may be considered as an asymptotically valid approximation for large values of T by neglecting the 4th-order cumulant of x(t).

If $S(\omega)$ does not change significantly within the bandwidth of the window, equation (29) can be written approximately as

$$Var(S^{(E)}(\omega)) \simeq S^2(\omega)\frac{1}{P}\times\sum_{k=-(P-1)}^{P-1}(1-\frac{|k|}{P})W^2(\tau=ks) \tag{30}$$

Using equations (25) and (30), the following ratio can be formed (Nuttall, 1971)

$$\eta_S = \frac{\hat{S}^2(\omega)}{Var(S^{(E)}(\omega))} \simeq \frac{P}{\sum_{k=-(P-1)}^{P-1}(1-\frac{|k|}{P})\left(\frac{W(\tau=ks)}{W(\tau=0)}\right)^2} \tag{31}$$

which can be interpreted as half of the equivalent number of degrees of freedom for each frequency point in the PSD estimate, and is to be maximized for the optimum degree of segment overlap under the constraint of equation (8) (with the sign of equality).

A similar expression, asymptotically valid for large values of T, can be derived for the variance of $B^{(E)}(\omega_1,\omega_2)$, considering the 6th-order covariance function of x(t). Using the results by cumulant expansion techniques (Leonov and Shiryaev, 1959) and neglecting cumulants of order 4 and higher (Rosenblatt and Van Ness, 1965; Alekseev, 1983) we obtain

$$\begin{aligned} Var(|B(\omega_1,\omega_2)|) \simeq\ & \frac{1}{(2\pi)^3P^2T^2}\sum_{p=1}^{P}\sum_{q=1}^{P}\iiint_{-\infty}^{+\infty} d\omega_1' d\omega_2' d\omega_3' \\ & S(\omega_1')S(\omega_2')S(\omega_3')e^{i(-\omega_1'+\omega_2'+\omega_3')(q-p)s} \\ & \Big\{|w^{(B)}(\omega_1+\omega_2-\omega_1')|^2|w^{(B)}(\omega_1-\omega_2')|^2|w^{(B)}(\omega_2-\omega_3')|^2 + \\ & |w^{(B)}(\omega_1+\omega_2-\omega_1')|^2 w^{*(B)}(\omega_1-\omega_2')w^{(B)}(\omega_2-\omega_2')\times \\ & w^{*(B)}(\omega_2-\omega_3')w^{(B)}(\omega_1-\omega_3')\Big\} \end{aligned} \tag{32}$$

The second term in equation (32) gives negligible contributions as long as ω_1 differs from ω_2 by more than the window bandwidth b, and doubles the variance for $\omega_1 = \omega_2$. It should be noted that equation (32) is not valid (also not even approximately) in the frequency regions around the frequency axes. The variance shows highly singular behaviour in these regions, even in the absence of any DC component of x(t). If we restrict the frequency range to the sector area $\omega_1 > \frac{b}{2}, \frac{b}{2} < \omega_2 < \omega_1 - b$ and apply the same approximation as used for equation (30), we can define a ratio to be maximized in the same way as equation (31)

$$\eta_B = \frac{T\hat{S}(\omega_1 + \omega_2)\hat{S}(\omega_1)\hat{S}(\omega_2)}{Var(|B^{(E)}(\omega_1, \omega_2)|)} \simeq \frac{P}{\sum_{k=-(P-1)}^{P-1}(1 - \frac{|k|}{P})\left(\frac{\xi_B^2 W(\tau=ks)}{\xi_S^2 W(\tau=0)}\right)^3} \tag{33}$$

Nuttall (1971) has considered the covariance of the PSD estimate. He has shown that the covariance is significantly larger than zero for two neighbouring spectrum points only when the main lobe of the spectral SWs overlaps. A corresponding statement holds for $Cov(|B(\omega_1, \omega_2)|, |B(\omega_3, \omega_4)|)$ which is a lengthy expression and exhibits a strongly peaked behaviour for $\omega_3 \rightarrow \omega_1, \omega_4 \rightarrow \omega_2$.

V. WINDOW SELECTION

For the characterization of an SW there are many parameters which describe the merits mainly in the frequency domain and are obviously related to the consequences in the PSD estimation. Due to inherent constraints the selection of an SW can never be made from the viewpoint of looking for the best one, but only for suitable ones. We are concerned here with the problem that the SW should be suitable for the PSD estimation as well as for the BSD estimation.

The SW w(t) has been defined in the interval (0,T). Because of the postulated symmetry condition by equation (14), its Fourier transform can be written as $w(\omega) = w_0(\omega)exp(-i\omega T/2)$, where $w_0(\omega)$ is a real-valued function of frequency. The phase factor is of no interest, since it is cancelled in all formulae of sections III and IV where $w(\omega)$ appears. One requirement for a good window is that $w_0(\omega)$ should not produce negative side lobe effects. According to equation (23) any type of SWs, whether having negative side lobe characteristics or not, will never produce negative side lobe phenomena in the PSD estimate. But equation (24) shows that negative side lobe characteristics will be forwarded to the BSD estimate. This limits drastically the use of known SWs to a few types and excludes, for example, the application of the very popular Hanning SW. Among the various SWs summarized in the paper of Harris (1978) we selected 4 types. They and their most interesting characteristics are listed in table 1.

Table 1. Signal Window Characteristics

Type [1]	$w_0(\omega)/w_0(\omega = 0)$ [2]	SW bandwidth in units 1/T [3]	Half-power band width in units 1/T [4]
Triangle	$sin^2(\frac{T\omega}{4})/(\frac{T\omega}{4})^2$	1.78	1.28
Bohman	$8\pi^4(1 + cos(\frac{T\omega}{2}))/(T^2\omega^2 - 4\pi^2)^2$	2.39	1.71
Parzen	$sin^4(\frac{T\omega}{8})/(\frac{T\omega}{8})^4$	2.56	1.71
Kaiser-Bessel $\alpha = 2.5$	$\frac{\pi\alpha}{sinh(\pi\alpha)} \times \frac{sin((\frac{T\omega}{2})^2-(\pi\alpha)^2)^{\frac{1}{2}}}{((\frac{T\omega}{2})^2-(\pi\alpha)^2)^{\frac{1}{2}}}$	2.19	1.82

Table 1. Signal Window Characteristics Cont.

Highest side lobe level db	ξ_S	ξ_B	Optimum fractional overlap %		Fractional overlap %	Fraction % of	
			PSD	BSD		max η_S	max η_B
[5]	[6]	[7]	[8]	[9]	[10]	[11]	[12]
-13.0	1.732	1.587	62.1	77.0	50	95	80
-23.0	1.846	1.608	73.2	78.3	65	95	90
-26.8	1.926	1.654	75.1	80.3	68	97	90
-29.5	1.769	1.561	70.1	76.8	60	95	90

Their spectral characteristics relative to the peak value are given in column 2. The first three SWs, the triangle SW, the Bohman SW, and the Parzen SW have positive side lobes only. The fourth SW, the Kaiser-Bessel SW with the choice of the parameter $\alpha = 2.5$, has also negative side lobes but its side lobe level is very low. In column 3 the SW bandwidths are given in units of 1/T. The SW bandwidth is defined as the frequency range over which $w_0(\omega)$ is greater than half of its peak value. Values of the usually specified half-power bandwidth in units of 1/T are listed in column 4. The half-power bandwidth is defined as the frequency range over which $w_0^2(\omega)$ is greater than half of its peak value. In column 5, values of the highest side lobe level of $|w_0(\omega)|$ in db units are given ($db = 10 \times \log(|w_0(\omega)/w_0(\omega = 0)|)$.

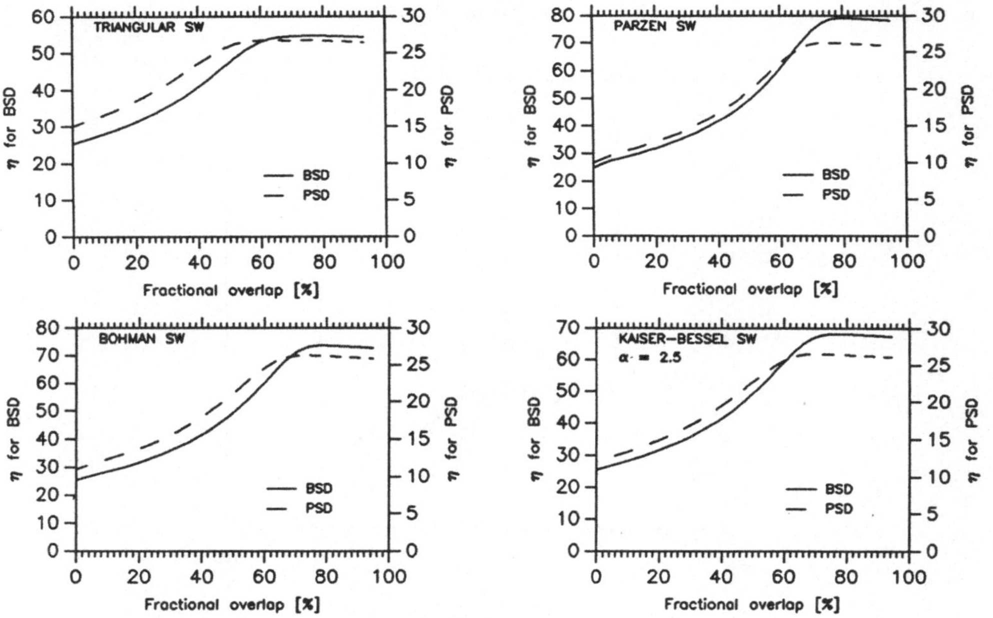

Fig. 1. Asymptotic values of η for different windows as function of the fractional segment overlap. Solid line : for BSD. Dashed line : for PSD.

The values of ξ_S and ξ_B (equations (27) and (28)) are shown in columns 6 and 7. They refer to the normalization w(T/2)=1 in the time domain. The numbers have been checked for the 4 SWs with a pair of quadratically phase-coupled harmonic sinusoids and by comparing their nominal amplitude values with those obtained from the integrated peaks in the PSD and BSD estimates. Columns 8 and 9 represent almost asymptotic values of the optimum fractional segment overlap (transform size in the digital FFT (number of signal data points in

T) of 1024 points, $P \geq 20$). Corresponding plots of η_S and η_B (equations (31) and (33)) are given in Fig.1 for each SW. For maximizing η_S and η_B the same procedure as described by Nuttall (1971) has been used. Optimal BSD estimation requires, in general, a higher degree of segment overlap than optimal PSD estimation. One can also observe the trend that an increase in the SW bandwidth is associated with a higher optimum fractional segment overlap, but with a decrease in the side lobe level. With regard to the frequency resolution and the low optimum fractional segment overlap, the triangular SW would be the most suitable. On the other hand, it shows the highest side lobe level between the 4 SWs. A good compromise seems to exist in the Bohman SW and the Kaiser-Bessel SW. Finally, in the last 3 columns of table 1, fractional overlap values are suggested for the combined spectra estimation and the corresponding fractions of the maximum η-values are indicated. There are other SWs which have positive side lobe characteristics, like the Bartlett-Priestley SW (Priestley, 1981), but their merits for the BSD estimation have not yet been explored.

VI. APPLICATION TO HARMONIC ANALYSIS

In a recent paper by Piñeyro and Behringer (1987) a new method, called displaced spectra techniques, has been presented for distinguishing between sinusoidal components and narrowband random noise contributions in otherwise random noise data. Since the power of higher harmonics usually decreases rapidly, the method is not suitable for the identification of higher harmonics. If the type of a peak in the PSD at the fundamental frequency can be well identified by that method, the use of the BSD is considered to be the complementary method for harmonic analysis because of more suitable amplitude relationships.

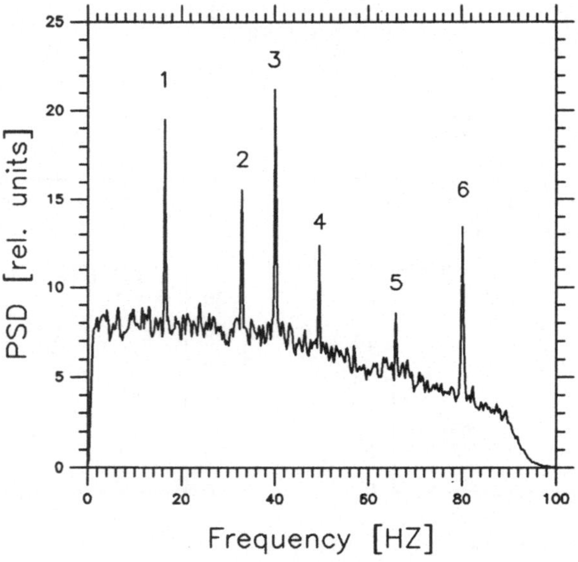

Fig. 2. PSD function

The example which we would like to present refers to a reanalysis of data used in the aforementioned paper. The data come from an experimental circuit which was made up in the 100 Hz frequency region for the generation of stationary analogue noise records containing Gaussian broadband random noise components, sinusoidal components and Gaussian narrowband random noise contributions including a squared portion. The PSD is shown in Fig.2. Four signals were generated and mixed together by a summer. The broadband random noise signal was obtained from a Gaussian white noise generator and a low-pass filter. Harmonic sinusoids (peaks 1,2,4,5) were generated by periodic rectangular pulses (fundamental frequency at 16.4 Hz, duty cycle $\simeq$ 20%) from a pulse generator. The signal from a second independent Gaussian white noise generator was fed to an extremely narrowband second-order resonance filter (peak 3, center frequency at 40 Hz, quality factor Q= 300). The filter output signal was branched to a squarer which gives non-Gaussian narrowband random noise contributions (peak 6, center frequency at 80 Hz). The mixed signal was high-pass filtered to remove the DC component and a sharp antialiasing filter was inserted before the AD converter. The sampling frequency was 208 Hz. The PSD shown in Fig.2 and the BSD shown in Fig.3 were estimated by 190 averages with 60% fractional segment overlap, using the Kaiser-Bessel SW ($\alpha = 2.5$) and an FFT transform size of 1024 data points.

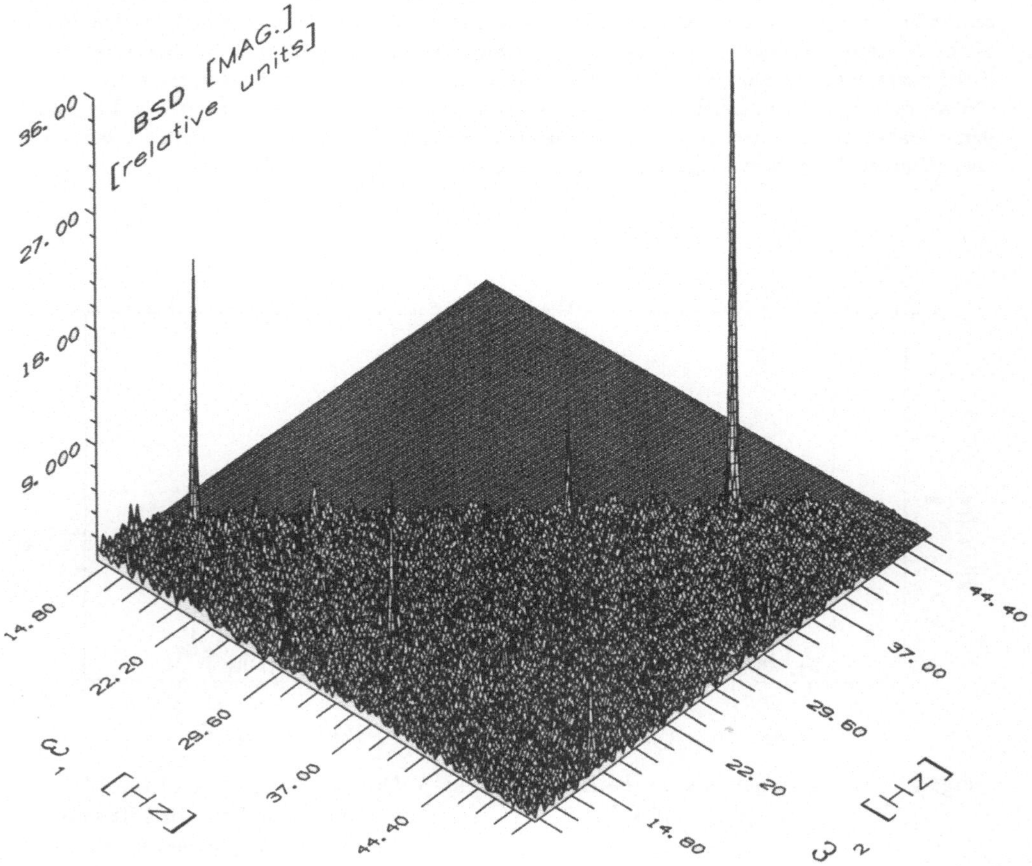

Fig. 3. BSD function (magnitude) in the interesting frequency region

If a signal y(t) is composed of a Gaussian random signal x(t) with the PSD $S(\omega)$ and a small squared portion, so that

$$y(t) = x(t) + \beta x^2(t) \tag{34}$$

the BSD of y(t) is given by (Huber et al.,1971; Nikias and Raghuveer, 1987)

$$B(\omega_1, \omega_2) = 2\beta \left[S(\omega_1)S(\omega_2) + (S(\omega_1) + S(\omega_2))S(\omega_1 + \omega_2)\right] + O(\beta^3) \tag{35}$$

According to equation (35), peaks 3 and 6 in the PSD combine into one peak in the BSD at $\omega_1/2\pi = \omega_2/2\pi = 40$ Hz, which is clearly seen in Fig.3.

If x(t) is a periodic signal function which is assumed to be different from a pure sinusoid and represented by its Fourier series with the DC component removed

$$x(t) = \sum_{\nu=-\infty}^{\nu=+\infty} c_\nu e^{i\nu\omega_0(t-t_0)} \tag{36}$$

where ω_0 is the fundamental frequency, t_0 is an arbitrary starting time point, and $c_0 = 0, c_{-\nu} = c_\nu^*$, the theoretical BSD is given in the sector $\omega_1 > 0, 0 < \omega_2 \leq \omega_1$ by

$$B(\omega_1, \omega_2) = (2\pi)^2 \sum_{2\mu \geq \nu > \mu = 1}^{\infty} c_\nu c_\mu^* c_{\nu-\mu}^* \delta(\omega_1 - \mu\omega_0)\delta(\omega_2 - (\nu - \mu)\omega_0) \tag{37}$$

where δ denotes the Dirac-delta function. The most intensive peaks in the BSD appear on the line $\omega_2 = \omega_0$, where the amplitudes of the higher harmonics combine with the amplitude of the fundamental wave. Fig.4 shows a cross-sectional plot at that line. On the other hand, if the same data are analyzed under the same conditions, but without using segment overlap (75 averages), a significantly lower quality estimation results. This is shown in Fig.5 and demostrates the importance of proper segment overlap. Estimations of inferior quality are also obtained if uniform weighting (without segment overlap) is applied.

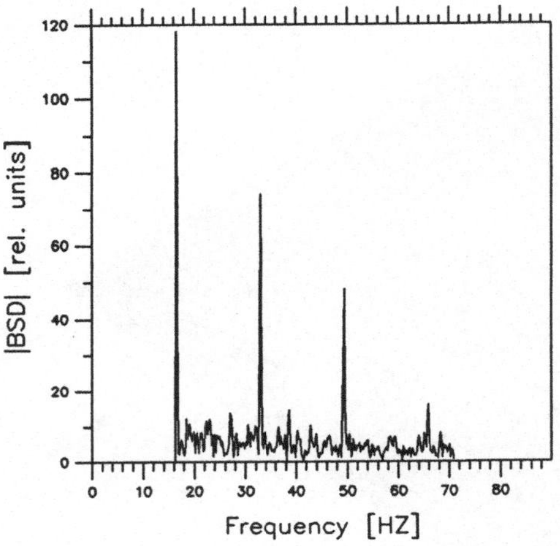

Fig. 4. Cross-sectional view of the BSD function at $\omega_2/2\pi = 16.4$ Hz

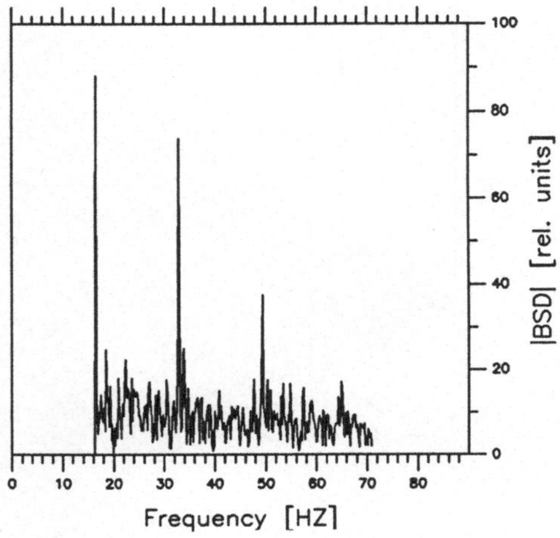

Fig. 5. Cross-sectional view of the BSD function at $\omega_2/2\pi = 16.4$ Hz estimated without segment overlap

We would like to mention that cases can occur which require up-decimation of the signal data by a factor 2 for the BSD estimation. The up-decimation procedure, described by Otnes and Enochson (1978), works well but requires doubling the FFT transform size to obtain the same frequency resolution.

VII. FINAL REMARKS

A method is proposed which combines the simultaneous estimation of the PSD and the BSD in a common procedure. The algorithm is a straightforward extension of Welch's

PSD estimation method. We limited the considerations to a few SWs. Since single FFT techniques only are involved and the BSD must be determined only in the triangular region $\omega_1 > 0, 0 < \omega_2 \leq \omega_1, \omega_1 + \omega_2 \leq \omega_c$ where ω_c is the Nyquist cutoff frequency, a large transform size (of up to 2048 data points) can be used to obtain good frequency resolution. We did not observe any problems in implementing the method on a VAX 11/785 and in specifying the required array dimension for the BSD. The FORTRAN code that we developed is relatively fast. For 100 averages and a FFT transform size of 1024 data points, about 6 min. CPU time is required for estimating the PSD and the BSD. We have not yet made comparisons with the indirect BSD estimation procedure or with parametric signal modelling methods, as described in the paper of Nikias and Raghuveer (1987). In the indirect BSD estimation method, however, the use of double FFT techniques is required. The attainable frequency resolution depends on the storage capacity of the available computer. The same problem may arise in the recently proposed lag reshaping method by Nuttall and Carter (1982), which can be considered as a combination of the direct and indirect estimation procedure and should also be applicable to the BSD estimation. However, the profit which could be achieved by the application of this method to the BSD estimation requires investigation.

REFERENCES

Alekseev V.G., 1983, Some aspects of estimation of the bispectral density of a stationary stochastic process, Probl. Inf. Transm. (USA) 19, 204:214.

Bendat J.S. and Piersol A.G., 1982, Spectral analysis of nonlinear systems involving square laws operations, J. Sound and Vibration 81, 199:212.

Brillinger D.R., 1965, An introduction to polyspectra, Ann. Math. Statist. 36, 1351:1374.

Brillinger D.R. and Rosenblatt, 1967, Asymptotic theory of estimates of k-th order spectra, in Spectral Analysis of Time Series, (B. Harris ed.) 153:188, Wiley, New York.

Harris F.J., 1978, On the use of windows for harmonic analysis with the discrete Fourier transform, Proc. IEEE 66, 51:83.

Hasselmann K., Munk W., Macdonald G., 1963, Bispectral of ocean waves, Proc. Symp. Time Series Analysis (M. Rosenblatt, ed), Wiley, New York, 125:139.

Helland K.N., Itsweire E.C.,and Lii K.S., 1985, A program for computation of bispectra with application to spectral energy transfer in fluid turbulence, Adv. Eng. Software 7, 22:27.

Huber P.J., Kleiner B., and Gasser T., 1971, Statistical methods for investigating phase relations in stationary stochastic processes, IEEE Trans. Audio Electroacoust. AU-19, 78:86.

Leonov V.P. and Shiryaev A.N., 1959, On a method of calculation of semi-invariants, Theor. Prob. Appl. 4, 319:329.

Lii K.S. and Helland K.N., 1981, Cross-bispectrum computation and variance estimation, ACM Trans. Math. Software 7, 284:294.

Lohmann A.W. and Wirnitzer B., 1984, Triple correlations, Proc. IEEE 72, 889:901.

Nuttall A.H., 1971, Spectral estimation by means of overlapped fast Fourier transform processing of windowed data, Naval Underwater Systems Center Technical Report 4169.

Nuttall A.H. and Carter G.C., 1982, Spectral estimation using combined time and lag weighting, Proc. IEEE 70, 1115:1125.

Nikias C.L. and Raghuveer M.R., 1987, Bispectrum estimation: A digital signal processing framework, Proc. IEEE 75, 869:891.

Otnes R.K. and Enochson L., 1978, Applied Time Series Analysis John Wiley and Sons, New York.

Piñeyro J. and Behringer K., 1987, Displaced spectra techniques as a tool for peak identification in PSD-analysis, Paper presented at SMORN V, Oct. 12-16, Munich (to be

published in Prog. Nucl. Energy); EIR-report 633.
Priestley M.B., 1981, Spectral Analysis and Time Series, Vol. 1, Academic Press.
Rao T.S. and Gabr M.M., 1984, An Introduction to Bispectral Analysis and Bilinear Time Series Models, Springer-Verlag.
Rosenblatt M. and Van Ness J.W., 1965, Estimation of the bispectrum, Ann. Math. Statist. 36 1120:1136.
Saito K. and Tanaka T., 1985, Exact analytic expression for Gabr-Rao's optimal bispectral two-dimensional lag window, J. Nucl. Sci. and Technology 22, 1033:1035.
Sasaki K., Sato T., and Yamashita Y., 1975, Minimum bias windows for bispectral estimation, J. Sound and Vibration 40, 139:148.
Sato T. and Sasaki K., 1977, Bispectral holography, J. Acoust. Soc. Amer. 62, 404:408.
Tukey J.W., 1959, An introduction to the measurement of spectra, in Probability and Statistics, (U. Grenander ed.), 300-330, Wiley, New York.
Wang M.C. and Uhlenbeck G.E., 1945, On the theory of the Brownian motion II, Rev. Modern Physics 17, 323:342.
Welch P.D., 1967, The use of FFT for the estimation of power spectra: A method based on time averaging over short modified periodograms, IEEE Trans. Audio Electroacoustics 15, 70:73.

NON LINEAR AR-NOISE INVESTIGATED VIA A KALMAN FILTER

G. Cojazzi†, M. Marseguerra† and C. M. Porceddu‡

† Politecnico di Milano, Milano, Italy

‡ ENEA, Centro "E. Clementel", Bologna, Italy

Abstract

The use of autoregressive (AR) models for extracting the information contained in the fluctuating component of a stationary signal represents a well established technique, widely used in the nuclear reactor engineering field. The possibility of treating non stationary fluctuations via a time varying autoregressive ($TVAR$) model is here considered in the framework of the Kalman filters theory. Firstly, the vanishing of the covariance matrix of the state vector is prevented by slightly increasing the matrix at each time step. Moreover the forecasting guess of the state vector is improved by the use of a least squares technique. The results of numerical experiments, made on AR models of increasing order up to four are encouraging and suggest the convenience of further investigating the features of the present approach to this non linear noise problem.

AR noise investigation

In nuclear reactor engineering, the noise analysis methods represent an important tool for the surveillance of components and for the early detection of malfunctions. Indeed the diagnostic value of the information carried out by the noise has long been appreciated and many reactors are now equipped with noise measuring devices [1].

Concerning the data analysis, the autoregressive (AR) and autoregressive moving average ($ARMA$) models became very popular both in the uni- and multi-variate cases.

Many methods are available for estimating the model coefficients, such as the original Yule-Walker equations [2,3] or the Young least squares algorithm.[4] Also the Kalman filters have been quite successfully proven to be able to estimate the coefficients of a stationary AR model, together with their statistical errors which are asymptotically vanishing[5,6].

A research line followed in the Institute of Nuclear Engineering of the Politecnico di Milano aims at investigating the potentiality of the Kalman filters also for dealing with non stationary noise, i.e. noise whose characteristics change in time. Such a situation easily occurs in a running plant not only when something goes wrong but also when something changes e.g. the power level or the state of a valve. Assuming that the noise may be described by an $AR(p)$ model, the coefficients correspondingly vary in time and also the model order itself may change since some coefficients may vanish or new ones may appear. In such cases it turns out that the coefficients delivered by a Kalman filter built for dealing with the stationary state, actually tend to follow the variations but so slowly that the results are not of use.

In a paper [7] presented at the 19th IMORN (Informal Meeting On Reactor Noise) held in Rome in 1986, some of the presents authors attributed this coefficients sluggishness to the decrease of the elements of their covariance matrix which is continuously updated in the course of the procedure, and proposed to get rid of it by restarting this matrix to unity every fixed number of steps. The coefficients computed after each exhibit wild oscillations, but soon they reach a sort of plateau about the true values. Throwing away the first data, these preliminary results were rather satisfactory. Fig. 1 reports a study case in which the covariance matrix was reinitialized to unity every 200 points, and plotted every 5 points. An obvious improvement of this procedure would consist in considering a set of M filters, each acting for a time T, to be restarted one after the other delayed by a time T/M. However in the course of the investigation it was clear that, instead of performing the above mentioned sudden enlargements of the covariance matrix every fixed bin of time steps, better results could be achieved by slightly amplifying the covariance matrix at each step.

In the present work the covariance matrix $\mathbf{P}_{t,t-1}$, once updated at time $t-1$ as usual in Kalman filtering, is also multiplied by an amplification factor:

$$q = 1 + \epsilon$$

where ϵ is a small quantity whose order of magnitude will be discussed later.

The proposed procedure also contains another feature. We remind that at each step the Kalman filter estimates the optimal coefficient values by suitably weighting:

(a) Their values as forecasted from the assumed dynamical model before taking the measurement.
(b) The innovation, i. e. the difference between the forecasted and the measured value of the time series.

To forecast the coefficient vector $\hat{\mathbf{x}}_t$ at time t, one should know the transition matrix $\mathbf{\Phi}_{t,t-1}$ which describes the evolution of the system from time $t-1$ to time t, according to the model:

$$\mathbf{x}_t = \mathbf{\Phi}_{t,t-1}\, \mathbf{x}_{t-1}$$

since $\mathbf{\Phi}_{t,t-1}$ is actually not known at each time step, we assumed a linear evolution of the coefficients based on a least square fit of the last few N data. In Fig. 2 the Kalman algorithm here implemented is reported.

In an $AR(p)$ model with time varying coefficients the measurement equation reads:

$$z_t = a_{1t}\, z_{t-1} + a_{2t}\, z_{t-2} + \ldots + a_{pt}\, z_{t-p} + v_t$$

where v_t is the $AR(p)$ model white noise with zero mean value and variance R, here set to unity; in a matrix formalism we have:

$$z_t = \mathbf{H}_t\, \mathbf{x}_t + v_t$$

where $\mathbf{H_t}$ is the $1 \times p \cdot N$ measure matrix:

$$\mathbf{H}_t = (z_{t-1}\, 0 \ldots 0, z_{t-2}\, 0 \ldots 0, \ldots, z_{t-p}\, 0 \ldots 0)$$

and the state vector $\mathbf{x}_t$ is defined as (superscript T denotes transpose):

$$\mathbf{x}_t = \left(a_{1t}\, a_{1,t-1} \ldots a_{1,t-N} \ldots a_{pt}\, a_{p,t-1} \ldots a_{p,t-N}\right)^{T}$$

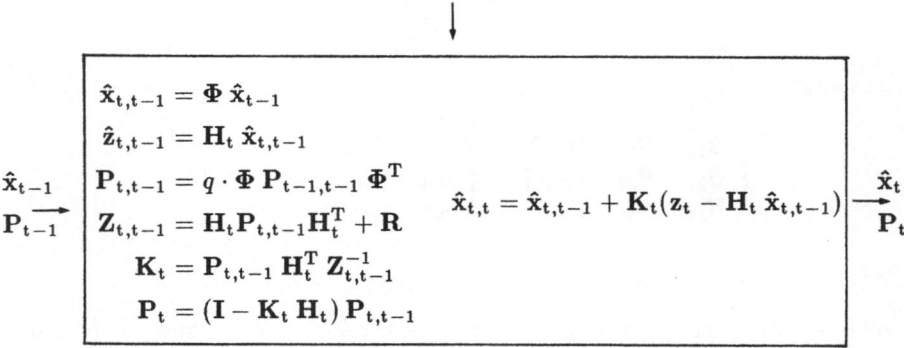

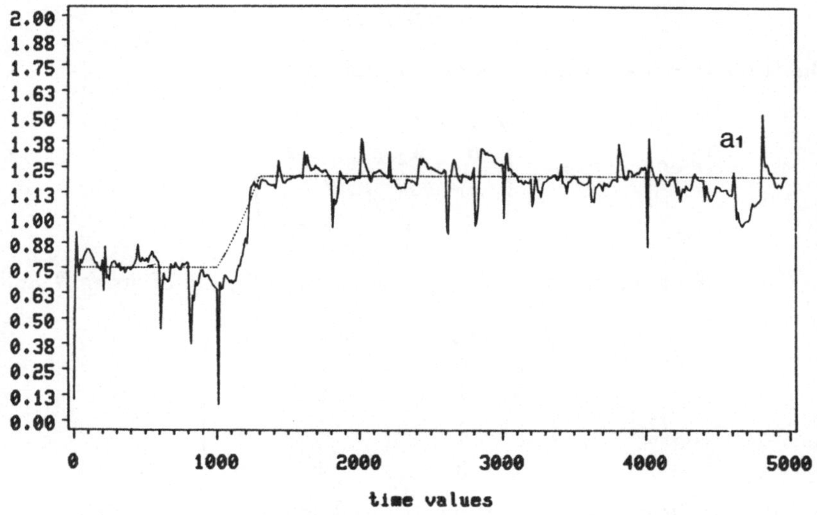

Simulated Time Series

Fig. 1 Estimation of the coefficient a_1 of an $AR(2)$ process.

SYSTEM EQUATION

$$\mathbf{x}_t = \mathbf{\Phi}\, \mathbf{x}_{t-1}$$

MEASUREMENT EQUATION

$$\mathbf{z}_t = \mathbf{H}_t\, \mathbf{x}_t + \mathbf{v}_t$$

ALGORITHM

$\mathbf{z}_t$ ↓

Input: $\hat{\mathbf{x}}_{t-1}$, $\mathbf{P}_{t-1}$ →

$$\hat{\mathbf{x}}_{t,t-1} = \mathbf{\Phi}\, \hat{\mathbf{x}}_{t-1}$$
$$\hat{\mathbf{z}}_{t,t-1} = \mathbf{H}_t\, \hat{\mathbf{x}}_{t,t-1}$$
$$\mathbf{P}_{t,t-1} = q \cdot \mathbf{\Phi}\, \mathbf{P}_{t-1,t-1}\, \mathbf{\Phi}^T$$
$$\mathbf{Z}_{t,t-1} = \mathbf{H}_t \mathbf{P}_{t,t-1} \mathbf{H}_t^T + \mathbf{R}$$
$$\mathbf{K}_t = \mathbf{P}_{t,t-1}\, \mathbf{H}_t^T\, \mathbf{Z}_{t,t-1}^{-1}$$
$$\mathbf{P}_t = (\mathbf{I} - \mathbf{K}_t\, \mathbf{H}_t)\, \mathbf{P}_{t,t-1}$$

$$\hat{\mathbf{x}}_{t,t} = \hat{\mathbf{x}}_{t,t-1} + \mathbf{K}_t(\mathbf{z}_t - \mathbf{H}_t\, \hat{\mathbf{x}}_{t,t-1})$$

→ Output: $\hat{\mathbf{x}}_t$, $\mathbf{P}_t$

Fig. 2 Kalman algorithm implemented in this work.

In this particular case the $\mathbf{\Phi}$ transition matrix is calculated by the least squares procedure and is time-independent; the state equation is consequently:

$$\mathbf{x}_t = \mathbf{\Phi}\,\mathbf{x}_{t-1}$$

where the transition matrix has a block diagonal form:

$$\mathbf{\Phi} = \begin{pmatrix} \mathbf{A}_1 & \mathbf{0} & \dots & \mathbf{0} \\ \mathbf{0} & \mathbf{A}_2 & \dots & \mathbf{0} \\ \vdots & \vdots & \ddots & \vdots \\ \mathbf{0} & \mathbf{0} & \dots & \mathbf{A}_p \end{pmatrix}$$

of order $N \cdot p \times N \cdot p$, in which all the square matrices $\mathbf{A}_j$ are of order N and equal, viz:

$$\mathbf{A}_j = \begin{pmatrix} \alpha_1 & \alpha_2 & \dots & \alpha_{N-1} & \alpha_N \\ 1 & 0 & \dots & 0 & 0 \\ 0 & 1 & \dots & 0 & 0 \\ \vdots & \vdots & \ddots & \vdots & \vdots \\ 0 & 0 & \dots & 1 & 0 \end{pmatrix}$$

and the scalar coefficients α_i come from the least square fit procedure.

The α_i values for the case of a fit with a constant are:

$$\alpha_i = 1/N \qquad \text{for} \qquad i = 1, 2, \dots, N \qquad \text{and} \qquad N \geq 1$$

The α_i values for the case of a straight line fit are:

$$\alpha_i = \frac{6}{N(N-1)}\left(\frac{2N+1}{3} - i\right) \qquad \text{for} \qquad i = 1, 2, \dots, N \qquad \text{and} \qquad N \geq 2$$

In case of a parabolic fit, we define:

$$u = \frac{N+1}{2}$$
$$v = \frac{2N+1}{3}$$
$$w = \frac{3N^2+3N-1}{5}$$

then defining:

$$\begin{pmatrix} B_{11} & B_{12} & B_{13} \\ B_{21} & B_{22} & B_{23} \\ B_{31} & B_{32} & B_{33} \end{pmatrix} = \begin{pmatrix} 1 & -u & uv \\ -u & uv & -Nu^2 \\ uv & -Nu^2 & uvw \end{pmatrix}^{-1}$$

we have:

$$\alpha_i = \frac{1}{N}(B_{11} - iB_{12} + i^2 B_{13}) \qquad \text{for} \qquad i = 1, 2, \dots, N \qquad \text{and} \qquad N \geq 3$$

In the present paper we selected straight line fits through $N = 4$ points. However good results may also be obtained by fitting with a constant and $N = 1$. This last case resembles that of a stationary state equation with added noise.

The formal expressions describing the generic Kalman filter, summarized in the Appendix, may be found in [5].

The results of some numerical experiments are presented in the Figs from 3 to 6.

All the numerical results here given refer to time series lasting 4000 time lags with $E[v_t] = 0$, $E[v_t^2] = 1$. In all cases the first guess values for the components of the state vectors $\hat{\mathbf{x}}_0$ were set to 0.1 and the initial covariance $\mathbf{P}_{0,0}$ was assumed to be a unit matrix. Moreover in all figures the dashed lines indicate the time behaviour of the true parameters a_i $(i = 1, 2, \ldots, p)$.

The results of the Kalman estimate relate to $\epsilon = 5 \cdot 10^{-3}$ and to a forecast based on a least squares straight line fit based on $N = 4$ points. These numerical values seem suitable also for other cases.

In Fig. 3 we consider the case of an $AR(1)$ process whose coefficient at $t = 2000$ suddenly jumps from the initial value -0.8 to a final value 0.8. Notice that the two halves of the time series are quite different, being representative of two distinct physical situations. Indeed in the first case the autocorrelation function exhibits dumped oscillations, while in the second one it goes to zero exponentially.

In Fig. 4 the time series relates to an $AR(2)$ model in which the first coefficient a_1 drops to zero at time $t = 1000$.

In Fig. 5 the AR model underlying the time series is initially of order $p = 2$ and at time $t = 800$ increases to $p = 3$. From this time on, all the coefficients a_1 $(i = 1, 2, 3)$ vary as indicated by the dashed lines. This case shows that the procedure here sketched may also be used to find the order of an AR model. This remark is also supported by the results shown in Fig. 6. Here we have a fourth order AR model which at time $t = 2000$ diminishes to the third order.

Concerning the results we obtained, it turned out that by increasing ϵ the filter follows more rapidly the sudden coefficient variations but the oscillations around the true coefficient values increase. On the other side, by decreasing ϵ the response is slower but the oscillations smooth out. In all the examined cases the procedure seems to work satisfactorily with the values $\epsilon = 0.5\%$, $N = 4$ and a straight line fit, which therefore represent an acceptable compromise. Alternatively, since the actual cases are often constituted by a sequence of stationary states represented by different AR models, one could be only interested in identifying the transition times from one state to another. In such a situation a greater ϵ value and a smaller N value, says 1.% and $N = 1$ (fit with a constant) could be in order. This last situation may occur in a safety monitoring.

A couple of final remarks concerning the results here presented is now in order.

Firstly, it appears that the time behaviour of almost all the estimated parameters follows the step variations of the true coefficients with a delay of about 10^2 time lags. Some improvements might be obtained utilizing by the Kalman filter in the context of a smoothing problem, i.e. by estimating $\mathbf{x}_t$ from the measurements taken up to the time $t + s$. Work along this line is in progress.

Secondly, when the true coefficients are constant, the estimated ones exhibit rather large oscillations. This is due to the amplification factor $q > 1$ and it seems intrinsic to the procedure. Were $q = 1$ the Kalman filter would deliver more stable results but then it would be unable to follow the sudden variations.

The conclusion of the investigation here presented is in favour of a more extensive utilization of the Kalman filters in time series analysis.

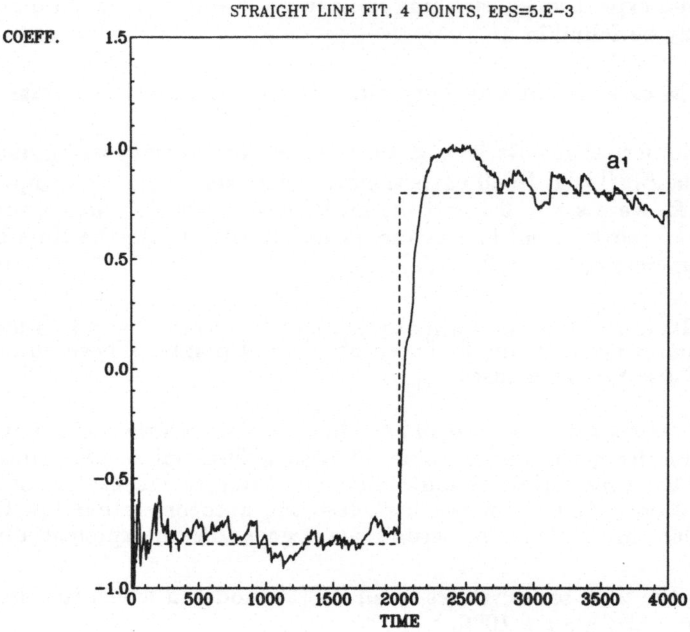

Fig. 3 Estimation of the coefficient a_1 of an $AR(1)$ process

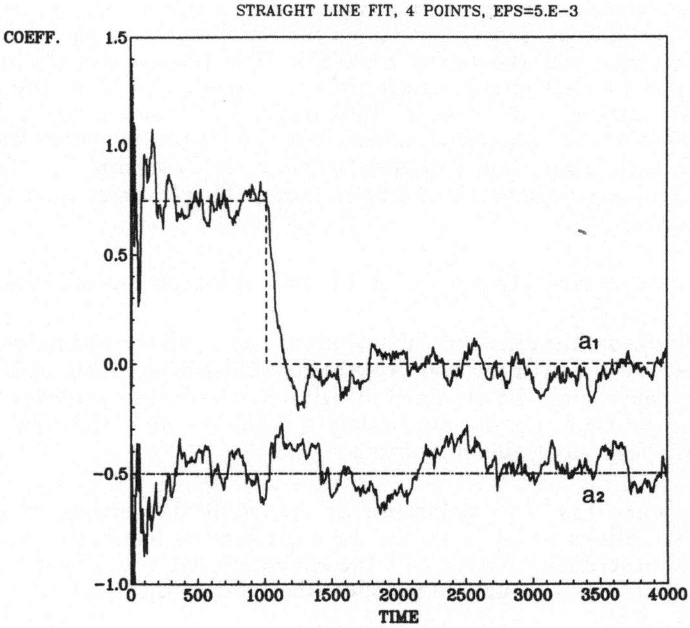

Fig. 4 Estimation of the coefficients a_i of an $AR(2)$ process

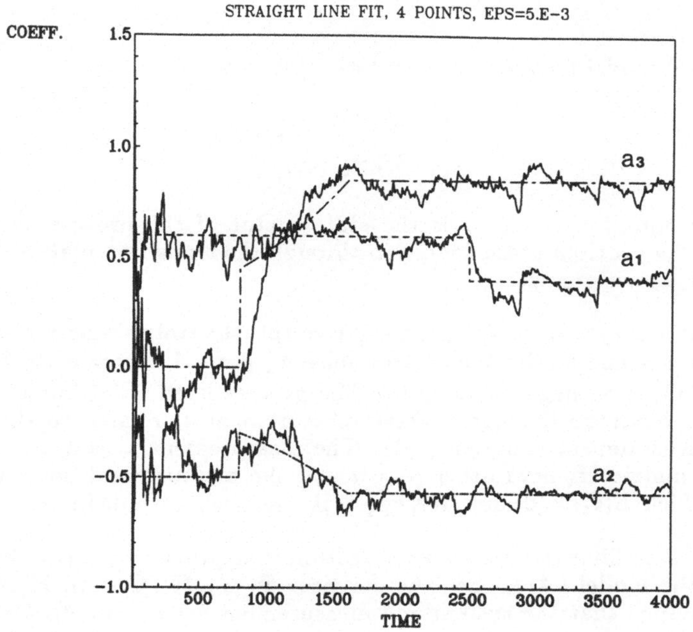

Fig. 5 Estimation of the coefficients a_i of an initially $AR(2)$ process which changes to $AR(3)$ at $t = 800$

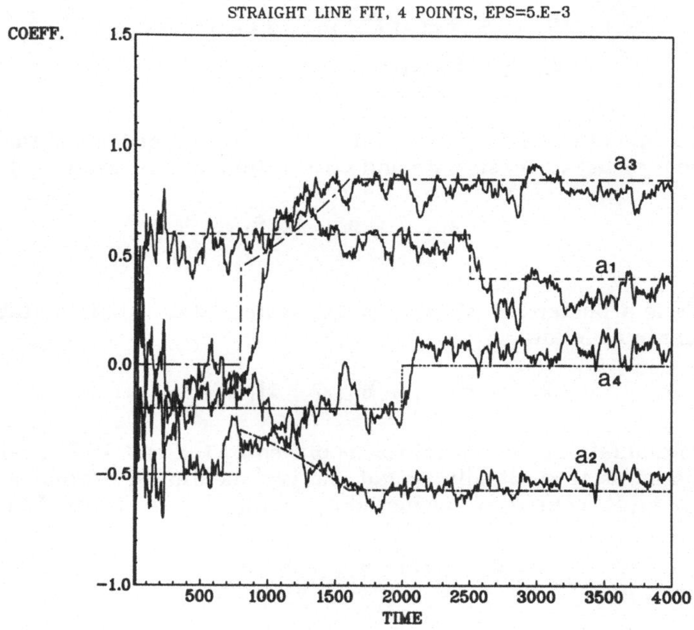

Fig. 6 Estimation of the coefficients a_i of an initially $AR(4)$ process whose order drops to 3 at $t = 2000$

Appendix

A general model for a linear stochastic system is:

$$\mathbf{x}_t = \mathbf{\Phi}_{t,t-1}\,\mathbf{x}_{t-1} + \mathbf{w}_t \tag{A1}$$

$$\mathbf{z}_t = \mathbf{H}_t\,\mathbf{x}_t + \mathbf{v}_t \tag{A2}$$

In the state equation (A1), $\mathbf{x}_t$ is the state vector of the process at time t which depends on the previous state condition through the transition matrix $\mathbf{\Phi}_{t,t-1}$ and on the system noise vector $\mathbf{w}_t$.

In detail, the system state $\mathbf{x}_t$ is a $n \times 1$ vector, the state transition matrix $\mathbf{\Phi}_{t,t-1}$, is of order $n \times n$ and finally the system noise $\mathbf{w}_t$ a $n \times 1$ vector with $E[\mathbf{w}_t] = \mathbf{0}$ and $E[\mathbf{w}_t\mathbf{w}_t^T] = \mathbf{Q}_t$, non negative definite. The process itself is not directly detectable: instead, it is observed through a noisy measurement $\mathbf{z}_t$ related to the state vector $\mathbf{x}_t$ by the measurement equation ($A2$). The measurement $\mathbf{z}_t$ is a $m \times 1$ vector, the observation matrix $\mathbf{H}_t$ is of order $m \times n$ and the measurement noise $\mathbf{v}_t$ is a $m \times 1$ vector for which $E[\mathbf{v}_t] = \mathbf{0}$ and $E[\mathbf{v}_t\mathbf{v}_t^T] = \mathbf{R}_t$, non negative definite.

The Kalman filter is a recursive algorithm that provides a way of estimating the state $\mathbf{x}_t$ of the model ($A1$), once the matrices $\mathbf{\Phi}_{t,t-1}, \mathbf{H}_t, \mathbf{Q}_t$ and $\mathbf{R}_t$ are all known and it is assumed that the system and measurement noise are both Gaussian White Noise processes.

Starting from an initial value $\mathbf{x}_0$ with covariance matrix $\mathbf{P}_{0,0}$, the procedure is recursive: from the estimates of $\hat{\mathbf{x}}_{t-1}$ and its covariance $\mathbf{P}_{t,t-1}$ at time $t-1$ and from the measurements $\mathbf{z}_t$ at time t, new estimates of $\hat{\mathbf{x}}_t$ and its covariance $\mathbf{P}_{t,t}$ may be attained. Indeed at time $t-1$ the following a priori estimates of state and measurement along with their accuracies could be forecasted for one step ahead:

$$\hat{\mathbf{x}}_{t,t-1} = \mathbf{\Phi}_{t,t-1}\,\hat{\mathbf{x}}_{t-1}$$

$$\mathbf{P}_{t,t-1} = \mathbf{\Phi}_{t,t-1}\,\mathbf{P}_{t-1,t-1}\,\mathbf{\Phi}_{t,t-1}^T + \mathbf{Q}_t$$

$$\hat{\mathbf{z}}_{t,t-1} = \mathbf{H}_t\,\hat{\mathbf{x}}_{t,t-1}$$

At the subsequent instant t , the $\mathbf{z}_t$ value is actually recorded and the innovation, i.e. the discrepancy between estimate and observation, can be evaluated, viz:

$$\mathbf{z}_t - \hat{\mathbf{z}}_{t,t-1} = \mathbf{z}_t - \mathbf{H}_t\,\hat{\mathbf{x}}_{t,t-1}$$

A reasonable a posteriori estimate of the system state would suitably combine these two estimate as follows:

$$\hat{\mathbf{x}}_{t,t} = \hat{\mathbf{x}}_{t,t-1} + \mathbf{K}_t(\mathbf{z}_t - \mathbf{H}_t\,\hat{\mathbf{x}}_{t,t-1})$$

in this linear combination, the central role is played by the matrix $\mathbf{K}_t$, called Kalman gain, to be chosen as to optimally weight the two mentioned estimates.The matrix $\mathbf{K}_t$, of order $n \times m$, is recursively calculated by minimizing the trace of the covariance matrix of $\hat{\mathbf{x}}_{t,t}$ i. e.:

$$E[(\hat{\mathbf{x}}_{t,t} - \mathbf{x}_t)^T(\hat{\mathbf{x}}_{t,t} - \mathbf{x}_t)] = \text{min.}$$

It turns out that:

$$\mathbf{K}_t = \mathbf{P}_{t,t-1}\,\mathbf{H}_t^T\,\mathbf{Z}_{t,t-1}^{-1}$$

where:

$$\mathbf{Z}_{t,t-1} = \mathbf{H}_t \mathbf{P}_{t,t-1} \mathbf{H}_t^T + \mathbf{R}_t$$

and the covariance matrix is updated this way

$$\mathbf{P}_t = (\mathbf{I} - \mathbf{K}_t \mathbf{H}_t) \mathbf{P}_{t,t-1}$$

Note that the Kalman gain $\mathbf{K}_t$, as well the matrices $\mathbf{P}_{t,t}, \mathbf{\Phi}_{t,t-1}, \mathbf{H}_t, \mathbf{Q}_t$ and $\mathbf{R}_t$ are deterministic quantities.

References

1. SMORN V, 12 – 16th October 1987, Munich, FRG.
2. G.E.P. Box and G. M. Jenkins, *Time Series Analysis: Forecasting and Control,* Holden Day Inc., San Francisco ,USA, (1976).
3. M. B. Priestley, *Spectral Analysis and Time Series*, Academic Press, London, (1981).
4. P. C. Young, *A Recursive Approach to Time Series Analysis,* Bull. Inst. Math. Appl., 10 : 209 (1974).
5. A. H. Jazwinski, *Stochastic Processes and Filtering Theory*, Academic Press, N. Y., (1970).
6. R. F. Kalman and R. S. Bucy, *New Results in Linear Filtering and Prediction Theory,* ASME Jour. Basic Eng. Series D, 83:85 (1961).
7. M. Marseguerra and C. M. Porceddu, *An Example of Utilization of Kalman Filters in Time Series Analysis,* 19th IMORN, 4 – 16th June 1986, Rome, Italy.

CHARACTERISTICS OF ADAPTIVENESS AND SELF-ORGANIZING AS APPLIED TO THE NON LINEAR NUCLEAR SYSTEMS

Kohyu Fukunishi

Advanced Research Laboratory,
Hitachi Ltd.,
P.O.Box 2, Kokubunji, Tokyo, Japan

ABSTRUCT

Supervising a operating nuclear reactor accurately with as short as time delay as possible is becoming more important as the demands increase for operating safety. The non linear and non stationary phenomena in a reactor may be one of the technical barriers to build a dynamic model permitting supervision and control. It appears that the characteristics of adaptiveness and self-organizing could extend the applicability of the conventional linear model to non linear and non stationary dynamics. The autoregressive (AR) and the adaptive digital filter (ADF) modeling techniques have been introduced for this study. The ADF is simple enough to permit real time modeling and can respond to non stationary and non linear state variations, but the AR technique needs complex time consuming computation. Application of ADF to anomaly detection and diagnosis of nuclear systems is discussed. Further, the stability monitoring method is discussed. Both of these have been confirmed by feasibility studies. Finally, adaptiveness of the reactor control method is discussed to assure its usefulness for operating efficiency.

INTRODUCTION

A quarter century has past since the Florida Symposium on noise in nuclear systems where the most significant contribution was the awareness of the role played by special effects upon noise measurement[1). Since the symposium, the importance of noise analysis has been confirmed and indeed increased in monitoring, supervising and furthermore the controlling of a operating nuclear reactor. In particular all the topics presented at the specialists' meetings of reactor noise, SMORNs, which have held four times since 1972, have described the vast possibility of reactor noise analysis for improvement of safety and suitability for nuclear power stations.

During these years, the methods adopted in reactor noise analysis have been more than just spectral analysis for a single-input and single-output system. Time series analysis for a multi-inputs multi-outputs system has recently occupied an important position with the increase in power reactors. This new method is eminently suitable for the detection of events due to reactivity perturbations. Fluctuations of the neutron flux, neutron noise, are interlinked with reactivity perturbations of physical systems such as the coolant and fuel temperature and steam bubble generation, etc..

This new approach has been expected to revealing the complex feedback relations between reactor power and the reactivity perturbations[2]. Particularly, the time series analysis with a multivariable autoregressive (MAR) model has proved useful in these investigations regardless of the controversial points discussed in bench mark tests presented in all SMORN meetings since 1981[3].

The reactor dynamics are known to be inherently non linear. However, the inherent non linearity does not caused any conspicuous characteristics in reactor noise if the reactor is operated sufficiently steady. Apart from the inherent non linear property, the occurrences of non linear phenomena in power reactors appear to be due to different external reactivity fluctuations, such as listed in Ref.4. The description of a general model for these non linear phenomena seem to be complex and impractical because of the many different physical mechanisms. With the recent emphasis on increasing plant safety, it is reasonable to expect a concurrent emphasis on the integrated method for supervising and controlling a nuclear system. The contribution of reactor noise analysis to the integrated control of nonlinear nuclear system is expected to rapidly increase.

The objective of this paper is to discuss the technical possibilty of utilizing noise analysis for control operations of non linear nuclear systems in order to meet the increasingly strict requirements for reactor operations. At the present stage, because non linear mathematical and physical representations are less practical from a viewpoint of real time execution, effective control can be made putting simple linear methods into practical properly. Adaptiveness and self-organizing may turn out to be important and realistic solutions, because they could extend the applicability of linear models to describe non linear reactor systems.

For this presentation I will concentrate on two areas that have potential for use in supervising and controlling a non linear nuclear system. The first is self-organizing modeling. Basic techniques are discussed, and MAR and ADF models are presented together with feasibility studies an anomaly detection/diagnosis and reactor stability of BWRs. Second, adaptive control systems are discussed and their efficiency is confirmed.

SELF-ORGANIZING MODELING

Modeling reactor dynamics, is generally complex, difficult and a tiresome task due to multiple feedback effects, non linear phenomena and, furthermore, the space dependent effects. assuming that a mathematical and/or physical dynamic model could be precisely accomplished, it will become a necessarily complex non linear representation and may not be suitable for the real time problems of reactor supervision and control. Therefore, it can be said that the simple model with linear terms should be extended to be cover non linear nuclear dynamics. Modeling using self-organization may be a practical method to extend the application of linear dynamic models to non linear nuclear phenomena and also to non stationary phenomena.

It can be considered that self-organizing is a remarkable characte-

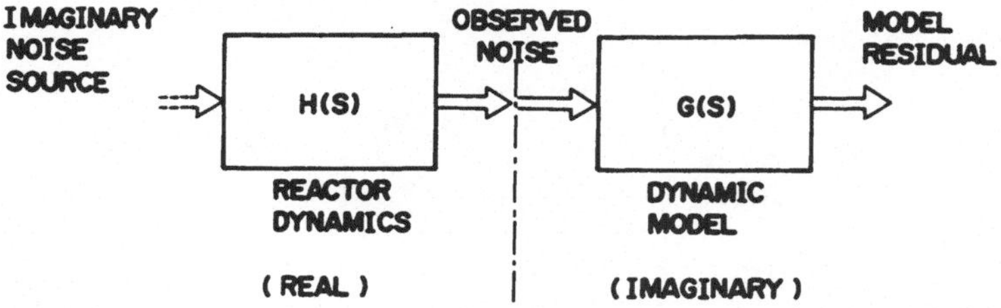

Fig.1 Modeling of reactor dynamics from a viewpoint ot the inverse system concept

ristics that is endued with noise analysis. That is, it is thought that noise analysis by time series analysis may play the major role in extracting some form of parametric model from the actual dynamic system. This can be explained in terms of the inverse system concept as illustrated in Fig.1. That is, considering the observed noise to be the output of the dynamic system perturbated by Gaussian white noise, the dynamical model can be revealed from the observed noise by noise analysis.

Supposing that an automatic or semi-automatic process is established for the extraction of linear parametric models, the dynamic model can be developed by itself from the observed noise. If the linear parametric model could be tuned to adapt to dynamic changes, such as non stationary variations, it has the possibility to be extended to non stationary and also non linear phenomena.

This approach to the modeling, that is self-organizing modeling, of non linear nuclear systems may be far from elegant and still imprecise, but it could be useful given the current insufficiency of the non linear dynamic models and methods for real time processing. In this section, the linear parametric model representation by MAR and ADF are discussed. Sample applications to anomaly detection, diagnosis and reactor stability monitoring are reviewed and presented.

Autoregressive Modeling

Multivariable autoregressive (MAR) modeling has been studied widely for supervising nuclear reactors since the suggestive investigations of reactor feedback effects [2,5]. A MAR model as well as a scalor AR model can be extracted semi-automatically by solving the generalized Yule-Walker equation with a criterion for evaluating the model order such as the AIC [6].

The MAR model is expressed as

$$\mathbf{x}_t = \sum_{i=1}^{p} \mathbf{A}_i \mathbf{x}_{t-i} + \mathbf{z}_t \,. \qquad (1)$$

Matrix $\mathbf{A}_i$ can be determined from the m-dimensional time series data $\{\mathbf{x}_t\}$, where $\mathbf{A}_i$, $i=1,2,...,p$ are m x m matrices, $\{\mathbf{z}_t\}$ is uncorrelated residual sequence and p means the model order. The matrix $\mathbf{A}_i$, $i=1,2,...,p$ and the value p are determined as follows. The Yule-Walker equations for matrix $\mathbf{A}_i$s.

$$\sum_{i=0}^{p} \mathbf{A}_i \mathbf{R}_{j-i} = 0 \,, \; j=1,2,\, \cdots,\; p, \qquad (2)$$

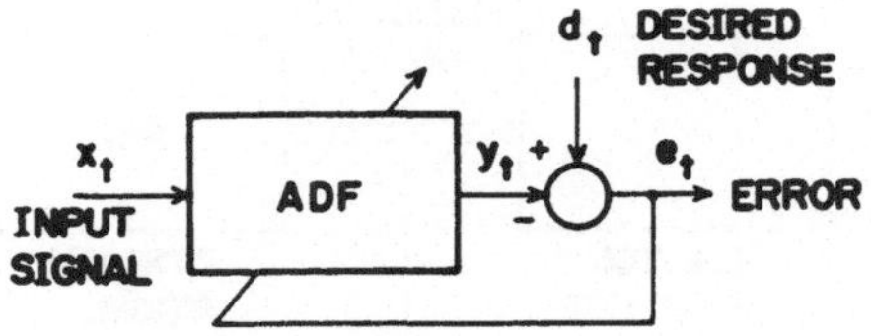

Fig.2 A schematic concept of ADF, adaptive digital filter.

are recursively solved for the tentatively fixed value p. Here, $\mathbf{R}_j$ means the correlation between $\mathbf{x}_t$ and $\mathbf{x}_{t+k}$ and $\mathbf{A}_0$ is the unit matrix. The optimum value p, can be evaluated by the AIC criteria from maximum likelihood obtained by Eq.(2) for the assumed p as

$$AIC_p = -2ln(maximum\ likelihood) + 2m(1+mp). \qquad (3)$$

When the reactor operating condition has changed, the MAR model can be revised automatically to correspond with the new state using the above procedure repeatedly. These calculation, however, require time consuming recursive computation for solving Eq.(2), this is a problem to be overcome for the real time processing. Detailed descriptions of the applications of the MAR model for the diagnosis and monitoring of nuclear system as discussed in Ref.(2,5,7) are not repeated here.

ADF Modeling

The adaptive digital filter (ADF) introduced by Widrow[8] has accurately self-organizing modeling structure and is drawn schematically in Fig.2. The filter comprises a tapped delay line and adjustable weights whose values are adapted by an optimal impulse response. The ADF maintains stationary and also non stationary learning characteristics by tuning the adaptation speed as described later. Thus ADF is applicable to not only stationary but also non stationary environments by slow and fast adaptation respectively.

In addition to the usual input signal x_t another input signal, the desired response d_t, must be supplied to the adaptive filter during the adaptation process as shown in Fig.2. When the observed data is applied to the ADF for the input signal and the desired response, the unknown dynamic system can be modeled automatically so as to extract the dynamic characteristics concealed in the observed signal as illustrated in Fig.3. It should be noted that this modeling method is one of the realizations of the inverse system concept described in Fig.1.

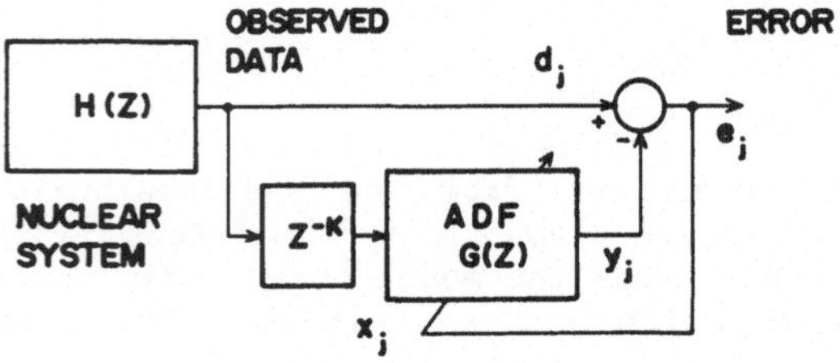

Fig.3 Self-organizing model generation of nuclear reactor dynamics using ADF, adaptive digital filter.

Widraw's ADF, with White's adaptation procedure[9], is briefly reviewed. The input and output relationship of the ADF is:

$$y_i = \sum_{k=0}^{p} b_{k,j}\, x_{j-k} + \sum_{k=1}^{q} a_{k,j}\, y_{j-k}\,, \tag{4}$$

where parameters $a_{k,j}$, $b_{k,j}$ can be adjusted adaptively so as to minimize the mean square of the error e_j between the desired response d_i and the ADF response y_i. The adaptation algorithm is based on the following steepest descent method using the least mean square error,

$$\mathbf{W}_{j+1} = \mathbf{W}_j - \mathbf{s}\nabla_{\mathbf{W}_j} e_j^2$$

where $\mathbf{W}_j = \mathrm{col}[b_{0j}, b_{1j}, \cdots, b_{pj}, a_{1j}, a_{2j}, \cdots, a_{qj}]$,

$$\mathbf{s} = \mathrm{diag}[s_{b1}, s_{b2}, \cdots, s_{bp}, s_{a1}, \cdots, s_{aq}], \tag{5}$$

$$e_j = d_j - y_j\,.$$

and $s_{b1}, \ldots s_{bp}, s_{a1}, \ldots, s_{aq}$, are constant values. After transformation, the least mean square algorithms for adaptively adjusting the parameters can be derived:

$$a_{k,j+1} = a_{k,j} + u\, e_j \frac{\partial y_j}{\partial a_{kj}}\,, k = 1, 2, \cdots, q,$$

$$b_{k,j+1} = b_{k,j} + v\, e_j \frac{\partial y_i}{\partial b_{kj}}\,, k = 0, 1, \cdots, p, \tag{6}$$

where $u = 2s_{b0} = 2s_{b1} = \cdots = 2s_{bp}$, $v = 2s_{a1} = \cdots = 2s_{aq}$.

In case of p=2 and q=0, the non recursive ADF can be drawn as shown in Fig.4. Thus, the ADF consists of simple and restricted procedures with some delay, summations and multiplications, it is very promising modeling method for real time execution in the reactor supervision and control.

The model order p and q can be evaluated by the AIC criteria as well as the MAR modeling. The ADF model is not as sensitive to the model order as the MAR model, the value p and q can be fixed in non stationary change to some extent by pretesting.

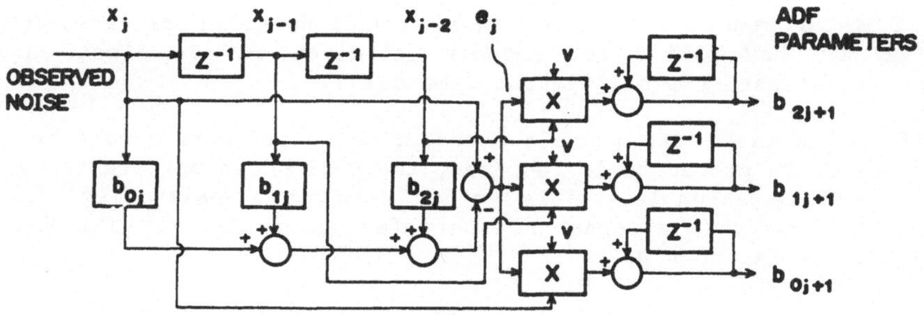

Fig.4 Block diagram of non recursive ADF modeling with the order p=2, q=0.

ADF Applications

The self-organizing modeling procedures using ADF have been applied to investigate real time supervision of nuclear reactors for anomaly detection, diagnosis and stability monitoring. Here, several experimental studies are discussed briefly.

A. Anomaly Detection and Diagnosis

The state value, the observed noise, x_t can be predicted from the past observed noise sequence $\{x_{t-1}\}$ using the ADF modeling. The predicted state value is defined by $x_{t/t-1}$. Here, the prediction error is expressed as

$$e_t = x_t - x_{t/t-1}. \tag{7}$$

This prediction error is designated "the innovation process" because it represents new information created by the latest observation[10].

The innovation process randomly fluctuates but in small quanties under normal conditions since all the systematic trends can be eliminated by ADF with slow adaptation speed. However, for occurrence of any faulty conditions, the innovation process may provoke severe systematic errors because the model deviates from the new state undergoing anomalous variations.

In normal reactor operating circumstances, the innovation derived represents Gaussian statistics with zero-mean and variance:

$$V_t = \frac{1}{N}\sum_{j=t-N+1}^{N} e_j^2. \tag{8}$$

The chi-square random variable with N degrees of freedom for the innovation process is expressed as

$$l_t = \frac{1}{N}\sum_{j=t-N+1}^{N} (e_j / \sqrt{V_t})^2. \tag{9}$$

The above variable means that the innovation variance is normalized with the Gaussian variance.

Statistical testing with the chi-square random variable can be utilized for anomaly occurrence detection by the following discrimination:

$$\begin{aligned} &l_t < \xi, \quad \text{normal}, \\ &l_t \geqq \xi, \quad \text{anomaly or non stationary}, \end{aligned} \tag{10}$$

where ξ is a prefixed value. Thus, the innovation process with ADF modeling may be applied for anomaly detection of non linear and non stationary nuclear systems, without time delay.

If diagnosis of the anomalous and/or non stationary condition detected by the above procedure is required, the well known pattern recognition method could be valuable. This method uses the Mahalanobius' distance between the observed patterns and the reference patterns. The Mahalanobius' distance is known as

$$D_t^k = \frac{1}{N}\sum_{j=t-N+1}^{N} (\alpha_j - \beta_j{}^k)\,(\mathbf{T}^k)^{-1}(\alpha_j - \beta_j{}^k)', \tag{11}$$

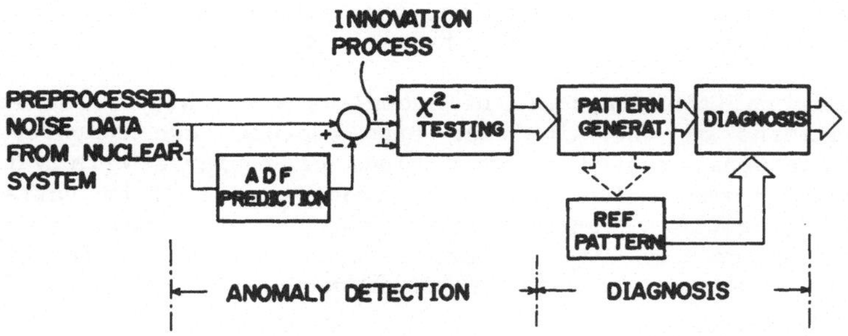

Fig.5 Real-time reactor anomaly detection and diagnosis procedures using chi-square testing with innovation process and pattern recognition.

where the super fix k means the k-th kind of anomaly, α_j represents the extracted features, that is the statiscal values such as the chi-square variable and/or the statiscal momentum of the innovation process and the observed data, and β_j^k represents the corresponding reference values with the covariance T^k. The specific anomaly may be diagnosised by selecting the anomaly which minimizes the Mahalanobis' distance.

The flow diagram of anomaly detection and subsequent diagnosis is illustrated schematically in Fig.5. In this study, prediction by the ADF is given as

$$x_{t/t-1} = \sum_{i=1}^{q} a_{i,t} x_{t-i/t-i-1} + \sum_{i=1}^{p} b_{i,t}\, x_{t-i}. \tag{12}$$

For the feasibility study of this method, nonlinear and non stationary phenomena in a nuclear system were simulated assuming a hypothetical loss of coolant flow event on a complex dynamic BWR simulator modified with the generation of stochastic random variations. Multiple variables such as reactor pressure, reactor water level, averaged neutron flex and dry well pressure, etc., were fed into the anomaly detection procedure. The innovation process which was derived automatically by the ADF predictor and statistical testing was executed for each variable. The points of special interest are that the anomaly detection has completely automatic procedure by the self-organization of the dynamic modeling after predetermining the model order and the parameter for the adaptation speed, and also that multivariable detections can be executed completely in parallel.

As example, the statistical testing result is shown in Fig.6. An explosion was detected in the chi-square value of the reactor pressure just after a hypothetical very small pipe rupture had been created in the simulator. And the result suggested that the threshold ξ of the chi-square testing can be determined easily beforehand. Furthermore, these results showed that the ADF was able to adjust itself to the new transition state in about 50 seconds after anomaly detection. This result also confirms the very important idea that anomaly detection by ADF modeling can be applied to non stationary and non linear variations without any additional labor. Here, the degrees of freedom for the chi-square testing were N = 20. The model order was selected as p, q = 4, but the innovation process was not so sensitive to the values. Therefore, frequent determination of the model order may be avoided in using ADF modeling. In addition, the value of u and v, the adaptation speed, was predetermined to be 0.2 by evaluating the AIC values.

Furthermore, the diagnosis of anomalies was tested. The statistical feature patterns of the simulated pipe rupture were compared with the reference patterns created by many kinds of ruptures with various size and types or forms. The innovation processes of three variables such as reactor pressure, reactor water level and dry well pressure were taken into account for the pattern recognition. As a result, simulated pipe rupture was estimated to have occurred at the gas flow level

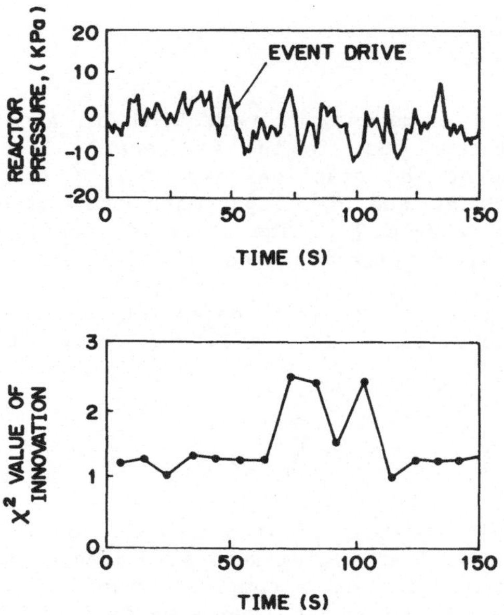

Fig.6 Feasibility study of anomaly detection using innovation process with ADF for non stationary and non linear transient of hypothentical loss of coolant flow event in BWR simulator.

with a break area of about 2.7 cm^2 as shown in Fig. 7. This estimated event agreed with the assumed one. The results, however, show that the sensitivity of the diagnosis by the pattern recognition is not so sharp in the large break area as in the small break area. This insufficiency can be improved by adding the momentum of the observed noise data to the elements of the feature vector.

This anomaly detection and diagnosis methods was adopted successfully to the diagnosis function in the prototype of a computerized operator guidance system for post-trip transient control of BWRs[11].

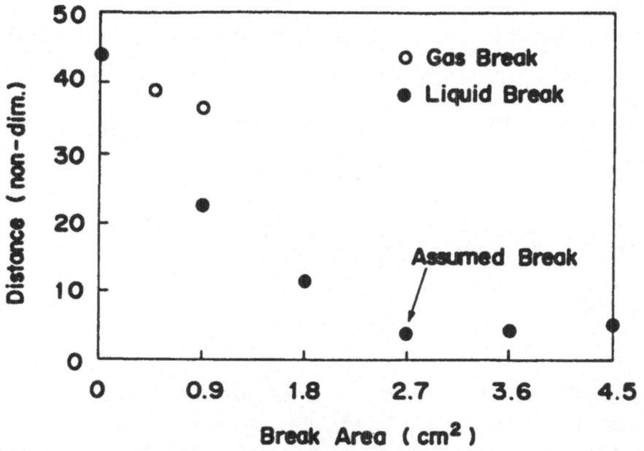

Fig.7 Simulation study of anomaly diagnosis after chi-square testing using pattern recognition method for hypothetical loss of coolant flow event in BWR simulator.

B. Stability Monitoring

The self-organizing modeling with ADF has been proposed because it has good potential to act as a real time stability monitor[12]. This stability monitoring method is based on the principle of the inverse system as illustrated in Fig. 3.

The power spectral density of the neutron flux in the frequencies band between about 0.3 Hz and about 0.7 Hz may represent the resonance fluctuation of a BWR reactor core which directly reflects on the reactor stability[7]. The normal spectral pattern of the neutron flux over the above frequency interval has only one peak as illustrated in Fig. 8, so that it can be described by the second-order oscillation model as

$$H(s)=\frac{c_0\, s+c_1}{s^2+2\zeta\omega_0 s+\omega_0{}^2}\,. \tag{13}$$

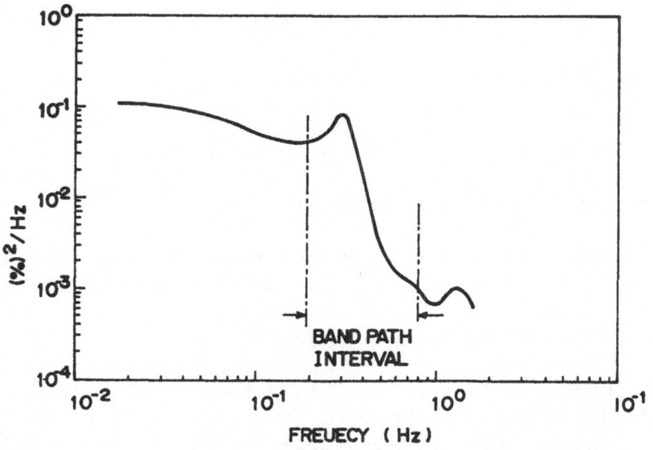

Fig.8 Power spectral density of averaged neutron power, APRM, noise and stability related frequencies.

where $H(s)$ denotes the transfer function of the second-order oscillation system.

The decay ratio, DR, and resonance frequency, f_0, of the above system are defined as

$$DR = \exp[-2\pi\xi/\sqrt{1-\zeta^2}],$$
$$f_0 = \omega_0\sqrt{1-\zeta^2}/2\pi, \quad (14)$$

where c_0 is assumed to be zero. The decay ratio of the second-order system corresponds to the empilical decay ratio which usually designated the amplitude ratio between two successive oscillation periods of the reactor dynamic response to an external impulse disturbance.

The inverse system relation illustrated in Fig. 3 can be expressed in this case by the following equation:

$$H(z) = (1-G(z))^{-1}, \quad (15)$$

where $H(z)$ is the discrete representation of $H(s)$ and is written as

$$H(z) = \frac{b_2 z^{-2}}{1+a_1 z^{-1}+a_2 z^{-2}}. \quad (16)$$

The above parameters a_1, a_2 and b_2 become the functions of the parameters ζ and ω_0 in Eq. (13). The characteristics of the ADF can be written by the expression:

$$G(z) = \frac{1+a_1 z^{-1}+(a_2-b_2)z^{-2}}{-b_2 z^{-2}} \quad (17)$$

The input-output relation of the ADF in the time domain can be denoted simply as

$$y_j = B_0 x_{j+2} + B_1 x_{j+1} + B_2 x_j, \quad (18)$$

where B_0, B_1 and B_2 are functions of a_1, a_2 and b_2. As a result, the decay ratio and resonance frequency in the neutron noise can be obtained according to the following computation:

(1). Determine parameters B_0, B_1 and B_2 by ADF procedure,
(2). Transform parameters ω_0 and ζ^2 through values of a_1, a_2 and b_2,
(3). Obtain DR and f_0 by Eq. (14).

The most prominent feature of this method is that real time stability monitoring of the whole reactor core through more than one hundred local-neutron-flux sensors may be possible using micro-processors because of algorithm simplicity and parallel calculation. Indeed, this approach offers significant potential for monitcring reactor stability in non stationary conditions by using a fast adaptation speed in the ADF convergency procedure. The decay ratio in non stationary states means only a approximate value and not have to be exactly determined.

A fesibility test result of this stability monitoring method is shown in Fig. 9, where the resonance frequency and the decay ratio revealed from the averaged neutron flux, APRM, fluctuations were indicated to be about 0.3 Hz and 0.23 respectively. These values were obtained after ADF convergence required about 30 seconds from zero set initial values of all the ADF parameters ezcept of the parameter of the adaptation speed. Here, the adjusting parameter v was chosen to be 15.0 and the sampling period was 0.1 second. And test data was under steady state at a natural circulation core flow operation. Those value of the stability evaluation also agreed with the values estimated by the completely different method with MAR modeling[13]. Thus, it is hoped that this method may prove useful in evaluating reactor core stability of BWRs.

ADAPTIVENESS

It was previously stated that self-organizing models, that is, adaptive and automatic modeling methods could to the application if linear models to non linear and non stationary phenomena. In addition to this, it can be argued that adaptiveness in the non linear phenomena such as adaptive control methods may also be effective to extend and improve the conventional linear control system in the nuclear plants.

The control systems of a nuclear plant are designed mainly on the linear control principle in the complex frequency domain. The optimality of the regulator in the linear control system generally is preserved only near the designed condition because of the non linear phenomena in the controlled nuclear system, therefore it is desirable to extend the application range of the linear regulator in order to improve control efficiency. Non linear control and adaptive control in the exact meaning such as the model reference adaptive control system MRAS may be optimum solutions for controlling the non linear nuclear system in the future, but they seems not to be feasible right now. The general principles of both control methods have problems for their utilization in the control of nuclear systems, that is, complexity, computatational difficulty and less reliability of the control modeling. However, it may be understood that the non linear problems in the control system can be overcome to a major extent by a simple adaptive modification to the conventional linear regulator of the today's nuclear control systems.

This approach is different from the usual reactor noise analysis which is based on the passive reactor noise, but utilizes transient data during start up testing and commissioning. This modification can be expressed in a series of operations using a pre auto-tuning method[14] as follows. At first, the dynamics of the controlled system which includes the non linear characteristics is identified using the actual step response data in a nuclear system. Generally, the model which corresponds to the identified and controlled system can be represented as

$$G(z) = \frac{b_1 z^{-1} + \cdots + b_p z^{-p}}{1 + a_1 z^{-1} + \cdots + a_q z^{-q}} z^{-d} , \tag{19}$$

where z^{-k}, $k=1,2,\ldots,p$ and q means the k-th time delay operator, and p and q show the model order, and d means a non linear time delay factor. The least square estimation technique and the AIC criterion are repeatedly applied to fit the above pseudo linear model to a step response transient output of the controlled system. That is, a set of the parameters $(a_1, a_2 \ldots, a_q, b_1, \ldots b_p)$ is determined so as to minimize the square of differences between the outputs of the model and the actual step response over the transient duration for a tentatively fixed model order. Then,

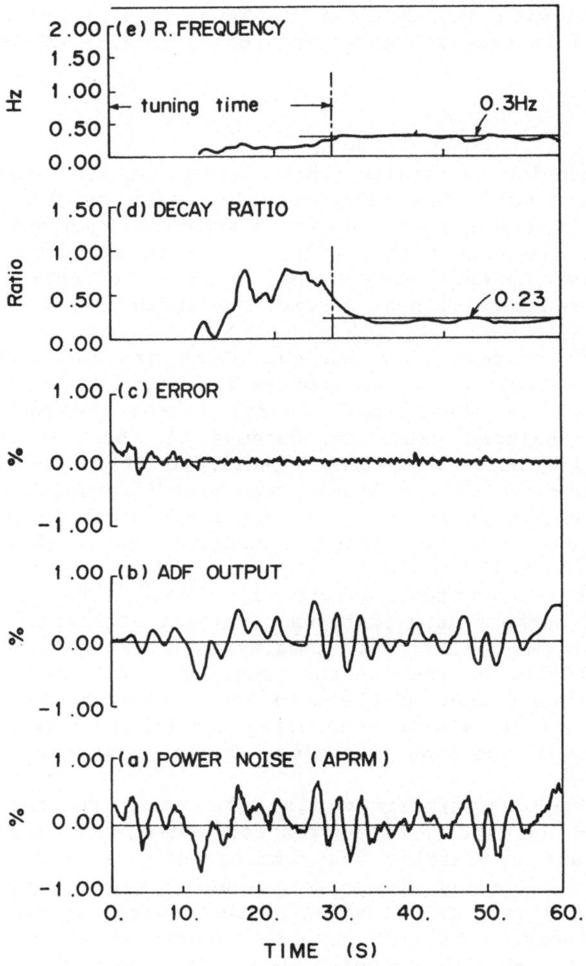

Fig.9 Feasibility test of reactor stability monitor using ADF modeling from averaged neutron power, APRM, noise at natural-circulation-core-flow operating condition of BWR.

the best fitted model is automatically selected by evaluating the AIC criteria on the tuned models.

Next, the optimal values of P,I,D (Proportional, Integral and Differential) regulator in the closed loop control system consisting of the generated model $G(z)$ given in Fig. 10, is evaluated through repeating the step response simulations of the control system. The criterion for optimum tuning is defined to be the direct engineering figures concerning to dynamic features such as the damping ratio, the peak response value and the settling time, etc.. Therefore, the performance index for the optimization of the PID regulator is defined as

$$Q = \min \max_{i} [r_1, r_2, \cdots, r_i, \cdots, r_l], \quad (20)$$

where $r_i, (i=1,\ldots l)$ means the tuning criterion normalized with each limiting conditions. The direct search simplex method of nonlinear programming[15] is introduced to the optimization algorithm for deciding the optimal P,I,D values. Using the optimization procedure, optimal sets of P,I,D values are prepared for each generated model corresponding to a different operational condition of a nuclear system.

Finally the functional relationship between each regulation parameter and related operational condition are fitted to each other as

$$g_\eta (\eta^{k_1}, \eta^{k_2}, \cdots, \eta^{k_n}, k_1 \; k_2, \cdots, k_n) = 0 , \quad (21)$$

where η is each P,I,D value and ki is the operational condition (the reactor power level). The optimal P,I,D values at an arbitrary power level can be obtained conversely from the Eq. (20) by selecting the appropriate power level during the reactor operation.

Thus an adaptive linear regulator is designed as follows:

(1) Model identification from an actual reactor transient,

(2) Optimization of P,I,D values by simulating the generated controlled system with the model identified,

(3) Fitting the optimal P,I,D values and the power level.

Feasibility testing of the adaptive regulator was studied using the transient data of a primary loop control system in a BWR simulator. The actual primary loop control system contains a flow coupler with non linear dynamics. Therefore, the control system with the linear PID regulator probably could not avoid the effects of non linearity.

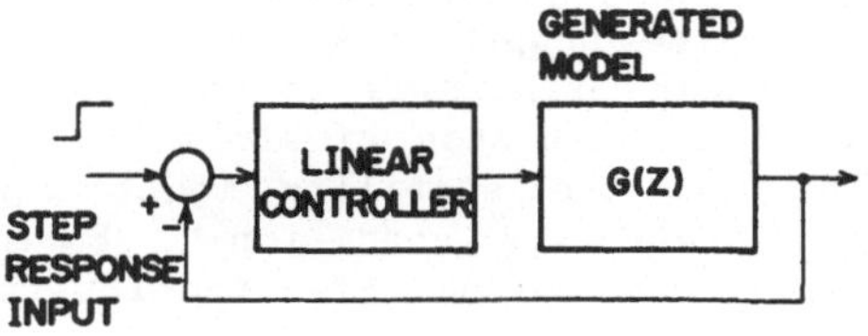

Fig.10 Simulation of optimal regulation design for adaptive control of closed loop control system with generated model as controlled object.

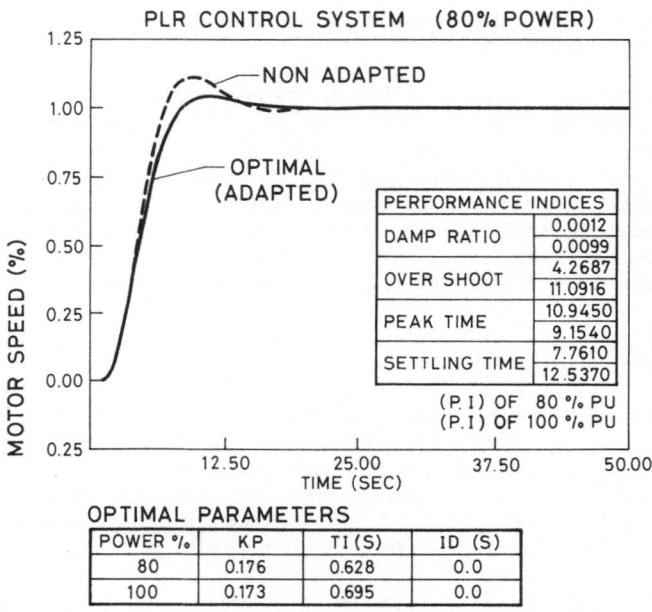

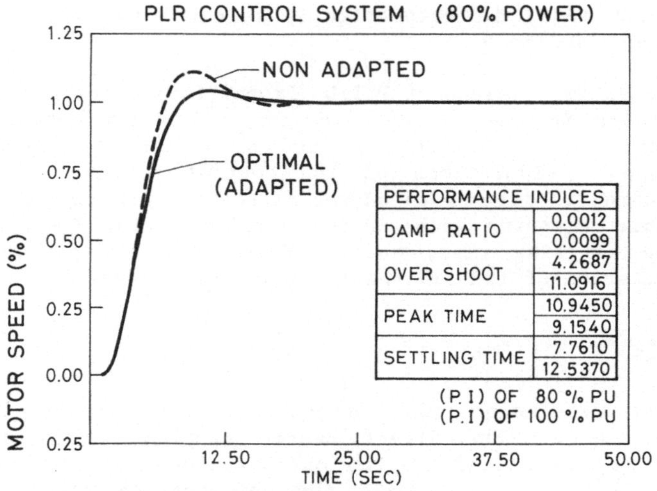

OPTIMAL PARAMETERS

POWER %	KP	TI (S)	ID (S)
80	0.176	0.628	0.0
100	0.173	0.695	0.0

Fig.11 Feasibility study of adaptive regulator by simulating optimal design for non linear primary loop control system of BWR using simulator (direct copy from CRT).

The simulated results shown in Fig. 11 confirmed the efficiency of the adaptive regulator for overcoming non linear effects. That is, the damping ratio of the moter speed at 80% power operation could be decreased almost zero by selecting the optimum P,I (Differential was eliminated here), values which had been chosen beforehand.

Thus the adaptiveness in the control system has the possibility to compensate for non linear dynamics of the controlled system and to improve the operating efficiency and availability.

CONCLUSION

The methods for supervising and controlling nuclear systems exhibiting non linear and non stationary phenomena have been discussed. The self-organizing modeling and the adaptiveness in control systems may extend the applicability of linear principles to non linear and non stationary systems. Especially, the usefulness of ADF modeling has been suggested for anomaly detection and the stability monitoring on the real time in particular to non stationary condition. Furthermore, the simple adaptive model has been shown to be effective for control of non linear systems.

REFERENCES

1. Uhrig, R.E. ed. (1964), Noise Analysis in Nuclear Systems, USAEC Symposium Series No. 4.
2. Fukunishi, K. (1976), Neutron Noise Analysis of BWR Using time Series Analysis, Proc. of IAEA/NPPCI Specialists' Meeting, MRR 160, 53.
3. Williams, M.M.R and McCormick, N.J. ed., Prog. Nucl. Energy vol. 9 (1982, SMORN III), vol. 15 (1985, SMORN IV), (1988, SMORN V).
4. Konno, H. (1987), Presented at SMORN V, 13.4. and ibid. (1988).
5. Fukunishi, K. (1977), Nucl. Sci. Eng., 62, 215.
6. Akaike, H. (1974), IEEE Trans. Automatic Control, AC19,716
7. Fukunishi, K. (1978), Nucl. Sci. Eng., 67, 296.
8. Widrow, B., McCool, J. M. and Larimore, M.G. and Johnson, C.R. (1976), Proc. IEEE, 64, 1151.
9. White, S.A. (1975), An Adaptive Recursive Digital Filter, Proc. 9th Asilomar Conf. Circuits, Systems and Computers, 21
10. Mehra, R.K. and Peschon, J. (1971), Automatica, 7, 637.
11. Fukunishi, K., Ohga, Y., Tanji., Kishi, S., Murata, F. and Hashimoto, S. (1984), A Computerized Operation Guidance Method for Post-Trip Transient Control of BWRs, Proc. IAEA Seminar, IAEA-SR-105/34.
12. Fukunishi, K. and Suzuki, S. (1987), Nucl. Tecnol., 78, 132.
13. Suzuki, S., Fukunishi, K., Kishi, S., Yoshimoto, Y. and Kishimoto, K. (1986), Nucl. Technol., 74, 132.
14. Suzuki, S. and Fukunishi, K. (1982), Nucl. Technol., 58, 379.
15. Jacoby, S.L.S., Kowalik, J.S. and Pizzo, J.T. (1972), Iterative Methods for Nonlinear Optimization Problems, Chap. 4, Prentice-Hall, Inc., New York.

THE MONTE CARLO SIMULATION OF THERMAL NOISE IN FAST REACTORS

G Hughes and R S Overton

Central Electricity Generating Board
Berkeley Nuclear Laboratories
Berkeley, Glos GL13 9PB, UK

INTRODUCTION

A safety concern for fast reactors is the possibility that a local flow blockage in a fuel pin bundle could lead to a more serious fault because of the high power density in the core. Experimental studies carried out in a number of countries (Weinkoetz, Krebs & Martin, 1982; Girard & Buravand, 1982; Firth & Conroy, 1986; Overton, Wey & Hughes, 1984) indicate that the monitoring of temperature fluctuations near the outlet of a fuel subassembly is one of the most sensitive ways of detecting a local flow blockage. These have been supported by theoretical modelling (Krebs & Bremhorst, 1983; Firth, 1977). In particular, a Monte Carlo computer program, STATEN, has been developed in the UK by the CEGB (Overton, Wey & Hughes, 1984; Hughes, Overton & Wey, 1986), and used to model the temperature fields in the outlet regions of reactor subassemblies under normal and fault conditions. This paper outlines the development and validation of the code and illustrates its use for the development of detection algorithms based on pattern recognition methods. In addition the modelling of the Superphenix-1 reactor at Creys-Malville, performed under the European Fast Reactor Agreement (Girard et al., 1987) is used to illustrate the comparison of its predictions against good experimental measurements.

THE STATEN CODE

The STATEN computer code, developed by CEGB, uses a Monte Carlo method to model turbulent flow in the open wrapper of a fast reactor subassembly between the pin bundle exit and the plane of the thermocouple. The degree of turbulent mixing is controlled by the cross-stream turbulence intensity specified to the code. This is used to allocate instantaneous cross-stream velocities to marker particles released in coplanar batches

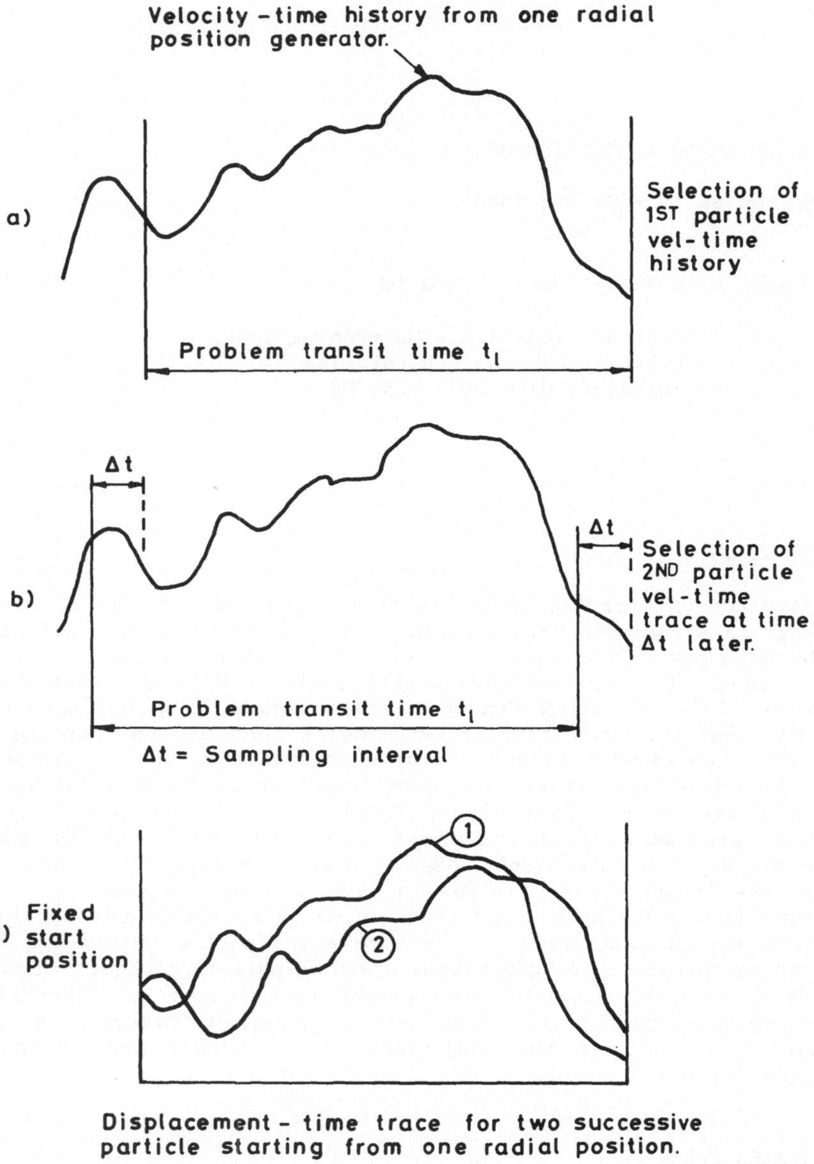

Figure 1. Generation of Displacement - Time Histories of Successive Particles.

into the flow. These velocities can be mutually weighted both spatially and temporally to achieve the desired turbulence length scales and spectral form of the velocity distribution. The particles carry representative temperatures downstream from an initial specified temperature profile, and explicit heat transfer calculations are performed at each of a large number of discrete steps between the initial plane and the final (thermocouple) plane, some distance beyond the subassembly outlet. A simulated temperature signal of full bandwidth can thus be built up at any specified point in the flow, and in particular at the thermocouple location. The details of the Monte Carlo simulation method are presented in Section 2.1 below.

To facilitate comparison with measured temperatures from limited bandwidth measurements, the simulated signals can be digitally low-pass filtered before computation of parameters such as r.m.s. noise levels and the skewness and kurtosis of the temperature distribution. For axially non-uniform geometries, such as the Superphenix subassembly open wrapper, STATEN is used in linked axial stages, which permits axial variation in turbulence conditions caused by changes in the geometry (grid, venturi, etc.) to be accommodated.

Details of the STATEN method

For a single point in the starting reference plane, the correlated velocity-time sequences for the x and y directions are used to define the trajectories of successive particles leaving that point. The two velocity components are independent owing to the assumption of isotropy; henceforth one velocity component is adequate to explain the motion. The velocity-time history, covering the problem transit time, t_L, for the first particle leaving the reference plane at time t_0 is selected as shown in Fig 1a. The second particle, immediately following in the flow, at time Δt later, has a different velocity-time history displaced by Δt from the original, and this is shown in Fig 1b. These two velocity-time histories are applied to two successive particles leaving the same point; thus the corresponding displacement-time histories for successive particles are as shown in Fig 1c. This is a simple way of maintaining the correct statistical characterisation of velocities at the starting plane and performing the Eulerian-Lagrangian transformation, the validity of which is a necessary assumption in such a representation of turbulent flow.

It is perhaps worth noting that whilst the velocity-time histories of successive particles only differ slightly by virtue of the temporal displacement, the successive particles move with different velocities at each timestep. Also, the Lagrangian velocity distributions, defined at the starting plane, are preserved at all axial positions downstream. This is the correct situation for the assumed homogeneous flow.

It is assumed that the Eulerian x and y radial components have the spectral form shown in Fig 2 (Lawn, 1977). This Figure gives the salient break frequencies at values derived from empirical correlations in terms of mean flow velocity U, pipe diameter D, and Kolmogorov length scale η. The -5/3 decay slope in the 'inertial-convective subrange' quoted by Lawn is approximated by -2 to make the filter design tractable. The slope in the 'inertial-diffusive subrange' is assumed to be -4.

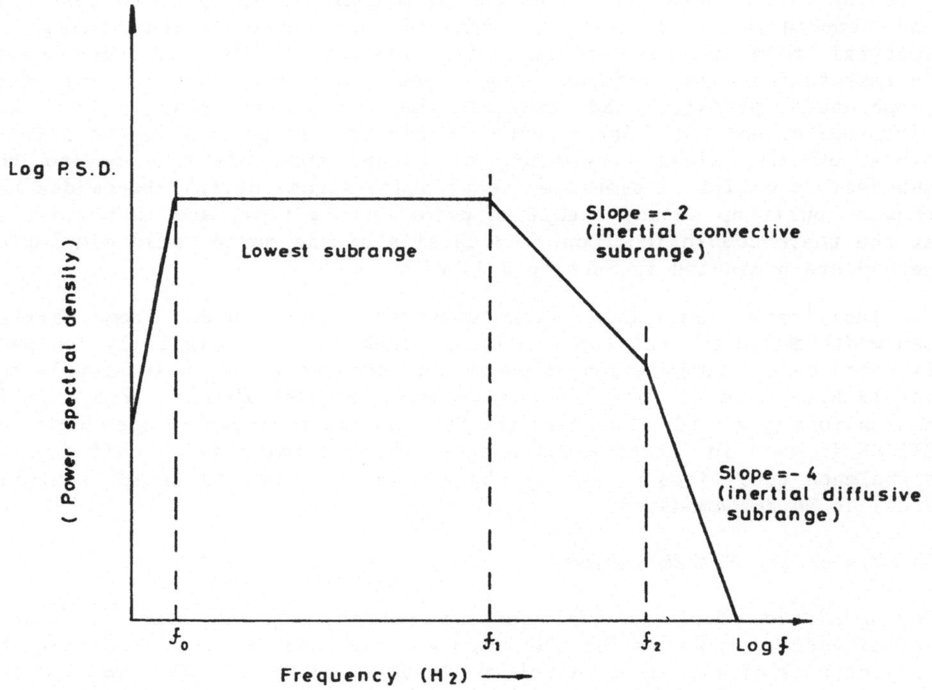

Figure 2. Asymptotic Approximation of Spectral Form of Velocity, Fluctuations.

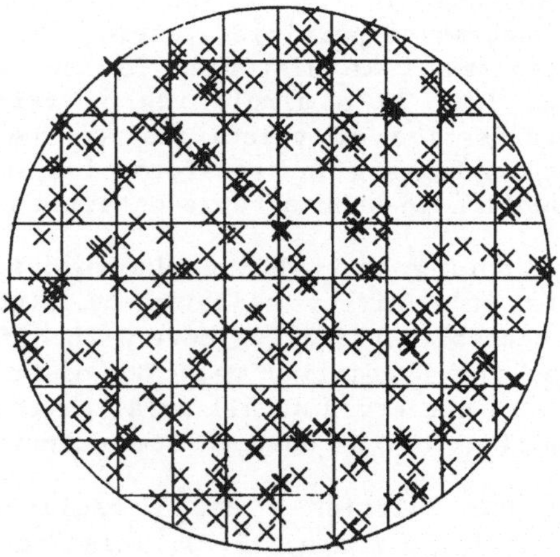

Figure 3. Typical Scattering of a Particle Batch Showing Heat Transfer Grid.

It is further assumed (Hinze, 1976) that the turbulent radial velocity macroscale (integral length scale) can be expressed as $\Lambda = D/8$, and thus the lower end of the inertial convective subrange is :

$$f_1 = \frac{3U}{16\pi\Lambda} = \frac{3U}{2\pi D}\, Hz$$

The upper limit of this subrange can be written as

$$f_2 = \frac{0.1U}{2\pi\eta} = \frac{0.1U}{2\pi}\left[\frac{2 \times 5 \times 10^{-3}U^3}{D\nu^3}\right]^{0.25} Hz$$

To minimise filter stabilising times it is desirable to keep the value of the low-frequency cut-off f_0 as high as possible. In practical systems it is likely that the low-frequency fluctuations will be dominated by external varaibles such as pump speed, pressures etc. and it is difficult to quantify the lowest frequency relevant to this study. An ad hoc method of assessing an appropriate value is to equate one-quarter of a sinusoidal wavelength of frequency f_0 to the transit length L. Thus $4L.f_0 = U$ and an upper bound for f_0 can be taken as $f_0 \leq U/4L$. Thus a particle moving with a frequency $f \geq f_0$ can reach its maximum displacement within the problem transit time. A particle moving at lower frequencies $f < f_0$ cannot achieve its maximum displacement within the duct length and thus the effect of these components reduces with reducing frequency.

A linear filter:

$$v(t) = \sum_{i=1}^{I_1} a_i v(t - i\Delta t) + \sum_{i=1}^{I_2} b_i w(t - i\Delta t)$$

is postulated, and the weights a_i and b_i can be calculated (Overton, Wey & Hughes, 1982) as functions of f_0, f_1, f_2 and the decay slopes of the power spectral density (PSD). It is found necessary to weight the previous five output values v and the previous four input values w in order to give the current output value.

The input values are provided by a Gaussian white noise generator; linear filtering preserves this distribution and thus the velocities used in the model are Gaussian, in accordance with Ohlmer and Schwalm, 1975.

The method described above produces velocity-time sequences which have no spatial correlation with each other. Correlation can, however, be achieved by using a weighted sum of independent velocities from surrounding cross-stream positions within a certain 'correlation area' R outside which correlation can be neglected. A correlation function is defined up to a maximum distance r_{max} :

$$c(r) = 0 \quad \text{for } r > r_{max}$$

$$\neq 0 \quad \text{almost everywhere on } [0, r_{max}]$$

At present the following experimental form is used (Hinze,1976):

$$c(r) = \exp\left(\frac{-r}{\Lambda}\right) \quad \text{for } 0 \leq r \leq r_{max}.$$

Defining the correlated velocity as

$$v_n^c(t) = \sum_{k \in R} \sigma_{nk} v_k(t),$$

(thus linearly preserving the Gaussian distribution of v) the weights σ can be calculated by solving a quadratic system of equations derived by considering equations of the form:

$$Corr(v_m^c, v_n^c) = Corr\left\{ \sum_{k \in R_1} \sigma_{mk} v_k, \sum_{l \in R_2} \sigma_{nl} v_l \right\}$$

$$= \sum_{k \in R_1 \cap R_2} \sigma_{mk} \sigma_{nk} = c(r) \quad (specified)$$

Details of the derivation and solution of these equations are also given in Overton, Wey & Hughes, 1982.

There are, however, considerable problems in modelling particle movements near a pipe wall. This is a region of anisotropic and imhomogeneous turbulence due to the radial velocity gradients in the region with possible fluid slowing and retention near the wall. Furthermore the limited spatial resolution of the present model makes it impossible to consider the turbulence transition region and laminar boundary layer in detail. For the present, particles incident on the wall are simply reflected back normal to the wall a distance equivalent to its hypothetical extramural movement. However, it is also relatively simple to incorporate other regimes, e.g. retention of a fraction of the particles on the wall, which may produce a better overall result without considering the details of the mechanisms involved.

At the reference plane the particles acquire temperatures from a fixed spatial mean temperature profile, with the optional addition of a specified r.m.s. temperature profile. The use of multiparticle batches enables a cross-stream heat transfer calculation to be carried out as follows. Each coplanar batch at each timestep consists of irregularly scattered particles carrying representative temperatures. (Typical coverage of a circular pipe is shown in Fig 3.) To make the heat transfer computation feasible a square grid is superimposed on the circular pipe cross section with a mesh coarser ($\geq \times 2$) than the grid from which the particles are released. At each timestep each particle in each batch is assigned to its nearest grid node. With all particles grouped temporarily in this way, the mean temperature of each nodal group is found and assigned to that nodal position. Linear interpolation is used to fill any empty nodes. Heat transfer between nodes is then computed for the duration of a timestep. The change in nodal temperature resulting is applied to each particle assigned to that node, before each particle is returned to its correct position.

After a coplanar batch of particles has been involved in the simulation for a time t_L corresponding to its axial movement through the given pipe length, it is necessary to extract a temperature at a specified temperature-measuring position. The positions and temperatures of all particles are known but it is highly improbable that one particle will be exactly coincident with the desired measuring point. Some form of temperature interpolation between nearest particles would immediately come to mind, but this would tend to attenuate the high frequency end of the noise spectrum. The temperature of the nearest particle may be taken to represent the temperature at the measuring point at each timestep

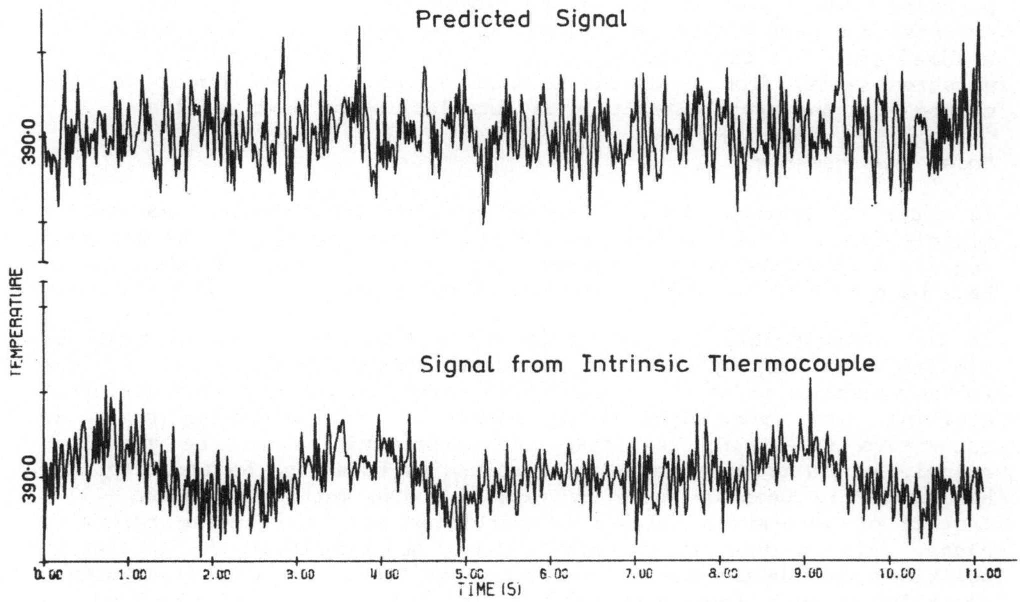

Figure 4. Typical Predicted and Measured Temperature Signals.

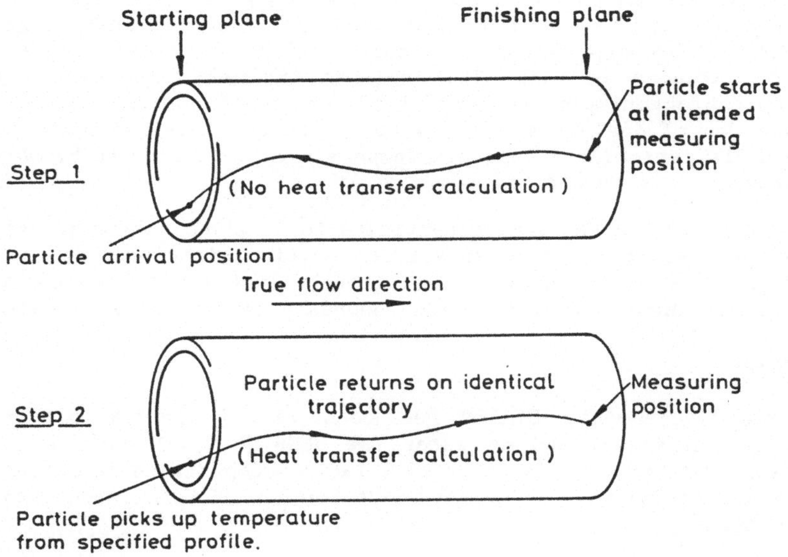

Figure 5. A Simple Illustration of the "Reverse Interpolation," Method for Deriving a Temperature Sample from STATEN.

provided there are sufficient particles per batch to avoid large cross-stream separations. A typical derived temperature-time history at a fixed point in the flow is shown in fig 4, together with a typical measured signal from an intrinsic thermocouple. A refinement to this method of temperature extraction is detailed in the next section.

Model Enhancements

In order to improve the statistical quality of predicted temperature signals from STATEN and to enable it to be used in geometries and flow regimes more varied than homogeneous pipe flow, a number of enhancements have been made to the basic model described above.

In the basic model, the extraction of a temperature signal from the simulation is done by selecting the temperature of the nearest particle to the measuring point. This inevitably leads to particles from different starting plane generators being nearer to the measuring point at successive timesteps, resulting in discontinuities in the simulated signal. Such a problem cannot be resolved by reducing the timestep, and has therefore been overcome by extending the method of Firth, 1977. Instead of generating batches of particles at the starting reference plane, this is done at the final plane, ensuring that the generating positions include the desired measuring positions. From the final plane, particles travel backwards to the initial plane, but carrying no temperature information. Here the particles are assigned temperatures from the specified initial profile, and return along their previous trajectories with a cross-stream heat transfer calculation being performed. This 'reversed path' technique thus ensures that particles will always arrive at the desired measuring positions. The scheme is summarised in Fig 5.

The basic STATEN method outlined assumes homogeneity in the flow, a justifiable assumption for fully developed pipe flow. In non-homogeneous conditions, STATEN may be run in a series of linked axial stages, with homogeneity assumed in each. This permits the mean flow velocity and turbulence characteristics to vary within the region to be modelled. The multistage use of STATEN can be combined with the 'reverse path' method to allow a statistically acceptable temperature simulation to be obtained under inhomogeneous conditions.

Another change to STATEN permits particles to be given an r.m.s. velocity dependent on their current cross-stream position, irrespective of axial position. Regions with radially varying turbulence intensity can thus be tackled, providing a way of modelling boundary layers and wall effects.

Verification

During its development, STATEN has been tested against out-of-pile measurements in sodium and in water, at CEGB (Wey, Overton & Hughes, 1982), UKAEA and KfK (Dubuisson et al., 1987). Comparisons in sodium have been most successful when a detailed distribution of turbulence intensity was available.

The first experiment against which the code was tested was in pipe flow in sodium (Hughes & Overton, 1980) at a comparatively low Reynolds number ($Re \simeq 4 \times 10^4$) in which hot sodium was injected continuously along the central axis of a cooler flow. A second pipe flow experiment in sodium (Wey, Overton & Hughes, 1982) was designed to simulate reactor flow rates ($Re \simeq 10^6$) and necessitated the use of a transient injection method. STATEN predictions have also been compared with results from an expanding jet experiment (Wey, Hughes & Overton, 1981) in which a hot sodium jet emerged into a cooler pool. To model these experiments successfully, certain assumptions were made for the turbulence parameters. However, agreement was particularly good for the expanding jet experiment for which turbulence data were obtained from Laser-Doppler measurements in an analogous water experiment.

An essential step in modelling the flow between the pin bundle and the subassembly outlet is the accommodation of the turbulent structure downstream of the pin bundle, where subchannels merge resulting in complex flow patterns. In order to study such conditions, experiments have been carried out at KfK (FRG) and at UKAEA Risley. These experiments have been modelled with STATEN to justify further its applicability to a subassembly geometry.

The KfK work (Weinkoetz, Krebs & Martin, 1982) comprised analogous sodium and water experiments ('TEFLU'). For these, turbulence intensity and macroscale values were available, along with wide-band temperature noise measurements. Detailed cross-stream temperature profiles were available at 20cm and at 70cm downstream of the 'jet block' which simulated the pin bundle, so STATEN was initially applied to that section of the flow. Measured axial turbulence intensity and macroscale data were averaged over the section and scaled by one half to give an estimate of the required radial parameters. This led to the use of the values of 0.013 for turbulence intensity ratio and 2.6mm for radial macroscale.

Since the heat transfer mesh size (see section 2.1) has to be of the order of the turbulence macroscale (and preferably much smaller), the number of nodes and hence the number of particles was significantly larger for the TEFLU simulation than for the larger macroscale pipe/jet flow experiments. In order to reduce the computing time and storage required, a 2D version of STATEN was developed (axial direction plus radial direction), taking advantage of the azimuthal symmetry of the temperature profiles in TEFLU.

THE DEVELOPMENT & TESTING OF DETECTION ALGORITHMS

Early detection of cooling anomalies has important safety and operational advantages: it reduces the need to understand the subsequent development of the fault, and it prevents significant fuel cladding failures which could lead to reduced plant availability. The main difficulty, which is common to all early diagnostic techniques, is that parameter changes generated by incipient fault conditions can often be masked by normal plant variations. Because of this, it is likely that decision-making will be based on some form of pattern recognition method, which can combine the multiple characteristics of the system into a simple indicator and learn normal and fault states over the range of operating conditions.

The initial use of STATEN was to generate a benchmark dataset to allow the performance of the temperature noise technique to be assessed and to enable the comparison of different detection methods. Such a dataset is

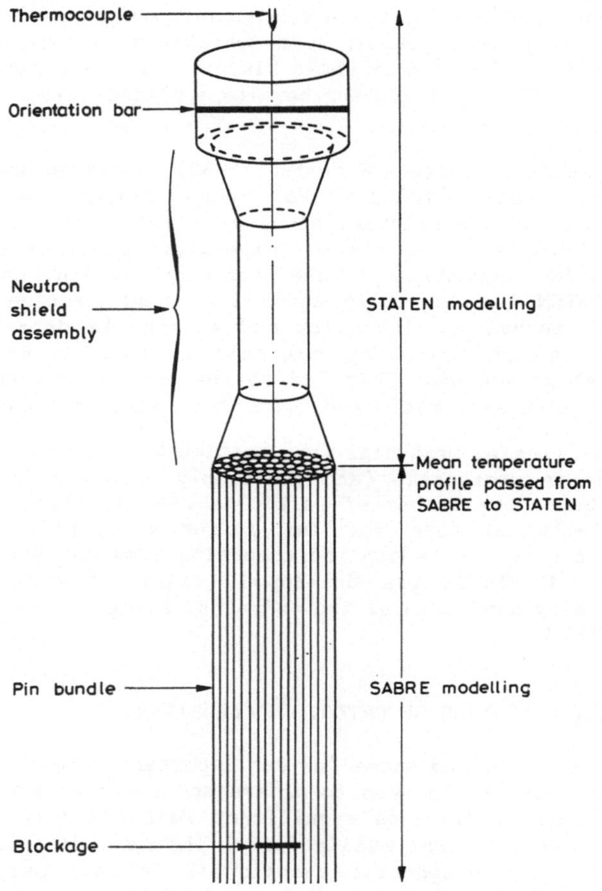

Figure 6. SABRE and STATEN Modelling Domains for a CDFR Subassembly

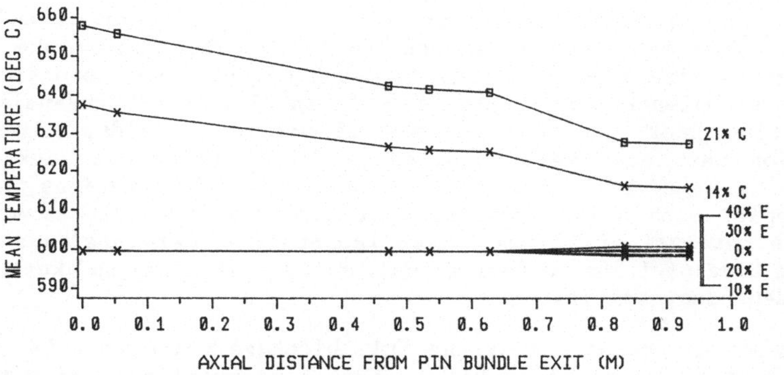

Figure 7a. STATEN Predictions of Axial Development of Mean Temperature at the Centre Line

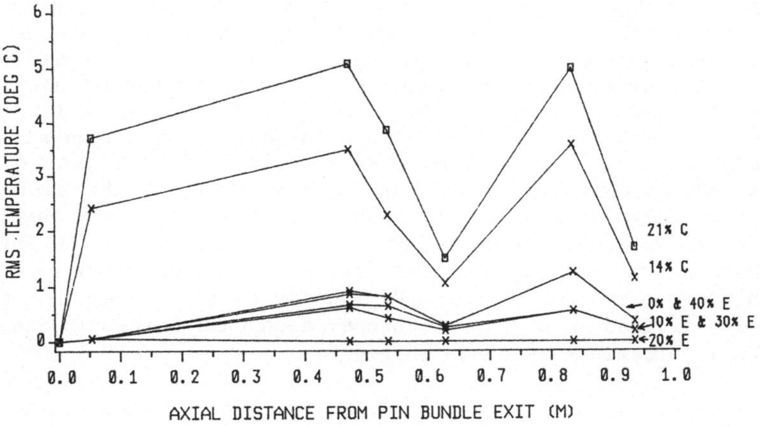

Figure 7b. STATEN Predictions of Axial Development of RMS Temperature at the Centre Line

seen as a potentially important aid to train reactor systems to recognise fault states before their actual occurrence. The generation of the dataset was based on a model of the CDFR (UK) subassembly (Hughes, Overton & Wey, 1986) using the calculational route outlined in Fig 6. The mean temperature profile at the outlet of the pin bundle is a function of the local blockage, the radial power profile and the normally cooler edge channels. This profile is mixed by the turbulence field downstream of the pin bundle. The mean profiles were estimated with the aid of the SABRE code (MacDougall, 1986) for blockage sizes of 1% to 40% at three different positions across the subassembly (left, centre, right), and under different cross-subassembly temperature tilts. The temperature signals on the centre line were predicted for each of the input conditions. The different turbulence conditions, from the small interpin channels to the large outlet duct and the presence of neutron shields, mixers, and orientation bars etc. were modelled by STATEN using six linked axial stages, to generate real-time temperature fluctuations at the thermocouple. Each signal consisted of discrete temperatures separated by a time interval of 2.5 ms. These temperatures were analysed over an averaging time of 1.28s to derive mean, r.m.s., skewness and kurtosis for eight independent samples.

A summary of the results over the full blockage range to 40%E is shown in Figs 7a and 7b, and in Table 1. The 'E' symbolises a ring blockage around the edge of the assembly which reduces edge channel cooling and provides a difficult detection requirement. The results clearly show the benefit of temperature noise detection compared with mean temperature monitoring.

For each blockage size, location and tilt the eight realisations of the vector (mean, r.m.s., skewness, kurtosis) were used to test two different pattern recognition techniques (Dubuisson et al.,1986). These were an Adaptive Learning Network (ALN) technique previously used for ultrasonic inspection, and a development of a Cluster Analysis (CA) method. The study concentrated on the incipient blockage (1% - 6%) stages and the ALN outputs for the different blockages are illustrated in Fig 8. A detection failure probability was determined from the separation of the assumed Gaussian distribution of ALN outputs from the distribution obtained in the no-blockage state. For the CA technique, these probabilities were deduced by counting the number of points for a given blockage size which fell into the zero blocakge class. The limited amount of test data did not permit the estimation of the low probabilities in the tail of the distribution (large blockage size). The detection failure probabilities for both techniques are shown in Table 1. A second comparison involving the derivation of correlation coefficients to measure the overlaps between different output classes. To facilitate this procedure the ALN outputs were divided into discrete bands, each band being redefined as a class. This comparison is shown in Fig 9.

COMPARISON WITH MEASUREMENTS IN SUPERPHENIX

Despite the experimental and theoretical work outlined above, the use of temperature noise as a diagnostic technique has not been widely accepted to date because of the lack of long term reactor measurements using sensors with a sufficiently broad bandwidth. However, the commissioning

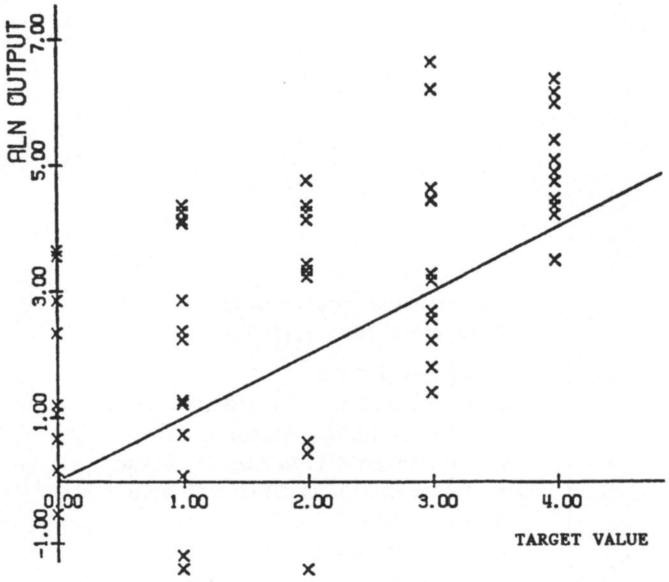

Figure 8. ALN Outpust 0-4% Blockage (STATEN I Data).

Table 1. Detection Failure Probabilities for Each Abnormal Condition Estimated for the ALN and CA Techniques.

% blockage	position	Detection failure probability	
		ALN	CA
1	L	0.57	0.79
1	R	0.39	0.25
2	L	0.36	0.42
2	C	8.8×10^{-2}	0.21
2	R	0.20	0.0
3	L	0.20	0.21
3	C	4.7×10^{-2}	4.2×10^{-2}
3	R	7.2×10^{-2}	0.0
4	L	0.14	4.2×10^{-2}
4	C	2.2×10^{-2}	8.3×10^{-2}
4	R	1.7×10^{-2}	0.0
5	L	6.5×10^{-2}	0.0
5	C	1.6×10^{-2}	0.0
5	R	2.1×10^{-2}	0.0
6	L	2.9×10^{-2}	0.0
6	C	3.2×10^{-3}	0.0
6	R	8.3×10^{-3}	0.0

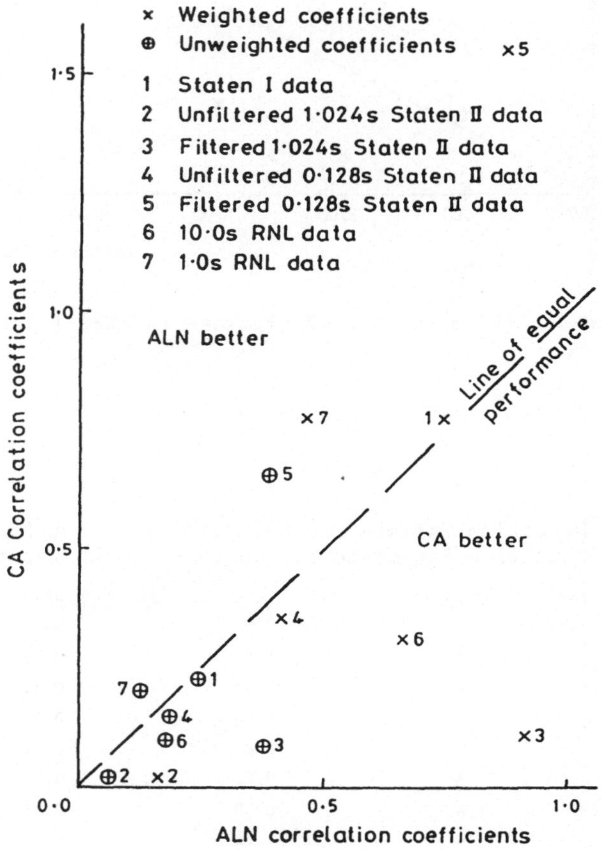

N.B. CA technique applied after analysis of data into principal components.

Figure 9. Comparison of ALN and CA Methods Using Averaged Correlation Coefficients

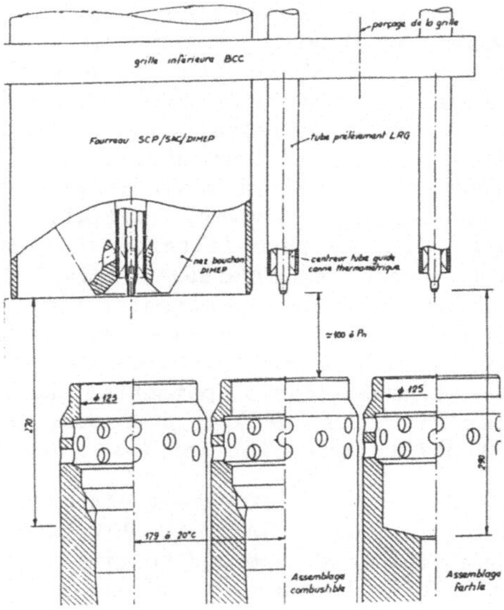

Figure 10. SPX 1 Above-Core Geometry

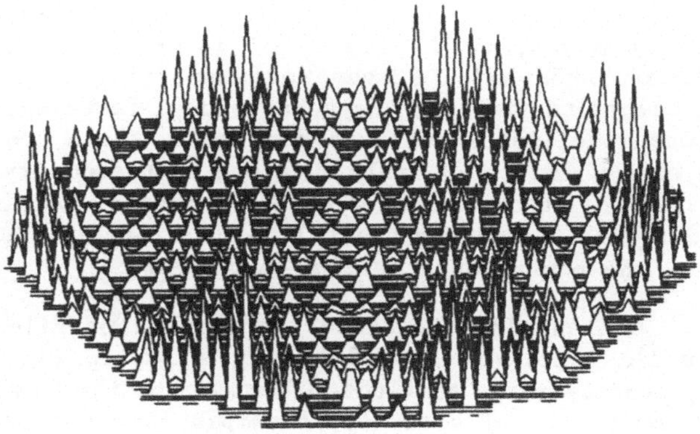

Figure 11. Superphenix Core Map - RMS at 80% Power

of the Superphenix-1 reactor at Creys-Malville, with its comprehensive array of fast response intrinsic thermocouples and associated data acquisition system, has provided an opportunity to extend the assessment of temperature noise, both from an experimental and a theoretical point of view. Under the European Fast Reactor Agreement, the UK (BNL and NRL, Risley) have been collaborating with the Federal Republic of Germany (KfK/IRB, Karlsruhe) and France (CEA, Cadarache) to obtain and analyse noise data, and to carry out theoretical modelling in support of the experimental programme. An initial summary report (Girard et al., 1987) has been presented to the SMORN-V meeting. Each of the core subassemblies in Superphenix, and each of the inner ring of breeder subassemblies (a total of 448) is monitored by two chromel/alumel thermocouples (bandwidth 0.2Hz) together with an intrinsic thermocouple of much greater bandwidth, potentially up to 100Hz (Fig 10 shows their above-core location). During the experimental programme, however, signals from the intrinsic thermocouples needed to be low-pass filtered at 10Hz by the ANABEL data acquisition system to remove extraneous voltages. The filtered signals were then recorded in analogue form on magnetic tapes, which were returned to the UK for analysis. During the experiment, further filtering of selected signals to below 0.2 Hz, followed by synchronisation to the corresponding chromel/alumel signals, permitted the time constants of the latter to be determined for safety purposes, and also enabled a calculation of the sensitivity of the intrinsic device.

Recordings were made for a number of steady state power levels (15%, 60%, 80%, 90% and 100%). In particular, at 80% power, signals for all instrumented subassemblies were recorded to give an idea of the core-wide variation of noise levels. Further recordings were made for a smaller number of subassemblies during power raising and selected control rod movement tests. The aim of this last set of tests was to induce a significant change in the radial temperature profile of the sodium emerging from the pin bundle, thus permitting an assessment to be made of the profile as a temperature noise generator. An example of the distribution of r.m.s. values across the core is shown in Fig 11 for 80% power.

Modelling the Superphenix Subassembly

For the purpose of modelling the region of interest of the Superphenix subassembly, five linked axial stages were used (Fig 12), with no heat flux modelling across the wall of the subassembly. Turbulence intensities based on previous modelling experience were used; these are presented as a ratio to the bulk velocity in Table 2, along with geometry and velocity details. As an initial study two cases were considered: namely subassembly 3230 at 80% power with the nearby control rod 1 at each of its two extreme positions during the experiment (z=228mm and z=0mm).

For these cases, CEA supplied detailed temperature profile calculations for the pin bundle exit, one example being shown in Fig 13. The near symmetry of the profile meant that the 2D version of the STATEN code could be used without much loss of accuracy. A sample length of 2048 x 1ms was generated by the code for the thermocouple location, and, as stated above, was low-pass filtered (10Hz, 1st order) before analysis for mean, r.m.s.,skewness and kurtosis. This ensured that the assessed frequency response of the measurement system was correctly incorporated into the

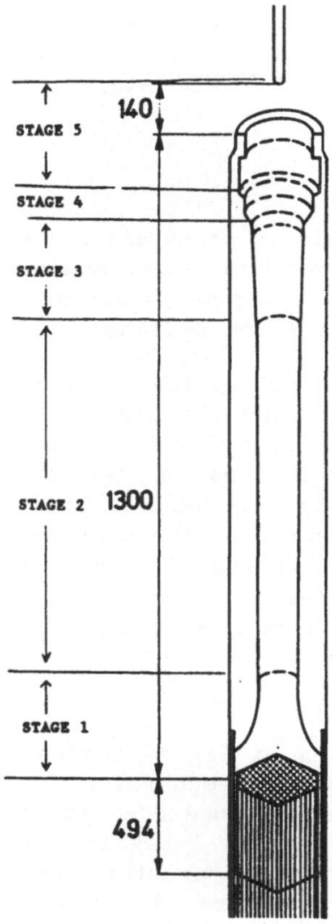

Figure 12. SPX 1 Geometry Modelled by STATEN

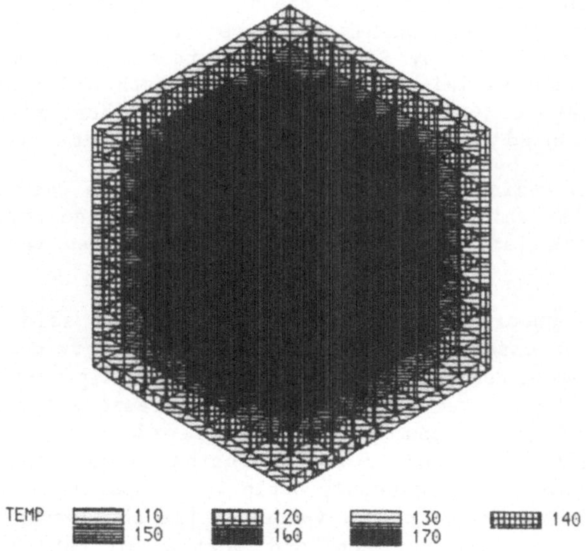

Figure 13. Pin Bundle Subchannel Exit Temperatures

model. In order to ascertain the spatial distribution of these parameters, simulated measurement positions were also placed across a diameter at the end of each of the five linked STATEN stages, and data were obtained at each position.

Results of the Simulation

The results for the two simulated cases are given in Table 3 together with the analysis of the corresponding experimental results. Since the two cases do not differ greatly, the results are presented graphically mainly for one case (control rod position z=228mm).

Fig 14 shows the calculated initial mean temperature rise profile at the pin bundle exit (cf Fig 12), and the prediction of mean temperature at the thermocouple plane. Figs 15, 16 and 17 show respectively the r.m.s., skewness and kurtosis predictions at the sensor plane derived from filtered and unfiltered signals. The analysed values from Superphenix measurements are also indicated, on the assumption that the thermocouple is located on the axis of the subassembly. Fig 18 compares the centre line axial development of filtered r.m.s. levels, and the Superphenix measurements, for each of the two control rod positions.

Discussion of Results

The mean temperature profiles of Fig 14 demonstrate the small degree of decay at the centre line (1.4°C) occurring during the transit time of 0.15s. This decay has, however, generated about 0.5°C r.m.s. (filtered) noise at the centre line (Fig 15). Significantly more is generated at the positions of maximum mean temperature gradient (up to 5°C r.m.s. filtered). The corresponding centre line value measured in Superphenix is 0.75°C r.m.s. From Fig 18 it can be seen that for the two different cases considered, the different levels of r.m.s. are echoed by the STATEN predictions. It should be noted, however, that there is a tolerance of several millimetres on the radial positioning of the thermocouple, and it can be seen that the Superphenix measurement is within the predictions at, say, ± 10mm. The Superphenix skewness value of -0.44 is nearer to zero than that predicted by STATEN, but kurtosis is closely predicted despite the erratic radial profile of this fourth order statistic. The lack of smoothness of the predicted skewness and kurtosis profiles may in fact be attributed to the limited length of the simulated data.

The variation in radial thermocouple location may be seen as a possible explanation of the variations of parameters across the core seen in Fig 11 except for the large and systematic change across the core/breeder boundary.

Considering the uncertainties involved, it may be said that, in the limited number of cases predicted so far, STATEN provides reassuringly close agreement with reactor measurements. However, it must be remembered that a great deal more verification will be necessary for subassemblies at different core positions and powers in order to ensure that the modelling is sufficient, and to assess whether unmodelled aspects such as cross-flow above the subassembly outlet are important. The predicted radial profiles of mean and r.m.s. values at the thermocouple plane could have important implications and are worthy of future experimental verification.

Table 2 . Geometry/Flow Details for the STATEN SuperPhenix Simulation

STAGE	STAGE LENGTH (mm)	INITIAL DIAMETER (mm)	FINAL DIAMETER (mm)	AV BULK VELOCITY (m/s)	TURBULENCE INTENSITY RATIO	RADIAL MACROSCALE (mm)
1	263	172	70	9.86	0.10	1.5
2	568	70	70	12.68	0.05	2.9
3	168	70	92	9.64	0.04	3.6
4	75	92	92	7.34	0.04	3.8
5	255	92	125	5.40	0.04	4.1

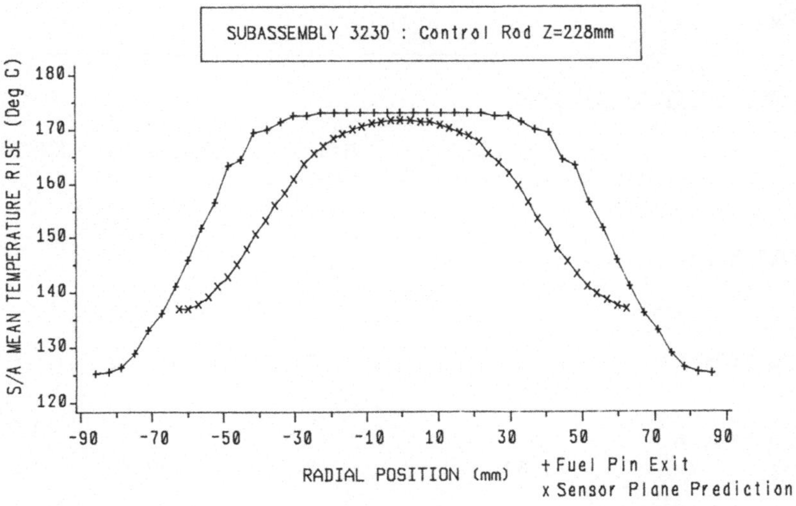

Figure 14. Predicted Mean Temperature Development

Table 3 . STATEN Predictions & Superphenix Analysis of the Temperature Distribution at the Thermocouple for s/a 3230 at 80% Power

CONTROL ROD POSITION Z=0MM

	MEAN TEMP RISE FROM INLET (DEG C)	RMS TEMP (DEG C)	SKEWNESS	KURTOSIS
PIN BUNDLE EXIT	154.90	-	-	-
UNFILTERED PREDICTED SIGNAL	153.81	0.78	-1.49	5.83
PREDICTED SIGNAL (1 - 10Hz)	153.81	0.39	-1.11	4.56
SUPERPHENIX ANALYSIS (1 - 10Hz)	-	0.66	-0.44	3.84

CONTROL ROD POSITION Z=228MM

	MEAN TEMP RISE FROM INLET (DEG C)	RMS TEMP (DEG C)	SKEWNESS	KURTOSIS
PIN BUNDLE EXIT	173.0	-	-	-
UNFILTERED PREDICTED SIGNAL	171.63	1.13	-1.78	7.82
PREDICTED SIGNAL (1 - 10Hz)	171.63	0.56	-0.82	4.16
SUPERPHENIX ANALYSIS (1 - 10Hz)	-	0.74	-0.39	4.03

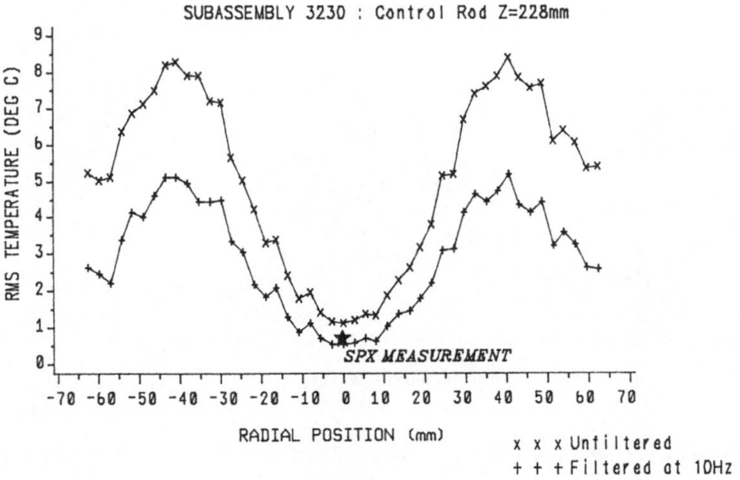

Figure 15. Predicted RMS Temperatures at the Termocouple Plane

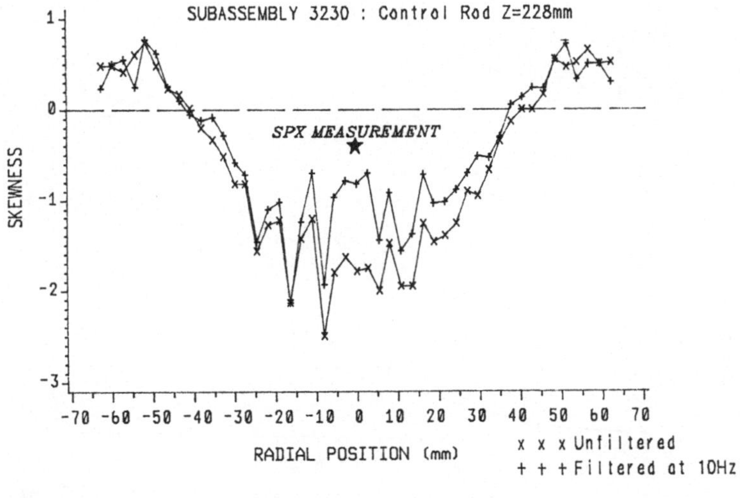

Figure 16. Predicted Temperature Skewness at the Thermocouple

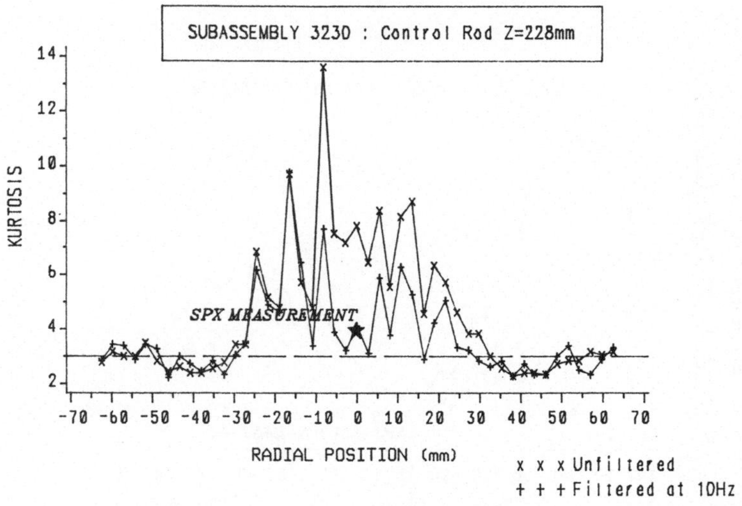

Figure 17. Predicted Temperature Kurtosis at the Thermocouple

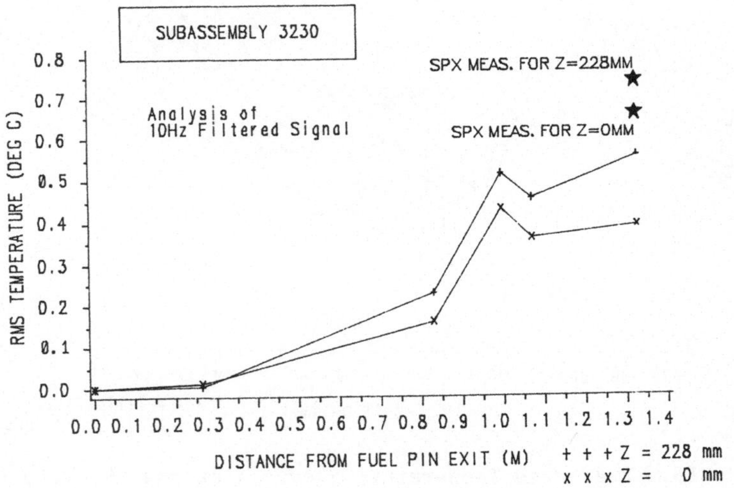

Figure 18. Predicted RMS Temperatures for Two Rod Positions

Extension to a Hypothetical Blocked Subassembly

On the premise that STATEN adequately modelled the two cases considered, it was decided to examine the effect on the predicted signal of a blockage occurring in the pin bundle. To comply with the requirements of the 2D model, the blockage had to be assumed to be located on the axis of the subassembly. Furthermore, it was assumed to be of a size and axial location that would give rise to a peak temperature, on the axis at the pin bundle exit, of 50°C greater than for the unblocked case. Also, the temperature disturbance was taken to extend over ± 15mm from the axis, at the pin bundle exit. The resulting profile at that plane, is admittedly arbitrary, but is used merely to illustrate the effect of a central blockage on the noise level at the thermocouple compared with that resulting from normal operation at 80% power (viz the z=0mm case considered previously).

It was assumed that the presence of a blockage would not alter the flow conditions downstream of the pin bundle; accordingly, the turbulence conditions and flow velocities remained as in Table 2. As before, a sample length of 2048 x 1ms was generated, filtered and analysed for each of a large number of radial positions at the end of each modelling stage.

A comparison of the predicted profiles of filtered r.m.s. temperature at the thermocouple plane for the unblocked and blocked cases is given in Fig 19. It can be seen that the centre line r.m.s. temperature undergoes a sevenfold increase because of the blockage, from 0.4°C to 2.8°C. Because of the small scale turbulence, the effect of the blockage is to increase r.m.s. temperatures only in a region about the centre line. The thermocouple, if positioned away from the centre line, would measure a less pronounced increase in r.m.s. level. Similarly, a less pronounced increase would be expected if there were a temperature tilt present (see Hughes, Wey & Overton, 1982). Finally, the effect of the blockage is to make the negative skewness values generally closer to zero (see Fig 20).

On the assumption that the modelling, as described earlier, is adequate, an immediate conclusion can be drawn: namely that a fairly small central blockage occurring at high power/flow conditions in a subassembly well away from the core/breeder boundary should be detectable by r.m.s. temperature monitoring. In this respect, r.m.s. temperature is a much more sensitive indicator of a blockage than mean temperature changes which may be subject to low frequency variation.

The case of off-central or corner blockages needs 3D STATEN modelling, after suitable verification against Superphenix measurements. It may nevertheless be stated that, because of the small scale turbulence, the perturbation of the mean temperature profile due to the blockage may not be transported radially as far as the thermocouple. On the other hand, the perturbation of the r.m.s. and skewness profiles is propagated radially to a greater extent than mean temperature changes (Hughes, Wey & Overton, 1982), and these parameters may well have more potential as a blockage indicator.

CONCLUSIONS

A modelling code based on Monte Carlo techniques has been developed to study the generation and dissipation of temperature fluctuations in flowing sodium. It has been tested against experimental data covering the range of turbulence intensities and macroscales found in the axially changing turbulence fields associated with the outlet regions of the present designs of LMFBR fuel subassemblies.

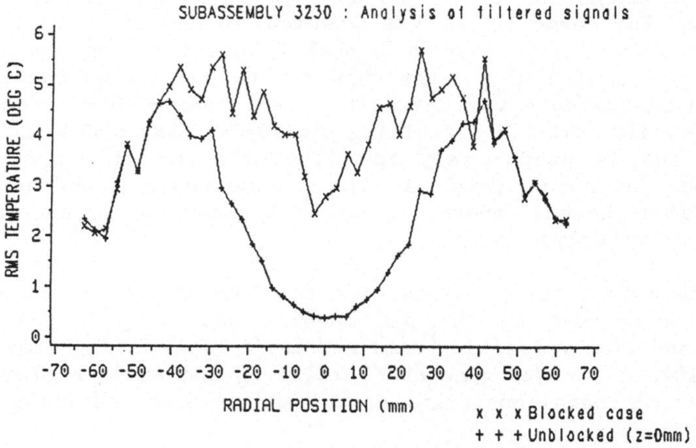

Figure 19. STATEN Prediction of the RMS Temperature Profile at Thermocouple Plane for an Hypothetical Blocked Subassembly.

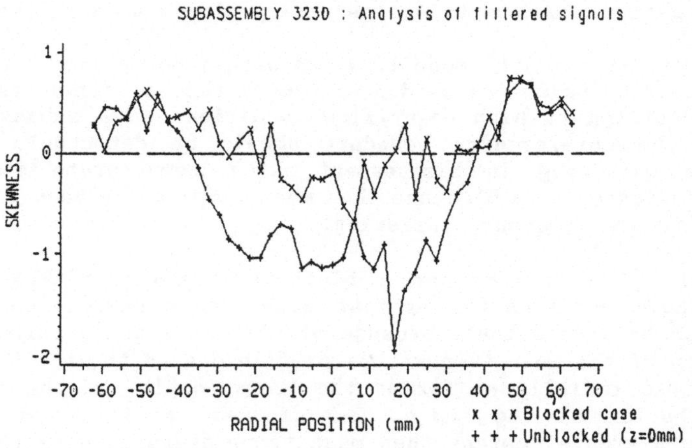

Figure 20. STATEN Prediction of the Temperature Skewness Profile at Thermocouple Plane for an Hypothetical Blocked Subassembly.

The code has been used to study the variation of temperature noise signals with the formation of blockages in the fuel bundle. It has demonstrated its usefulness by simulating data to test the sophisticated detection methods that are likely to be used in a reactor environment to distinguish fault conditions from normal plant variations. The provision of such test data is seen as an essential requirement to train the detection systems to recognise fault states before their actual occurrence in the reactor.

Recently, the code has been used for initial modelling of the high frequency temperature noise signals obtained from intrinsic sodium/steel thermocouples on the Superphenix-1 reactor during its commissioning. The results compare well with the measured values, given the sources of uncertainty in both modelling and measurement. The predictions indicate a number of important aspects which require future study, but in the meantime a speculative extension of the modelling to a blocked Superphenix subassembly has indicated the value of temperature noise monitoring as a means of blockage detection.

ACKNOWLEDGEMENTS

This paper is published by permission of the Central Electricity Generating Board.

REFERENCES

Dubuisson B., Richard P., Krebs L., Weinkoetz G., Girard J.P., Firth D., Conroy P.J., Wey B.O., Overton R.S., Hughes G. & Diamantidis Z. (1986) 'Recent European Collaborative Work on Temperature Noise Monitoring and the Generation of Benchmark Tests'. Paper 569 at Conference on Science & Technology of Fast Reactor Safety, Guernsey.

Dubuisson B., Richard P., Overton R.S., Wey B.O. & Hughes G. (1987) 'The Diagnostic Potential of Pattern Recognition Methods applied to Temperature Noise Data from Simulated Fast Reactor Fuel Subassemblies'. European Appl.Res.Rept - Nucl Sci Tech Vol 7(6) P929

Firth D. (1977) Progress in Nucl Energy. 1. p527.

Firth D. & Conroy P. (1986) 'Basic Studies of Temperature Noise in a Simulated Fast Reactor Subassembly'. Paper 584 at Conference on Science & Technology of Fast Reactor Safety, Guernsey.

Girard J.P. & Buravand Y. (1982) 'Temperature Field Downstream of a Heated Bundle Mock-up: Results for Different Power Distributions' Paper presented at the 10th LMBWG Meeting, Karlsruhe.

Girard J.P., Recroix H., Beesley M., Weinkoetz G., Krebs L., Overton R.S. & Hughes G. (1987) 'Detection of Coolant Temperature Noise in SPX-1 Using Intrinsic High Frequency Thermocouples'. Paper presented at the fifth Specialists' Meeting on Reactor Noise (SMORN-V). Munchen.

Hinze J.O. (1976) Turbulence (2nd Edn). McGraw Hill.

Hughes G. & Overton R.S. (1980) CEGB Report RD/B/N4933

Hughes G., Overton R.S. & Wey B.O. (1986) 'Modelling of Subassembly Outlet Region Flow and Temperature Distributions'. Paper 570 at the Conference on Science & Technology of FR Safety. Guernsey.

Hughes G., Wey B.O. & Overton R.S. (1982) 'A Simulation of Fluctuating Temperature Signals at the Outlet of a Blocked Subassembly'. Paper presented at the 10th Liquid Metal Boiling Working Group, Karlsruhe.

Krebs L. & Bremhorst K. (1983) 'Verification of Extended Gradient Diffusion Model by Measurements of the Mean and Fluctuating Temperature Fields in a Sodium Flow Downstream of a Multi-bore Jet Block'. Proceedings of the Fourth Symposium on Turbulent Shear Flows, Karlsruhe.

Lawn C.J. (1977) Int. J. Heat Mass Transfer. 20. pp1035-1044.

MacDougall J.D. (1986) 'Thermohydraulic Calculations for Fast Reactor Fuel Elements using the SABRE Code'. Paper 559 at the Conference on Science & Technology of FR Safety. Guernsey.

Ohlmer E. & Schwalm J. (1975) Int. J. Heat Mass Transfer. 19. p 765.

Overton R.S., Wey B.O. & Hughes G. (1982) 'A Multiparticle Monte Carlo Simulation of Temperature Noise and Heat Transfer in Turbulent Flow'. Ann. Nucl. Energy. 9. P 297

Overton R.S., Wey B.O. & Hughes G. (1984) 'The Temperature Noise Simulation Code STATEN and its Experimental Verification'. Paper presented at an IAEA/IWGFR Specialists' Meeting on Methods and Tools to Detect Thermal Noise in Fast Reactors. Bologna.

Weinkoetz G., Krebs L. & Martin H. (1982) 'Measurement and Analysis of Temperature Fluctuations at the Outlet of an Electrically Heated 28 Rod Bundle without and with Flow Blockage'. Paper presented at the 10th Liquid Metal Boiling Working Group Meeting, Karlsruhe.

Wey B.O., Hughes G. & Overton R.S. (1981) 'Modelling of Temperature Noise in a Divergent Jet' CEGB Report RD/B/5033N81

Wey B.O., Overton R.S. & Hughes G. (1982) 'The Decay of Temperature Fluctuations in Sodium Pipe Flow at High Reynolds Numbers' CEGB Report TPRD/B/0029/N82

PUMP INDUCED LINEAR AND NON LINEAR VIBRATIONS AT P.W.R

R. Sanchis Arnal,
G. Verdú Martín,
and J.L. Muñoz-Cobo

Universidad Politécnica de Valencia
Departamento Ingeniería Química y Nuclear PO. Box 22012
46071 Valencia

1. INTRODUCTION

The fluctuations about the average in detector signals from a nuclear power plant contain information, about their origin and about the dynamic transmission properties of the reactor. Such fluctuations can be represented by noise descriptors, such as power spectral densities (PSD) and cross power spectral densities (CPSD), which display features, peaks and valleys, that are related to a specific causative mechanism such as fuel-assembly vibrations, core-barrel motion, thermal shield vibrations, pump vibrations, reactivity feedback effects and so on [1].

Malfunctions of structural components in nuclear power plants can produce linear and nonlinear vibrations which induce changes in the signatures PSD and CPSD [2]. Therefore, it is very important to know the existing correspondence between the noise-source profiles and those of neutron noise. These kind of studies have great interest in pattern recognition of malfunctions and anomalies by neutron-noise analysis methods.

This paper studies the local vibrations of reactor components driven by Gaussian random and pump induced forces, when nonlinear vibrations arise. Also we study the problem of fitting neutron-noise experimental data to perform pattern recognition analysis.

2. LINEAR AND NONLINEAR VIBRATIONS DRIVEN BY SINUSOIDAL AND RANDOM FORCES

To calculate the spectrum, x(w), and the power spectral density, PSD(w), of the displacement, x, of given reactor internal, under coloured Gaussian random and sinusoidal deterministic forces, one must solve the system of coupled equations (1) and (2), where we have assumed that U(x) has the polynomial form and $F_b(t)$ is a white Gaussian noise.

$$\ddot{x}(t) + c.\dot{x}(t) + U(x) = F(t) + B.\cos(\,w_{pt}+\sigma) \qquad (1)$$

$$\dot{F}(t) + \tau^{-1}\,.F(t) = \tau^{-1}\,.F_b(t) \qquad (2)$$

$$U(x) = \sum_{i=1}^{n} a_i \cdot x^i \tag{3}$$

The numerical simulation of a stochastic differential equation is quite different from that of an ordinary differential one [3]. What one does, is to simulate a particular trajectory which is consistent with the stochastic equation. In obtaining a particular trajectory, a set of pseudo-random numbers is used. To obtain the statistical descriptors one must compute the average over many of these simulated trajectories.

The first step is to integrate equation (2) from t to t+Δ, where Δ is the integration step. In this way it is obtained:

$$F(t+\Delta) = F(t) - \tau^{-1} \int_t^{t+\Delta} F(t').dt' + \tau^{-1} \int_t^{t+\Delta} F_b(t').dt' \tag{4a}$$

from equation (4a) we compute F(t'), which is given for $t \le t' \le t+\Delta$, and the direct substitution into equation (4a) leads to:

$$F(t+\Delta) = F(t)\ (1-\Delta/\tau) + \tau^{-2} \int_t^{t+\Delta} dt' \int_t^{t'} dt''.F(t'') - \tag{4b}$$

$$- \tau^{-2} \int_t^{t+\Delta} dt' \int_t^{t'} dt''.F(t'') + \tau^{-1} \int_t^{t+\Delta} F_b(t').dt'$$

This process can continue up to the desired order of approximation. The $0(\Delta^{3/2})$ approximation, where $\Delta^{3/2}$ terms have been neglected, is obtained by setting F(t')=F(t) in equation (4a). In the same way the $0(\Delta^{5/2})$ approximation can be obtained setting F(t")=F(t) in equation (4b). The results for both approximations are:

$$F(t+\Delta) = F(t).(1-\Delta/\tau) + \tau^{-1}.X_1(t)+0(\Delta^{3/2}) \tag{5}$$

$$F(t+\Delta) = F(t).[1-\Delta/\tau+1/2(\Delta/\tau)^2]+\tau^{-1}.X_1(t)-\tau^{-2}.X_2(t)+0(\Delta^{5/2})$$

Where $X_1(t)$ and $X_2(t)$ are linear transformations of Gaussian random numbers given by:

$$X_1(t) = \int_t^{t+\Delta} dt'.F_b(t') \tag{6}$$

$$X_2(t) = \int_t^{t+\Delta} dt' \int_t^{t'} dt''.F_b(t'')$$

Therefore, it follows that $X_1(t)$ and $X_2(t)$ are also Gaussian random numbers of zero mean as $F_b(t)$.

Next we write equation (1) in the form:

$$[d\dot{x}(t)/dt] = -c.\dot{x}(t) - U(x) + F(t) + B.\cos(w_p t+\sigma) \qquad (7)$$

$$[dx(t)/dt] = \dot{x}(t) \qquad (8)$$

Integrating both equations from t to t+Δ, it is obtained that:

$$\dot{x}(t+\Delta) = \dot{x}(t).[1-(c\Delta)+(c\Delta)^2/2]+c.\sum_i a_i.x(t)^i.\Delta^2/2-\sum_i a_i[x(t)^{i-1}. \qquad (9)$$

$$.\dot{x}(t).\Delta^2/2+x(t)^i.\Delta] + B.F1(t)-c.B.F2(t) + F(t)[\Delta-\tau^{-1}.\Delta^2/2]+$$

$$+ \tau^{-1}.X_1(t)-c.F(t).\Delta^2/2+ O(\Delta^{5/2})$$

$$x(t+\Delta) = x(t) + \dot{x}(t).\Delta -c.\dot{x}(t).\Delta^2/2-\sum_i a_i.x(t)^i.\Delta^2/2 + \qquad (10)$$

$$+ F(t).\Delta^2/2+ B.F2(t)+ O(\Delta^{5/2})$$

where the sinusoidal terms are:

$$F1(t) = \int_t^{t+\Delta} dt'.[\cos(w_p t'+\sigma)]= \qquad (11)$$

$$= 1/w_p\{sen(w_p t+\sigma)[\cos(\Delta w_p)-1]+\cos(w_p t+\sigma).sen(\Delta w)\} =$$

$$= \{-sen(w_p t+\sigma).(\Delta w)^2/2+ \cos(w_p t+\sigma)[\Delta w-(\Delta w)^3/6]\}\ 1/w_p + O(\Delta^{5/2})$$

$$F2(t) = \int_t^{t+\Delta} dt' \int_t^{t'} dt''.[\cos(w_p t''+\sigma] = \qquad (12)$$

$$= 1/w^2{}_p\{\cos(w_p t+\sigma)[1-\cos(w_p\Delta)]+ sen(w_p t+\sigma)sen(\Delta w_p)\}-$$

$$- (\Delta w_p/w^2{}_p).sen(w_p t.\sigma)=$$

$$= \{\cos(w_p t+\sigma)(w\Delta)^2/2 - sen(w_p t+\sigma)(w_p\Delta)^3/6\}.1/w^2{}_p + O(\Delta^{5/2})$$

In obtaining equations (6), (9) and (10), we neglected the terms of order equal or greater than $\Delta^{5/2}$. This set of equations form the basis to simulate the trajectories. The computer code called NOLPOS [4], has been developed to perform the numerical simulation.

Each particular trajectory consists of a set of values of x(1+KΔ). x(1+KΔ) and F(1+KΔ) for K=1,2..,N) consistent with the stochastic equations, which correspond to a particular realization of the stochastic term, i.e. to a given sequence of random numbers. The statistical descriptors in which one is interested are then obtained by taking the average over the set of simulated trajectories.

In order to compute the trajectories, one must choose some initial boundary conditions, but if at a given time we fix some boundary conditions, then several points at the beginning of each particular trajectory are dependent on the intial conditions.

Because in the stationary ergodic state the statistical descriptors are independents on the initial conditions, we then need to know when the stationarity has been reached. So, the statistical descriptors can be calculated from the trajectories neglecting all points below the stationarity time, t_e. To compute t_e we have applied the cumulant neglect closure technique [5].

Finally we extend the numerical results to the frequency domain. Because we deal with stationary ergodic systems, instead of calculating the displacement power spectral density as the Fourier transform of the autocorrelation function, it is better from a numerical point of view, to compute the displacement PSD, by means of the formula:

$$PSD(w) = E[x(w).x^*(w)] \tag{13}$$

where x(w) is the Fourier transform of x(t), computed over one trajectory, and $x^*(w)$ is the complex conjugate of x(w), and E[ ] is the expectation operator. The Fourier transform were calculated using the FFT (fast Fourier transform) algorithm.

3. HARMONIC OSCILLATIONS WITH UNSYMMETRICAL NONLINEAR CHARACTERISTIC

We consider the following case:

$$\ddot{x}(t) + k.\dot{x}(t) + a_1.x + a_2.x^2 + a_3.x^3 = b.\cos wt \tag{14}$$

This equation is readily transformed to alternative form [6]:

$$\ddot{x}(t) + K\,\dot{x}(t) + c_1.x + c_3.x^3 = B.\ \cos Wt + B_o \tag{15}$$

where:

$$c_1 + c_3 = 1 \tag{16}$$

$$c_1 = (a_1 - a^2{}_2/3a_3)/(a_1 - a^2{}_2/3a_3 + a_3)$$

$$c_3 = a_3/(a_1 - a^2{}_2/3a_3 + a_3)$$

$$K = k/(a_1 - a^2{}_2/3a_3 + a_3)^{1/2}$$

$$B = b/(a_1 - a^2{}_2/3a_3 + a_3)$$

$$B_o = (a_1a_2/3a_3 - 2.a^3{}_2/27.a^2{}_3)/(a_1 - a^2{}_2/3a_3)$$

$$W = w/(a_1 - a^2{}_2/3a_3 + a_3)^{1/2}$$

So, we see that the nonlinear characteristic is symmetrical, but the external force is unsymmetrical, since it contains the unidirectional component B_o.

In the case of harmonic oscillations in which the fundamental component having the period $(2\pi/w)$ predominates over the higher harmonics, the periodic solution of equation (15) way be assumed by the form:

$$X(t) = z + m.sen(Wt) - n.cos(Wt) \tag{17}$$

Substituting equations (17) into (15) and equating the constant term and the coefficients of the terms containing sin(Wt) and cos(Wt) separately to zero yields:

$$K\, W\, n + A\, m = 0 \tag{18}$$

$$A = W^2 - (3/4)\, c_3\, r^2 - c_1 - 3\, z^2\, c_3$$

$$K\, W\cdot m - A\, n = B$$

$$c_3\, z^3 + (3/2)\, z.r^2.c_3 = B_o$$

or eliminating m and n from the equations (18) yields:

$$K^2\, W^2\, r^2 + A^2\, r^2 = B^2 \tag{19}$$

Stability condition of the periodic solutions

The periodic states of equilibrium determined by equations (18) and (19) are not always realized, but are actually able to exist only so long as they are stable. We shall investigate the stability of the equilibrium states and find the periodic solutions which are sustained in the stable state. To do this, we consider the small variation ξ from the equilibrium states and substitute $(x_0+\xi)$ in place of x in equation (15). We obtained the variational equation:

$$d^2\xi/dt^2 + K\, d\xi/dt + c_1.\xi + 3.X^2.c_3.\xi = 0 \tag{20}$$

By use of the transformation:

$$\xi = \exp\,(-Kt\eta/2) \tag{21}$$

and substituting the periodic solution (17), we ultimately obtain the Mathieu's equation:

$$d^2\eta/dt^2 + [\sigma_0 + 2 \sum_i \sigma_i.\cos(iwt - \epsilon_i)].\eta = 0 \tag{22}$$

where: $\sigma_0 = c_1 - (1/4)\, K^2 + 3\, c_3\, (z^2 + r^2/2)$

$$\sigma_1 = 3\, c_3\, r\, z$$

$$\sigma_2 = 3\, c_3\, r^2/4$$

$$\epsilon_1 = \text{arc.tg}\, (x/y)$$

$$\epsilon_2 = \text{arc.tg}\, (2xy/y^2-x^2)$$

We shall particulary investigate the stability condition for the unstable regions in order to test the stability against buildup of an unstable oscillation having a frequency the same as that of the periodic solution. Thus, we write:

$$[\sigma_0-(n/2)^2.W^2]^2+ 2[\sigma_0+(n/2)^2W^2](K^2/4)+ (K^4/16) > \sigma^2_n \qquad n=1,2 \tag{23}$$

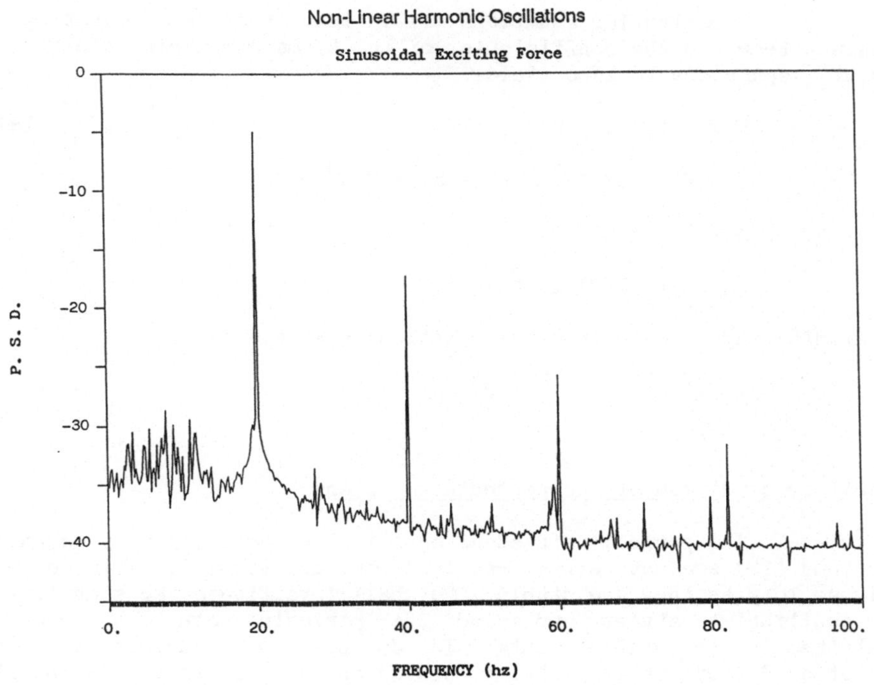

Figure 1

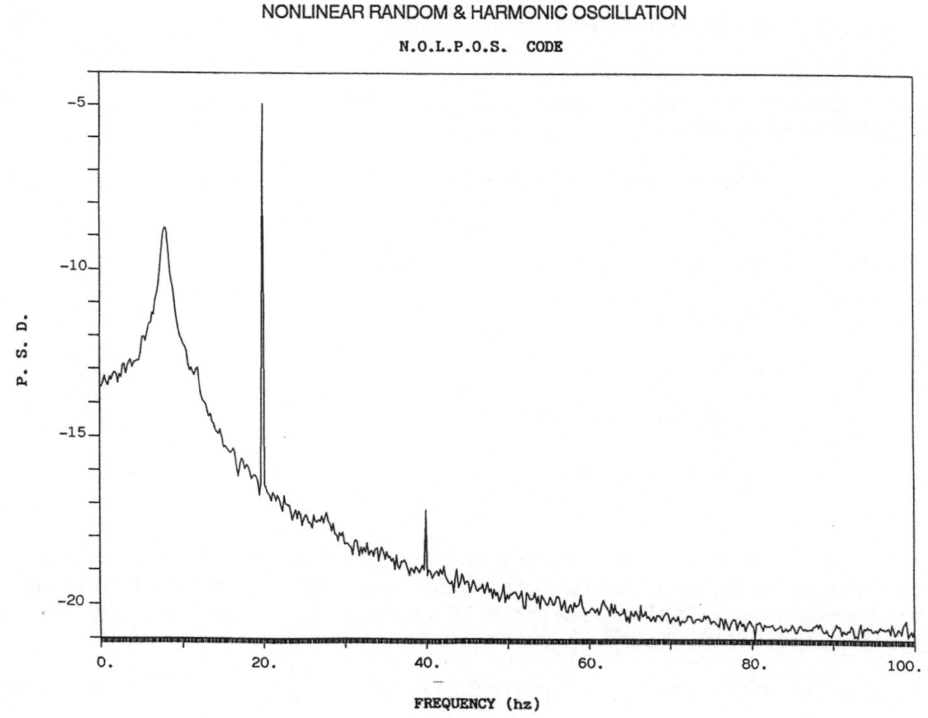

Figure 2

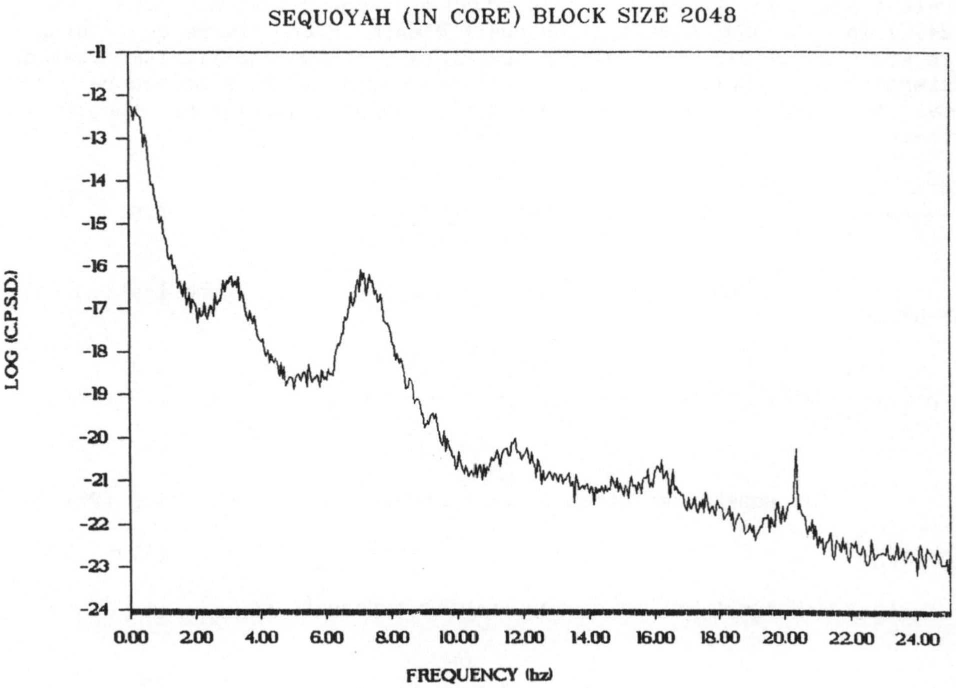
SEQUOYAH (IN CORE) BLOCK SIZE 2048
-11
-12
-13
-14
-15
-16
-17
-18
-19
-20
-21
-22
-23
-24
LOG (C.P.S.D.)
0.00
2.00
4.00
6.00
8.00
10.00
12.00
14.00
16.00
18.00
20.00
22.00
24.00
FREQUENCY (hz)

Figure 3

Higher-harmonic oscillation

In the preceding section the stability of harmonic oscillations was investigated by making use of variational equations of the Hill type. But when the amplitude B of the external force is very large, oscillations under this condition are worthy of consideration, since anomalous excitation of higher-harmonic components results if the above stability condition is not satisfied.

In the case in which B is large, a periodic solution for equation (15) may preferably be chosen as:

$$x(t) = z + m \text{ sen } (wt) + n \cos (wt) + m_3 \text{ sen } (3wt) + n_3 k \cos(3wt) + m_2 \text{ sen}(2wt) + n_2 \text{ sen}(2wt) \qquad (24)$$

Terms of harmonics higher than the third are certain to be present but are ignored to this order of approximation. Substituting (24)) in (15) and equating the coefficients of the terms containing sin wt, cos wt, sin 3wt, cos 3wt, separately of the equilibrium states, determine the stability of the equilibrium states, by a procedure like that used before. So we obtain the varitional equation of the Hill, type:

$$\frac{d^2\eta}{dt^2} + [\sigma_0 + 2 \sum_{n=1}^{6} \sigma_n \cos (nwt - \epsilon_n)] \eta = 0 \qquad (25)$$

and the stability condition (to a first approximation) is obtained:

$$(\sigma_0 - (\frac{n}{2} w)^2)^2 + (\sigma_0 + \frac{n}{2} w)^2 \frac{k^2}{2} + \frac{k^4}{16} > \sigma^2_n \qquad n = 1, 2, 3, 4. \qquad (26)$$

The general solution to extended form Hill's equation (26) is:

$$\eta = C_1 e^{\mu t} \phi (\tau, g_1) + C_2 e^{-\mu t} \phi (\tau, g_2) \qquad (27)$$

Where μ is the characteristic exponent, so

$$\mu^2 = - (\sigma_0 + \frac{nw}{2} + \sqrt{} 4 \frac{mw}{2} \sigma_0 + \sigma^2_n \qquad (28)$$

and

$$\phi (\tau, g) \approx \sin (\frac{nw}{2} \tau - g)$$

C_1, and C_2 are arbitrary constants, and the parameters g_1, g_2, will be determined from the expansions of σ_0.

Consequently, by virtue of the stability condition a periodic solution becomes unstable if the point (σ_o, σ_n) lies in the nth unstable region, furthermore, if (σ_0, σ_1) lies in the first unstable region an unstable oscillation of subharmonics builds up in this region. To summarize the above considerations, we say that if the amplitude B of the external force is very large and the nonlinearities are important, the periodic solution is featured by the ocurrence of even and odd harmonics, and would be possible the occurrence of subharmonics if the first stability condition (n = 1) is not satisfied.

The figure-1 display the PSD obtained running the NOLPOS Code, using the following values:

$K=4.5$, $a_1=2500$, $a_2=-3000$, $a_3=1600$, $b=1000$, $w_p=20$ hz.

The picture shows the following features:

- The peak related to the pump frequency is clearly
- There is a peak related to the characteristic frequency of the oscillator, $w_o=50\ s^{-1}$.
- We observe the occurrence of the a subharmonic region, below 20 hz.
- The higher harmonics, 2w, 3w,... are excited because the amplitude B of the external force is very large. If we decrease B, the peak height of the 2w, 3w, harmonics becomes smaller.

The figure-2 display the PSD obtained with NOLPOS Code. In this case b=1000 and the strength of the noise $D_F=100$. The features are very similar to the previous case, but we note that the noise hiddens the posible subharmonic zone, due to the sinusoidal exciting force. However the picture shows the 2w harmonic, but its peak is very low.

4. FITTING OF NEUTRON NOISE EXPERIMENTAL DATA

The next step was to try to fit neutron noise experimental data. We chose the frequency range from 0 to 25 Hz, and we used neutron noise data from the Instumentation and Control Division Noise Data Bank (ORNL). These data were obtained from Sequoyah-1 Nuclear Power Plant.

The figure 3 display the PSD fits obtained for the In-Core detector neutron noise data. Obviously the In-Core detector did not see the core-barrel peak, and the pump peak appears as a delta-function when the block size increase and the frequency resolution is better, but we work with few blocks.

In this figure is clearly that the pump resonance is the superposition of the modes of reactor internals, and one delta function of the sinusoidal deterministic force of the pump.

5. CONCLUSIONS

Our model, developed above, provides valuable information about the characteristics of the local vibrations of the reactor components driven by gaussian random and pump induced frorces. So we mention the following features:

- The pump resonance is the superposition of the modes of a reactor internals and a delta function related to the Fourier transform of the sinusoidal deterministic force of the pump.

- The pump induces higher harmonics only if there are nonlinearities, but its peaks are very low.

- The pump induces subharmonics, sometimes (only if there are nonlinearities), but if the oscillator is driven by random noise the subharmonic region is hidden.

Therefore, the main conclusion is that if we observe in the PSD higher harmonics, two, three times the pump frequency , there are internal components that show non linear behavior.

6. REFERENCES

[1] Williams M.M.R. (1974), "Random Processes in Nuclear Plants". Pergamon Press, Oxford.

[2] Thie J.A. (1971), Reactor Technol. **14**, 354.

[3] Sancho J., San Miguel M. et al. (1982), Phys. Rev. **A3**, 1588.

[4] Sanchis R., Verdú G., Muñoz-Cobo J. (1986), Ann. Nucl. Energy **15**,275

[5] Wu W., Lin Y. (1984), Int. J. Non-Linear Mechan. **4**,349.

[6] Hayashi C. (1964), "Nonlinear Oscillations in Physical Systems". McGraw-Hill Inc.

CHAPTER VI

NONLINEAR DYNAMICS AND TRANSITION TO CHAOS IN NUCLEAR SYSTEMS.

PART I

VARIATIONAL METHODS FOR ANALYZING LIMIT CYCLES IN B.W.R.

G. Verdu Martin, P. Jimenez and J. Pena

Departamento de Ingeniería Química y Nuclear
Departamento de Matemática Aplicada

Universidad Politecnica de Valencia
P.O. Box 22012. 46071 Valencia

In this paper we analize the phenomenological nonlinear model of BWR dynamics (lumped model). To study the model we develop two general techniques which introduce coordinate systems in which computations are more easily carried out : Center Manifolds method and Normal forms method. The analytical study reveals that if the equilibrium point have a pair pure imaginary eigenvalues, a Hopf bifurcation arises and the solutions are stable limit cycles.

I. INTRODUCTION

There are many commercial boiling water reactors (BWRs) in operation or under construction in the World. Therefore, a great effort has been devoted to the study of BWR systems under a wide range of plant operating conditions. This work represents a contribution to this effort; its object is to study the non linear dynamic processes in BWR reactors.

The BWRs have been experimentally found to behave as linear systems under normal operating conditions. However, recent stability tests have shown that this type of reactor is susceptible to instabilities when operated at low-flow conditions, [1], [2]. In this case, limit cycles are observed [3], indicating that these instabilities cause a non linear operating regime.

The analytical and physical understanding of the transition to non linear regimes is important to the commercial operation of BWR at low flow conditions.

This paper is organized as follows:

In Section II we describe a lumped phenomenological model for studying the non linear dynamic behavior of a BWR. In Section III we study the bifurcation problems to describe the splitting of equilibrium solutions in a family of differential equations. In particular we focus our attention in Hopf bifurcations [5] and [6] derive the stability formula to predict limit cycles and calculate its period.

In Section IV we develop an algorithm by means of variational method to display and locate limit cycle consuming short computational times [7]. Finally in Section V we show the conclusions and summary.

II.- NON LINEAR BWR DYNAMICS

II.A Phenomenological model

In the reference 3 it has been shown that the simplest phenomenological model that contemplates the essential physical processes relative to the dynamic behavior of a BWR contains a one point representation of the reactor kinetics, a one node representation of the heat transfer in the fuel, and a two node representation of the channel thermal hydraulics that includes the void reactivity feedback. In mathematical form, the processes can be represented by the following system of ordinary differential equations:

$$dn/dt = [\rho(t)-\beta]/\Lambda\, n + \lambda c + \rho/\Lambda \quad (1)$$

$$dc/dt = \beta/\Lambda\, n - \lambda c \quad (2)$$

$$dT/dt = An - BT \quad (3)$$

$$d^2\rho_\alpha/dt^2 + a_1\, d\rho_\alpha/dt + a_2\rho_\alpha = a_3 T \quad (4)$$

$$\rho(t) = \rho_\alpha(t) + D_F T \quad (5)$$

where

$n(t)$ = excess neutron density normalized to the steady state neutron density.

$c(t)$ = excess delayed neutron precursors concentration, also normalized to the steady state neutron density.

$T(t)$ = excess average fuel temperarure.

$\rho_\alpha(t)$ = excess void reactivity feedback.

D_F = Doppler coefficient.

II.b Analytical study

The first step to study the model is to obtain the equilibrium points by setting the time derivatives in Equations (1) through (5) to zero. Straightforward algebra yields two equilibrium points, designated P_A and P_B, where

$$P_A = [n = c = T = \rho_\alpha = 0] \quad (6)$$

and

$$P_B = [n = -1, c = -\beta/\lambda\Lambda, T = -A/B, \rho\alpha = -a_3\, a/B\, a_2] \qquad (7)$$

Equilibrium point P_A corresponds to normal reactor operation, while point P_B corresponds to a shutdown condition. Suppose then that we have a fixed point and we wish to characterize the behavior of solutions near the fixed point. If we represent the system of equations (1).(5) by:

$$\dot{x} = f(x) \qquad (8)$$

and the fixed point by $\bar{x}$, the linearized system near $\bar{x}$ is:

$$\dot{\xi} = Df(\bar{x})\xi \qquad (9)$$

where $Df = [\partial f_j/\partial x_j]$ is the Jacobian matrix of first partial derivatives of the function. Note that in our case $x = 0$ (origin)

The important question is, what can we say about the solutions of (8) based on our knowledge of (9). The answer is provided by the Hartman-Grobman theorem, it say that if Df(x) has no zero or purely imaginary eigenvalues, then there is a homeomorphism h defined on some neighborhood U of x locally taking orbits of the nonlinear flow $\varnothing_t$, of (8), to those of the linear flow $e^{tDf(x)}$. Furthermore, there exist local stable and unstable manifolds W^s_{loc} (x), W^u_{loc} (x) of the same dimensions n_s, n_u as those of the eigenspaces E^s, E^u of the linearized system (9), and tangent to E^s, E^u at x. W^s_{loc} (x), W^u_{loc} (x) are as smooth as the function f.

It follows from the above discussion that the trajectories should appear as shown in fig.1 when they are close to the equilibrium points (P_A and P_B). The situation shown in fig. 1.a display the trajectories when equilibrium point P_A is stable, while the fig. 1.b represents the situation when the linear stability threshold is crossed, points P_A becomes unstable.

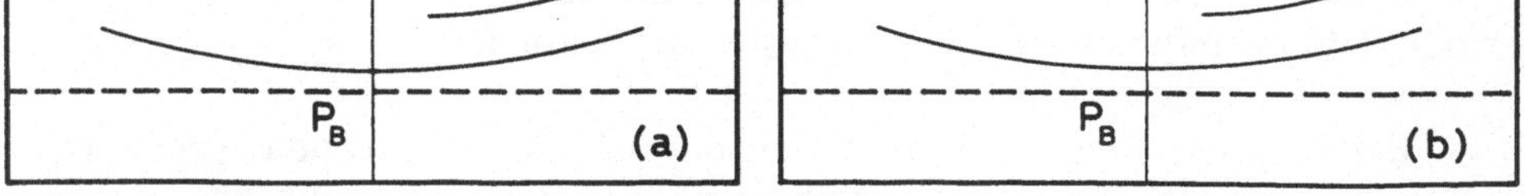

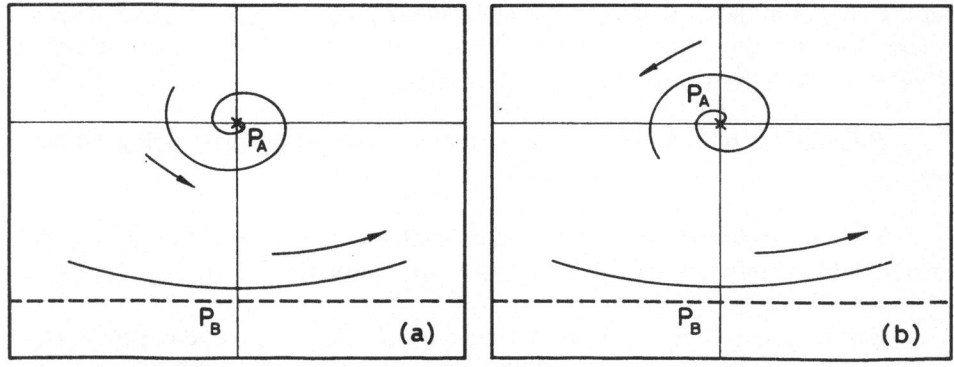

Figure 1. Equilibrium points

Away from the two equilibrium points, possible behaviors of the trajectories are presented in fig. 2. Case a display a globally unstable system where the trajectory continually spirals away from point P_A. Case b describes a situation where the trajectory departs from the equilibrium point but eventually stabilizes itself due to the nonlinearities in the system; thus the trajectory remains bounded and converges to a closed line, a limit cycle, which corresponds to a periodic solution of fixed amplitude. Case c is similar to case b, but the trajectory is repelled by equilibrium point a and remains bounded due to the nonlinearities. The difference between cases b and c is that no periodic solution exists, although the trajectory stays within a bounded regim, that is called a strange attractor.

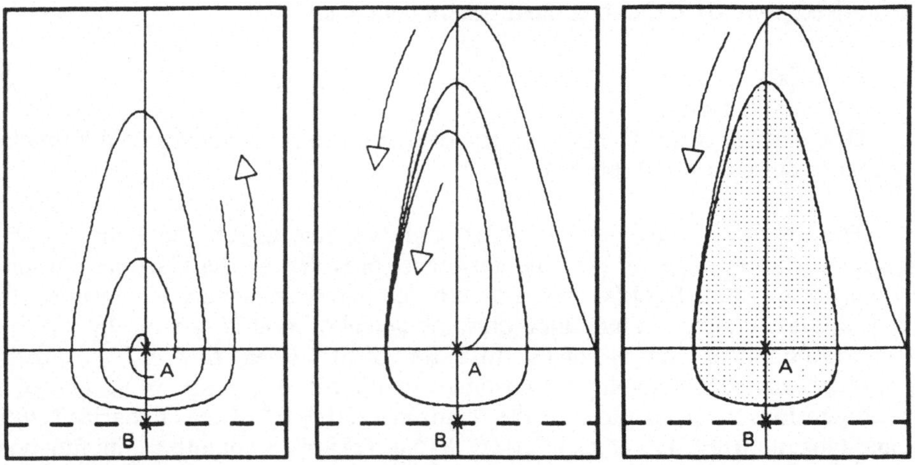

(a) Divergent (b) Limit Cycle (c) Strange Attractor

Figure 2.

III. BIFURCATION PROBLEMS

The term bifurcation was originally used by Poincare to describe the splitting of equilibrium solutions in a family of differential equations. We note that the vectorial field depends on mainly, the feedbaak gain, a_3, that we denote by μ. Therefore a value μ_0 of equation (8) for which the flow is not structurally stable is a bifurcation value of μ.

Bifurcations of equilibria usually produce changes in the topological type of flow.

Like as shown in reference 3, the feedback gain μ produces a cascade of period-doubling bifurcations, which leads to an aperiodic regime.

Before proceeding with a discussion of the bifurcations problems, we develop two general techniques which have the effect of introducing coordinate systems in which computations are more easily carried out.

III.1 Center Manifolds

The Center manifold theorem, provides a means for systematically reducing the dimension of the state spaces which need to be considered when analyzing bifurcations.

We say before that if the linearization of f at the fixed point has no pure imaginary eigenvalues, then it states that the number of eigenvalues with positive and negative real parts determine the topological equivalences of the flow near x, but if there are eigenvalues with zero real parts, then the flow near the fixed point (origen) can be quite complicated. In general, the center manifold method isolates the complicated asymptotic behavior by locating an invariant manifold tangent to the subspace spanned by the eigenspace of eigenvalues on the imaginary axis.

The center manifold method implies that the bifurcating system is locally topologically equivalent to

$$\dot{x} = Bx + f(x,y)$$

$$\dot{y} = Cy + g(x,y) \qquad (x,y) \in R^2 x R^3 \tag{10}$$

where B and C are 2x2 and 3x3 matrices whose eigenvalues have, respectively, zero real parts and negative real parts, and f and g vanish, along with their first partial derivatives, at the origin.

Note that in our case the unstable manifold is empty (spanned by the eigenspace of eingevalues with positive real parts).

Since the center manifold is tangent to E^c (the y= 0 space). We can represent it as a local graph

$$W^c = \{(x,y)\ y = h(x)\};\ h(0) = Dh(0) = 0$$

where h: U--R^3 is defined on some neighborhood U C R^2 of the origin. Fig.3.

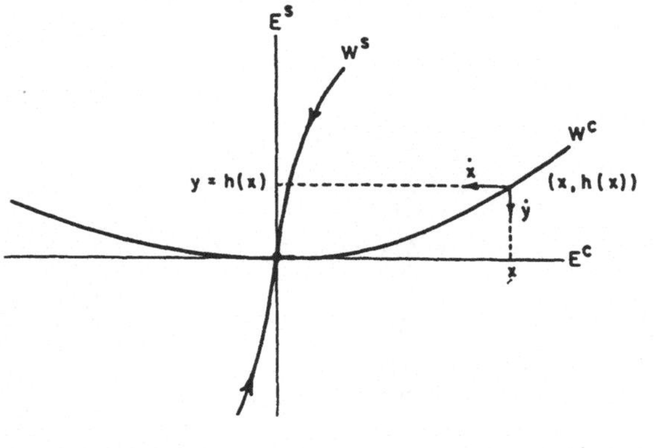

Figure 3.

We now consider the projection of the vector field on y= h(x) onto E^c:

$$\dot{x} = Bx + f(x, h(x)) \tag{11}$$

Since h(x) is tangent to y= 0, the solutions of equation (11) provide a good approximation of the flow solution of equation (10) restricted to W^c.

We now show how h(x) can be calculated, or at least approximated. Substituting y= h(x) in the second component of (10), we obtain:

$$\dot{y} = Dh(x)\,\dot{x} = Dh(x)\,[Bx + f(x, h(x))] = Ch(x) + g(x, h(x)) \tag{12}$$

with boundary conditions

$$h(0) = Dh(0) = 0 \tag{13}$$

This partial differential equation for h cannot, of course, be solved exactly in most cases, but its solution can be approximated arbitrarily closely as a Taylor series at x = 0.

III.2 Calculus of the graph h

Before we obtain the graph we calculate the eigenvalues of the linearized system at origin (negative real parts).

Note that we study the system when it becomes unstable, in this case, the eigenvalues are:

$\delta_1, \delta_2, \delta_3$ (negative real parts) and $\pm\,\omega i$.

Using a classic transformation matrix, the system (8) is transformed to a differential equations system like as (10), so:

$$\dot{u} = \omega v + f_1(u,v,w,y,z)$$

$$\dot{v} = -\omega u + f_2(u,v,w,y,z)$$

$$\dot{w} = \delta_1 w + g_1(u,v,w,y,z)$$

$$\dot{y} = \delta_2 y + g_2(u,v,w,y,z)$$

$$\dot{z} = \delta_3 z + g_3(u,v,w,y,z) \tag{14}$$

$$f_1(u,v,u,y,z) = a_{10} u^2 + a_{11} uv + a_{12} uw + a_{13} uy + a_{14} uz + $$
$$+ a_{15} v^2 + a_{16} vw + a_{17} vy + a_{18} vz + \text{terms } w^2 y^2 z^2 \; wy \; wz \; yz \tag{15}$$

$$f_2(u,v,w,yz) = a_{20} u^2 + a_{21} uv + a_{22} uw + a_{23} uy + a_{24} uz + $$
$$+ a_{25} v^2 + a_{26} vw + a_{27} vy + a_{28} vz + \text{terms } w^2 y^2 z^2 \; wy \; wz \; yz \tag{16}$$

where the coefficients can be complex, and g_1, g_2, g_3 vanish with their first partial derivatives at the origin.

Now, we consider the projection of the vector field on

$$w = h_1(u,v),\; y = h_2(u,v),\; z = h_3(u,v) \tag{17}$$

onto E^c:

We set

$$h_j(u,v) = \alpha_j u^2 + \beta_j v^2 + \gamma_j uv \qquad (18)$$

We seek Taylor series of second order because it suffices to analyze the qualitative behaviour near 0.

Substituting (18) into (12), we have:

$$\partial h_1/\partial u\, \dot{u} + \partial h_1/\partial v\, \dot{v} = \delta_1 h_1(u,v) + g_1(u,v,h_1(u,v), h_2(u,v), h_3(uv)) \qquad (19)$$

$$\partial h_2/\partial u\, \dot{u} + \partial h_2/\partial v\, \dot{v} = \delta_2 h_2(u,v) + g_2(u,v,h_1(uv), h_2(u,v), h_3(uv)) \qquad (20)$$

$$\partial h_3/\partial u\, \dot{u} + \partial h_3/\partial v\, \dot{v} = \delta_3 h_3(u,v) + g_3(u,v,h_1(uv), h_2(u,v), h_3(uv)) \qquad (21)$$

equating the terms u^2, v^2, uv, we find the unknow coefficients α_j, β_j, γ_j. Doing so, we obtain:

$$\gamma_j = \frac{\text{coefficient}_j(v^2) - \text{coeffcient}_j(u^2) - \delta_j \text{coefficient}_j(u,v)}{\delta_j + 4\omega^2} \qquad (22)$$

$$\alpha_j = \frac{\gamma_j \omega - \text{coefficient}_j\,(u^2)}{\delta_j} \qquad (23)$$

$$\beta_j = \frac{\gamma_j \omega - \text{coefficient}_j\,(v^2)}{\delta_j} \qquad (24)$$

coefficient$_j$ $(v^2, u^2, u\,v)$ is the coefficient that multiply to term (v^2, u^2, uv) relative to vectorial field $g_j\,(u,v,w,y,z)$.

The next step is to substitute the graph obtained into equation (15) and dropping the terms of order four and major $(u^4, u^3 v)$. So, we obtain

$$\begin{aligned}\dot{u} = {} & \omega v + a_{10} u^2 + a_{11} uv + a_{12} [\alpha_1 u^3 + \beta_1 uv^2 + \gamma_1 u^2 v] \\ & + a_{13} [\alpha_2 u^3 + \beta_2 uv^2 + \gamma_2 u^2 v] + a_{14} [\alpha_3 u^3 + \beta_3 uv^2 + \gamma_3 u^2 v] \\ & + a_{15} v^2 + a_{16} [\alpha_1 u^2 v + \beta_1 v^3 + \gamma_1 uv^2] + a_{17} [\alpha_2 u^2 v + \beta_2 v^3 + \gamma_2 uv^2] \\ & + a_{18} [\alpha_3 u^2 v + \beta_3 v^3 + \gamma_3 uv^2] \end{aligned} \qquad (25)$$

$$\dot{v} = -\omega v + a_{20} u^2 + a_{21} uv + a_{22} [\alpha_1 u^3 + \beta_1 uv^2 + \gamma_1 u^2 v]$$

$$+ a_{23} [\alpha_2 u^3 + \beta_2 uv^2 + \gamma_2 u^2 v] + a_{24} [\alpha_3 u^3 + \beta_3 uv^2 + \gamma_3 u^2 v]$$

$$+ a_{25} v^2 + a_{26} [\alpha_1 u^2 v + \beta_1 v^3 + \gamma_1 uv^2] + a_{27}[\alpha_2 u^2 v + \beta_2 v^3 + \gamma_2 uv^2]$$

$$+ a_{28} [\alpha_3 u^2 v + \beta_3 v^3 + \gamma_3 uv^2] \tag{26}$$

As will see the analysis of equations (25) and (26) suffices to know the behavior of the solution of equations (1)-(5).

III.3. Normal forms

In this section, we assume that the center manifold method has been applied to system and henceforth we restrict our attention to the flow within the center manifold. We shall try to find additional coordinate transformations which simplify the analytic expression of the vector field on the center manifold, these simplified vector fields are called normal forms and make some symmetry properties of the bifurcation apparent.

We are specifically in equilibria at which there are eigenvalues with zero real parts. At such equilibria, the linearization problem cannot be solved and there are nonlinear resonance terms which cannot be removed by coordinate changes. The normal form method formulates systematically how well one can do. We start with a system of differential equations:

$$\dot{x} = g(x) \tag{27}$$

If $L = Dg\,(0)\,x$ denotes the linear part of (27) at $x = 0$, then L induces a map adL on the linear space H_K of vector fields whose coefficients are homogeneous polynomials of degree K. The map ad L is defined by

$$adL(y) = [Y,L] = DLY - DYL \tag{28}$$

where [,] denotes the Lie bracket operation. In component form, we have

$$[Y,L]^i = \sum_{j=1}^{n} (\partial L^i/\partial y_j\, Y^j - \partial Y^i/\partial y_j\, L_j) \tag{29}$$

The normal form method say that if $\dot{x} = g(x)$ be a system of differential equations with $g(0) = 0$ and $Dg\,(0)x = L$. Then there is an analytic change of coordinates in a neighborhood of the origin which transforms the system to

$$\dot{y} = K(y) = g^{(1)}(y) + g^{(2)}(y) + \dots\, g^{(r)}(y) + R_r \text{ with } L = g^{(1)} \text{ and } g^{(K)} \in G_K$$

for $2 \le K \le r$, where G_K is a complement for ad $L(H_K)$ in H_K

$$(H_K = adL(H_K) + G_K) \qquad (R_r = 0(y^r) \tag{30}$$

After straightforward algebra is easy to demonstrate that the normal form method gives a coordinate transformation which transforms the system

$$\dot{x} = -wy + 0(u,v) \tag{31}$$

$$\dot{y} = wx + 0(u,v)$$

into the system

$$\dot{u} = -wv + (au - cv)(u^2 + v^2) + \text{higher-order terms}$$

$$\dot{v} = wu + (av + cu)(u^2 + v^2) + \text{higher-order terms} \tag{32}$$

for suitable constants a and b

III. 4 Hopf Bifurcations

A clue to what happens in the generic bifurcation problem involving an equilibrium with pure imaginary eigenvalues can be gained examining the system obtained after we applied the normal form method. By smooth changes of coordinates, the Taylor series of degree 3 for the general problem can be brought to the following form (see. Eq. 32)

$$\dot{u} = (d\mu + a(u^2+v^2))\,u - (w + c\mu + b(u^2+v^2))\,v$$

$$\dot{v} = (w + cu + b(u^2+v^2)\,u + (d\mu + a(u^2+v^2))\,v \tag{33}$$

in our case applied to equations (1)-(5)

$$d = d/da_3\, R(\lambda(a_3))\ a_3 = \dot{a}_{30} \tag{33.a}$$

$$\mu = a_3 - a_{30}$$

λ is the eigenvalue with real part non negative, and $R(\lambda(a_3))$ and $Im(\lambda(a_3))$ means real and imaginary part, respectively, of the eigenvalue

$$c = d/da_3\, Im(\lambda/(a_3))\ a_3 = a_{30} \tag{33.b}$$

the equations (33) are expressed in polar coordinates as

$$\dot{r} = (d\mu + ar^2)\, r$$

$$\dot{\theta} = (w + c\mu + br^2) \tag{34}$$

We see that there are periodic orbits of (34) which are circles, r= cte, obtained from the nonzero solutions of r= 0. If a= 0 and d= 0 these solutions lie along the paraboloid $\mu = -ar^2/d$. This implies that the surface of periodic orbits has a quadratic tangency with its tangent plane $\mu = 0$, furthermore the Hopf bifurcation theorem say that the qualitative properties near the origin remain unchanged if higher order terms are added to the system. If a < 0 then these periodic solutions are stable limit cycles. While if a > 0 the periodic solutions are repelling.

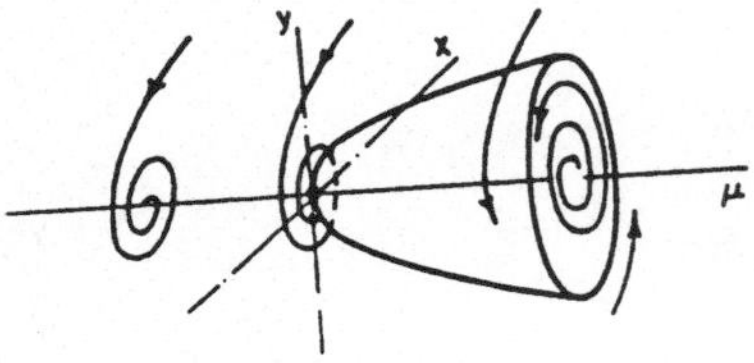

Hopf Bifurcation (Supercritical)

If we add higher order terms to (34), we obtain:

$$\dot{r} = (d\mu + ar^2 + \gamma r^4)\, r \tag{35}$$

The solutions r= cte are obtained equating r to 0

$$(d\mu + ar^2 + \gamma r^4) = 0 \tag{36}$$

$$r^2 = z \qquad d\mu + az + \gamma z^2 = 0 \text{ so}$$

$$z = [\, -a \pm a\sqrt{1-4\gamma d\mu/a}\,]/2\gamma \tag{37}$$

$$z = \begin{cases} -a/2\gamma - a/2\gamma + d\mu/a \\ -a/2\gamma + a/2\gamma - d\mu/a \end{cases} \tag{38}$$

when a<0 , the two stable solutions are $d\mu/a$ and $a/\gamma - d\mu/a$ but, if a>0, the solutions are unstable.

Therefore, the qualitative properties of equation (34) and (35) are the same.

III. 4.1 Derivation of stability formula

If the reduced (approximate) system equations (25) and (26), has a imaginary pair of eigenvalues, $\lambda, \bar{\lambda} = r \pm iw'$, $r = \mu d$, $w' = \omega + c\mu$, then it can be conveniently represented as a simple complex equation:

$$\dot{z} = \lambda z + h(z,\bar{z}) \tag{39}$$

however, the normal form becomes

$$\dot{w} = \lambda w + C_1 w^2\bar{w} + C_2 w^3\bar{w}^2 + ... + C_N w^{k+1}\bar{w}^k + 0(w^{2N+3}) =$$

$$= \lambda w + C_1 w^2\bar{w} = \lambda w + \hat{h}(w,\bar{w}) \tag{40}$$

where the complex coefficients are of the form $C_j = a_j + ib_j$ and an overbar denotes complex conjugation.

To transform (39) to (40), we use the near identity transformation

$$z = w + \Psi(w,\bar{w}). \qquad \Psi = 0(|w|^2) \tag{41}$$

substituting (41) in (39) and using (40) we obtain

$$\dot{w} + \Psi_w \dot{w} + \Psi_{\bar{w}} \dot{\bar{w}} = \lambda(w+\Psi(w,\bar{w})) + h(w+\Psi,\bar{w} + \bar{\Psi})$$

$$= \lambda w + \hat{h}(w,\bar{w}) + \Psi_w(\lambda w + \hat{h}(w,\bar{w})) + \Psi_{\bar{w}}(\overline{\lambda \bar{w} + \hat{h}(w,\bar{w})}) \tag{42}$$

therefore,

$$\lambda[\Psi_w w-\Psi] + \bar{\lambda}\bar{w}\Psi_{\bar{w}} = h(\Psi+w,\bar{w}+\bar{\Psi})-\hat{h}(w,\bar{w})(1+\Psi_w)-\overline{\hat{h}(w,\bar{w})}\Psi_{\bar{w}} \tag{43}$$

We now express Ψ as a Taylor, series so

$$\Psi(w,\bar{w}) = \Psi_{ww} w^2/2! + \Psi_{\bar{w}\bar{w}} \bar{w}^2/2! + \Psi_{w\bar{w}} w\bar{w} +$$

$$+ \Psi_{www} w^3/3! + \Psi_{\bar{w}\bar{w}\bar{w}} \bar{w}^3/3! + \Psi_{w2\bar{w}} w^2\bar{w}/2 +$$

$$+ \Psi_{\bar{w}2w} \bar{w}^2 w/2 + o(|w|^4) \tag{44}$$

where subscripts denote partial differentiation.

Substituting (44) into (43), we obtain to order two

$$w^2 [\lambda \Psi_{ww}/2] + w\bar{w}[\bar{\lambda} \Psi_{\bar{w}w}] + \bar{w}^2 [\bar{\lambda} \Psi_{\bar{w}\bar{w}} - \lambda \Psi_{\bar{w}.} \bar{w}/2] =$$

$$= h_{ww} w^2/2 + h_{w\bar{w}} w\bar{w} + h_{\bar{w}\bar{w}} \bar{w}^2/2 \tag{45}$$

so,

$$\Psi_{ww} = h_{ww}/\lambda, \Psi_{\bar{w}w} = h_{w\bar{w}}/\lambda, \Psi_{\bar{w}\bar{w}} = h_{\bar{w}\bar{w}}/(2\bar{\lambda}-\lambda) \tag{46}$$

We now carry out the expansion to one higher order and equate the coefficients of the normal form term $w^2\bar{w}$. Is easy to check that, for this terms the coefficient on the left hand side of (43) becomes

$\Psi_{ww\bar{w}}\,[\lambda+\bar{\lambda}]/2 \approx 0$

Now, we develop the right hand side of (44)

$$h(w+\Psi, \bar{w}+\bar{\Psi}) = h(w,\bar{w}) + \Psi h_w(w,\bar{w}) + \bar{\Psi} h_{\bar{w}}(w,\bar{w}) \qquad (47)$$

If we express $h(w,\bar{w})$ as a Taylor series, we have

$$h(w,\bar{w}) = w^2 h_{ww}/2! + w\bar{w}h_{w\bar{w}} + \bar{w}^2 h_{\bar{w}\bar{w}}/2!$$
$$+ h_{ww\bar{w}}\, w^2\bar{w}/2 + \ldots \qquad (48)$$

Therefore, the coefficient for the term $w^2\bar{w}$ in right hand side becomes

$$\Psi_{ww}\, h_{w\bar{w}}/2 + h_{ww}\,\Psi_{w\bar{w}} + \overline{\Psi_{\bar{w}\bar{w}}}/2\; h_{\bar{w}\bar{w}} + \overline{\Psi_{w\bar{w}}}\, h_{w\bar{w}} + h_{ww\bar{w}}/2 - C_1 \qquad (49)$$

so,

$$C_1 = \Psi_{ww}\, h_{w\bar{w}}/2 + h_{ww}\,\Psi_{w\bar{w}} + \overline{\Psi_{\bar{w}\bar{w}}}/2\; h_{\bar{w}\bar{w}} + \overline{\Psi_{w\bar{w}}}\, h_{w\bar{w}} + h_{ww\bar{w}}/2 \qquad (50)$$

Substituting (46) into (50), we obtain finally.

$$C_1 = h_{ww\bar{w}}/2 + h_{ww}/\lambda\; h_{w\bar{w}}/2 + h_{ww}h_{w\bar{w}}/\bar{\lambda} + h_{\bar{w}\bar{w}}\overline{h_{\bar{w}\bar{w}}}/2(2\lambda-\bar{\lambda}) + \overline{h_{w\bar{w}}}\, h_{w\bar{w}}/\lambda \qquad (51)$$

Recalling the equation (33) we obtain that

$$a = \text{Real}\,\{C_1\} \qquad (51.a)$$

$$b = \text{Imaginary}\,\{C_1\} \qquad (51.b)$$

Now, if in general,

$$h(z,\bar{z}) = f(u,v) + i\, g(u,v);\; z = u + iv \qquad (52)$$

where f y g are real functions.

We can express f(u,v) and g(u,v) as a Taylor series.

so,

$$f(u,v) = f_{uu}\, u^2/2 + f_{uv}\, uv + f_{vv}\, v^2/2 + f_{uuu}\, u^3/3! + f_{uuv}\, u^2v/2! +$$
$$+ f_{uvv}/2!\, u^2v^2 + f_{vvv}/3!\, v^3 + \ldots$$

$$g(u,v) = g_{uu}/2!\, u^2 + g_{uv}\, uv + g_{vv}\, v^2/2! + g_{uuu}/3!\, u^3 + g_{uuv}\, u^2 v/2! +$$

$$+ g_{uvv}\, u\, v^2/2! + g_{vvv}/3!\, v^3 + \ldots \quad (53)$$

And developing $h(w,\bar{w})$ the same form, we have

$$h(w,\bar{w}) = h_{ww}\, w^2/2 + h_{w\bar{w}}\, w\bar{w} + h_{\bar{w}\bar{w}}\, \bar{w}^2/2! + h_{ww\bar{w}}/3! + h_{w\bar{w}\bar{w}}/3! + \ldots \quad (54)$$

Substituting (53) into (52), bringing to mind (41), and equating to (54) we have,

$$h_{ww} = 1/4\, [f_{uu} + ig_{uu} - f_{vv} - ig_{vv} - 2if_{uv} + 2g_{uv}]$$

$$h_{\bar{w}\bar{w}} = 1/4\, [f_{uu} + ig_{uu} - f_{vv} + ig_{vv} + 2if_{uv} + 2g_{uv}]$$

$$h_{w\bar{w}} = 1/4\, [f_{uu} + f_{vv} + ig_{uu} + ig_{vv}]$$

$$h_{ww\bar{w}} = 1/8\, [f_{uuu} + f_{uvv} + g_{uuv} + g_{vvv} + ig_{uuu} + ig_{uvv} - if_{uuv} - if_{vvv}] \quad (55)$$

Introducing (55) into (51), finally we calculate the variable C_1, and therefore, a and b.

The variables a, b and d, c expressed by equation (33.a) and (33.b) characterize completely the system; if $a < 0$, the system is a limit cycle (paraboloid $\mu = -ar^2/d$) with period $w + c\mu - b\mu d/a$. If $a > 0$ the solutions are repelling.

IV. THE POINCARE MAP

To study the stability of closed orbits or periodic solutions we introduce a geometric tool equivalent in essence to Floquet multiplier method: The Poincaré map.

Let γ be a periodic orbit of some flow Φ_t in R^n arising from a nonlinear vector field f(x). We take a local cross section $\Sigma \subset R^n$, of dimension n-1; transverse to flow everywhere. Denote the point where γ intersects Σ by p and let $U \subseteq \Sigma$ be some neighborhood of p. Then the first return or Poincaré map $P: U \to \Sigma$ is defined for a point $q \in U$ by

$$P(q) = \Phi_\tau(q) \quad (56)$$

where $\tau = \tau(q)$ is the time taken for the orbit $\Phi_\tau(q)$ based at q to first return to Σ. τ generally depends upon q and need not be equal to $T = T(p)$, the period of γ, however, $\tau \to T$, when $q \to p$. p is a fixed point for the map P, and it is not difficult to see that the stability of p for P reflects the stability of γ for the flow Φ_τ. figure 4

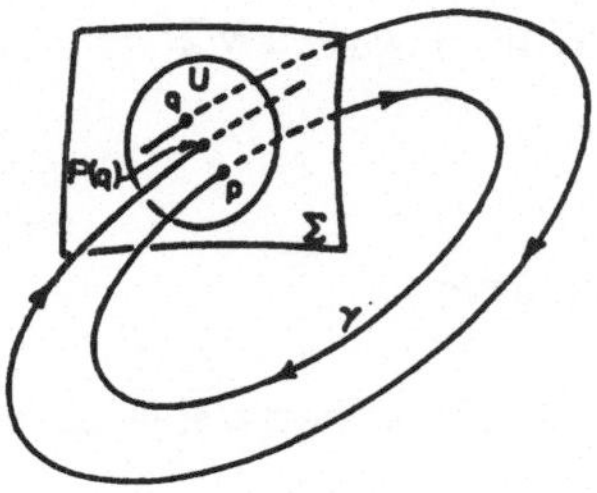

Figure 4 . Poincarè Map

Now we demonstrate the general relationship between Poincaré maps and linearized flows.

Let $x(t) = x(t+T)$ be a solution lying on the closed orbit γ based at $x(0) = p \in \Sigma$. Linearizing the differential equation about γ, we obtain

$$\dot{\xi} = Df(x(t))\xi \tag{57}$$

where $Df(x(t))$ is a T-periodic matrix. The fundamental solution matrix of such a T periodic system can be written in the form (Floquet theory)

$$X(t) = Z(t)\, e^{tR}; \qquad Z(t) = Z(t+T) \tag{58}$$

where X, Z and R are square matrices.

In particular, we can choose $X(0) = Z(0) = I$, so that

$$X(T) = Z(T)\, e^{TR} = Z(0)\, e^{TR} = e^{TR} \tag{59}$$

it then follows that the behavior of solutions in the neighborhood of γ is determined by the eigenvalues of the constant matrix e^{TR}. The eigenvalues $\lambda_1 \ldots \lambda_n$ are called the characteristic Floquet-Lyapunov multipliers. The multiplier associated with perturbations along γ is always unity. The moduli of the remaning determine the stability of γ.

Althoùg the matrix R is not determined uniquely by the solutions, the eigenvalues of e^{TR} are uniquely determined.

Therefore, to calculate the eigenvalues we solve the matrix differential equation (57) but we need the knowledge of a periodic solution; this step is very difficult because it implies to solve previously the initial non linear differential equation.

Another strategy more feasible is to develop an algorithm based on the variational equation of the non linear differential equation, so we represent the non linear equation

$$\dot{x} = f(x), \; x(t = t_0) = x_0$$

by

$$\dot{\Phi}_t(x_0, t_0) = f(\Phi_t(x_0,t_0)) \tag{60}$$

If now derive respect to x_0 (initial conditions) we obtain

$$D_{x0}\, \dot{\Phi}_t\,(x_0,t_0) = D_x\, f(\Phi_t(x_0,t_0))\, D_{x0}\, \Phi_t(x_0,t_0)$$

$$D_{x0}\, \Phi_{t0}\,(x_0,t_0) = I \tag{61}$$

If $D_{x0}\, \Phi_t(x',t_o)$ is represented by $\Phi_t\,(x',t_0)$

it follows that

$$\Phi_t(x_0,t_0) = D_x\, f(\Phi_t.\,(x_0,t_0))\, \Phi_t(x_0,t_0) \tag{62}$$

We note that if $x_0 \in \gamma$ (periodic solution of period T), the eigenvalues of Φ_T are the Floquet multipliers.

The algorithms for locating limit cycles require the solution of the variational equation. However $\Phi_t(X)$ is most easily calculated by appending the variational equation to the equations describing the dynamical system. The expressions can be written in compact form:

$$\begin{bmatrix} \dot{x} \\ \dot{\Phi}_t \end{bmatrix} = \dot{F}(x,\Phi_t) = \qquad \begin{bmatrix} f(x) \\ Df(x)\Phi_t \end{bmatrix} \begin{bmatrix} x(t_0) \\ \Phi_{t0} \end{bmatrix} = \begin{bmatrix} x_0 \\ I \end{bmatrix} \tag{63}$$

The obvious way to locate a limit cycle steady state solution using computer is what we call the brute-force method: integrate the dynamical system until the transient has died out. This method has the advantage of simplicity, but it suffers from several drawback, the most important of which is long simulation times. (An another problem is that there is no reliable way to detect the end of the transient).

The algorithm that we have developped uses the Newton-Raphson method. Consider a function H: R^n--R^n, with a zero in x (i:e. $H(x) = 0$). Newton-Raphson method attempts to calculate x by iterating from an initial guess $x^{(0)}$, as follows.

In our case, we define

$$H(x,T) = \Phi_T\,(x) - x$$

$$y^i = H(x^{(i)}, T^{(i)}) \tag{64}$$

If now we linearize (64) we obtain:

$$\Delta y^{(i)} = D_x\, H(x^{(i)}, T^{(i)})\Delta x^{(i)} + D_T\, H(x^{(i)})\, T^{(i)}\, \Delta T^{(i)} =$$

$$= [\Phi_{T(i)} (x^{(i)}) - I] \, \Delta x^{(i)} + f(\Phi_{T(i)} (x^{(i)})) \, \Delta T^{(i)} \tag{65}$$

where

$$\Delta x^{(i)} = x^{(i+1)} - x^{(i)}$$

$$\Delta y^{(i)} = y^{(i+1)} - y^{(i)}$$

$$\Delta T^{(i)} = T^{(i+1)} - T^{(i)} \tag{66}$$

Given $x^{(i)}$, $y^{(i)}$, and $T^{(i)}$, we want to choose $x^{(i+1)}$ and $T^{(i+1)}$ such that $y^{(i+1)} = 0$, therefore, the Newton-Raphson i teration is given by

$$-y^{(i)} = (\Phi_{T(i)}(x^{(i)}) - I) \, (x^{(i+1)} - x^{(i)}) + f(\Phi_{T(i)} (x^{(i)}))(T^{(i+1)} - T^{(i)}) \tag{67}$$

We note that (67) is a system of n equations in the $(n+1)$ unknowns $x^{(i+1)}$ and $T^{(i)}$. One more constraint must be found to make the system of equations uniquely solvable, and this is how method 1 and 2 differ.

Method 1

In this method, the correction term $\Delta x^{(i)}$ is restricted to be orthogonal to the trajectory

$$f(x^{(i)})^T \, \Delta x^{(i)} = 0$$

This constraint results in a system of $(n+1)$ equation in $(n+1)$ unknowns

$$\begin{bmatrix} -y^{(i)} \\ 0 \end{bmatrix} = \begin{bmatrix} \Phi_{T(i)} (x^{(i)}) - I & f(\Phi_{T(i)} (x^{(i)})) \\ f(x^{(i)})^T & 0 \end{bmatrix} \begin{bmatrix} x^{(i+1)} - x^{(i)} \\ T^{(i+1)} - T^{(i)} \end{bmatrix}$$

Method 2

In this method, T is fixed. We know its value from the equations (34) and (51). In this case we have a system of n equations and n unknowns

$$[-y^{(i)}] = [\Phi_T(x^{(i)}) - I \quad f(\Phi_T(x^{(i)}))] \, [x^{(i+1)} - x^{(i)}]$$

Comparing both methods, notably the second is better if we know the period of the limit cycle, it require short time of simulation and we reach the steady solution quickly.

IV.a Practical application

The specific numeric values for this practical application have been chosen, because a limit cycle was experimentally observed for these conditions in the Vermont Yankee reactor [3].

The model parameters are showed in table 1

Table 1.

$A = 25.04$	$a_{30} = -3.10 \cdot 10^{-3}$
$Df = -2.52 \cdot 10^{-5}$	$B = 0.23$
$\beta = 0.0056$	$a_1 = 2.25$
$\Lambda = 4.00 \cdot 10^{-5}$	$\lambda = 0.08$
$a_2 = 6.82$	

The parameter a_3, which is proportional to the void reactivity coefficient and the fuel heat transfer coefficient controls the gain of the feedback. The value a_{30} given in table is the critical value above which the model become unstable.

Now, we apply the techniques developed in this work to analyze the system represent by equations (1)-(5). First, we calculate the eigenvalues of the differential equations system... This serves us to split the system in two matrix differential equations whose eigenvalues have, respectively, zero real parts and negative real parts. The second step is the calculus of the graph that projects the vector field related to eigenvalues with negative real parts into the center manifold. Henceforth we try to find additional coordinate transformations which simplify the analytic expression of the vector field on the center manifold. We apply the expressions (54) and (55) to obtain the values a, and b. The d, c values are expressed by equations (33.a) (33.b).

The results that we have obtained are the followings:

$a = -24.$

$c = 141.24$

$d = 155.85$

This shows that the system follows periodic orbits, ie limit cycles of the period

$$\omega + c\mu + bd\mu/a = 2.7$$

Now we apply the algorithm to locate the periodic solution. The initial guess is arbitrarily. After short simulation time and apply Newton-Raphson method reiterately we obtain finally the limit cycle displayed in fig 5.

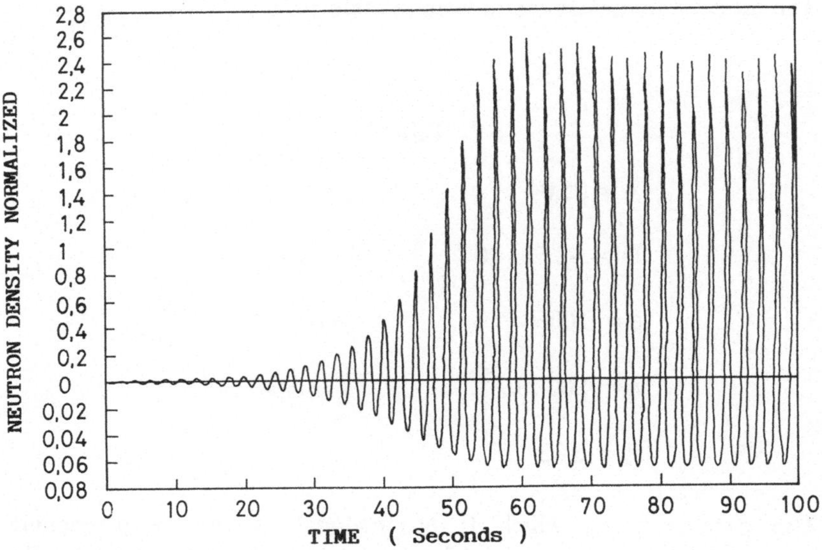

Figure 5. Example of Limit Cycle TIME (Seconds)

V. CONCLUSIONS

We analyze the phenomenological nonlinear model of BWR dynamics [lumped model] described in Section II. To study the model we develop two general techniques which introduce coordinate systems in which computations are more easily carried out: Center Manifolds method and Normal forms method.

The analytical study reveals that if the equilibrium point have a pair pure imaginary eigenvalues, a Hopf bifurcation arises and the solutions are stable limit cycles.

We have derived stability formula for the model and we obtain the parameters that describe the behavior of the periodic solution (periode amplitude...)

Finally we have developped a algorithm to find the limit cycles using variational methods. Thus, in short computational times we displayed the final solution when the transient died out. Furthermore the algorithm provides the Lyapunov-Floquet multipliers which determine the behavior of solutions in the neighborhood of the limit cycle.

VI. REFERENCES

1. Y. WAARANPERA and S. ANDERSON. Trans. Am. Nucl. Soc. 39, 868 (1981)

2. E. GIALDI, S. GRIFONI, C. PARMEGGIANI, and C. TRICOLI. " Core Stability in Operating BWR : Operational Experience ". Proc. Specialist Mtg. Reactor Noise SMORN IV. Dijon. France, Octuber. 15-19, 1984

3. J. MARCH-LEUBA, D. G. CACUCI, R. B. PEREZ. " Nonlinear Dynamics and Stability of Boiling Water Reactors". Part 1 : Qualitative Analysis. Nuclear Science and Engineering. 93. 111-123 (1986)

4. J. MARCH-LEUBA, D. G. CACUCI, R. B. PEREZ. "Nonlinear Dynamics and Stability of Boiling Water Reactors". Part 2 : Quantitive Analysis. Nuclear Science and Engineering. 93. 124-136 (1986)

5. J. GUCKENHEIMER AND P. HOLMES. " Nonlinear Oscillations, Dynamical Systems, and Bifurcations of vector fields". Springer Verlag. 1986

6. M. GOLUBITSKY and W. LANGFORD. " Classification and unfolding of degenerate Hopf Bifurcations ". Journal Of Differential Equations. 41. 375-415.1981

7. T. PARKER and L. CHUA. "Chaos : A tutorial for engineers ". Proceedings of the IEEE. August 1987

NONLINEAR DYNAMICS AND CHAOS IN BOILING WATER REACTORS

José March-Leuba

Instrumentation and Controls Division
Oak Ridge National Laboratory*
Building 3500
P.O. Box 2008, Oak Ridge, TN 37831-6010 (USA)

INTRODUCTION

There are currently 72 commercial boiling water reactors (BWRs) in operation or under construction in the Western world, 37 of them in the United States. Consequently, a great effort has been devoted to the study of BWR systems under a wide range of plant operating conditions. This paper represents a contribution to this ongoing effort; its objective is to study the basic dynamic processes in BWR systems, with special emphasis on the physical interpretation of BWR dynamics. The main thrust in this work is the development of phenomenological BWR models suited for analytical studies performed in conjunction with numerical calculations. This approach leads to a deeper understanding of BWR dynamics and facilitates the interpretation of numerical results given by currently available sophisticated BWR codes.

In BWRs, water acts both as moderator and as coolant. During normal operating conditions, both the density reactivity and Doppler reactivity coefficients are negative, so the reactor power is stable (i.e., the power level is maintained constant) without the intervention of the reactor's control system. For example, if the reactor's power level is increased, these negative reactivity coefficients induce negative feedback processes that ultimately reduce the power, thereby stabilizing the reactor.

It is known, though, that systems with negative feedbacks are susceptible to instabilities if the feedback gain or the phase lag is increased beyond some critical value. In the case of a BWR, a reactivity instability manifests itself as a diverging power oscillation. As recent stability experiments[1,2] have shown, BWRs are susceptible to reactivity instabilities when operated at low flow and relatively high power levels (e.g., 32% flow and 51% power). The importance of understanding reactivity instabilities is underscored by the fact that United States utilities are required by the Nuclear Regulatory Commission to evaluate the reactor stability for every reload core unless plant technical specifications provide for monitoring of neutron flux oscillations in the

*Operated by Martin Marietta Energy Systems, Inc., for the U.S. Department of Energy under Contract No. DE-AC05-84OR21400.

so-called limit-cycle "Detect and Suppress" (D&S) region. This region is defined by these specifications and commonly lies below the 40% flow line and above the 80% rod control line. Within this region, the reactor operator must monitor average and local power oscillations to detect instabilities. Should instabilities occur, they should be suppressed either by inserting control rods or by increasing recirculation pump speed.

It has been shown experimentally that BWRs behave as linear systems under normal operating conditions.[1] As discussed in the foregoing, however, BWRs are susceptible to instabilities. When the equilibrium (i.e., operating) point becomes unstable, small oscillations grow large enough so that nonlinearities become important. This phenomenon has been observed in recent experiments in which the reactor power was increased slowly while the flow rate was essentially kept constant. When a certain critical power level was reached, the reactor became unstable and the measured signals began displaying an oscillatory behavior. If the reactor were a linear system, these oscillations would have diverged exponentially. However, these experiments showed that the amplitudes of these oscillations grew only slightly, then stabilized (i.e., remained bounded) due to the appearance of a limit cycle. As is well known, the appearance of a limit cycle is a typical nonlinear phenomenon.

To analyze the effect of nonlinearities on the stability and dynamic behavior of a BWR, it is useful to summarize the main nonlinear process in a BWR. An obvious nonlinearity is associated with the neutron field. For example, in the point kinetics representation of the neutron field, the term "reactivity-times-neutron-density" (ρn) is nonlinear because, due to the inherent reactivity feedback, ρ depends on n. Physically, this means that reactivity perturbations are weighted by the neutron density, a fact that forces the neutron density to be positive at all times. In addition, the reactivity feedback introduces nonlinearities because the cross sections and the associated reactivity coefficients are nonlinear functions of the temperature and moderator density.

In general, the fuel heat-conduction equation is also nonlinear because the heat conductance, density, and heat capacity of the fuel depend on the temperature. The fluid energy and momentum conservation equations are also nonlinear. For example, nonlinearities in the momentum equation appear in both the kinetic-energy and friction terms. Nonlinearities are also found in the dynamics of the recirculation loop. Although the nonlinearities mentioned in the foregoing have little effect on the reactor's stability under small perturbations from steady-state operating conditions, they become the dominant factors governing the reactor's dynamic behavior beyond the regime of linear stability.

A PHENOMENOLOGICAL MODEL OF BWR DYNAMICS

A phenomenological model for studying nonlinear BWR dynamics must obviously contain the essential processes that control the dynamic behavior of a BWR, yet it must be sufficiently simple to allow extensive numerical simulation of a wide range of reactor operating conditions. The development of such a model has been the objective of the work described in Refs. 3 and 4. There it has been shown that the simplest phenomenological model that retains the essential physical processes controlling the dynamic behavior of a BWR contains a one-point representation of the reactor kinetics, a one-node representation of the heat transfer process in the fuel, and a two-node representation of the channel thermal hydraulics to account for the void reactivity feedback. Mathematically, these processes can be represented by the following system of ordinary differential equations:

$$\frac{dn(t)}{dt} = \frac{\rho(t) - \beta}{\Lambda} n(t) + \lambda c + \frac{\rho}{\Lambda} \quad , \tag{1}$$

$$\frac{dc(t)}{dt} = \frac{\beta}{\Lambda} n(t) - \lambda c \quad , \tag{2}$$

$$\frac{dT(t)}{dt} = a_1 n(t) - a_2 T(t) \quad , \tag{3}$$

$$\frac{d^2 \rho_\alpha(t)}{dt^2} + a_3 \frac{d\rho_\alpha(t)}{dt} + a_4 \rho_\alpha = kT(t) \quad , \tag{4}$$

and

$$\rho(t) = \rho_\alpha(t) + DT(t) \quad , \tag{5}$$

where

$n(t)$ = excess neutron density normalized to the steady-state neutron density,

$c(t)$ = excess delayed neutron precursors concentration, also normalized to the steady-state neutron density,

$T(t)$ = excess average fuel temperature,

$\rho_\alpha(t)$ = excess void reactivity feedback.

Note that the nonlinearity appears in the neutronic equation through the parametric feedback introduced by the reactivity.

As detailed in Ref. 4, the parameters in the phenomenological model represented by Eqs. (1) through (5) were obtained by functionally fitting the transfer function, as calculated by the LAPUR code, of the Vermont Yankee reactor with operating conditions equivalent to those of stability test 7N. These test conditions have been chosen for the present nonlinear study because a limit cycle was experimentally observed during this test. The specific numerical values for the resulting model parameters are presented in Table I.

The parameter k, which is proportional to the void reactivity coefficient and the fuel heat transfer coefficient, controls the gain of the feedback and, thus, defines the linear stability of this reactor model. The value of k_0 given in Table I is the critical value above which the model becomes unstable. By artificially increasing the value of k above k_0, the model can be made unstable and can be used to study BWR dynamic behavior in the nonlinear regime.

The validation of the above model has been detailed in Ref. 4. The model accurately represents the dynamic behavior of the reactor. For example, Fig. 1 shows the remarkable agreement between the model's reactivity-to-power transfer and the LAPUR-calculated transfer function for the conditions of test 7N.[2]

ANALYTICAL STUDY

Taking advantage of the low order of the present model, analytical work can be performed to understand the behavior of the model's solutions. To start the transient solutions reported in this work, a step perturbation Δ in the neutron population was applied to the heat

Table 1

Model Parameters for Vermont Yankee Test 7N

Parameter	Value	Units
a_1	25.04	$K \cdot s^{-1}$
a_2	0.23	s^{-1}
a_3	2.25	s^{-1}
a_4	6.82	s^{-2}
k_o	-3.70×10^{-3}	$K^{-1} \cdot s^{-2}$
D	-2.52×10^{-5}	K^{-1}
β	0.0056	---
Λ	4.00×10^{-5}	s^{-1}
λ	0.08	s^{-1}

generation term in Eq. (3). (This perturbation has the same effect as changing the reactivity and was chosen to facilitate subsequent computations.) The equilibrium points were obtained by setting the time derivatives in Eqs. (1) through (5) to zero. Straightforward algebra yields two equilibrium points, designated A and B, where

$$A = [n = -\Delta,\ c = -\beta\Delta/(\lambda\Lambda),\ T = 0,\ \rho_\alpha = 0] \quad , \tag{6}$$

and

$$B = [n = -1,\ c = -\beta/(\lambda\Lambda),\ T = a_1(\Delta - 1)/a_2,\ \rho_\alpha = ka_1(\Delta - 1)/(a_2 a_4)] \quad . \tag{7}$$

Equilibrium point A corresponds to normal reactor operation, while point B corresponds to a shutdown condition. Note that the definition $n = (N - N_0)/N_0$ implies that point $n = -1$ corresponds to $N = 0$, where N is the absolute neutron density. For a perturbation Δ of this kind, point B is always unstable while point A is stable as long as $k < k_0$. For $k > k_0$, both equilibrium points are unstable.

An examination of Eq. (1) reveals that neglecting the delayed neutron effects gives

$$\frac{dn}{dt} = \frac{\rho(n + 1)}{\Lambda} \quad . \tag{8}$$

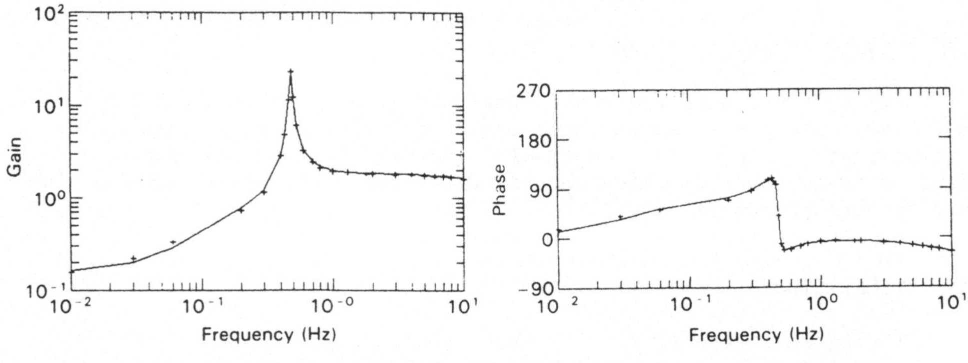

Fig. 1. Comparison between the reactivity-to-power transfer functions calculated by the code LAPUR (+++) and the phenomenological model (---).

Equation (8) highlights the fact that dn/dt will always tend to zero when n approaches $n = -1$, regardless of the value of ρ. Physically this means that the absolute neutron population N cannot be negative. Therefore, not just point B, but the whole $n = -1$ line in phase space is unstable in the sense that it repels trajectories (i.e., temporal model solutions). This is in contrast to the linear models customarily used for BWR dynamic modeling; in such models, ρ does not induce a parametric reactivity feedback, so the absolute neutron population can become mathematically negative.

It follows from the above discussion that the phase-space trajectories should appear as shown in Fig. 2 when they are locally close to the equilibrium points (A and B). The situation shown in Fig. 2(a) depicts the trajectories when equilibrium point A is stable. All the trajectories spiral toward the stable equilibrium point. This situation represents normal BWR operation. Once the linear stability threshold is crossed, point A becomes unstable, so the trajectories spiral away from it. This scenario is depicted in Fig. 2b. In both cases, the trajectories are asymptotically tangent to the $n = -1$ because they can neither cross it nor reach the unstable equilibrium point B.

Away from the two equilibrium points, possible behaviors of the phase-space trajectories are presented in Fig. 3. Case (a) depicts a globally unstable system where the trajectory continually spirals away from point A. Case (b) describes a situation where the trajectory departs from the equilibrium point but eventually stabilizes itself due to the nonlinearities in the system; thus the trajectory remains bounded and converges to a closed line. This closed line defines a limit cycle, which corresponds to a periodic solution of fixed amplitude. Case (c) is similar to case (b) insofar as the trajectory is repelled by equilibrium point A and remains bounded due to the nonlinearities. The difference between cases (b) and (c) is that no periodic solution (i.e., a limit cycle or closed line in phase space) exists in case (c); although the trajectory stays within a bounded region, it never converges to a closed curve or to an equilibrium point. This bounded region is called a strange attractor, and the solution of a system of equations with a strange attractor is said to be aperiodic.

DETERMINISTIC NUMERICAL ANALYSIS

For the deterministic numerical analysis presented below, a 10% step-type perturbation was introduced in the neutron population in

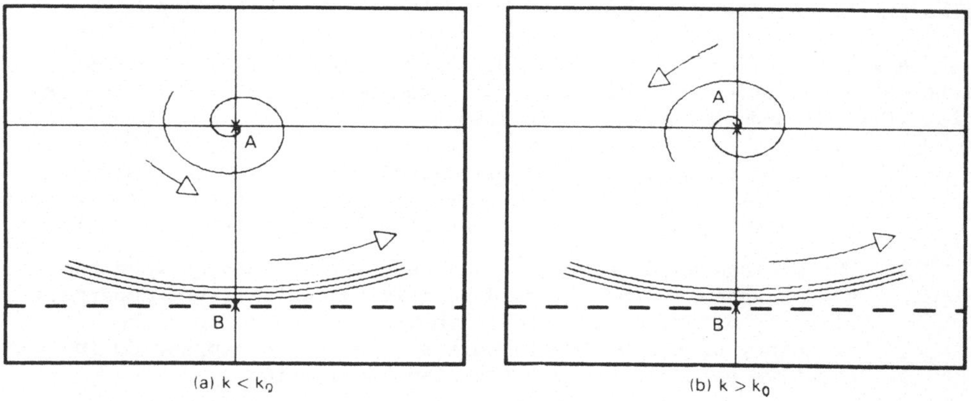

Fig. 2. Phase-space trajectory of the solution close to the two equilibrium points.

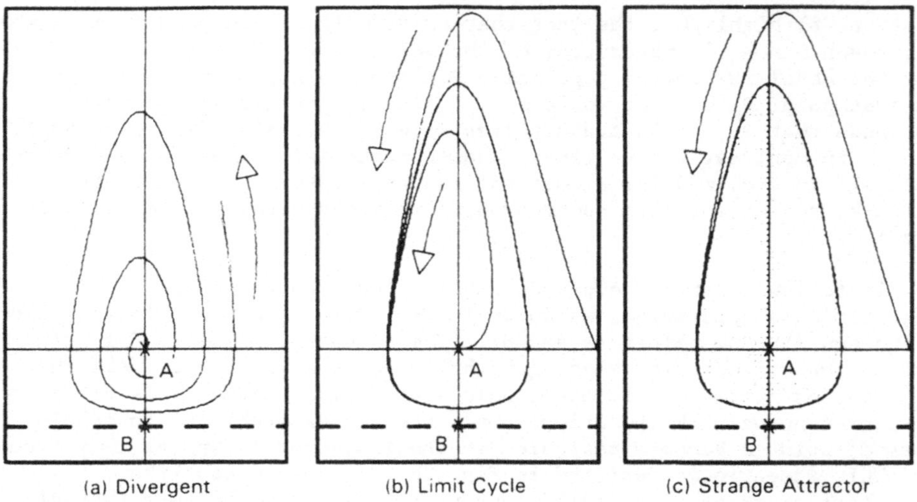

Fig. 3. Possible behavior of the phase-space trajectories away from the equilibrium points.

Eq. (3) at time $t = 0^{+}$. The solution was then allowed to converge freely to its final state. This numerical procedure was used because the solution converged quickly and smoothly to its final state.

The numerical solution of the model shows that limit cycles appear when feedback gain k is increased past critical value k_0. A limit cycle corresponds to a periodic and bounded solution of a system of equations that describes a closed trajectory in phase space. The phase space for the present model has five dimensions: the neutron density n, the delayed neutron precursors concentration c, the fuel temperature T, and the void reactivity feedback ρ_α and its derivative $d\rho_\alpha/dt$. The trajectory is parameterized in phase space by time. As an illustrative example, Fig. 4 presents the neutron density time trace when the system is perturbed from the unstable equilibrium point by inserting a $\Delta = 10\%$ perturbation. This time trace shows the development of a typical limit cycle; it also shows that once the limit cycle is reached, the signal becomes periodic and oscillates with an amplitude of ~20%. The amplitude of this limit cycle has the same order of magnitude as the one experimentally observed at Vermont Yankee. This fact increases confidence in the fact that this phenomenological model represents the general qualitative dynamic behavior of BWRs not only in the linear but also in the nonlinear regime.

A limit cycle of large amplitude is presented in Figs. 5 and 6. The main characteristics of this limit cycle can be studied in the time trace plot (Fig. 5) and summarized as follows:

1. The neutron density undergoes a series of periodic pulses of large magnitude. Between pulses, the neutron population remains at a low level close to the unstable shutdown equilibrium point.
2. The average void fraction, obtained as the ratio between the void reactivity and the void reactivity coefficient, undergoes a slightly deformed sinusoidal behavior, which defines the frequency of the neutronic pulses. The pulses appear during the negative part of the void fraction oscillation.
3. The fuel temperature rises sharply during the neutron pulse and then decays exponentially with the fuel time constant. The

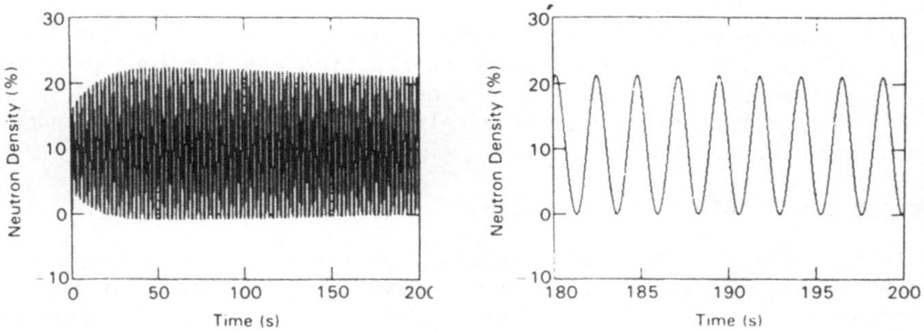

Fig. 4. The development of a typical limit cycle in the time domain

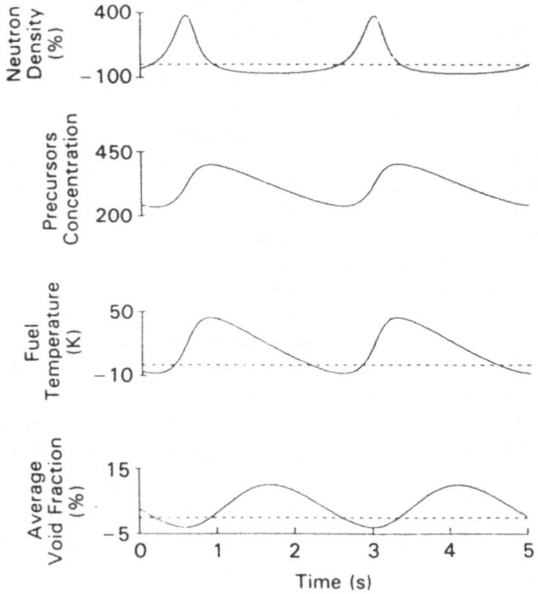

Fig. 5. Time domain representation of a large-amplitude limit cycle

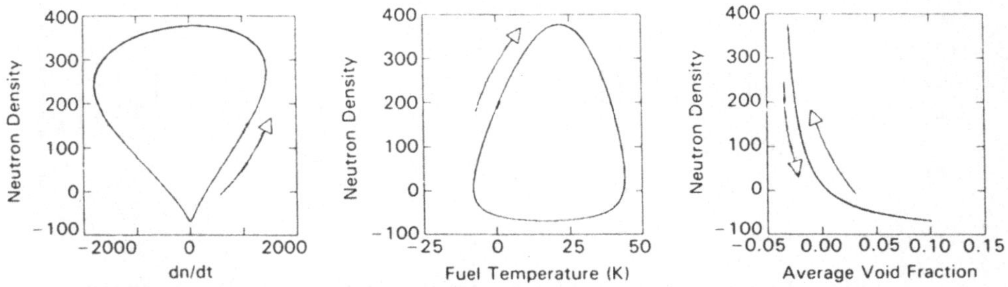

Fig. 6. Phase-space representation of a limit cycle

relative timing of the neutron, temperature, and void fraction pulses indicates that the void fraction oscillations provide the mechanism for generating the neutron pulses, while the temperature oscillations merely follow the evolution of the neutron pulses.

LIMIT CYCLE STABILITY: BIFURCATIONS

In the linear region, the stability of the system is quantified in terms of an *asymptotic decay ratio*;[4] however, in the nonlinear regime, the asymptotic decay ratio is always equal to 1.0, due to the appearance of limit cycles. Therefore, the decay ratio is not a good descriptor of the reactor dynamic state in this regime. A better dynamic descriptor in the nonlinear regime is the amplitude of the limit cycle oscillations. These two descriptors (i.e., the decay ratio and the limit cycle amplitude) are mutually complementary. Thus, the decay ratio defines the stability of the reactor in the linear regime where the amplitude of the limit cycle oscillations is zero. In the nonlinear regime, the decay ratio becomes constant and equal to 1.0, but the amplitude of the oscillations defines the dynamic behavior of the reactor.

The previous discussion indicates that the concern in the nonlinear regime is the stability of the amplitude of the oscillations. This fact is highlighted in Fig. 7. This figure shows the development of the limit cycle for three different values of the feedback gain: (a) $k = 1.2$, (b) $k = 1.4$, and (c) $k = 1.5$. We observe clear differences among the ways the limit cycle is reached in the three cases. In case (a), the amplitude of the oscillation (which is equal to the maximum value of the pulses, i.e., the signal envelope) follows a smooth curve and promptly converges to the final amplitude. In case (b), the amplitude oscillates around the final value and eventually converges to it. In case (c), however, the amplitude oscillates but never converges to a final value; the amplitude itself describes an undamped periodic oscillation. Thus the amplitude of the limit cycle in case (c) has become unstable and is following a new limit cycle of its own with twice the original period. This fact causes the original signal to periodically exhibit two pulses

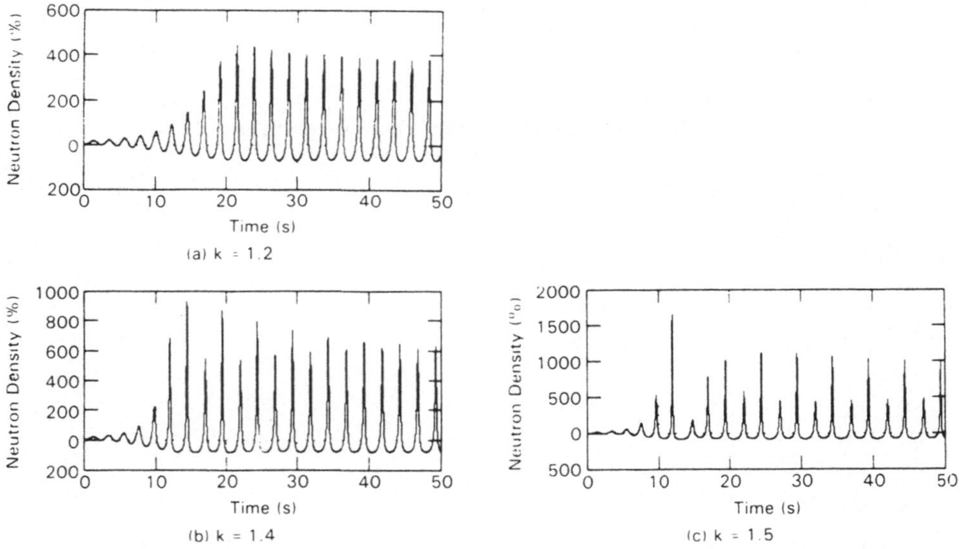

Fig. 7. Development of an instability of the limit cycle amplitude

of different magnitude. This process is customarily called a "period-doubling pitchfork bifurcation."[5]

The bifurcation process can be seen more clearly in phase space. For instance, in the $n - T$ phase-space plane (Fig. 8), the original limit cycle curve splits (bifurcates) into two new curves. The trajectory described by the solution follows first the inside curve and then the outside curve. The cycle is then repeated periodically. The small and large curves in Fig. 8 correspond to the small and large pulses in Fig. 7c.

Further increasing the feedback gain k produces a cascade of period-doubling bifurcations, which leads to an aperiodic regime. The transition to aperiodicity is governed by a set of universal constants predicted by Feigenbaum's universality theory.[5] The bifurcation diagram shown in Fig. 9 is constructed by plotting the extrema of the stationary solution $n(t)$ as a function of the heat transfer coefficient k. A cascade of period-doubling pitchfork bifurcations arises as k is increased past k_0. The critical bifurcation values k_j converge to the accumulation point $k_\infty = 1.61811k_0$; at k_∞ the BWR model enters into an aperiodic self-pulsing mode. Increasing k past k_∞ leads to regions of aperiodic behavior intermixed with windows of periodicity. Since this BWR model is not driven externally, its period is determined by the parameters appearing in Eqs. (1) through (5), and this period varies continuously with k except when $k = k_j$, $j = 0, 1, \ldots$.

Table II presents calculated values of k_j/k_0, the ratios $\delta_j = (k_j - k_{j-1})/(k_{j+1} - k_j)$, and the ratios α_j of the successive pitchfork splittings, for $j = 0$ through 6. Feigenbaum's theory predicts that, as $j \to \infty$, δ_j and α_j tend to the universal constants $\delta = 4.6692\ldots$ (the convergence ratio) and $\alpha = 2.5029\ldots$ (the pitchfork scaling parameter) shown in the last row of Table II. Also presented in this row is the value, calculated by extrapolation, of the accumulation point k_∞. (In the literature, the "accumulation point" is occasionally called the "critical point.") Overall, there is good agreement between these theoretically predicted values for δ and α and the calculated values δ_j and α_j.

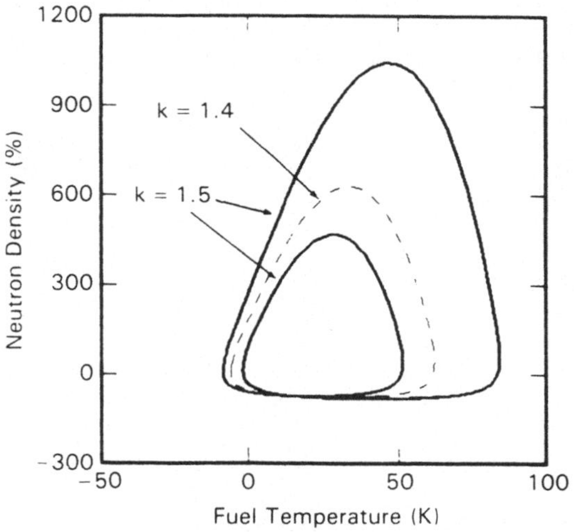

Fig. 8. Illustration of a period-doubling bifurcation in phase space.

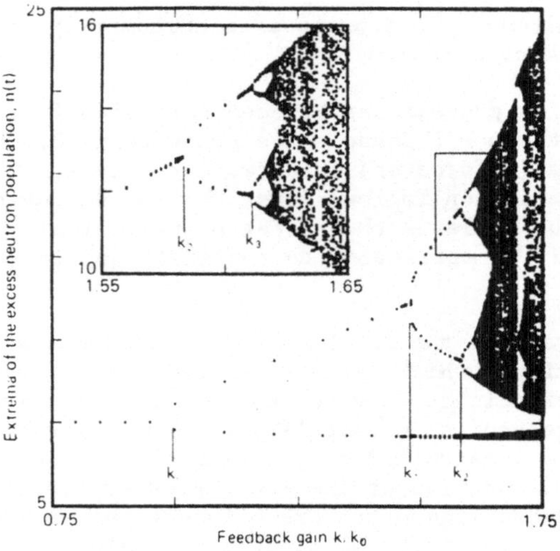

Fig. 9. Bifurcation diagram for $n(t)$.

Table II

Estimation of Universal Constants δ and α; Aperiodic Behavior Commences at k_∞

Cycle j	Period 2^j	Critical Bifurcation Values, k_j/k_o	δ_j	α_j
0	1	1.0	---	---
1	2	1.470 ± 0.002	---	---
2	4	1.584 ± 0.001	4.123 ± 0.126	1.095
3	8	1.6103 ± 0.0001	4.335 ± 0.295	2.207 ± 0.097
4	16	1.6165 ± 0.0001	4.242 ± 0.314	2.391 ± 0.332
5	32	1.61775 ± 0.00001	4.960 ± 0.596	2.465 ± 0.154
6	64	1.618025 ± 0.000005	4.545 ± 0.648	2.517 ± 0.094
∞	∞	1.61811 (extrapolated)	4.6692...(Ref.2)	2.5029...(Ref.2)

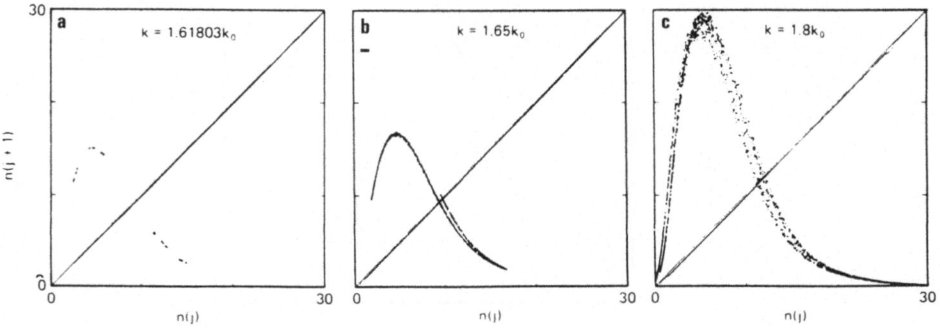

Fig. 10. Plot of the $(j + 1)$'th maximum of the excess neutron population, $n(t)$, versus the j'th maximum of $n(t)$, for $j = 0,1,2,\ldots$: (a) periodic region, $k = 1.61803k_0$, (b) aperiodic region, $k = 1.650k_0$, and (c) aperiodic region, $k = 1.80k_0$.

Figure 10 shows one-dimensional mappings obtained when the Poincare map for this BWR model is reduced by plotting the j'th maximum of $n(t)$ versus the $(j + 1)$'th maximum of $n(t)$. The plot in Fig. 10a is obtained for $k = k_7 = 1.61803k_0$, when the BWR model has just undergone its seventh bifurcation to a 128-cycle; to the numerical accuracy in this work, $k = k_7$ delineates the limits of the periodic region since the slightest increase over k_7 triggers aperiodic behavior. Figure 10a indicates that in the periodic region, the behavior of this BWR model is essentially that of a one-dimensional noninvertible map with a quadratic extremum, a result expected from the previously discussed calculations of δ and α.

In the aperiodic region, the one-dimensional maps shown in Fig. 10b ($k = 1.650$) and in Fig. 3c ($k = 1.80$) display unexpected foldings, indicating that a double-valued relation might exist between successive maxima. This double valuedness, though, is illusory: An examination of the dynamic evolution of the relation between successive maxima reveals that the BWR model evolves either on the lower or on the upper branch according to whether the magnitudes of preceding maxima form an increasing or a decreasing sequence. This hysteresis-like folding displayed by the reduced Poincaré map indicates that a many-term recursion relation is needed to investigate the behavior of this BWR model in its aperiodic region.

STOCHASTIC NONLINEAR ANALYSIS

In contrast with the previous section, which dealt exclusively with a deterministic analysis of the nonlinear behavior of BWRs, this section analyzes the effects of nonlinearities on the BWR behavior under stochastic (random) excitations (sources). For this purpose, the phenomenological model was externally driven with a band-limited Gaussian white noise, and the equations were solved numerically in the time domain.

Two parameters were varied for this stochastic analysis: the feedback gain k and the variance of the driving noise source. The traces generated for $n(t)$ were fast-Fourier transformed to obtain power spectral densities (PSDs). The development of a limit cycle in the presence of stochastic sources is shown in Fig. 11, where the envelopes (maxima and minima) of the oscillation are plotted as a function of time for three different values of the noise-source variance. To generate these plots, the system was held originally at the unstable equilibrium point. At time $t = 0$, a zero-mean white noise was applied. The amplitude of the oscillations increased until it reached a limit cycle. As expected, the amplitude of these limit cycle oscillations was found to be independent of the magnitude of the driving noise variance.

Studying the effects of the feedback gain on the system behavior shows that, as predicted by linear studies, for stable systems (when $k < k_0$) the neutron PSD exhibits a single peak at the reactor characteristic frequency of oscillation. However, as k approaches k_0 while maintaining the driving-source variance constant, the PSD develops peaks at the harmonics of this fundamental frequency. For $k > k_0$, the power oscillations increase in time and eventually reach a limit cycle, with an enhancement of the harmonic components of the PSD as seen in Fig. 12. This figure shows three PSDs for different values of k. In case (a) the model is barely stable and only the fundamental peak is clearly discernible at ~0.4 Hz. Case (b) represents a small amplitude limit cycle for which the value of k is only slightly above the critical value k_0. Case (c) corresponds to a fully developed large amplitude limit cycle. The main difference between the stable and the unstable

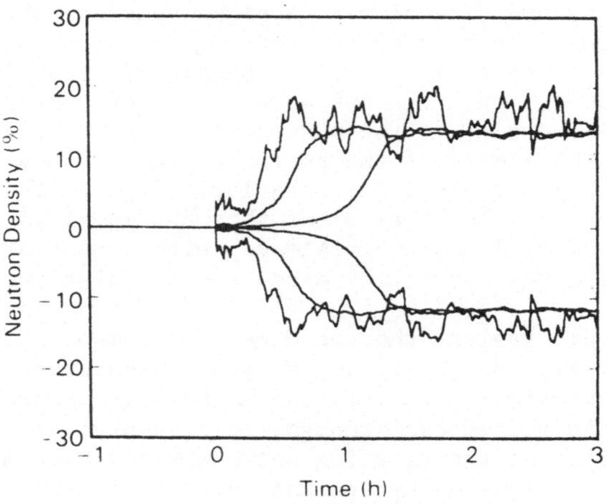

Fig. 11. Envelopes of the development of a limit cycle in the presence of noise.

PSDs is the appearance of higher harmonics. These harmonics have a large magnitude, so they should be measurable in real-life experiments even in the presence of measurement and process noise.

One of the consequences of the appearance of a limit cycle in a reactor is an increase in the variance (noisiness) of the neutron density as seen by the in-core neutron detectors. An increase in neutron noise variance, however, could also be due to an increase in the noise of other variables, such as flow, that influence the neutronics. It is of current interest to be able to distinguish between these two kinds of noise increases because different corrective action might need to be taken, depending on the cause. For example, if the increase in noise is due to an instability (i.e., the appearance of a limit cycle), the reactor can be made more stable by increasing the flow rate through the core; thus, a flow increase eliminates the extra noise. If the increase in noise is due to a pump malfunction, however, an increase in flow would probably only worsen the problem.

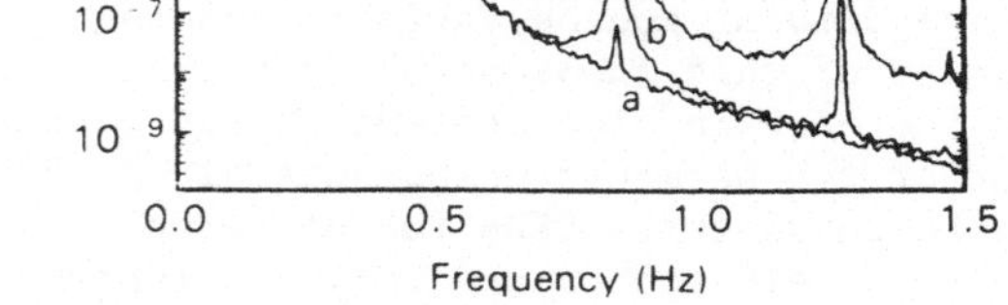

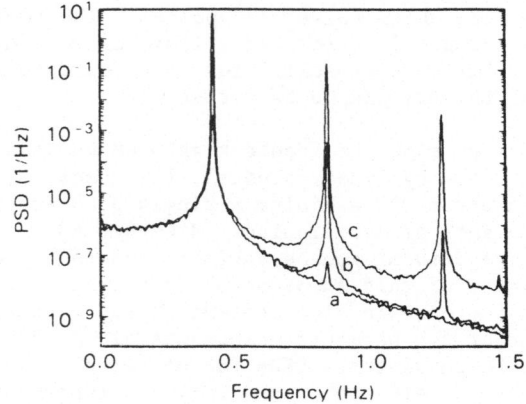

Fig. 12. The PSDs before and after the development of a limit cycle: (a) slightly stable, (b) slightly unstable, and (c) fully developed limit cycle.

Distinguishing between different causes leading to a noise increase is a current concern regarding BWR single-loop operation (SLO):[6] Operating some reactors with a single recirculation pump above 40% of rated flow causes neutron noise increases of 300%. This increase could be caused by a hypothetical instability induced by the special mode of operation, or it could be caused by increased flow noise due to crossflow in the downcomer between the inactive and active jet pump banks. This highlights the need to develop a technique for differentiating between these two scenarios.

The present model was used to generate neutron density time traces for both stable and unstable conditions. The variance of the input noise source was adjusted so that the variance of the output neutron noise was of the same order of magnitude in both cases. The resulting time traces are presented in Fig. 13. Although there are obvious differences between the unstable reactor condition (a) and the stable one (b), it is not easy to determine if case (a) is really a limit cycle or not. A more sensitive technique is required to differentiate these two cases. Figure 14 presents the PSDs of the time traces for the above two cases.

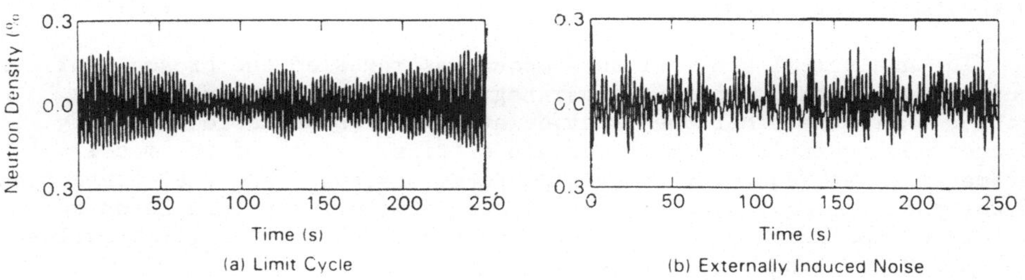

Fig. 13. Comparison between limit cycle oscillations and externally induced noises in the time domain.

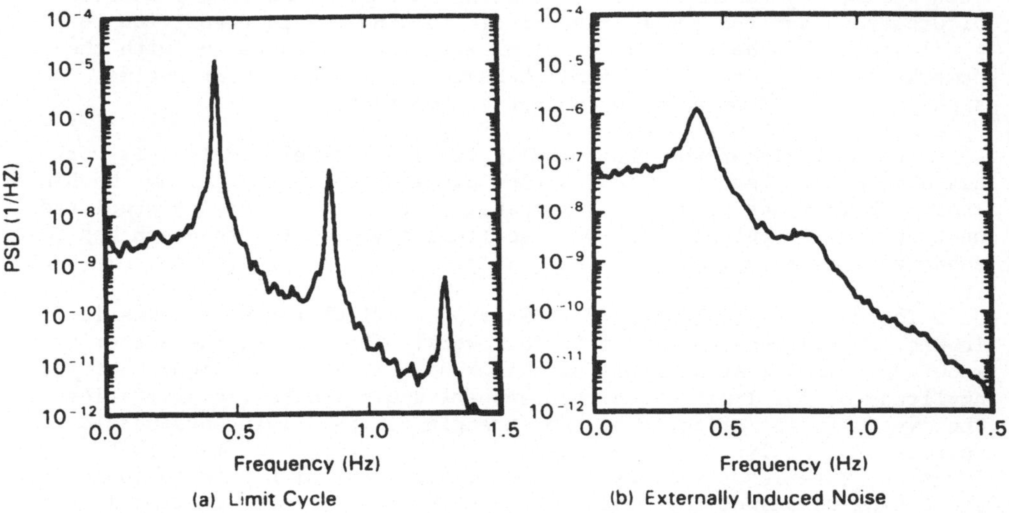

Fig. 14. Comparison between limit cycle oscillations and externally induced noise in the frequency domain.

Here, the differences are more obvious. In case (a), where the reactor is unstable, the characteristic peak at ~0.4 Hz is very sharp. The main difference, however, is the appearance of higher harmonics in the neutron PSD. Case (b), i.e., the stable condition, also shows harmonic contamination due to the large amplitude of the noise, but since the peak is wider, this contamination is not as obvious as in case (a). Furthermore, in a real-life measurement it would be impossible to distinguish this small harmonic contamination from the measurement and process noise. Therefore, if large-amplitude oscillations occur, nonlinearities appear as a harmonic contamination in the neutron PSD. If the oscillations are due to the appearance of limit cycles, the harmonic contamination is easily distinguishable from the background noise because it appears as sharp peaks at harmonic frequencies of the characteristic peak.

SUMMARY AND CONCLUSIONS

A phenomenological nonlinear model of BWR dynamics has been developed. In the linear regime, the results given by this model closely agree with the results from large, detailed BWR codes. This model also predicts the appearance of experimentally observed limit cycle behavior in the nonlinear regime.

The analytical study of this model has revealed the presence of two equilibrium points. In the linear regime, one equilibrium point is stable, while the other one is associated with an unstable shutdown reactor configuration. Above certain critical values of the model parameters, both equilibrium points become unstable, and the interaction of the phase-space trajectories with two equilibrium points leads to a variety of nonlinear phenomena, such as the appearance of limit cycles and period-doubling bifurcations.

The predictions of the phenomenological model have been analyzed under both deterministic and stochastic stimulations. The deterministic analysis was performed by increasing the feedback gain. In the linear sense, the effect of this increase would be to make the reactor more unstable. In the nonlinear sense, however, the effect is to change the amplitude of the limit cycle. Additional increases in the feedback gain induce limit cycle instabilities leading to period-doubling pitchfork bifurcations and the onset of aperiodic pulsing. This transition to aperiodicity has been found to follow Feigenbaum's scenario, with the remarkable result the BWR dynamic behavior in the nonlinear regime appears to depend on a set of universal constants.

The analysis of the phenomenological model developed in this work has also shown that the void fraction oscillations arising from the BWR's thermal hydraulics provide the mechanism for the onset of the predicted neutron density pulses. The Doppler (fuel temperature) feedback has been shown to play only a minor role.

The nonlinear behavior of a BWR system stimulated by stochastic inputs has also been analyzed in this work. This is, to the author's knowledge, the first analysis of this kind. It has been found that the amplitude of the limit cycle that appears under stochastic excitation of the system is not affected by the strength of the driving stochastic source. The analysis of stochastic descriptors, such as the PSD function, revealed the appearance and enhancement of higher harmonics of the 0.5-Hz resonance that characterizes the BWR dynamic behavior in the linear regime. This finding has led to the development of diagnostic rules for differentiating between limit cycle oscillations and externally induced increases in the system's noise level. These diagnostic rules have been successfully applied to analyze the problem of abnormal noise increases that have been observed in BWRs under single-loop operating conditions.

To conclude, the work reported in this paper, in addition to its academic value, provides a basis for scoping calculations to determine the viability of limit cycle operation of commercial BWRs.

REFERENCES

1. Y. Waaranpera and S. Anderson, "BWR Stability Testing: Reaching the Limit Cycle Threshold at Natural Circulation," *Trans. Am. Nucl. Soc.*, 39:868 (1981).
2. S. A. Sandoz and S. F. Chen, "Vermont Yankee Stability Tests During Cycle 8," *Trans. Am. Nucl. Soc.*, 45:727 (1983).
3. J. March-Leuba, D. G. Cacuci, and R. B. Perez, "Nonlinear Dynamics and Stability of BWRs: Part 1—Qualitative Analysis" and "Part 2—Quantitative Analysis," *Nucl. Sci. Eng.*, 93:111-136 (1986).
4. J. March-Leuba, "Dynamic Behavior of Boiling Water Reactors," PhD Dissertation, The University of Tennessee, Knoxville, Tennessee (1984).
5. M. J. Feigenbaum, "The Transition to Aperiodic Behavior in Turbulent Systems," *Commun. Math. Phys.*, 77:65 (1980).
6. J. March-Leuba, R. T. Wood, P. J. Otaduy, and C. O. McNew, "Stability Tests at Browns Ferry Unit 1 Under Single Loop Operating Conditions," *Nucl. Technol.*, 74:38-52 (1986).

NONLINEAR DYNAMICS OF THREE-DIMENSIONAL POINT FISSION REACTORS

A.A. Harms and Z. Bilanovic

McMaster University
Hamilton, Ontario, Canada

INTRODUCTION

Fission reactors are devices which sustain specified nuclear reactions at controlled rates. These reactions combine to form a linked network consisting of single-body decays and two-body collisions for which some of the reaction rate parameters are also functions of the medium temperature. Particular nonlinear dynamical forms thus result.

Our objective here is to examine selected dynamical features of fission reactors as determined by some nonlinear extensions of point reactor kinetics characterizations. Particle densities and medium temperatures are the variables of interest in the formulation of the dynamical equations. Subsequently, we employ some analytical and numerical means to elucidate selected nonlinear dynamical respresentations.

SOURCES OF NONLINEARITY

The microscopic variations of particle densities such as fuel atoms, chain carriers, reaction products, and precursors -- all as a function of time -- constitute the fundamental description of common interest because these densities relate directly to macroscopic reactor operational features such as power generation. A particularly convenient and useful dynamic in the so-called point-kinetics formulation[1,2] commonly written as

$$\frac{dN_n}{dt} = \left(\frac{\rho - \beta}{\Lambda}\right) N_n(t) + \sum_i \lambda_i N_i(t) \tag{1a}$$

$$\frac{dN_i}{dt} = \left(\frac{\beta}{\Lambda}\right) N_n(t) - \lambda_i N_i(t), \quad i = 1, 2, \ldots \tag{1b}$$

Here $N_n(t)$ is the neutron density, $N_i(t)$ is the i'th delayed neutron precursor density and the remaining symbols are parameters of the fission system.

The most "sensitive" parameter in Eqs. (1) is the reactivity ρ. This is of particular importance herein because its dependence on medium temperature -- either fuel temperature $T_f(t)$ or coolant temperature $T_c(t)$ or both -- may significantly effect the form of the dynamical equations. This becomes evident if we write

$$\rho = g_\rho[T_f(t),\ T_c(t)] \tag{2}$$

for which, by energy conservation, dynamical equations for these temperatures take on forms such as the following.

$$\frac{dT_j}{dt} = q_{in}H_{j,in} - q_{out}H_{j,out} + \sum_k \xi_{jk}R_{jk}Q_{jk} \tag{3}$$

Here $q_{(\,)}$ are heat conduction parameters, $H_{(\,)}$ are cross-boundary energy flow rates, $Q_{(\,)}$ are reaction energy Q-values, and $\xi_{(\,)}$ are thermal parameters. The complete nonlinear fission dynamic of interest is therefore given in the form

$$\frac{dN_i}{dt} = f_i(N_i, T_j) \; ; \qquad i = 1, 2, ..., I \tag{4a}$$

$$\frac{dT_j}{dt} = f_j(N_i, T_j) \; ; \qquad j = 1, 2, ..., J \tag{4b}$$

We next consider several reductions of Eq. (4) to specially recognizable form for which $I+J = 3$.

SELECTED THREE-PHASE-SPACE REDUCTIONS

Our interest in three coupled nonlinear dynamical equations stems from the recognition that the necessary -- though not sufficient -- conditions for interesting dynamical features such as chaos and strange attractors demands at least three such equations[3-7].

As a first dynamic of interest, we begin with the point kinetics description for the neutron density $N_n(t)$ and one effective group of delayed neutron precursors $N_d(t)$, Eq. (1). We next use the reactivity taken to depend upon the fuel temperature $T_f(t)$ according to

$$\rho(T_f) = \alpha_f[T_f(t) - T_{fo}] \tag{5}$$

Here α_f is the positive or negative temperature coefficient of reactivity and T_{fo} is the nominal operating fuel temperature. Then we take the fuel temperature to be given by Newton's Law of cooling

$$\frac{dT_f}{dt} = \gamma_f N_n(t) - \frac{[T_f(t) - T_{co}]}{\tau_f} \tag{6}$$

where the first term is the fission rate heat generation with γ_f constant, τ_f is the mean energy residence time in the fuel and T_{co} is a constant coolant temperature. Substituting Eq. (5) into Eq. (1) and combining with Eq. (6) yields therefore the following nonlinear dynamical system:

$$\frac{dN_n}{dt} = \left[\frac{\alpha_f T_{fo} - \beta}{\Lambda}\right] N_n(t) + \left(\frac{\alpha_f}{\Lambda}\right) T_f(t) N_n(t) + \lambda_d N_d(t) \tag{7a}$$

$$\frac{dN_d}{dt} = \left(\frac{\beta}{\Lambda}\right) N_n(t) - \lambda_d N_d(t) \tag{7b}$$

$$\frac{dT_f}{dt} = \gamma_f N_n(t) - \frac{[T_f(t) - T_{co}]}{\tau_f} \tag{7c}$$

For reasons of textual clarity we will call this the Case I "Fuel Temperature Feedback" dynamics.

A ready extension of Eq. (1) valid for longer time intervals, follows from the suppression of delayed neutron precursors and admitting the coolant temperature to be a function of time and affecting the reactivity by superposition of temperature effects:

$$\rho = \alpha_f[T_f(t) - T_{fo}] + \alpha_c[T_c(t) - T_{co}] \tag{8}$$

Here α_c is the coolant temperature coefficient of reactivity. Employing further Newton's Law of cooling for both the fuel-coolant boundary and for the coolant-sink boundary, then yields the following system with the notation self-evident,

$$\frac{dN_n}{dt} = \left[\frac{\alpha_f T_{fo} + \alpha_c T_{co}}{\Lambda}\right] N_n(t) + \left(\frac{\alpha_f}{\Lambda}\right) N_n(t)\, T_f(t) + \left(\frac{\alpha_c}{\Lambda}\right) N_n(t)\, T_c(t) \tag{9a}$$

$$\frac{dT_f}{dt} = \gamma_f N_n(t) - \frac{[T_f(t) - T_c(t)]}{\tau_f} \tag{9b}$$

$$\frac{dT_c}{dt} = \frac{[T_f(t) - T_c(t)]}{\tau_f} - \frac{[T_c(t) - T_{sink}]}{\tau_c} \tag{9c}$$

By analogy to the Case I "Fuel Temperature Feedback" dynamic, Eqs. (7), this second nonlinear formulation, Eqs. (9), will be labeled the Case II "Fuel-Coolant Temperature Feedback" dynamic.

The fuel temperature coefficient of reactivity α_f will invariably be negative while the coolant temperature coefficient of reactivity α_c can be positive or negative. Hence, Case I will be restricted to $\alpha_f < 0$ but Case II will be either Case II+ (for $\alpha_c > 0$) or Case II− (for $\alpha_c < 0$).

ANALYTICAL EXAMINATIONS

We have undertaken an analytical examination for the two fission reactor dynamics of interest, Eqs. (7) and Eqs. (9): the "eigenvalue characterization"(8) about the point(s) of equilibria. This examination has provided some indications on questions of parameter-range implications or the local-global flow of the trajectories. The eigenvalues are of particular relevance.

Equilibrium points X* for a nonlinear dynamic of the form of Eq. (1a) are simultaneous solution of

$$\left.\frac{dX_i}{dt}\right|_{X=X^*} = 0 \tag{10}$$

Sufficiently small perturbations about these equilibrium points are introduced and the eigenvalues for the consequent linear dynamic -- which here involves a cubic polynomial characteristic equation -- found.

For the fuel temperature dynamic, Eqs. (7), two equilibria exist. One set and its corresponding eigenvalues for typical parametric reactor values yields

$$N^*_{n,1} = 0 \qquad \lambda_{1,1} = 16.87$$

$$N^*_{d,1} = 0 \qquad \lambda_{1,2} = -8.53$$

$$T^*_{f,1} = T_c\ ^\circ C \qquad \lambda_{1,3} = -8.35$$

and evidently represent an unstable point. The second equilibria and eigenvalue set are

$$N^*_{n,2} = \frac{T_{fo} - T_c}{\gamma \tau_f} \qquad \lambda_{2,1} = -75.07$$

$$N^*_{d,2} = \frac{\beta(T_{of} - T_c)}{\gamma \Lambda \lambda_d \tau_f} \qquad \lambda_{2,2} = -0.017 + i0.028$$

$$T^*_{f,2} = T_{fo} \qquad \lambda_{2,3} = -0.017 - i0.028$$

and clearly suggest a stable spiral.

A similar analysis was undertaken for the fuel-coolant temperature feedback dynamic with the result displayed in Table II.

NUMERICAL ANALYSIS

Figures 1 to 8 provide a graphical depiction of some of the dynamical patterns associated with the nonlinear descriptions developed here. These solutions have been obtained for the parameter values listed in Table I which suggests a "typical" fission reactor.

Based on the analytical and numerical investigations, the parameters which appear to exert the most profound influence on the reactor behaviour are the mean energy residence times τ_f and τ_c; the other parameters alter the quantitative but not the qualitative behaviour of the solutions.

For Case I, the "Fuel Temperature Feedback" system, two equilibria points exist. The first at zero neutron density is unstable under a small perturbation while the second is a stable spiral. The "tightness" of the spiral is governed by the magnitude of τ_f. For a comparatively large τ_f, that is long mean energy residence time, slow fuel heat removal, the time to reach equilibrium is longer than for a smaller τ_f. Physically, this means that the high fuel temperatures will be retained for a longer time inhibiting fission reactions. For a small τ_f, that is rapid fuel heat removal, the imaginary parts of the eigenvalues vanish and the equilibrium point becomes stable. Thus, a small τ_f minimizes the length of time the fuel retains high temperatures and decreases the adverse effect on neutron production.

For Case II+, the effect of varying τ_f and τ_c are the same as for Case I.

Case II–, however, is significantly different. As can be seen from Table II, the first attractor is a saddle point and the second attractor possesses one positive, one negative and one zero eigenvalue. Figures 5 and 6 show the neutron density as a function of fuel and coolant temperature respectively. From this figure, it can be seen that the solution is periodic (limit cycle) due to the complex interplay of the two unstable attractors. The magnitude of τ_f and τ_c again determine the rapidity with which the solution approaches the limit cycle. In addition to the magnitude, the ratio, τ_f/τ_c, is also important. For t_f/τ_c small (for our case $\tau_f/\tau_c < 2$), the solution becomes unstable and for τ_f/τ_c large ($\tau_f/\tau_c > 2$) the solution becomes stable as shown in Figures 7 and 8. This suggests that the two parameters, τ_f and τ_c, as well as their ratio, can alter the behaviour of a fission system from a stable reaction to a periodic limit cycle mode of behaviour.

CONCLUDING COMMENT

It is evident that a nonlinear characterization of fission reactor provides dynamical information of considerable interest and relevance. Further research is evidently required in order to clarify both the mathematical description and the methodology of reactor parameter specifications.

TABLE I

List of parameter values used in calculations

$\alpha_f = 10^{-5}\ °C^{-1}$	$T_{of} = 850°C$
$\alpha_c = 3\times10^{-5}\ °C^{-1}$	$T_{oc} = 250°C$
$\beta = 0.0075$	$T_{SINK} = 200°C$
$\Lambda = 10^{-4}\ s^{-1}$	τ_f – variable
$\gamma = 3.4\times10^{-6} cm^3\ °C\ s^{-1}$	τ_c – variable

TABLE II

Equilibria and eigenvalue characterizations

Case I: Fuel temperature feedback dynamic, Eqs. (7).

$N^*_{n,1}=0$	$\lambda_{1,1}=16.87$
$N^*_{d,1}=0$	$\lambda_{1,2}=-8.53$
$T^*_{f,1}=T_c$	$\lambda_{1,3}=-8.35$
$N^*_{n,2}=\dfrac{T_{fo}-T_c}{\gamma\tau_f}$	$\lambda_{2,1}=-75.07$
$N^*_{d,2}=\dfrac{\beta(T_{fo}-T_c)}{\gamma\Lambda\lambda_d\tau_f}$	$\lambda_{2,2}=-0.017+i0.028$
$T^*_{f,2}=T_{fo}$	$\lambda_{2,3}=-0.017-i0.028$

TABLE II continued

Case II +: Coolant-fuel temperature feedback dynamic, Eqs. (9).

$N^*_{n,1}=0$	$\lambda_{1,1}=16.74$
$T^*_{f,1}=T_s$	$\lambda_{1,2}=-33.38$
$T^*_{c,1}=T_s$	$\lambda_{1,3}=16.64$
$N^*_{n,2}=\dfrac{(T^*_{c,2}-T_s)}{\tau_f\gamma}$	$\lambda_{2,1}=-0.06$
$T^*_{f,2}=T^*_{c,2}(1+\dfrac{\tau_f}{\tau_c})-\dfrac{T_s\tau_f}{\tau_c}$	$\lambda_{2,2}=-0.053+i1.29$
$T^*_{c,2}=\dfrac{[\alpha_f(T_{fo}+\dfrac{T_s\tau_f}{\tau_c})-\alpha_cT_{co}]}{[\alpha_f(1+\dfrac{\tau_f}{\tau_c})-\alpha_c}$	$\lambda_{2,3}=-0.053-i1.29$

Case II −: Coolant-fuel temperature feedback dynamics, Eqs. (9):

$N^*_{n,1}=0$	$\lambda_{1,1}=30.17$
$T^*_{f,1}=T_s$	$\lambda_{1,2}=-66.0$
$T^*_{c,1}=T_s$	$\lambda_{1,3}=35.8$
$N^*_{n,2}=\dfrac{(T^*_{c,2}-T_s)}{\gamma\tau_f}$	$\lambda_{2,1}=0.0$
$T^*_{f,2}=T^*_{c,2}(1+\dfrac{\tau_f}{\tau_c})-\dfrac{T_s\tau_f}{\tau_c}$	$\lambda_{2,2}=0.24$
$T^*_{c,2}=\dfrac{[\alpha_f(T_{fo}-\dfrac{T_s\tau_f}{\tau_c})+\alpha_cT_{co}]}{[\alpha_f(1+\dfrac{\tau_f}{\tau_c})+\alpha_c}$	$\lambda_{2,3}=-8.24$

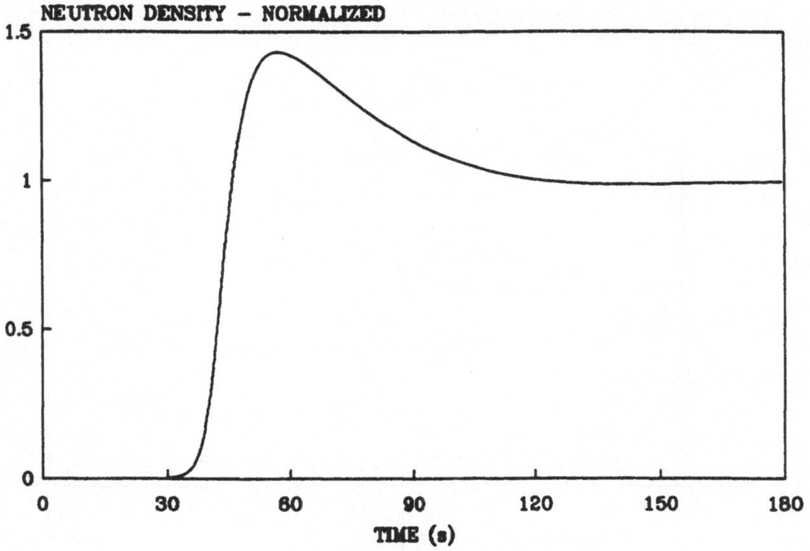

Figure 1 – Case I : Neutron Density vs Time

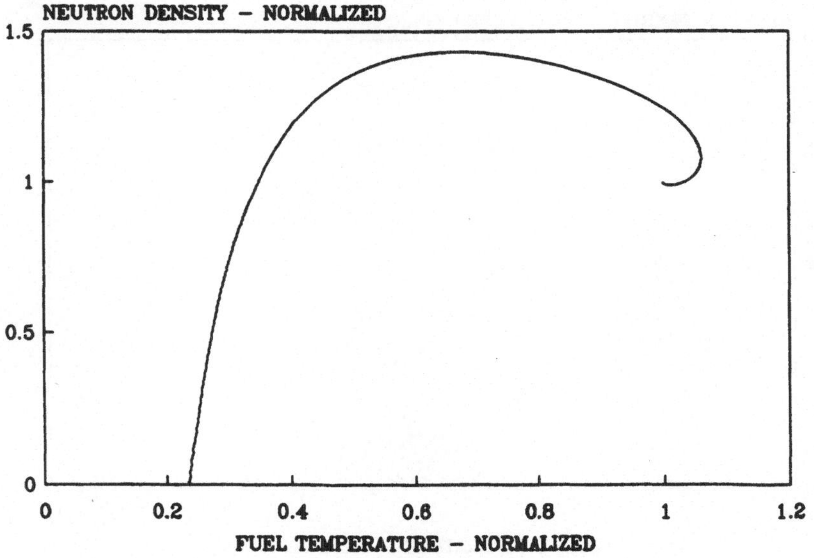

Figure 2 – Case I : Neutron Density vs Fuel Temperature

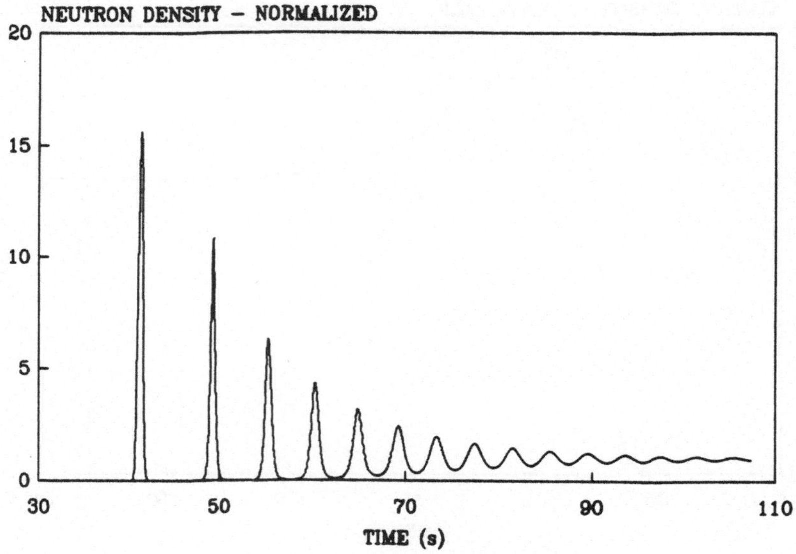

Figure 3 – Case II+ : Neutron Density vs Time

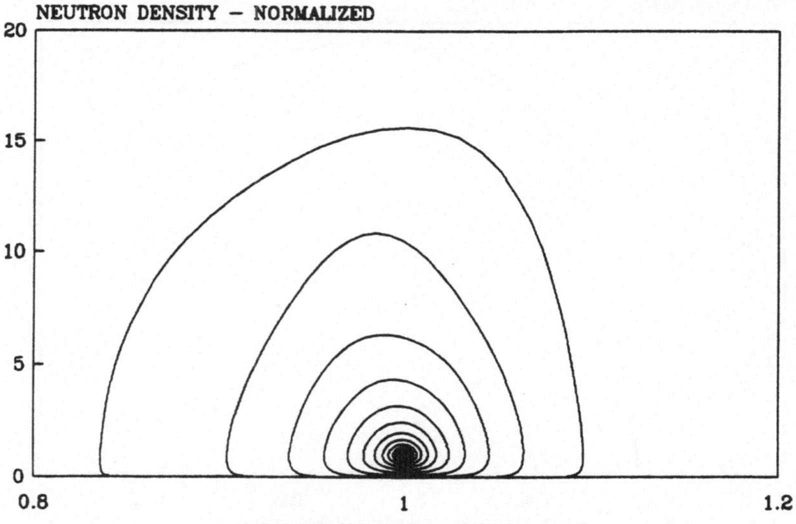

Figure 4 – Case II+ : Neutron Density vs Fuel Temperature

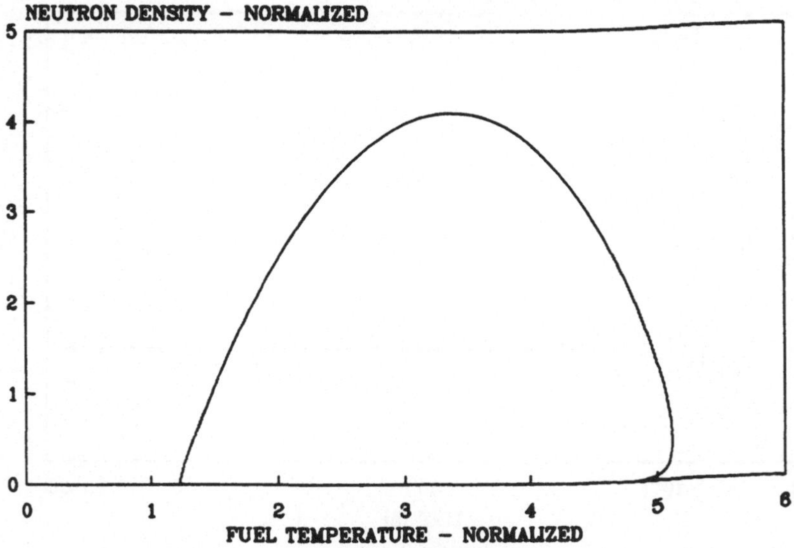

Figure 5 – Case II– : Neutron Density vs Fuel Temperature

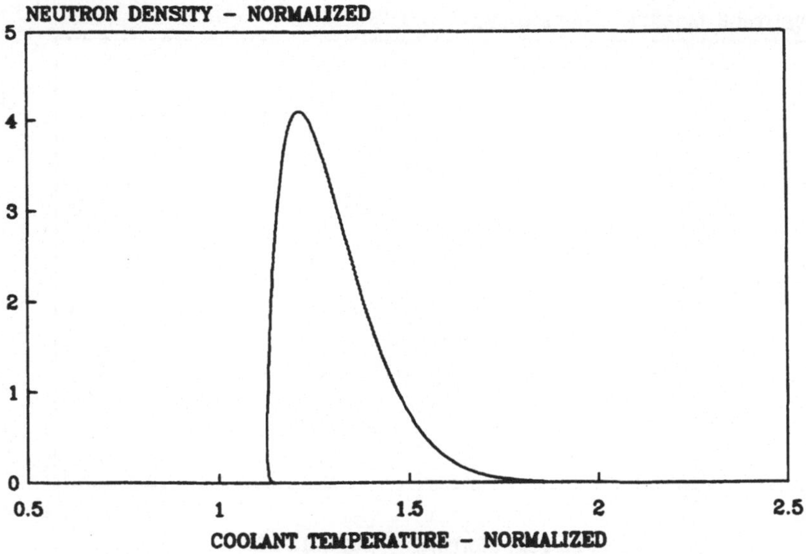

Figure 6 – Case II– : Neutron Density vs Coolant Temperature

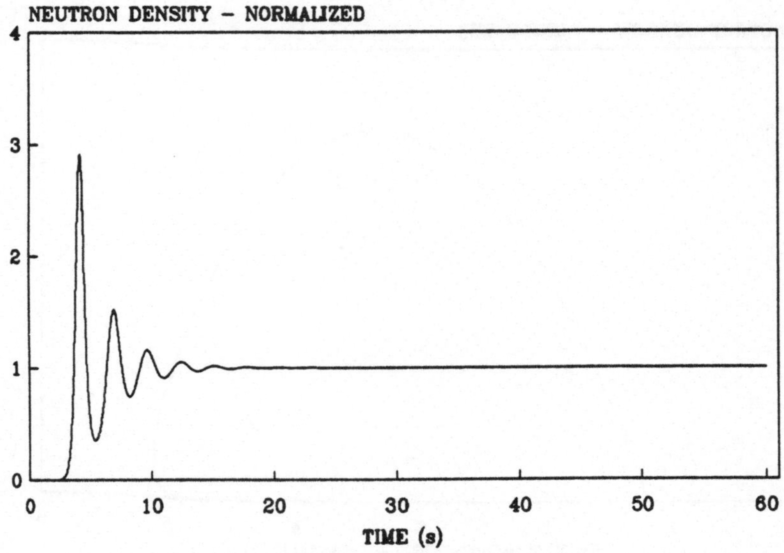

Figure 7 – Case II– : Long mean energy residence times

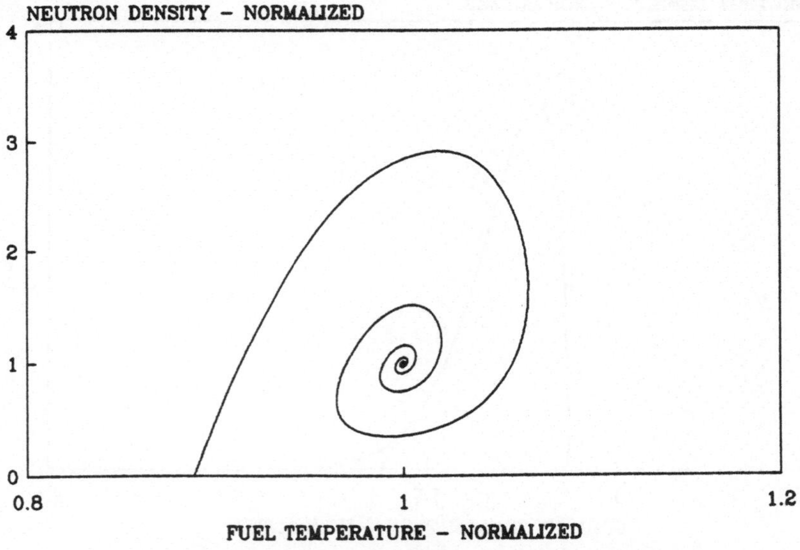

Figure 8 – Case II– : Long mean energy residence times

REFERENCES

1. D.L. Hetrick, Dynamics of Nuclear Reactors, University of Chicago Press, Chicago, USA (1971).
2. J. Lewins, Nuclear Reactors Kinetics and Control, Pergamon Press, Oxford, UK (1978).
3. J. Guckenheimer and P. Holmes, Nonlinear Oscillations, Dynamical Systems, and Bifurcation of Vector Fields, Springer-Verlag, Heidelberg, FRG (1983).
4. H.G. Schuster, Deterministic Chaos, Physik Verlag, Weinheim, FRG (1984).
5. J. March-Leuba, D.G. Cacuci and R.B. Perez, "Universality and Aperiodic Behaviour of Nuclear Reactors", Nucl. Sci. Eng. 86, 401 (1984).
6. A.A. Harms, Z. Bilanovic and H.K.Y. Leung, "The Existence of Fission-Driven Nonlinear Maps", Ann. Nucl. Energy 13(6), 341 (1986).
7. J. March-Leuba, D.G. Cacuci and R.F. Perez, "Nonlinear Dynamics and Stability of Boiling Water Reactors (Parts 1 and 2)", Nucl. Sci. Eng. 93, 111 (1986).
8. H. Haken, Advanced Synergetics, Springer-Verlag, Heidelberg, FRG (1983).

PHASE PLANE ANALYSIS OF NUCLEAR-COUPLED DENSITY-WAVE OSCILLATIONS

John C. Lee and Anozie Onyemaechi

Department of Nuclear Engineering
The University of Michigan
Ann Arbor, Michigan 48109-2104

INTRODUCTION

Due to a tight coupling between thermal-hydraulic and neutronic performance of boiling water reactors (BWRs), there exists the potential for nonlinear limit cycle oscillations in coolant flowrate and core power. Physical understanding of the limit cycle oscillations is of considerable importance for safe and reliable operation of BWRs, especially in regard to anticipated transient without scram (ATWS) events. Although considerable attention[1-8] has been given to these potential instability problems, most of the earlier investigations[1-3] have focused mainly on frequency-domain linear models for prediction of the instability threshold, assessment of the stability margins, and establishment of the stability boundary. While these linear analyses are intended for use as a design tool to avoid the inception of flow oscillations, they are not capable of predicting the severity of oscillations in an unstable configuration. More recently,[6-8] the potential for large-amplitude limit cycle oscillations has been explicitly studied with nonlinear BWR models for linearly unstable operating regimes.

In modern commercial BWRs, the predominant mode of interest[1,6-10] for the system stability is the density-wave oscillation (DWO). This low-frequency oscillation occurs as a result of regenerative interactions among the overall channel pressure drop, flowrate, and vapor generation rate. If the pressure drop across the boiling channel remains approximately constant during the oscillation, as is the case in BWR coolant channels subject to a large recirculation flowrate, a perturbation in inlet flowrate will cause an immediate change in outlet flowrate in the opposite direction. In addition to this momentum feedback effect, the inlet flowrate perturbation will also result in perturbations in vapor generation rate, boiling boundary, and void fraction in the two-phase region, with a time delay associated with the fluid motion. These delayed effects will eventually be propagated to the outlet, which will, in turn, cause an immediate reversal of the initial inlet perturbation due to the momentum feedback effect. In BWR channels, the DWO behavior will be reflected in neutronic power and flux oscillations due to void reactivity feedback. Thus, the characteristics of the nuclear-coupled density-wave oscillations (NCDWOs), including the stability and oscillation period, are determined essentially by the DWO phenomena. In unstable NCDWOs, the oscillation amplitude grows as heat generation exceeds dissipation and fuel temperature increases. With the resulting increase in the heat transferred to the coolant channel, the coolant density decreases thereby reinforcing the ongoing DWOs. The limit cycle is reached when the heat generation equals dissipation over each cycle.

The current generation of BWRs has been experimentally found[3] to behave as linear systems under normal operating conditions. Recent stability tests[4,5] have, however, shown that this type of reactor is susceptible to NCDWO instabilities when operated at low flow conditions. In fact, a recent transient event[11] at the LaSalle Unit 2 plant has realistically indicated the safety implications of these nuclear-coupled thermal-hydraulic instabilities. In this March, 1988, event, large-amplitude oscillations in neutron flux and power occurred following a recirculation pump trip. The reactor was operating at a steady-state condition with 84% of the rated power and 76% of the rated flow, when a valving error was made by instrumentation personnel. This resulted in a pressure pulse which tripped both recirculation pumps and rapidly reduced the power to 45% at natural circulation flow. The feedwater controller was unable to handle the large magnitude of the resulting steam flow and load reduction, causing the feedwater temperature to decrease by 45 ^{o}F and thereby inserting a positive reactivity. The plant went through hundreds of unstable oscillatory cycles before the reactor scrammed automatically on a high neutron flux level of 118%. Although the associated temperature oscillations were considerably smaller in magnitude and no damages to the plant equipment occurred, this event has essentially simulated an ATWS event and raised concerns regarding the potential limit cycle oscillations in BWRs.

Among the concerns regarding the BWR system stability is the sensitivity of the limit cycle amplitude of the power oscillations to initial reactor conditions. This is evident from the recent NCDWO studies[6-8] performed with BWR models of varying degrees of complexity and nonlinearity. March-Leuba et al. studied the nonlinear BWR stability with both a phenomenological[6] and a detailed physical[7] model for the boiling channel, with an emphasis on the potential bifurcation phenomena. Ward and Lee[8] used a two-region boiling channel representation, which is simple yet based on first principles. Coupled with a form of the point-kinetics equation, this thermal-hydraulic model was used to investigate the behavior of large amplitude NCDWOs through conventional time-integration solution and through a phase-space solution of the relaxation oscillations. One major limitation of this semi-analytic adaptation of Mishchenko's singular perturbation technique,[12] is the need to anchor a point in the trajectory through a separate numerical solution.

This paper presents a new method to determine phase-plane trajectories during the approach to a limit cycle. The method can serve as an efficient means to anchor the limit cycle trajectory in the singular-perturbation framework or as a short cut in conventional time-domain integration of the NCDWO equations. A simplified, nonlinear NCDWO model[8] has been used to demonstrate the algorithm which utilizes both the phase-plane and time-domain solutions to locate the extremum points of void reactivity K. This is accomplished by combining the energy and momentum conservation equations for the boiling channel in the phase plane of K and fuel temperature T. With these extremum points anchored, normal time-domain solution or phase-plane analysis of the system equation can be performed to obtain the limit cycle trajectory. This method of approach to the limit cycle is considerably faster than conventional numerical solution of the system equations, which may take hundreds of cycles before the magnitude of the limit cycle oscillations can be determined.

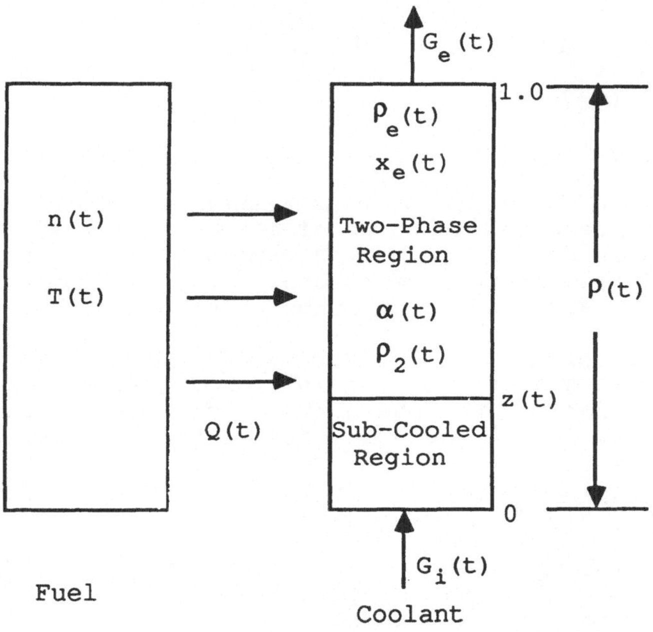

Fig. 1. The Simple NCDWO Model Structure

SIMPLIFIED NCDWO MODEL

The BWR model we have used in our NCDWO analysis is a single channel, lumped parameter representation of Ref. 8, shown schematically in Fig. 1. In order to clarify the physics behind the model and the NCDWO phenomena, the basic assumptions of the model are summarized here. The governing equations are also repeated here although with a simpler notation. In developing this coupled-nuclear thermal-hydraulic model, an effort has been made to make maximum use of first-principle descriptions of the key phenomena but with the least possible details. A key feature of the model is an explicit coupling between the neutronic and thermal hydraulic models through the void reactivity feedback and heat flux dependency on fuel temperature.

Neutronics and Core Energy Balance

The neutronic model is based on the point-kinetics equation with the infinite delayed approximation. With void reactivity K(t) and linear reactivity feedback coefficient C_1 associated with fuel temperature T(t), the normalized core power n(t) is calculated from:

$$\Lambda\frac{dn}{dt} = (K + C_1T - \beta)n + \beta, \tag{1}$$

where Λ is the neutron lifetime and β is the delayed neutron fraction. Core energy balance is written as

$$\frac{dT}{dt} = (n - 1 - T), \tag{2}$$

with the heat flux Q(t) to the coolant channel approximated as

$$Q(t) = Q_o(1 + T).$$

Thermal Hydraulic Model

The fluid dynamics and heat transfer characteristics of BWR cores are represented in terms of a two-region, single-channel model shown in Fig. 1. A moving boundary located at normalized distance z(t) separates the subcooled region from the two-phase region, with saturation properties assumed constant in time.

Mass Conservation. The subcooled region is assumed incompressible with normalized inlet mass velocity $G_i(t)$. Mass conservation for the two-phase region can then be represented in terms of K(t), $G_i(t)$, and exit mass velocity $G_e(t)$:

$$\frac{dK}{dt} = C_2(G_i - G_e), \tag{3}$$

where C_2 is a lumped constant including the void reactivity coefficient C_3 and channel length H. Here we make use of the relation between K(t) and channel average density $\rho(t)$:

$$K(t) = C_3[\rho(t)-\rho(0)]/\rho(0), \tag{4}$$

where

$$\rho = \alpha(1-z)\rho_{fg} + \rho_f \tag{5}$$

is given in terms of two-phase average void fraction α, boiling length z, and saturation properties ρ_{fg} and ρ_f.

Energy Conservation. For the subcooled region, with the inlet enthalpy assumed constant in time, the energy conservation equation is integrated over the boiling length z(t) to yield:

$$\frac{dz}{dt} = C_4[G_i - (1 + T)z / z_o], \tag{6}$$

where z_o = z(0) and C_4 is another constant parameter. The two-phase energy conservation relation is represented in terms of a time derivative of two-phase average void fraction $\alpha(t)$, which is eliminated through the corresponding two-phase continuity equation. We then obtain an algebraic form of the two-phase energy conservation relation:

$$G_e(1 - C_5 x_e) = G_i - C_6(1 + T)(1 - z), \tag{7}$$

where $x_e(t)$ is the channel exit quality, and C_5 and C_6 are constants dependent on saturation properties, $x_e(0)$, and z_o. Based on the assumption of spatially uniform heat flux $Q(t)$, α can be related to x_e:

$$\alpha = \frac{\gamma}{\gamma - 1}[1 - \frac{1}{\gamma - 1}\frac{\ln\{1 + x_e(\gamma - 1)\}}{x_e}], \tag{8}$$

where Thom's slip factor γ is calculated[13] at a given channel pressure.

Combining Eqs. (4), (5), and (8), we can express $K(t)$ in terms of $x_e(t)$ and $z(t)$, or equivalently represent $x_e(t)$ as a function of $K(t)$ and $z(t)$. Thus, we may write the two-phase energy conservation equation (7) as a nonlinear algebraic equation of the form:

$$f_1(G_i, G_e, T, K, z) = 0 . \tag{9}$$

Momentum Conservation. A key assumption of our NCDWO model is that the pressure drop ΔP across the channel remains constant during the transient. This condition is approximately satisfied in boiling channels with large recirculation flowrates. In stability analysis of U-tube steam generators, Kornfilt and Lee[14] observed that the variation in the overall pressure drop of a boiling channel in the steam generator secondary-side is considerably less than the corresponding variations in the single- and two-phase pressure drops. The constant pressure drop assumption is likewise expected to be a reasonable approximation for NCDWO analysis of BWR systems. Apart from this assumption, we represent as explicitly as possible various pressure drop terms including the gravitational term, two-phase frictional and acceleration terms, and inlet and exit form loss terms:

$$\Delta P = \frac{K_i}{2\rho_f}G_i^2 + \frac{K_e}{2\rho_e}G_e^2 + \frac{G_e^2}{\rho_2}\{F_o G_e^{-0.25} + A_o\}(1 - z)H + gH\rho \tag{10}$$

where

K_i, K_e = inlet and exit form loss coefficients, respectively,
ρ_e = exit fluid density = function of $x_e(t)$ and saturation properties,
ρ_2 = average two-phase density = $(1 - \alpha)\rho_f + \alpha\rho_g$, and
F_o, A_o, g = constants.

This accurate representation of ΔP is necessary because the interaction between single- and two-phase pressure drops drives the DWOs as discussed in the introduction. Similar to the formal representation of Eq. (9) for two-phase energy conservation, the momentum conservation equation (10) can be written as

$$f_2(G_i, G_e, K, z) = 0 \ . \tag{11}$$

Our NCDWO model can now be considered as four ordinary differential equations (1), (2), (3), and (6) for four system variables n(t), T(t), K(t), and z(t), respectively. The flowrates $G_i(t)$ and $G_e(t)$ appearing in these differential equations are obtained through two nonlinear algebraic equations (9) and (11). Although our model is based on a number of simplifying assumptions, it represents key physical phenomena consistently and efficiently. For a boiling channel with a significant two-phase pressure drop component, we readily observe from Eq. (10) that an increase, for example, in inlet flowrate G_i results in an essentially instantaneous decrease in outlet flowrate G_e. This momentum feedback effect is a direct consequence of the constant pressure drop boundary condition imposed by Eq. (10). Eq. (6) clearly illustrates the delayed effect of the inlet flow perturbation reflected in the boiling boundary perturbation with a time constant $C_4(1+T)/z_o$. Through the channel mass conservation relation of Eq. (3) and the two-phase energy conservation relation of Eq. (7), the initial increase in G_i and the delayed increase in z are reflected in exit flowrate G_e and exit flow quality x_e.

TIME-DOMAIN NCDWO SOLUTION

Time-domain solution[8] of the NCDWO model was obtained with the Crank-Nicolson scheme[15] with a time step of 100 μs and compared with the Vermont Yankee test data.[5] Although the comparison for the amplitudes of power and flow oscillations was reasonable, the calculated oscillation period of 1.4 s was considerably less than the measured value of 2.3 s. Through comparison of the DWO part of the model with the multi-cell finite-difference steam generator model of the TRANSG code,[16] it was concluded that the discrepancy in the oscillation period is a limitation of the two-region fluid channel model, which tends to propagate too rapidly perturbations at the channel inlet to the outlet.

A high sensitivity of the limit cycle amplitude to initial operating conditions was also noted,[8] as is discussed later in more detail. The numerical results also show that large relaxation oscillations can be expected. For a transient starting with a steady-state operation at 66% of the rated power and natural circulation flow, limit cycle amplitudes of 400% in power and 17 K in fuel temperature were calculated. It is worth noting that the fuel temperature oscillations are rather modest even with large power oscillations because the fuel heat capacity plays the role of a filter and dampens the effect of the power oscillations.

The numerical solution also shows that the overall transient is driven by DWOs with the associated characteristic behavior discussed in the introduction. For unstable BWR operations, very likely at natural circulation flow conditions, heat generation during the power pulse initially exceeds heat dissipation during the cooling period and the excess energy is accumulated resulting in a gradual increase in fuel temperature. With the resultant increase in the heat transferred to the coolant channel, the average channel density decreases and the negative void feedback increases, thereby reinforcing the ongoing DWOs. This, in turn, results in a larger positive reactivity feedback, when the oscillation resumes in the power pulse. This increase in the magnitude of the negative void feedback during the cooling period and in the positive void feedback during the power pulse transforms the initial sinusoidal oscillations to relaxation oscillations, which are characterized by a sharp power pulse over a short period followed by a slow recovery over a long period. These relaxation oscillations eventually can reach a limit cycle as the heat generated during the rapid power pulse is matched by the heat dissipated during the slow cooling period.

SINGULAR PERTURBATION ANALYSIS OF NCDWO

Based on the characteristic relaxation oscillation behavior observed in the NCDWO solutions, Ward and Lee[8] also adapted the singular perturbation techniques,[12] developed by Pontryagin, Mishchenko and other Russian mathematicians, for phase-plane analysis of the limit cycle NCDWO phenomena. The singular perturbation method is derived for a system described by:

$$\varepsilon dx/dt = f(x,y), \qquad dy/dt = g(x,y). \tag{12}$$

where ε is a small, dimensionless parameter. In a relaxation oscillation, system variable x, called fast variable, undergoes a large variation while variable y, called slow variable, remains essentially unchanged over certain portions of the cycle. These relaxation phases or fast-motion parts are characterized by finite values of f(x,y), where $O(dx/dt) = O(1/\varepsilon) \gg O(dy/dt)$. In contrast, during the recovery phases or slow-motion parts, we have $O(dx/dt) \cong O(dy/dt)$ corresponding to the condition $f(x,y) \cong 0$. Thus, the first phase-plane approximation, i.e., $O(\varepsilon^0)$ solution, consists of a trajectory following the degenerate curve $\Gamma = \{(x,y) \mid f(x,y) = 0\}$ for the slow parts and jumps between two points on Γ, called junction and drop points, for the fast parts. Higher order approximations may be obtained through different perturbation series on several distinct parts of the trajectory.

Ward and Lee[8,17] applied Mishchenko's singular perturbation method to phase-plane analysis of the Ergen-Weinberg model[18] and the NCDWO model. One key element in these applications has been variable transformations that are reminiscent of the Liénard's transformation.[19] Through these transformations, attempts have been made to represent relaxation oscillations in terms of Mishchenko's fast and slow variables. One major limitation of these applications has been the lack of junction points required for anchoring the phase-plane trajectory.

In the Ergen-Weinberg case, this problem was circumvented[17] by using the initial point in the trajectory corresponding to a step reactivity insertion K_o. With this approach to anchor the Ergen-Weinberg trajectory in the two-dimensional phase plane, the $O(\varepsilon)$ singular perturbation solution yielded the oscillation period and peak power with a 1~2 % accuracy for a case with K_o = 0.25 \$. The $O(\varepsilon^0)$ solution yields the well-known first-order estimate for the oscillation period equal to the time required to dissipate the total accumulated energy at the initial heat removal rate.

For the four-dimensional NCDWO problem,[8] one fast variable and three slow variables were selected through variable transformations and the $O(\varepsilon^{o})$ solution was anchored by using the dynamic average temperature obtained from a separate numerical solution. The $O(\varepsilon)$ singular perturbation solution yielded estimates of 1.21 to 1.28 s for the oscillation period, in good agreement with the direct numerical estimate of 1.20 s. The $O(\varepsilon)$ phase-plane trajectory, however, suffers discontinuities and lacks in accuracy in general.

The singular perturbation technique still holds the potential for efficient estimates of limit cycle trajectories for relaxation oscillations, including the NCDWO. This observation has provided the motivation to obtain a simple technique for anchoring limit cycle NCDWO trajectories, which will remove one of the limitations in the singular perturbation analysis of the limit cycle behavior. One such technique is presented in the next section.

LOCATING THE PHASE-PLANE TRAJECTORY THROUGH K_{min}

The semi-analytical algorithm we have developed to anchor the limit cycle NCDWO trajectory is based on determination of the extremum points of void reactivity, K_{min} and K_{max}, of the cycle. Once K_{min} is located for any cycle, a quick time-domain integration of the system equations is performed to determine the complete trajectory for the cycle, from which we can tell if the oscillation is converging or diverging. By iteratively applying this technique at a few different values of K_{min}, we can finally arrive at the limit cycle.

T-K-G Phase-Space Contour

From the mass conservation equation (3), the extremum points of void reactivity correspond to the points in a given trajectory where the inlet and exit flowrates are balanced, i.e., $G_i = G_e = G$. This allows us to rewrite Eq. (7) as

$$C_5 G x_e - C_6 (1+T)(1-z) = 0,$$

which can be combined with Eqs. (4), (5), and (8) to obtain a modified form of Eq. (9):

$$f_1(G,T,K,z) = 0. \tag{13}$$

A similar substitution of $G_i = G_e = G$ into Eq. (10) yields a modified form of Eq. (11):

$$f_2(G,K,z) = 0. \tag{14}$$

Combining Eqs. (13) and (14) to eliminate z, we finally obtain

$$f_3(G,T,K) = 0, \tag{15}$$

which describes contours of constant G in the T-K phase plane.

The goal here is to find the combinations of T, K, and G

that will satisfy Eq. (15) along the path toward a limit cycle. From our time-domain study, we note that, at a limit cycle, the flowrates G(t) at the extremum points of K are slightly lower than the steady state value G(0). This fact suggested construction of the contours of constant G in the T-K plane such that $G \le G(0)$. Fig. 2 shows such contour plots of extremum values of K corresponding to an initial Vermont Yankee operation at 65% of the rated power and natural circulation flow. Any point in the contour map of Fig. 2 corresponds to a solution of Eq. (15), which is, however, only a necessary condition for locating the extremum points. To determine the extremum K values uniquely in the T-K plane, we introduce an approximate phase-plane relation for each cycle:

$$T(K_{min}) = T(K_{max}). \tag{16}$$

As is shown later with actual numerical results, Eq. (16) is an accurate approximation for our NCDWO model and may well be a decent approximation even for more realistic BWR models. For each value of fuel temperature T during the approach to a limit cycle, a unique combination of G, K_{min}, and K_{max} is then obtained through a simultaneous solution of Eqs. (15) and (16). This is accomplished through an iterative numerical approach discussed below. Even in this numerical implementation, the constant-G contour plots of Fig. 2 serve as a very useful guide to possible ranges of extremum K points. This point may be appreciated when one recognizes that rather a small deviation from the initial

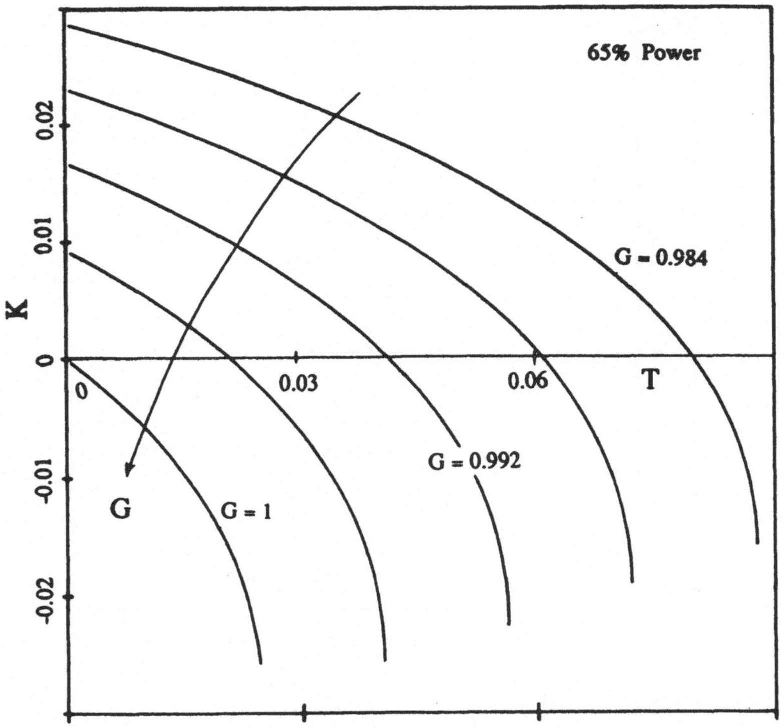

Fig. 2. T-K-G phase space contour

flowrate G(0) = 1.0 is all that is required to set up the extremum K points even for a large-amplitude limit cycle oscillation.

Numerical Implementation

The actual steps taken in the numerical implementation of the search for extremum points of void reactivity K are illustrated in Fig. 3. The simultaneous solution of Eqs. (15) and (16) requires a time-domain integration of the system equations (1), (2), (3), and (6) in an iterative manner:

(a) For an assumed value of fuel temperature T, estimate G and K_{min}, based on the constant-G contour plots of Eq. (15) in the T-K plane. Solve Eq. (13) for z with fixed T, K_{min}, and G. Check if G, K_{min}, and z satisfy Eq. (14). Adjust K_{min} as necessary until Eqs. (13) and (14) are simultaneously satisfied for fixed T and G. This step corresponds to an iterative solution of Eq. (15) for K_{min} with fixed T and G.

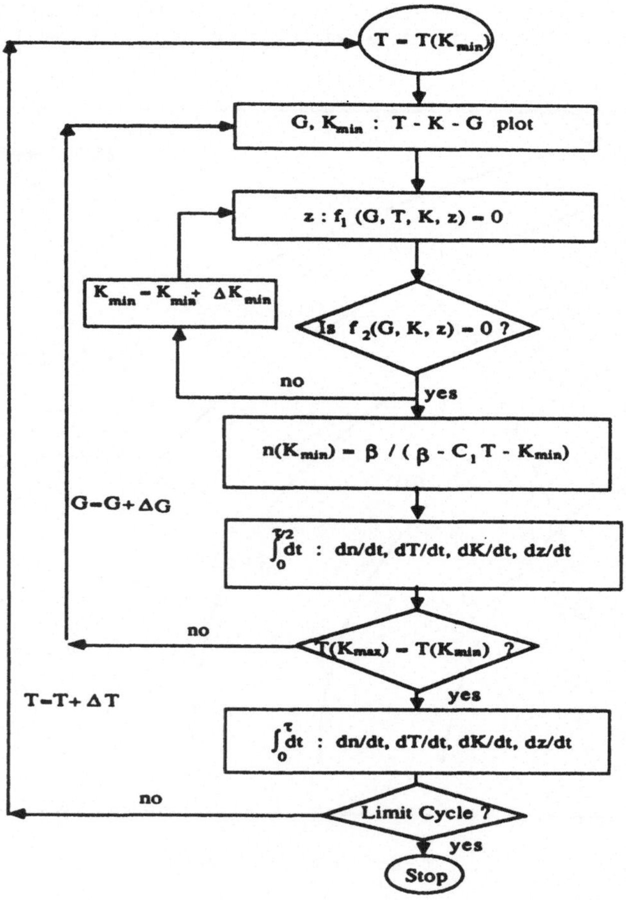

Fig. 3. Numerical implementation of K_{min} search

(b) With T, G, z, and K_{min} fixed, obtain n at K_{min} through the prompt jump approximation to Eq. (1):

$$n(K_{min}) = \beta/(\beta - C_1T - K_{min}). \tag{17}$$

(c) Integrate Eqs. (1), (2), (3) and (6) in time over half a cycle to determine T at K_{max}.

(d) Check if Eq. (16) is satisfied. If not, obtain a new estimate of G with $T(K_{min})$ fixed and repeat steps (a) through (c).

Once K_{min} and G satisfying Eqs. (15) and (16) are found for a given T, the time integration of step (c) is continued over a full cycle to determine the proximity to the limit cycle. If the trajectory diverges, the limit cycle is not yet reached; if the trajectory converges, it is possible the limit cycle is being approached from the outside. Fuel temperature T is then adjusted accordingly and steps (a) through (d) repeated to arrive at the limit cycle.

Numerical Results

We have applied the above algorithm to analyze BWR limit cycle behavior for natural circulation flow conditions of the Vermont Yankee test.[5] The algorithm yields, in a few iterations over steps (a) through (d), accurate estimates of K_{min} and G, which can then anchor the oscillatory cycle for an arbitrary value of T. Depending on how tightly Eqs. (15) and (16) are satisfied, two or three cycles may have to be followed, in practice, to determine accurately if the oscillation is converging or diverging for a particular value of T.

Fig. 4 shows the NCDWO trajectories as the limit cycle is approached from the inside for the initial power level of 65% of the rated. Fig. 5, with the same trajectories superimposed on the T-K-G contour of Fig. 2, further illustrates the utility of such a contour map. This figure demonstrates again the sensitive nature of the approach to a limit cycle. While the peak core power level n_{max} increases from a value of 1.86 to 3.13 over the four trajectories shown, $G(K_{min})$ varies from 1.0 to 0.997 and $T(K_{min})$ from 0.010 to 0.031.

In our phase-plane search for K_{min}, boiling boundary z is always obtained through Eq. (13), and indeed z could have been used instead of K in Eq. (15). This point is illustrated in Fig. 6 with the T-z trajectories superimposed on a T-z-G contour map. Here we also note that initial trajectories are not completely contained in subsequent trajectories as the oscillation grows to a limit cycle. Fig. 7 shows the T-K trajectories at 64% power as the limit cycle is approached from the outside. Fig. 7 also demonstrates that, irrespective of the initial estimate of T, the algorithm always yields the limit cycle trajectory. The actual limit cycle can be determined with only a few values of T arbitrarily tried.

Table 1 shows a comparison between the results for a cycle at 65% power obtained with our phase-plane scheme and through direct time integration. The trajectory compared here corresponds to a nearly converged limit cycle, and $T(K_{min})$ for the phase-plane scheme is chosen as an algebraic average of

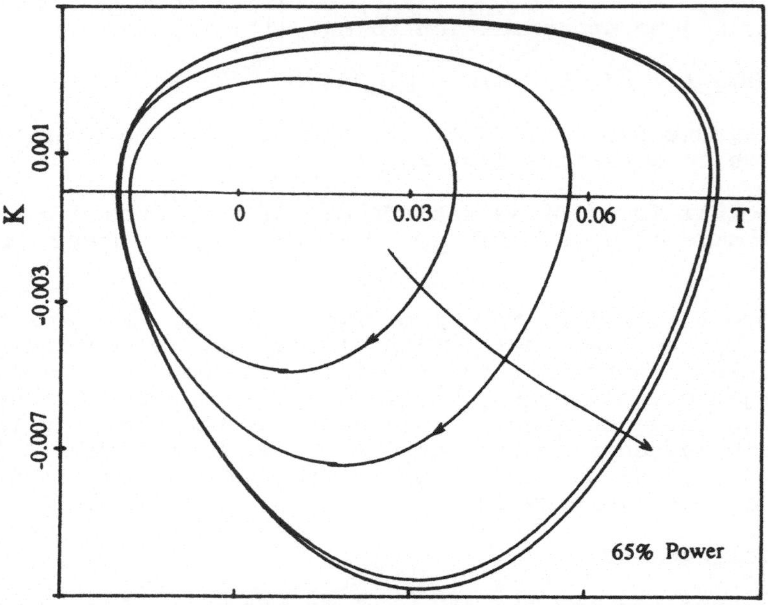

Fig. 4. T-K phase plane trajectory at 65% power approaching the limit cycle from inside.

$T(K_{min})$ and $T(K_{max})$ from the time-domain solution to minimize effects of less than full convergence. The comparison of the extremum values of the four system variables in Table 1 indicates that the phase-plane estimate of the trajectory is quite accurate and agrees with the time-domain solution with relative errors ranging from -0.6% to 0.6%. This provides a validation of the two key approximations, Eqs. (16) and (17), in our phase-plane algorithm. The extremum search solution was obtained with minimum computational effort, while the time-domain solution was obtained after following through hundreds of oscillatory cycles.

We finally show in Fig. 8 the results of parametric NCDWO studies performed with the proposed algorithm. The maximum core power in the limit cycle is calculated to be several times larger than the rated power and is a sensitive function of the initial power level and core flowrate. This sensitivity has serious reactor safety implications. We also note that the limit-cycle power amplitude is not a simple function of the power-to-flow ratio, which is a common measure of BWR stability. With various approximations introduced in our NCDWO model, however, the predicted limit cycle amplitudes should be considered only as an indication of the magnitude of potentially severe oscillations in operating BWRs.

CONCLUDING REMARKS

We have analyzed the limit cycle behavior of the nuclear-coupled density-wave oscillations in BWRs, using a simplified, nonlinear BWR model. We have developed a semi-analytic

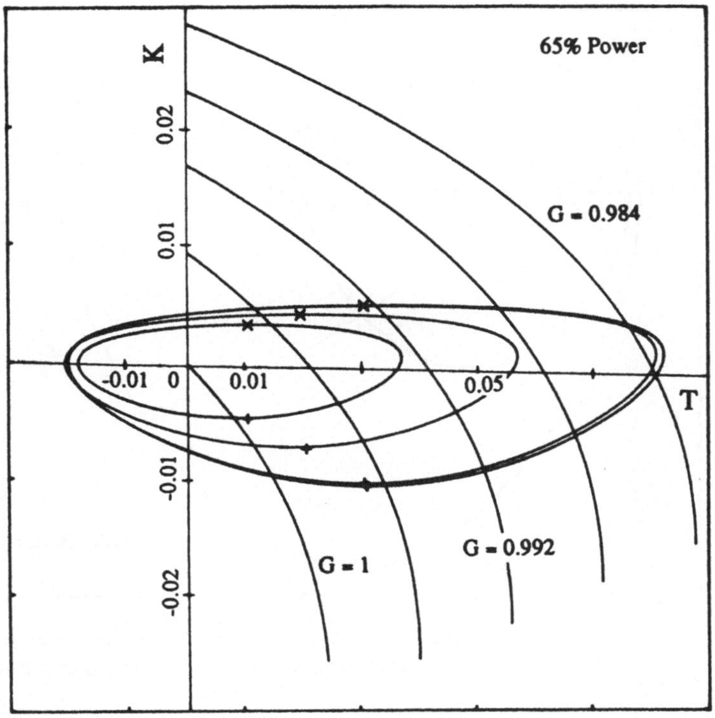

Fig. 5. T-K phase plane trajectory at 65% power with the T-K-G contour.

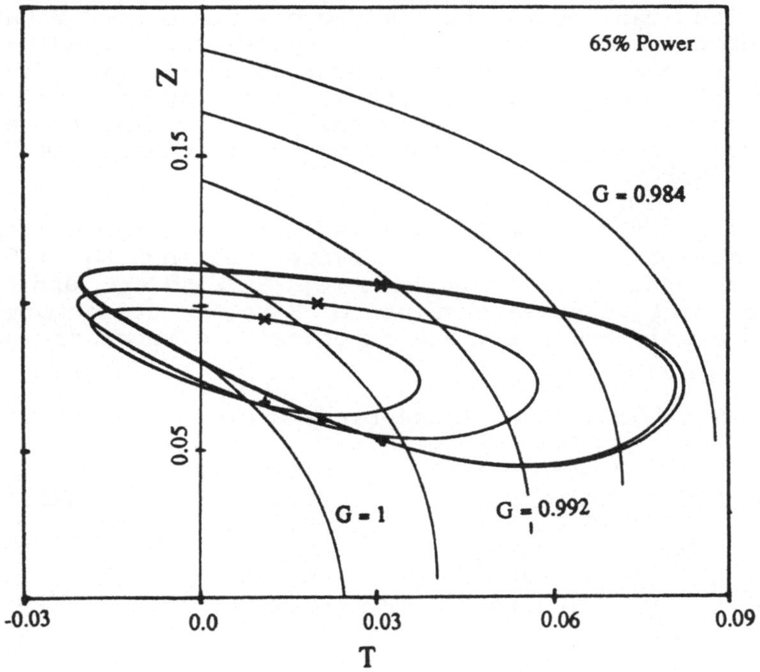

Fig. 6. T-z phase plane trajectory at 65% power with the T-z-G contour.

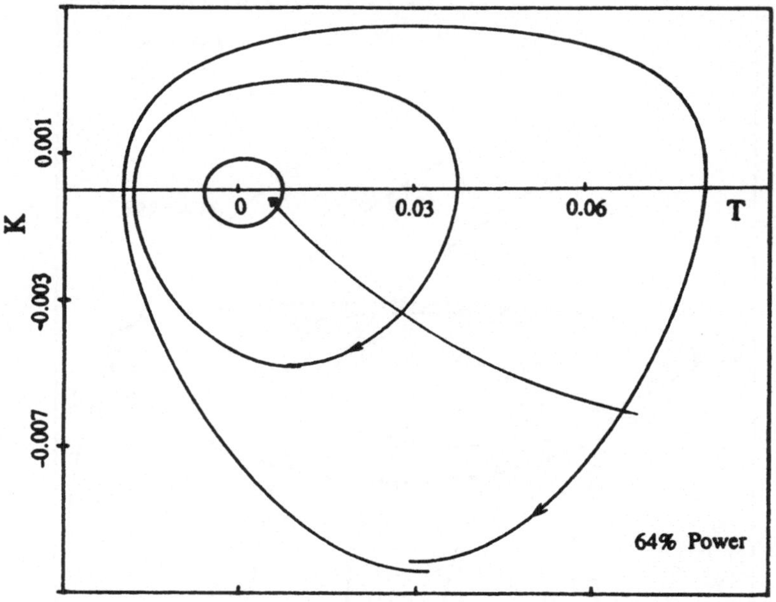

Fig. 7. T-K phase plane trajectory at 64% power approaching the limit cycle from outside

algorithm suitable for studying the approach to a limit cycle. This algorithm is based on a phase-plane search for extremum values of void reactivity K, which is accomplished through combining the energy and momentum conservation equations for the boiling channel in the phase plane of K and fuel temperature T. The algorithm can locate the limit cycle trajectory with time-domain integrations over only a few cycles corresponding to a few trial values of T, thereby providing an efficient means for obtaining NCDWO limit cycles with minimum computational effort. We have applied the algorithm to analyze BWR limit cycle behavior at natural circulation flow conditions of the Vermont Yankee reactor.

The algorithm we have developed for phase-plane location of the extremum points of void reactivity K through Eq. (16) suggests an analogy to Poincaré's map.[20] With K_{min} and K_{max} of a cycle providing successive values of K as the trajectory traverses a Poincaré's surface in opposite directions, we could

Table 1. Time-Domain vs. Phase-Plane Extremum Search Solution

Variables	Time-domain Solution	Phase-plane Solution	Relative Error (%)
K_{min}	-0.01056	-0.01050	0.6
K_{max}	0.00458	0.00457	0.2
$T(K_{min})$	0.03067	0.03086	-0.6
$T(K_{max})$	0.03104	0.03086	0.6
$z(K_{min})$	0.05348	0.05318	0.6
$z(K_{max})$	0.10542	0.10535	0.1
$n(K_{min})$	0.37876	0.38021	-0.3
$n(K_{max})$	3.13191	3.12404	0.3

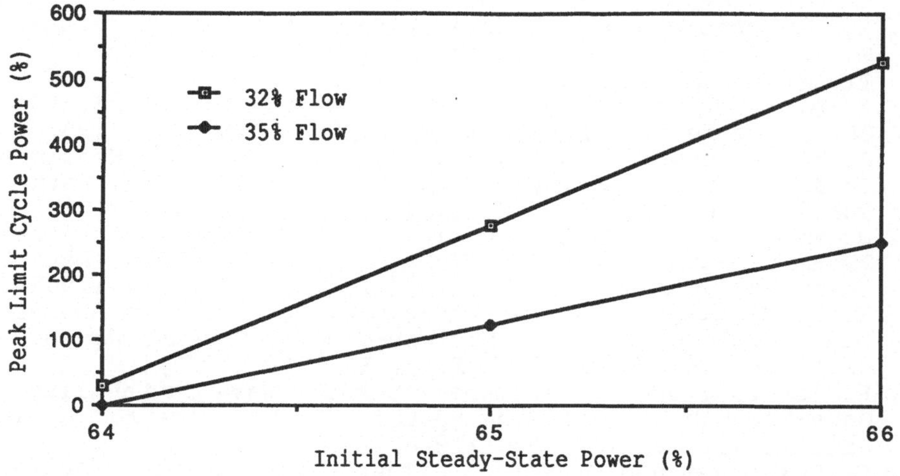

Fig. 8. Limit cycle amplitude vs. power. Power and flowrates are given in units of percent of the rated

consider the maximum void reactivity K_i of cycle i as a function of K_{i-1}, i.e., a map of the form $K_i = f(K_{i-1})$. Our phase-plane method to determine a limit cycle can then be considered equivalent to finding the attractor of a map without the need to follow successive points on the map one by one.

Application of this phase-plane approach to more detailed nonlinear BWR models using the RELAP5/MOD2 code[21] is under study. The method is expected to provide a means for obtaining simple but sufficiently accurate estimates of NCDWO characteristics. In light of the recent NCDWO event at the LaSalle Unit 2 reactor, an improved understanding of the nonlinear NCDWO behavior would be very much required for safe and reliable operation of BWR plants.

ACKNOWLEDGEMENTS

The authors appreciate the support provided by S. P. Kalra on this work through Electric Power Research Institute contract RP 1384-4.

REFERENCES

1. R. T. Lahey, Jr. and D. A. Drew, " An Assessment of the Literature Related to LWR Instability Modes," NUREG/CR-1414, U. S. Nuclear Regulatory Commission (1980).
2. G. Park, M. Z. Podowski, M. Becker, R. T. Lahey, Jr., and S. J. Peng, "The Development of a Closed-Form Analytical Model for the Stability Analysis of Nuclear-Coupled Density-Wave Oscillations in BWRs," Nucl. Eng. Des., 92, 253 (1986).
3. L. A. Carmichael and R. O. Niemi, "Transient and Stability Tests at Peach Bottom Atomic Power Station Unit 2 at End of Cycle 2," EPRI NP-564, Electric Power Research Institute (1978).

4. Y. Waaranpera and S. Anderson, "BWR Stability Testing: Reaching the Limit Cycle Threshold at Natural Circulation," Trans. Am. Nucl. Soc., 39, 868 (1981).
5. S. A. Sandoz and S. F. Chen, "Vermont Yankee Stability Tests During Cycle 8," Trans. Am. Nucl. Soc., 45, 727 (1983).
6. J. March-Leuba, D. G. Cacuci, and R. B. Perez, "Nonlinear Dynamics and Stability of Boiling Water Reactors: Part 1 -- Qualitative Analysis," Nucl. Sci. Eng., 93, 111 (1986).
7. J. March-Leuba, D. G. Cacuci, and R. B. Perez, "Nonlinear Dynamics and Stability of Boiling Water Reactors: Part 2 -- Quantitative Analysis," Nucl. Sci. Eng., 93, 124 (1986).
8. M. E. Ward and J. C. Lee, "Singular Perturbation Analysis of Limit Cycle Behavior in Nuclear-Coupled Density-Wave Oscillations," Nucl. Sci. Eng., 97, 190 (1987).
9. K. Takitani and T. Takemura, "Density Wave Instability in Once-Through Boiling Flow System, (I) Experiment," J. Nucl. Sci. Tech., 15, (5), 355 (1978).
10. R. Takahashi and M. Shindo, "Theoretical Study of Two-Phase Flow Oscillation in a Hot Channel, (I) Theoretical Study for Interpreting the Mechanism of Hydrodynamic Instability," J. Nucl. Sci. Tech., 8, (11), 637 (1971).
11. M. A. Ring, "Dual Recirculation Pump Trip Event of March 9, 1988, at the LaSalle County Station Unit 2," Augmented Inspection Team Report, U. S. Nuclear Regulatory Commission (1988).
12. E. F. Mishchenko and N. Kh. Rozov, Differential Equations with Small Parameters and Relaxation Oscillations, Plenum Press, New York (1980).
13. J. R. S. Thom, "Prediction of Pressure Drop During Forced Circulation Boiling of Water," Int. J. Heat Mass Transfer, 7, 709 (1964).
14. J. Kornfilt and J. C. Lee, "Stability Analysis of Boiling Water Systems and Steam Generators," Proceedings of the Second International Topical Meeting on Nuclear Reactor Thermal Hydraulics, Santa Barbara, CA (1983).
15. G. Dahlquist and A. Björck, Numerical Methods, Prentice-Hall, Englewood Cliffs, NJ (1974).
16. M. W. Crump and J. C. Lee, "Nonlinear Transient Analysis of Light Water Reactor Steam Generators Using an Implicit Eulerian Method," Nucl. Sci. Eng., 77, 192 (1981).
17. M. E. Ward and J. C. Lee, "Singular Perturbation Analysis of Relaxation Oscillations in Reactor Systems," Nucl. Sci. Eng., 95, 47 (1987).
18. M. Ash, Nuclear Reactor Kinetics, McGraw Hill, New York (1965).
19. N. Minorsky, Nonlinear Oscillations, Van Nostrand, Princeton, NJ (1962).
20. C. Grebogi, E. Ott, and J. A. Yorke, "Chaos, Strange Attractors, and Fractal Basin Boundaries in Nonlinear Dynamics," Science, 238, 632 (1987).
21. V. H. Ransom, R. J. Wagner, J. A. Trapp, L. R. Feinauer, G. W. Johnsen, D. M. Kiser, and R. A. Riemke, "RELAP5/MOD2 Code Manual," NUREG/CR-4312, U. S. Nuclear Regulatory Commission (1985).

CHAPTER VII

NONLINEAR DYNAMICS AND TRANSITION TO CHAOS IN NUCLEAR SYSTEMS.

PART II

NONLINEAR DYNAMICS OF THE NUCLEAR BREEDING PROCESS

V.C. Boffi*, G. Spiga**and P. Vestrucci*

* Nuclear Engineering Laboratory of the University of Bologna, Bologna, Italy
**Mathematics Department of the University of Bari Bari, Italy

INTRODUCTION

The time evolution of the nuclear breeding process, modeled as a mixture of three different species of interacting particles (neutrons, fissile and fertile atoms, respectively) diffusing in a host medium, is studied by means of the set of nonlinear integro-partial differential Boltzmann equations governing the relevant distribution functions f_1, f_2, f_3. It is shown that the Boltzmann system for the f_j's yields, under suitable assumptions, a system of conservation equations for the number densities ρ_1, ρ_2, ρ_3, which are defined by just integrating the corresponding f_j over the velocity domain. The zero-dimensional approach, for which the conservation system is a nonlinear first-order ordinary differential system, is first studied on the basis of the theory of dynamical systems, with particular regard to the stability of the fixed points. Curves representing some numerical results for the phase trajectories are presented. The space-dependent problem is then sketched, for which the conservation system can be instead of fully hyperbolic type. Finally, properties and open problems related to this latter case are shortly discussed on both mathematical and physical ground.

MATHEMATICAL SETTING

In some recent papers[1,2] an extended kinetic theory for an unbounded spatially homogeneous mixture of N different rarefied gases has been formulated in the frame of the so-called scattering kernel formulation, and including not only scattering (as usually done in classical kinetic theory), but also removal, creation and host medium effects. The N distribution functions $f_1, f_2, \ldots, f_N$ are governed by the set of nonlinear transport Boltzmann-like equations $(j=1,2,\ldots,N)$

$$\frac{\partial f_j}{\partial t} = - \sum_{l=1}^{N+1} f_j(\bar{v},t) \int_{R_3} g_{jl}(|\bar{v}-\bar{w}|)\, f_l(\bar{w},t)\, d\bar{w} +$$

$$\sum_{l=1}^{N+1} \int_{R_3}\int_{R_3} \overset{S}{g}_{jl}(|\bar{v}'-\bar{w}'|)\, \overset{S}{\Pi}_{jl}(\bar{v}',\bar{w}'\to\bar{v})\, f_j(\bar{v}',t)\, f_l(\bar{w}',t)\, d\bar{v}'d\bar{w}' + \qquad (1)$$

$$\sum_{k=1}^{N} \sum_{l=k}^{N+1} \int_{R_3}\int_{R_3} \overset{C}{g}_{kl,j}(|\bar{v}'-\bar{w}'|)\chi^{j}_{kl}(\bar{v}',\bar{w}'\to\bar{v}')f_k(\bar{v}',t)f_l(\bar{w}',t)d\bar{v}'d\bar{w}' +$$

$$+\ S_j(\bar{v},t) \qquad (\bar{v}\in R_3;\ t\in[0,\infty))$$

For the meaning of the symbols in Eq.(1) we recall shortly what follows.

- The index N+1 is used to label the background host medium, whose particles have a fixed distribution function, constant in time for simplicity.
- m_j and $S_j(\bar{v},t)$ stand for mass and external source for the particles of the species j.
- The function $(j=1,2,\ldots,N;\ l=1,2,\ldots,N+1)$

$$g_{jl}(|\bar{v}-\bar{w}|) = \overset{S}{g}_{jl}(|\bar{v}-\bar{w}|)+\overset{R}{g}_{jl}(|\bar{v}-\bar{w}|)+ \sum_{m=1}^{N} \overset{C}{g}_{jl,m}(|\bar{v}-\bar{w}|) =$$
$$= \overset{S}{g}_{jl}(|\bar{v}-\bar{w}|)+\overset{A}{g}_{jl}(|\bar{v}-\bar{w}|) \qquad (2a)$$

denotes the total (scattering plus removal plus creation) microscopic frequency of the collisions between particles j and particles l, under the assumption that creation by collision produces only one participating species.

- $\chi^{j}_{kl}(\bar{v}',\bar{w}'\to\bar{v})d\bar{v}$ is the number of particles j, created by a collision between a particle k with velocity $\bar{v}'$ and a particle l with velocity $\bar{w}'$, which are given a final velocity in $d\bar{v}$ at $\bar{v}$.

- $\overset{S}{\Pi}_{jl}(\bar{v}',\bar{w}'\to\bar{v})$ is the probability distribution for a particle j scattered by a particle l, and obeys the normalization condition

$$\int_{R_3} \overset{S}{\Pi}_{jl}(\bar{v}',\bar{w}'\to\bar{v})d\bar{v} = 1\ . \qquad (2b)$$

THE CONSERVATION SYSTEM

A peculiar feature of the Boltzmann system, Eq.(1), is that - if we assume that all the microscopic collision frequencies are nonnegative constants (maxwellian particles) - then, integrating both sides of Eq.(1) itself over the velocity domain, setting

$$\rho_j(t) = \int_{R_3} f_j(\bar{v},t)\, d\bar{v} \tag{3}$$

and recalling Eq.(2b), we get for the number densities ρ_j's the following nonlinear first-order ordinary differential set, namely

$$\rho_j(t) = -\sum_{l=1}^{N+1} g^A_{jl}\rho_j(t)\rho_l(t) + \sum_{k=1}^{N}\sum_{l=k}^{N+1} \eta^j_{kl}\, g^c_{kl,j}\rho_k(t)\rho_l(t) + Q_j(t) \tag{4}$$

to be integrated upon the initial condition $\rho_j(0)=\rho_{0j}>0$. In Eq.(4)

$$\eta^j_{kl} = \int_{R_3} \chi^j_{kl}(\bar{v}',\bar{w}'\to\bar{v}')\, d\bar{v}' \tag{5a}$$

is the mean number of secondary particles j created by collision between a particle k and a particle l, taken to be independent of both the velocities $\bar{v}'$ and $\bar{w}'$ before collision and possibly greater than unity, and

$$Q_j(t) = \int_{R_3} S_j(\bar{v},t)\, d\bar{v}\ . \tag{5b}$$

THE NONLINEAR DYNAMICS OF THE BREEDING PROCESS

Bearing in mind the scheme of principle of a fast breeder nuclear reactor, let us specialize the conservation system, Eq.(4), to the case N=3, when interactions to be accounted for in the mixture are those taking place typically in the breeding process. More precisely, if indeces 1,2,3,4 refer to neutron, fissile, fertile and background species, respectively, the following mechanisms of removal and creation will be considered:

-i) removal of 1 by 2,3,4; -ii) creation of 1 by fission of 2, and consequent removal of 2; -iii) creation of 2 by chemical reaction between 1 and 3, and consequent removal of 3; -iv) external sources of 2 and 3 taken to be constant.

With $\eta^2_{13}=1$, assuming that $\eta^1_{12} g^C_{12,1} > g^A_{12}$ and $\nu^A_1 = g^R_{14}\ \rho_4>0$ (as it is usually the case) and introducing the new time variable $\nu^A_1 t$ and the new dependent variables

$$x_1 = g^A_{13}\rho_1/\nu^A_1;\ x_2 = (g^A_{13})^2\rho_2/\nu^A_1\, g^C_{13,2};\ x_3 = g^A_{13}\ \rho_3/\nu^A_1, \tag{6}$$

from the general system, Eq.(4), we derive then the set of first-order ODE's with quadratic nonlinearity

$$\begin{aligned} \dot{x}_1 &= Ax_1x_2 - x_1x_3 - x_1 \ ; \\ \dot{x}_2 &= -Dx_1x_2 + x_1x_3 + F \ ; \\ \dot{x}_3 &= -x_1x_3 + G \ , \end{aligned} \tag{7}$$

where the four constant coefficients A,D,F,G are defined as

$$\begin{aligned} A &= (\eta^{1}_{12}\, g^{C}_{12,1} - g^{A}_{12,1})/g^{C}_{13,2} ; \quad D = g^{A}_{12}/g^{A}_{13} \ ; \\ F &= (g^{A}_{13})^2 Q_2/(\nu^{A}_{1})^2 g^{C}_{13,2} \quad ; \quad G = (g^{A}_{13})\, Q_3/(\nu^{A}_{1})^2 \end{aligned} \tag{8a}$$

and are such that

$$A>0 \quad ; \qquad D>0 \quad ; \qquad F \lesseqgtr 0 \quad ; \qquad G\geq 0. \tag{8b}$$

The inhomogeneous term F - which will be exploited as the main parameter in all the following discussion- may be also negative so that it simulates the operating conditions in which a breeder system not only produces energy, but also extra fuel to be extracted from the system itself.

It is remarkable that Eq.(7), derived here rigorously in a rarefied gas dynamics context, and in the hypothesis of spatial homogeneity and infinite extent, could be obtained also heuristically, on purely physical grounds, as overall balance equations for a finite heterogeneous breeder reactor in a point-kinetics-like approach. Coefficients should then be redefined as suitable averages of the underlying physical parameters, and would include leakage effects, depending on shape and size of the actual reactor. Of course, Eq.(7) provides anyway an oversimplified description of the various physical phenomena and operations occurring in a real breeding system (for instance, thermal-hydraulics effects and delayed neutrons are not taken into account), but deserves however our attention, for a better understanding of the dynamical behavior, and of its possible consequences and applications, related to the breeding process taking place in the reactor. The use of Eq.(7) is of particular interest when the dimensionless number densities x_j are of the same order of magnitude, since, then, the nonlinear coupling inhibits the usual linear approach of classical reactor physics.

ANALYSIS OF THE SYSTEM, EQ.7

From a mathematical point of view, the system, Eq.7, which -because of the appearance of the product x_1x_3- is not of Lotka-Volterra type, must be solved for nonnegative x_j's satisfying the initial condition $x_j(0)=x_{0j}>0$ ($j=1,2,3$). It is seen by inspection that x_1 cannot increase without any bound, because that would entail a decrease of x_3, and then of x_2 and of neutron generation rate. The analysis can be further performed according to the following steps.

Conditions upon which the nonnegative orthant of the phase space R_3 is positively invariant

Such a condition is easily checked to be $F\geq0$. If, instead, $F<0$, x_2 could become negative starting from positive, crossing the plane $x_2=0$ in the region defined by $x_1x_3<-F$. (The time evolution terminates thus at the time t for which $x_2=0$).

Divergence

If Φ_j denotes the quadratic form in the x_j's, appearing in the r.h.s. of the general equation of the system, Eq.(7), then the divergence

$$\sum_{j=1}^{3} \frac{\partial\Phi_j}{\partial x_j} = -1 - (D+1)x_1 + Ax_2 - x_3 \tag{9}$$

can be either positive or negative. The problem is thus neither conservative nor dissipative.

Fixed points

For

$$F > F_0 = (D-A)\ G/A \tag{10}$$

there exists a unique fixed point, whose coordinates are

$$x_1 = \frac{F-F_0}{D} A\ ; \quad x_2 = \frac{F+G}{A(F-F_0)}\ ; \quad x_3 = \frac{DG}{A(F-F_0)}\ . \tag{11}$$

If $F=G=0$, all points of the plane $x_1=0$ are fixed points. No other fixed points exist. (In the sequel we shall essentially concentrate on the case $G>0$, and thus only on the fixed point, Eq.(11)).

Linear stability of the fixed point, Eq.(11)

The eigenvalues of the Jacobian matrix relevant to Eq.(11) are the roots of the third order algebraic equation with real coefficients

$$\lambda^3 + \frac{D+1}{D} K\lambda^2 + (\frac{K^2}{D} + AF - G)\lambda + \frac{K^2}{D} = 0 \qquad (12)$$

with

$$K = A(F-F_0). \qquad (13)$$

It is worthwhile studying the existence of eigenvalues on the imaginary axis, as they could correspond to loss of stability when parameters are varied. The condition for existence of roots $\pm i\omega$ has admissible solutions only for $G>0$ and $A > D-1$, in which case it gives the unique critical value

$$F=F^*=F_0+[[D^4+4D(D+1)^2(A+1-D)G]^{1/2}-D^2]/2A(D+1). \qquad (14)$$

This is just a bifurcation value since the sum of the eigenvalues is always negative so that the third eigenvalue is real and negative, and moreover the transversality conditions are satisfied. We can thus conclude by stating that, when $G>0$ and $A>D-1$, the fixed point, Eq.(11), is stable for $F>F^*$, undergoes a Hopf bifurcation for $F=F^*$, and becomes unstable for $F_0<F<F^*$. If, instead, $G=0$ or $A\leq D-1$ $(D>1)$, the fixed point is stable for any value of $F>F_0$, since all eigenvalues have a negative real part.

In Fig.1 and Fig.2 we represent the solution (x_1,x_2,x_3) to the system, Eq.(7), as numerically obtained for the case of $A=D=1$, $G=3$ (thus $F_0=0$ and $F^*=1.5$), and F ranging from the negative value -0.10 up to the positive value 10. In the first row of each Figure the three-dimensional phase trajectory is plotted, whereas its three bi-dimensional projections $(x_1,x_2),(x_3,x_1),(x_3,x_2)$ are reported in the second, third and fourth row, respectively. Finally, the three curves, representing the general x_j $(j=1,2,3)$ as a function of time, are given. The curves are representative of most dynamical trends observed in all numerical experiments. The initial point is kept fixed at (1,2,2). For $F=-0.10$ there are no fixed points, the orbit is attracted by the invariant plane $x_1=0$, in whose neighborhood x_2 and x_3 behave linearly in time, and x_2 tends to vanish. At $F=1.40$ there is a fixed point, but not stable yet, and the trend is similar to the previous one, only with x_2 and x_3 both tending to diverge linearly on time. At $F=1.60$ the stability thresold is overcome, but the initial point is outside the basin of attraction of the stable fixed point; the dynamics is almost the same as in the previous column. Finally, the cases $F=2.00$, 3.00, and 10 of Fig.2 share the feature that the orbit is captured by the fixed point, but with oscillations becoming less important, and eventually disappearing for increasing F, in the approach to equilibrium.

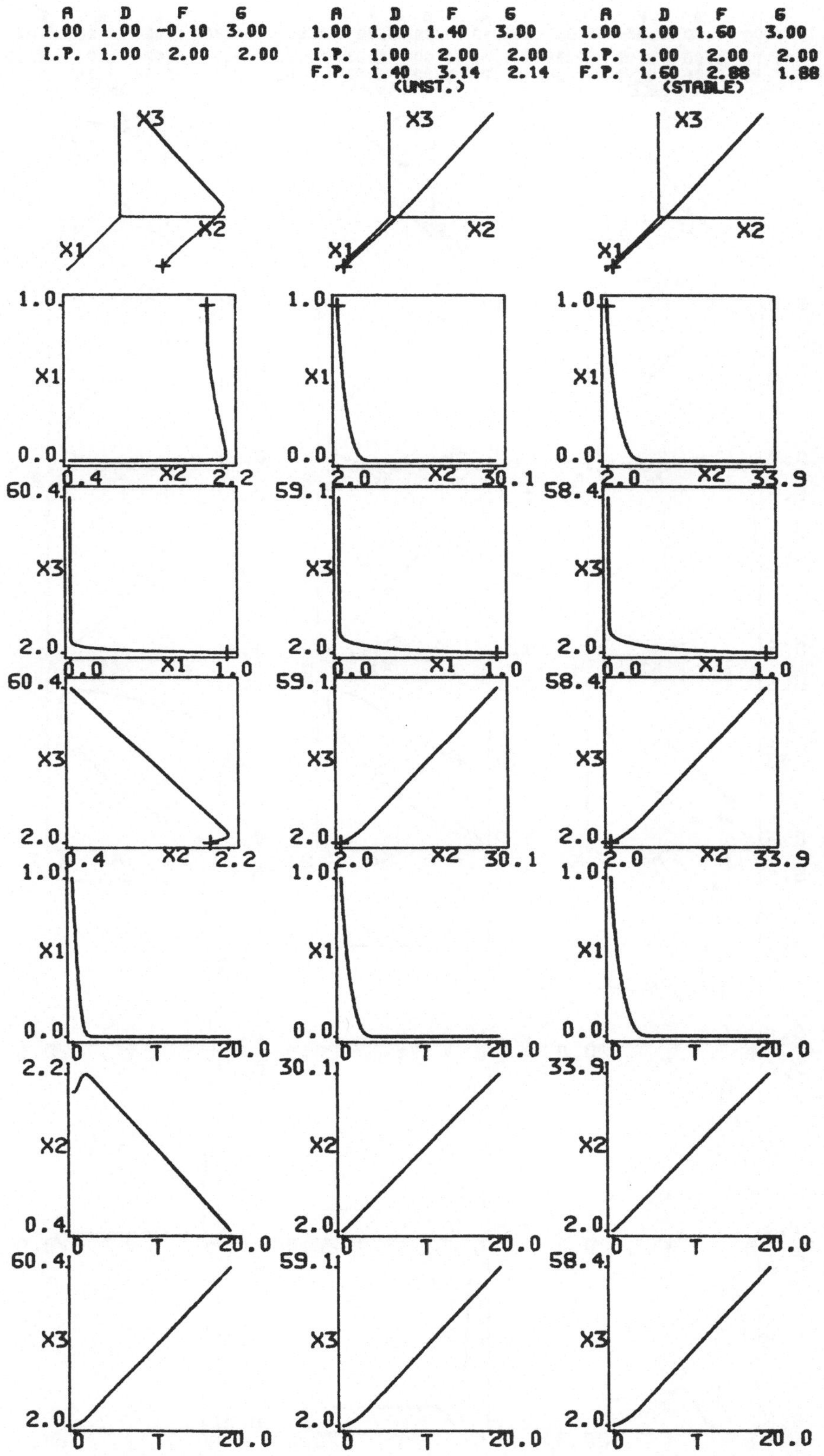

Fig. 1. Numerical results for the solution to Eq. (7)

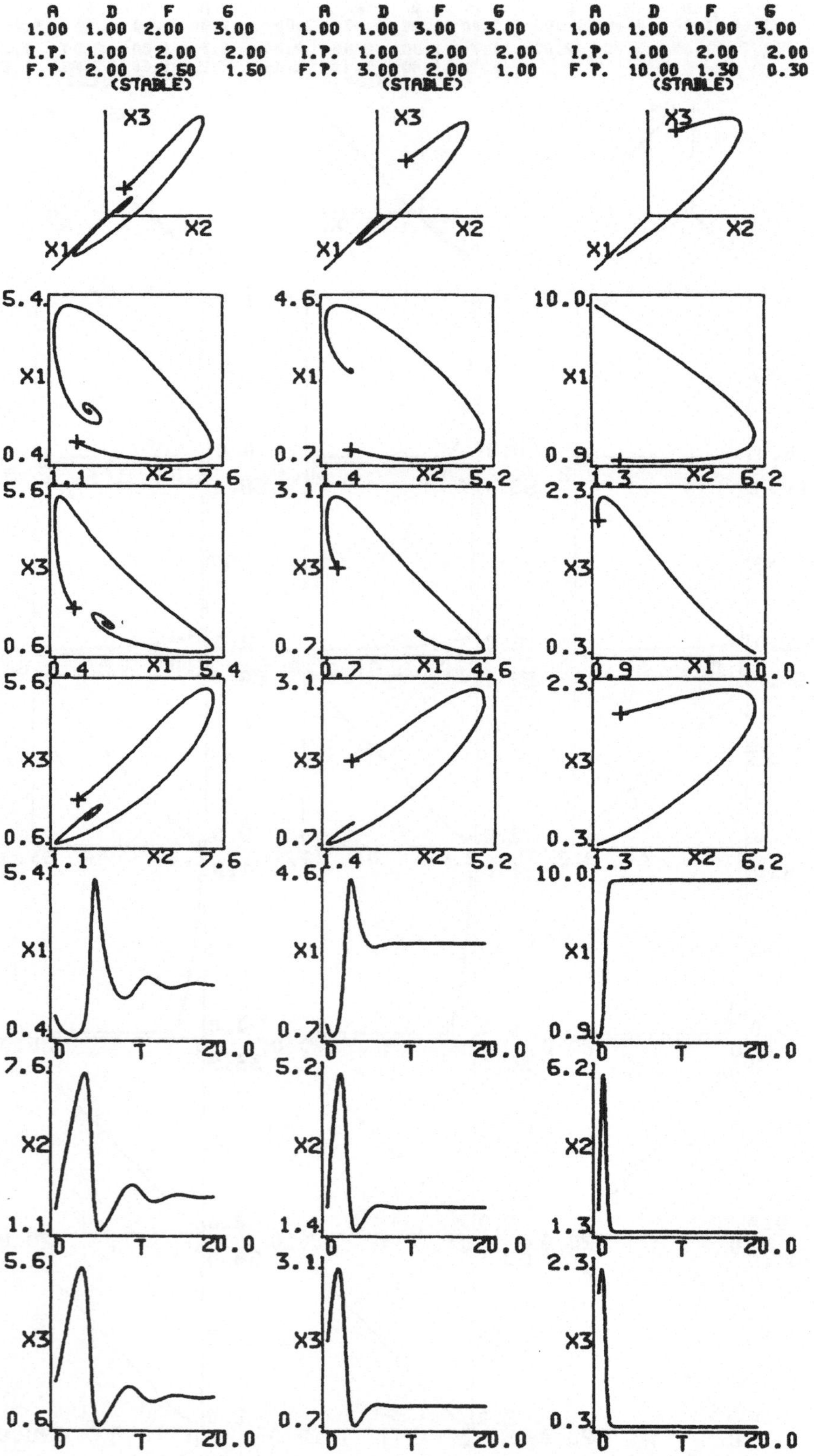

Fig. 2. Numerical results for the solution to Eq. (7)

If spatial dependence is to be accounted for, the Boltzmann system, Eq.(1), holding in the spatially homogeneous case, must be modified by adding in its l.h.s. the streaming term $\bar{v}\cdot\partial f_j/d\bar{x}$, and, of course, inserting $\bar{x}$, besides $\bar{v}$ and t, at the argument of f_j. We shall restrict, here, to a Boltzmann system, without inhomogeneous term, to be integrated upon the initial condition

$$f_j(\bar{x},\bar{v},0) = S_j(\bar{x},\bar{v}) \tag{15}$$

which corresponds to Eq.(1) with vanishing conditions and with pulsed source $S_j(\bar{x},\bar{v},t)=S_j(\bar{x},\bar{v})\delta(t)$, where $S_j(\bar{x},\bar{v})$ is a bounded positive function for both $\bar{x}$ and $\bar{v}$ in R_3.

It is spontaneous to ask then if also for the present spatially inhomogeneous case it is possible to get from the original Boltzmann system for the f_j's a "differential" system for the ρ_j's, as we did with the system, Eq.(4), in the spatially homogeneous case. The answer is positive only if, together with the assumption of constant collision frequencies, additional assumptions, concerning the scattering and creation probability distributions as well as the initial data are invoked[3,4]. In few words, an analytical treatment is possible by adopting for the functions $\Pi^{S}{}_{jl}$ and $\chi^{j}{}_{kl}$ some sinthetic model depending only on the final velocity $\bar{v}$ or some discrete velocity representation. As a test reference case, the limiting situation in which

$$\Pi^{S}_{jl}(\bar{v}',\bar{w}'\to\bar{v})=\delta(\bar{v}-\bar{v}_{jl}); \qquad \Pi^{S}_{j,N+1}(\bar{v}'\to\bar{v})=\delta(\bar{v}-\bar{v}_{j,N+1}) \; ; \tag{16}$$

$$\chi^{j}_{kl}(\bar{v}',\bar{w}'\to\bar{v})=\eta^{j}_{kl}\delta(\bar{v}-\bar{v}^{j}_{kl}); \quad \chi^{j}_{j,N+1}(\bar{v}'\to\bar{v})=\eta^{j}_{j,N+1}\delta(\bar{v}-\bar{v}^{j}_{j,N+1}); \tag{17}$$

$$S_j(\bar{v},\bar{v}) = Q_j(\bar{x}) \; \delta(\bar{v}-\bar{v}_{j0}) \; , \tag{18}$$

can be worked out in detail.
In Eqs.(16), (17) and (18) $\bar{v}_{jl}$, $\bar{v}_{j,N+1}$; $\bar{v}^{j}_{kl}$, $\bar{v}^{j}_{j,N+1}$; $\bar{v}_{j0}$ (j,k,l=1,2,.. ...,N) constitute a set of M=N+(N+1)!/2(N-1)!+3 assigned velocities attained by the particles j after being scattered, after being created and after being emitted by the external source at time t=0, respectively. The original, spatially inhomogeneous Boltzmann system becomes now

$$\frac{\partial f_j}{\partial t} + \bar{v}_{j0}\cdot \frac{\partial f_j}{\partial \bar{x}} = - \sum_{l=1}^{N+1} g_{jl}\rho_l(\bar{x},t)\, f_j(\bar{x},\bar{v},t) +$$

$$\sum_{l=1}^{N+1} g^S_{jl}\delta(\bar{v}-\bar{v}_{jl})\, \rho_l(\bar{x},t)\, \rho_j(\bar{x},t) +$$

$$\sum_{k=1}^{N} \sum_{l=k}^{N+1} g^C_{kl,j}\eta^j_{kl}\delta(\bar{v}-\bar{v}^j_{kl})\, \rho_k(\bar{x},t)\, \rho_l(\bar{x},t), \tag{19}$$

so that, by integrating both sides of this equation over the velocity domain, we get the system

$$\frac{\partial \rho_j}{\partial t} + \frac{\partial}{\partial \bar{x}}\cdot \bar{J}(\bar{x},t) + \Psi_j(\bar{x},t) = 0 , \tag{20}$$

where

$$\bar{J}_j(\bar{x},t) = \int_{R_3} \bar{v}\, f_j(\bar{x},\bar{v},t)\, d\bar{v} \tag{21}$$

is the current density and $\Psi_j(\bar{x},t)$ is nothing else but the r.h.s. of Eq.(4) with $\rho_j=\rho_j(\bar{x},t)$ and $Q_j(t)=0$. The system, Eq.(20), will become fully explicit if we were able to determine the current density $\bar{J}_j$ as a function of all the ρ_j's. That this is the case can be verified by integrating Eq.(19) along the characteristics, and then inserting the resulting expression for $f_j(\bar{x},\bar{v},t)$ (which is an explicit solution to the Boltzmann system in terms of all the ρ_j's) in Eq.(21). There follows that $\bar{J}_j(\bar{x},t)$ will be given by the constitutive equation

$$\bar{J}_j(\bar{x},t) = \bar{v}_{j0}\, \rho_j(\bar{x},t) + \bar{J}_j^{SC}(\bar{x},t), \tag{22}$$

which is also the closure condition for the problem under consideration. In Eq.(22) $\bar{J}_j^{SC}$ is the following cumbersome functional

$$\bar{J}_j^{SC}(\bar{x},t) = \sum_{l=1}^{N+1} g^S_{jl} \int_0^t (\bar{v}_{jl}-\bar{v}_{j0})\rho_l[\bar{x}-\bar{v}_{jl}(t-u),u]\rho_j[\bar{x}-\bar{v}_{jl}(t-u),u]$$

$$\exp\left\{ - \sum_{l=1}^{N+1} g_{jl} \int_u^t \rho_l[\bar{x}-\bar{v}_{jl}(t-u'),u']du' \right\}du +$$

$$\sum_{k=1}^{N} \sum_{l=k}^{N+1} g^C_{kl,j}\, \eta^j_{kl} \int_0^t (\bar{v}^j_{kl}-\bar{v}_{j0})\rho_k[\bar{x}-\bar{v}^j_{kl}(t-u),u]\rho_l[\bar{x}-\bar{v}_{kl}(t-u),u]$$

$$\exp\left\{ - \sum_{l=1}^{N+1} g_{jl} \int_u^t \rho_l[\bar{x}-\bar{v}^j_{kl}(t-u'),u']du' \right\}du. \tag{23}$$

With $\bar{J}_j(\bar{x},t)$ given by Eq.(22)+Eq.(23), the system, Eq.(20), for the ρ_l's can be classified as a semilinear functional-hyperbolic system. But, as we can observe from Eq.(23), there are two cases in which $\bar{J}_j^{SC}(\bar{x},t)$ vanishes.

-i) When no scattering and creation, but only removal events take place in the considered spatially inhomogeneous mixture. The system, Eq.(20), reads then as

$$\frac{\partial \rho_j}{\partial t} + \bar{v}_{j0}\cdot\frac{\partial \rho_j}{\partial x} + \sum_{l=1}^{N+1} \overset{R}{g}_{jl}\rho_l(\bar{x},t)\,\rho_j(\bar{x},t)=0;$$

$$\rho_j(\bar{x},0)=Q_j(\bar{x}) \tag{24}$$

which is a semilinear hyperbolic system of conservation type.

-ii) When $\bar{v}_{jl}=\bar{v}^j_{kl}=\bar{v}_{j0}$ $(l=1,2,\ldots,N+1)$. In this hypothetical case, in which removal, scattering and creation all occur at the velocity $\bar{v}_{j0}$, the system, Eq.(20), becomes

$$\frac{\partial \rho_j}{\partial t} + \bar{v}_{j0}\cdot\frac{\partial \rho_j}{\partial x} + \sum_{l=1}^{N+1} \overset{A}{g}_{jl}\,\rho_l(\bar{x},t)\,\rho_j(\bar{x},t) -$$

$$\sum_{k=1}^{N}\sum_{l=k}^{N+1} \overset{C}{g}_{kl,j}\,\eta^j_{kl}(\bar{x},t)\,\rho_k(\bar{x},t)\,\rho_l(\bar{x},t)=0\ ;$$

$$\rho_j(\bar{x},0)=Q_j(\bar{x}). \tag{25}$$

With $N=3$ and on the basis of the same removal and creation mechanisms as considered in the spatially homogeneous case, we obtain from Eq.(25) the 3x3 system

$$\frac{\partial \rho_1}{\partial t} + \bar{v}_{10}\cdot\frac{\partial \rho_1}{\partial x} + C_{12}\rho_1\rho_2 + C_{13}\rho_1\rho_3 + C_{14}\rho_1 = 0\ ;$$

$$\frac{\partial \rho_2}{\partial t} + \bar{v}_{20}\cdot\frac{\partial \rho_2}{\partial x} + C_{21}\rho_1\rho_2 + C_{23}\rho_1\rho_3 = 0\ ; \tag{26}$$

$$\frac{\partial \rho_3}{\partial t} + \bar{v}_{30}\cdot\frac{\partial \rho_3}{\partial x} + C_{31}\rho_1\rho_3 = 0$$

with coefficients expressed in terms of the physical parameter as

$$C_{12} = \overset{A}{g}_{12}-\eta^1_{12}\overset{C}{g}_{12,1}\ ;\quad C_{13} = \overset{A}{g}_{13} > 0\ ;\quad C_{14} = \overset{R}{g}_{14} > 0\ ;$$

$$C_{21} = \overset{A}{g}_{12}\ ;\quad C_{23} = -\overset{C}{g}_{13,2} < 0\ ;\quad C_{31} = C_{13} > 0\ . \tag{26a}$$

The study can be further restricted to the case when the fertile material can be regarded as part of the fixed background. In such a case the system, Eq.(26), reduces to

$$\frac{\partial \rho_1}{\partial t} + \bar{v}_{10} \cdot \frac{\partial \rho_1}{\partial x} + C_{12}\rho_1\rho_2 + C_{13}\rho_1 = 0 \; ;$$

$$\frac{\partial \rho_2}{\partial t} + \bar{v}_{20} \cdot \frac{\partial \rho_2}{\partial x} + C_{21}\rho_1\rho_2 + C_{23}\rho_2 = 0 \qquad (27)$$

with $C_{13}\rho_3 + C_{14} \rightarrow C_{13}$ and $C_{23}\rho_3 \rightarrow C_{23}$.

Equation (27) can be solved analytically, at least in several selected cases, when $\bar{v}_{10}=\bar{v}_{20}$[5]. For $\bar{v}_{10}\neq\bar{v}_{20}$ the problem could be faced by a Lie group analysis, leading to the formulation of similarity variables and, then, to the construction of invariant solutions. This part will be, however, object of a separate paper. Another important open problem is the investigation of the boundary value problem associated with the system, Eq.(27), arising when the spatial domain is finite. In this framework the formulation of appropriate boundary conditions will be required.

The possible use of differential systems like Eq.(20) in studying the nonlinear space-dependent dynamics of the nuclear breeding process in realistic physical conditions is under investigation now. The most delicate points are, of course, related to the choice of suitable functionals of the densities, replacing the addends $(\partial/\partial\bar{x})\cdot\bar{J}_j(\bar{x},t)$, for an adequate description of the streaming terms, as well as to the determination of proper boundary conditions. These and other issues will be object of future work.

ACKNOWLEDGEMENTS

The authors wish to thank the Italian Ministry of Education for the financial support in the frame of 40% and 60% Projects. Also the support of C.N.R.'s National Group of Mathematical Physics in the frame of the activities of PS-AITM is acknowledged.

REFERENCES

1. V.C.Boffi, V.Franceschini and G.Spiga, Dynamics of a gas mixture in an extended kinetic theory, Phys. Fluids 28:3232 (1985).

2. G.Spiga, Dynamical systems in nonlinear transport theory, Mathematical Department of the University of Bari, Report 7/87, Bari (1987).
3. V.C.Boffi and K.Aoki, A system of conservation equations arising in nonlinear dynamics of gas mixture, Il Nuovo Cimento D, to appear.
4. V.C.Boffi, Systems of conservation equations in nonlinear transport theory, in Proceedings of Wing's Conference on "Invariant Imbedding, Transport Theory and Integral Equations" (Santa Fe, NM, 20-22 January 1988), D.L.Seth, ed., to appear.
5. V.C.Boffi and G.Spiga, Spatially inhomogeneous nonlinear dynamics of a gas mixture, in Proceedings of XVI International Symposium on "Rarefied Gas Dynamics" (Pasadena, 10-16 July 1988), E.P.Muntz, ed., AIAA, New York, to appear.

NONLINEAR XENON OSCILLATIONS

David L. Hetrick

Department of Nuclear and Energy Engineering
The University of Arizona
Tucson, Arizona 85721, U.S.A.

ABSTRACT

Space-independent xenon oscillations are linearly unstable at high neutron flux, even with a negative power coefficient of reactivity, if the magnitude of the power coefficient is too small. These linearly unstable oscillations typically develop into stable limit cycles when nonlinear neutronics and xenon burnout are included. If the equilibrium flux is greater than about 2×10^{10}, some linearly stable cases exhibit an unstable limit cycle inside a stable limit cycle. An example of this was first described in 1961 by Jack Chernick as "relaxation oscillations within the domain of linear stability." This paper describes some aspects of the system in parameter space and state space.

INTRODUCTION

It has long been known that the interaction of xenon and other (e.g., thermal) reactivity effects can result in linear instability and sustained nonlinear oscillations (stable limit cycles). A linear analysis was given by Schultz (1961), and a more complete analysis, both linear and nonlinear, was given by Chernick et al (1961).

The controlling dynamic processes are extremely slow. The half-lives for the fission products iodine-135 and xenon-135 are about 7 hours and 9 hours respectively. Control is therefore relatively easy, although space-dependent xenon effects in large graphite or heavy-water reactors, which are not considered here, can present a difficult control design problem.

Our original motivation for re-examining space-independent xenon oscillations was to search for chaotic behavior of the nonlinear equations. The only chaotic behavior that was encountered was an artifact of the computer software. It would be interesting to search for chaotic behavior induced by external perturbations such as oscillating control rods, but this has not yet been attempted.

We found linearly stable examples for which an unstable limit cycle is nested inside a stable limit cycle. This confirms the original discovery of such behavior by Chernick et al (1961). The example is given in Figure 9 of that paper, where an unstable limit cycle can be inferred to exist between the growing and decaying oscillations.

DYNAMIC EQUATIONS

The following differential equations were employed for most of this study:

$$dI/dt = \gamma_I \Sigma_f \phi - \lambda_I I \quad (1)$$

$$dX/dt = \lambda_I I + \gamma_X \Sigma_f \phi - \lambda_X X - \sigma_X \phi X \quad (2)$$

$$d\phi/dt = \lambda R\phi \tag{3}$$

$$R = R_0 + R_X + R_T \tag{4}$$

$$R_X = -(\sigma_X/\nu p\epsilon\beta\Sigma_f)\,\delta X \tag{5}$$

$$R_T = -P_c(P_r/\phi_r)\,\delta\phi \tag{6}$$

The state variables are: iodine concentration I, xenon concentration X, and flux ϕ. Fission yields are $\gamma_I = 0.056$ and $\gamma_X = 0.003$. The macroscopic fission cross section is irrelevant because it cancels out in the normalized equations. Decay constants in sec^{-1} are $\lambda_I = 2.9 \times 10^{-5}$, $\lambda_X = 2.1 \times 10^{-5}$, and, for delayed neutron emitters, $\lambda = 0.0767$. The cross section for neutron capture in xenon is $\sigma_X = 3.5 \times 10^{-18}$ cm^2.

The other variables are the reactivity R and its feedback components R_X and R_T (xenon and temperature), where $\delta X = X - X_0$ and $\delta\phi = \phi - \phi_0$ and where the subscript zero refers to steady state with external reactivity $R_0 = 0$. Computations were for a typical thermal reactor with $\nu p\epsilon = 2.0$ and $\beta = 0.0065$ (usual notation). The power-to-flux ratio P_r/ϕ_r was chosen to illustrate a typical heavy-water reactor (2.954×10^{-11} MW cm^2 sec). The parameters to be varied are the equilibrium flux ϕ_0 and the power coefficient of feedback reactivity P_c (dollars/MW).

For convenience, the equations were normalized before programming. Normalized state variables are $RI = I/I_0$, $RX = X/X_0$ and $F = \phi/\phi_0$ where

$$I_0 = \gamma_I\,\Sigma_f\,\phi_0/\lambda_I \tag{7}$$

and

$$X_0 = (\gamma_X + \gamma_I)\,\Sigma_f\,\phi_0/(\lambda_X + \sigma_X\phi_0) \tag{8}$$

The normalized state equations are

$$d(RI)/dt = \lambda_I(F - RI) \tag{9}$$

$$d(RX)/dt = (\lambda_X + \sigma_X\phi_0)\,(\gamma_X F + \gamma_I RI)/(\gamma_X + \gamma_I) - (\lambda_X + \sigma_X\phi_0 F)RX \tag{10}$$

and, letting $F = e^U$,

$$dU/dt = \lambda R \tag{11}$$

The feedback equations become

$$R_X = -(\gamma_X + \gamma_I)\,\sigma_X\phi_0\,(RX - 1)/\nu p\epsilon\beta(\lambda_X + \sigma_X\phi_0) \tag{12}$$

and

$$R_T = -\phi_0 P_c(P_r/\phi_r)\,(F - 1) \tag{13}$$

This third-order system is easily integrated using a stiffly-stable integrator, but the large relaxation oscillations can be tracked only with very stringent error controls.

The models for neutron dynamics and thermal feedback can be improved by using the prompt-jump approximation with several delayed-neutron groups and by including a time lag for heat transfer. To be consistent, one should also include tellurium-135, the 11-sec precursor of iodine-135. We have experimented with several variations, and, as expected, none of this makes any difference because the xenon and iodine time constants are so large. Of course, results from the present model would be invalid if the reactivity were to approach prompt critical; this would signify a potential runaway and the need to adopt a more realistic model.

LINEAR STABILITY

The dynamic stability boundary for the associated linear system is shown in Figure 1. The parameter space is P_c vs ϕ_0, and the system is stable above the curve and unstable below it. Remember that P_c is the power coefficient of feedback reactivity, which is the negative of the power coefficient of the net reactivity R in Eqs. (3) and (4).

In other words, for ϕ_0 greater than 9.4 x 10^8, the system is unstable with a negative power coefficient if the magnitude of the power coefficient is too small. Systems below the curve have divergent oscillations, systems above the curve have decaying oscillations, and systems on the curve have sustained oscillations (vortex points in state space). For future reference we note that the stability boundary is at P_c = 0.00227 for $\phi_0 = 10^{10}$, at P_c = 0.00177 for ϕ_0 = 2 x 10^{10}, and at P_c = 0.000862 for $\phi_0 = 10^{11}$. See also Figure 2 of the paper by Chernick et al (1961).

It should be noted that the result quoted by Schultz (1961) is for a second-order system having static neutronics, i.e., reactor power proportional to net reactivity (constant forward-loop transfer function). The stability curve given by Schultz in his Figure 5-27 matches our Figure 1

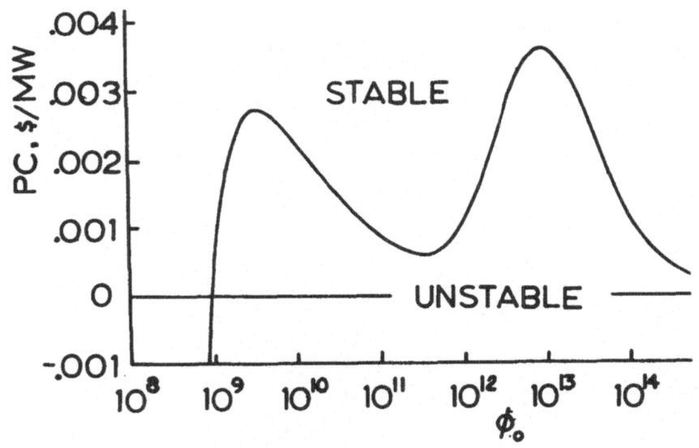

Fig. 1. Dynamic stability boundary (threshold of linear stability) for xenon oscillations, plotted as thermal feedback power coefficient vs equilibrium flux level.

at high flux, but the two curves diverge below $\phi_0 = 10^{12}$. Our curve has two peaks, while the curve given by Schultz has only one peak and predicts that the minimum flux for instability is 3.2 x 10^{11}. The model with static neutronics is invalid for smaller values of ϕ_0 and results in an overestimate of the stable domain. Note that Schultz plotted K_{TC} vs ϕ_0 where, in terms of our notation,

$$K_{TC} = \nu p \epsilon P_r P_c / \sigma_X \phi_r \tag{14}$$

NONLINEAR STABILITY

As expected, stable limit cycles were observed in the linearly unstable region. Figure 2 shows peak values of RI vs P_c for $\phi_0 = 10^{10}$. Note that stable limit cycles may appear even for P_c < 0 (positive power coefficient). The stable limit cycles disappear at the stability boundary P_c = 0.00227.

Figure 3 is a similar plot for $\phi_0 = 2 \times 10^{10}$, and Figure 4 is for $\phi_0 = 10^{11}$. The stable limit cycles persist into the stable region, and the additional points in the stable region represent unstable limit cycles nested within the stable limit cycles. It is concluded that unstable limit cycles can exist for values of equilibrium flux greater than about 2×10^{10}.

Another description of Figure 4 might be as follows: as the magnitude of the power coefficient is reduced, a linearly stable system undergoes a bifurcation into an unstable limit cycle inside a stable limit cycle. As the stability threshold is approached, the unstable limit cycles become smaller and the stable limit cycles become larger. At the threshold, the unstable limit cycles disappear, and there is a center (vortex point) inside a stable limit cycle. As one moves into the domain of linear instability the stable limit cycles continue to grow larger. Eventually a region of unbounded non-oscillatory solutions is reached.

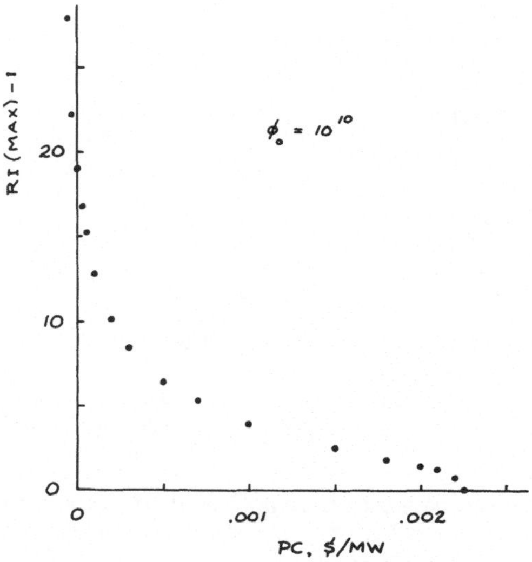

Fig. 2. Limit cycle peaks (maximum iodine-135 concentrations) vs power coefficient for equilibrium flux 10^{10}, critical $P_c = 0.00227$ dollar/MW.

An example of a stable limit cycle is shown in Figure 5. It corresponds to the stability threshold in Figure 4 where $P_c = 0.000862$ for $\phi_0 = 10^{11}$, and is therefore a stable limit cycle surrounding a vortex point. The shapes of the curves are typical of many of the large limit cycles that have been observed in stable and unstable regions of parameter space. These larger limit cycles are relaxation oscillations in which the flux undergoes sharp maxima and extremely small minima. Note how the reactivity curve mirrors the xenon curve except in the neighborhood of the power peak where the effect of temperature feedback briefly appears.

Figure 6 is a two-dimensional projection of the three-dimensional state space for $P_c = 0.0010$ and $\phi_0 = 10^{11}$ in the linearly stable region of Figure 4. The attracting set consists of the equilibrium point (1,1) and the dashed line (the stable limit cycle). The unstable limit cycle is between the small converging trajectory and the diverging trajectory that approaches the stable limit cycle.

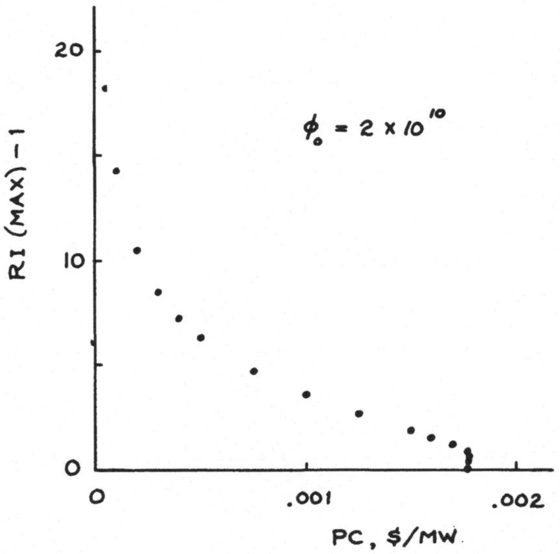

Fig. 3. Limit cycle peaks vs power coefficient for equilibrium flux 2 x 10^{10}, critical P_c = 0.00177 dollar/MW.

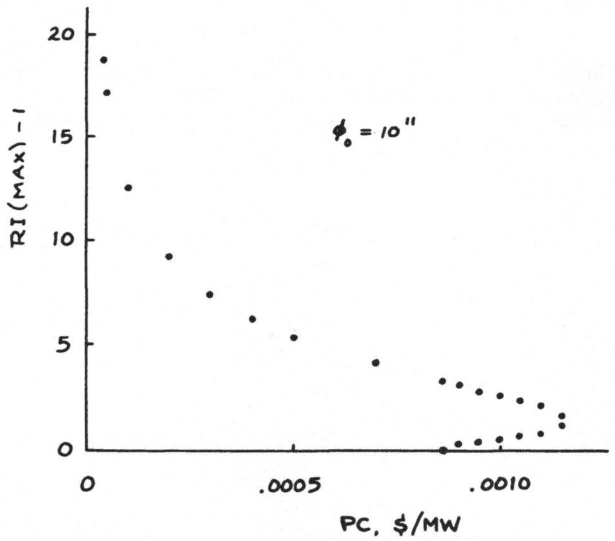

Fig. 4. Limit cycle peaks vs power coefficient for equilibrium flux 10^{11}, critical P_c = 0.000862 dollar/MW. Bifurcation into stable and unstable limit cycles occurs at P_c = 0.00115.

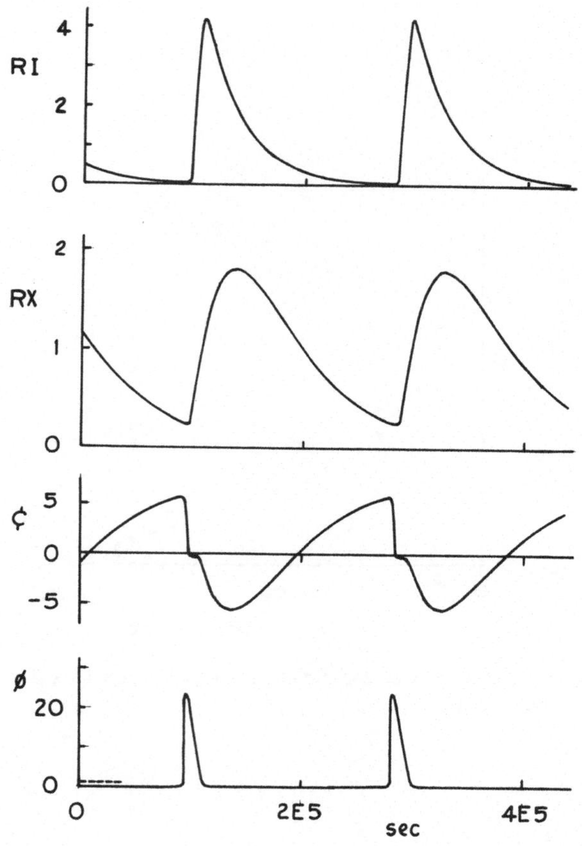

Fig. 5. Stable limit cycle surrounding vortex point for equilibrium flux 10^{11} and P_c = 0.000862 dollar/MW. RI = relative concentration of iodine-135; RX = relative concentration of xenon-135; ¢ = reactivity in cents; ϕ = relative neutron flux.

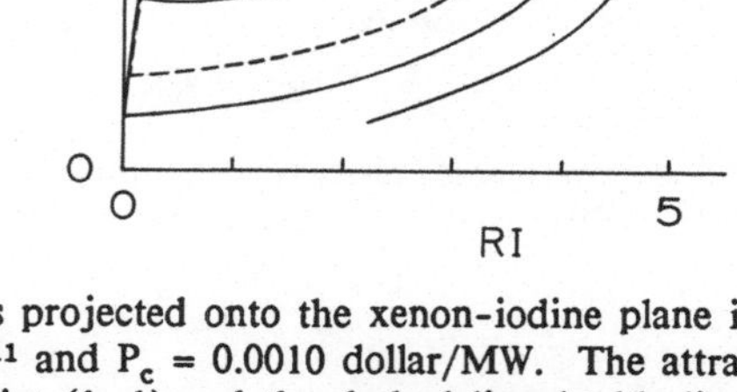

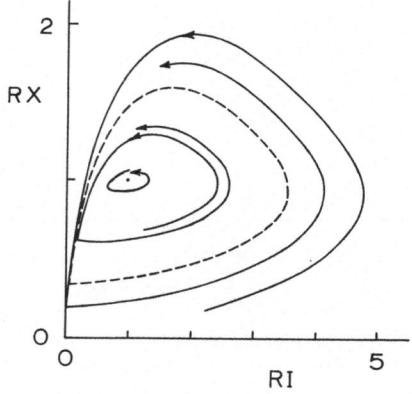

Fig. 6. Solution curves projected onto the xenon-iodine plane in state space for equilibrium flux 10^{11} and P_c = 0.0010 dollar/MW. The attracting set consists of the equilibrium point (1, 1) and the dashed line (stable limit cycle). The unstable limit cycle is just outside the small converging trajectory.

CONCLUSION

Nonlinear space-independent xenon oscillations have been re-visited, and some previous results (Chernick et al, 1961) have been confirmed and exhibited in a slightly different form. We hope to encourage more extensive numerical experiments and theoretical studies in this area.

DEDICATION

This paper is dedicated to the memory of Jack Chernick, 1911-1971.

REFERENCES

Chernick, J., Lellouche, G., and Wollman, W., 1961, The effect of temperature on xenon instability, Nucl. Sci. Engr., 10:120.

Schultz, M. A., 1961, "Control of Nuclear Reactors and Power Plants," 2nd ed., McGraw-Hill Book Company, Inc., New York.

NONLINEAR REACTOR STABILITY ANALYSIS WITH ARBITRARY FEEDBACK

Turan B. Enginol

Boğaziçi University
İstanbul, Turkey

INTRODUCTION

In this investigation a sufficient condition for the asymptotic stability of a reactor in terms of the equilibrium power level and delayed neutron parameters is presented. This criterion is somewhat more general than various stability criteria proposed by different authors previously, as is clarified later in this paper.

The integrodifferential form of the point kinetics equations for an autonomous reactor can be written as in Akçasu et al.[1]

$$\Lambda\dot{P}=\left[\delta_o+k_f(P)\right]P+\int_0^{\infty}du\ \left[P(t-u)-P(t)\right] \tag{1}$$

where,

$$D(u)\equiv\sum_{i=1}^{6}a_i\lambda_i\ \exp(-\lambda_i u),\qquad \sum_{i=1}^{6}a_i=1\quad,\qquad \Lambda\equiv\ell/\beta$$

and where P is the reactor power, δ_o the constant value of the external reactivity needed to overcome the feedback reactivity at the operating power level.

The feedback reactivity functional can be written as a functional series;

$$K_f(P)=\sum_{m=1}^{6}\int_{-\infty}^{t}du_1\dots\int_{-\infty}^{t}du_m G_m(t-u_1,\dots,t-u_m)P(u_1)\dots P(u_m) \tag{2}$$

which is time invariant, causal and is assumed to be stable. Here the feedback kernels $G_m(u_1,\dots,u_m)$ can be assumed to be symmetric[1] with respect to the arguments u_i without any loss of generality. Note also that the feedback functional has no constant term because it will be presumed to cancel out the constant external recativity δ_o.

After dividing Eq.(1) by P, introducing a new variable,

$$Q(t)=\ln p(t)/P_o \tag{3}$$

with P_o as the equilibrium power level, and taking the time derivative of the resulting equation gives[2]

$$Q= \frac{1}{\Lambda}\frac{d}{dt}K_f(Q)+\frac{1}{\Lambda}\frac{d}{dt}\{\exp(-Q)\int_0^\infty \exp\left[Q(t-u)\right] D(u)\,du\} \tag{4}$$

Here the derivative of the first order feedback kernel is,

$$\frac{d}{dt}\left[\int_0^\infty G_1(u)\,P_o\exp(Q(t-u))\,du\right]= -\int_0^\infty G_1(u)\,P_o\frac{d}{du}\exp(Q(t-u))\,du$$

which in turn gives, by partial integration

$$\frac{d}{dt}\left[\int_0^\infty G_1(u)\,P_o\exp(Q(t-u))\,du\right]= G_1(0)\,P_o\left[\exp(Q(t))-1\right]$$

$$+\int_0^\infty G_1'(u)\,P_o\left[\exp(Q(t-u))-1\right]du \tag{5}$$

Here we assumed $G_1(\infty)=0$ which is reasonable for physical systems.

Now, also assuming $G_1(0) < 0$ and defining

$$\omega^2\equiv\frac{P_o}{\Lambda}\left|G_1(0)\right|$$

Eq.(4) becomes through the use of Eq.(5)

$$\ddot{Q}+\omega^2(\exp(Q)-1) = \frac{1}{\Lambda}\int_0^\infty G_1'(u)\,P_0\left[\exp(Q(t-u_1)-1\right]du_1$$

$$+\frac{1}{\Lambda}\frac{d}{dt}\{\sum_{m=2}^{\infty}\int_0^\infty du_1\ldots\int_0^\infty du_m\,G_m(u_1,\ldots,u_m)P_o^m$$

$$\times\exp\left[Q(t-u_1)+\ldots+Q(t-u_m)\right]\}$$

$$+ \frac{1}{\Lambda} \frac{d}{dt} \{\exp(-Q(t)) \int_0^\infty \exp(Q(t-u))\, D(u)\, du\} \qquad (6)$$

Since Q(t) is a much slower varying function of time than P(t) is, for sufficiently small power variations, the functional power series expansion can be terminated after the first term and $G_1'(u)$ can be taken to be small. For this case of nearly constant power extraction, taking the delayed neutron concentration constant at the equilibrium value, the linearized form of Eq.(6) becomes

$$\ddot{Q} + \omega^2 Q = - \frac{1}{\Lambda} \dot{Q} , \qquad \omega^2 = \frac{P_o}{\Lambda} |G_1(0)|$$

which is a familiar form of linearized point kinetics equations.

We can thus write Eq.(6) as

$$\ddot{Q} + \omega^2 Q = \varepsilon f(Q,\dot{Q}) \qquad (7)$$

where

$$\varepsilon f(Q,\dot{Q}) = \omega^2 Q + \frac{1}{\Lambda} \frac{d}{dt} \sum_{m=1}^{\infty} P_o^m \int_0^\infty du_1 \ldots \int_0^\infty du_m\, G_m(u_1, \ldots, u_m)$$

$$\times \exp\left[Q(t-u_1)+\ldots+Q(t-u_m)\right]$$

$$+ \frac{1}{\Lambda} \frac{d}{dt} \{\exp(-Q) \int_0^\infty du\, D(u) \exp\left[Q(t-u)\right]\} \qquad (8)$$

Here ε serves to indicate that the right hand side of Eq.(7) is small under the specified conditions. Note that although the $\omega^2 Q$ term itself may not be small, on the right hand side it is also included in the linear term of the feedback series expansion with opposite sign and is therefore cancelled out.

THE METHOD OF KBM

Since the solution of the homogenous form of Eq.(7) is $Q = a \cos\omega t$ where a and ω are constants, it is assumed in the method of Krylov, Bogoljubov and Mitropolsky[3] that as a first approximation

$$Q(t) = a(t) \cos(\omega t + \theta(t)) = a(t) \cos\psi(t) \qquad (9)$$

can be written. Here a(t) and the additional phase term $\theta(t)$ are now time dependent but vary slowly because of the small nonlinearity $\varepsilon f(Q,\dot{Q})$.

Hence,

$$\frac{da}{dt} = \varepsilon A(a) \quad \text{and} \quad \frac{d\psi}{dt} = \varepsilon B(a) + \omega \tag{10}$$

describe, to a first order in **ε**, these time dependencies. Our aim is to find the function A so that stability of the amplitude a(t) can be studied[4].

For this purpose,

$$\dot{Q} = \frac{da}{dt}\cos\psi - a\frac{d\psi}{dt}\sin\psi = -\omega a\sin\psi + \varepsilon(A\cos\psi - aB\sin\psi)$$

$$\ddot{Q} = \frac{d^2a}{dt^2}\cos\psi - 2\,\frac{da}{dt}\,\frac{d\psi}{dt}\sin\psi - a\,\frac{d^2\psi}{dt^2}\sin\psi - a\left(\frac{d\psi}{dt}\right)^2\cos\psi$$

$$= -\,\omega^2 a\cos\psi + \varepsilon(-2\omega A\sin\psi - 2\omega a B\cos\psi) + \varepsilon^2\ldots$$

can be obtained from Eq.(9) and the terms up to the order ε be inserted into Eq.(7) so as to get,

$$\varepsilon(-2\omega\, A\sin\psi - 2\omega\, aB\cos\psi) = \varepsilon f(Q,\dot{Q})$$

Expanding $\varepsilon f(Q,\dot{Q})=\varepsilon f(Q_o+\Delta Q,\dot{Q}_o+\Delta\dot{Q})$ into Taylor's series around $(Q_o,\dot{Q}_o)$ where $Q_o=a\cos\psi$ and $\dot{Q}_o=-\omega a\sin\psi$, and equating the terms of first order in ε gives;

$$-2\omega A\sin\psi - 2\omega aB\cos\psi = \varepsilon f(a\cos\psi,-\omega a\sin\psi) \tag{11}$$

Now, in order to determine the function A, consider the Fourier series expansion of

$$F_o(a,\psi)=f(a\cos\psi,-\omega a\sin\psi)=\sum_{n=-\infty}^{\infty} F_{on}(a)\exp(ik\psi)$$

with

$$F_{on}=\frac{1}{2\pi}\int_0^{2\pi} F_o(a,\psi)\exp(-in\psi)\,d\psi$$

Inserting this expansion into Eq.(11) and equating the terms containing the first harmonics gives

$$-2\varepsilon A\omega\sin\psi - 2\varepsilon\omega Ba\cos\psi = \varepsilon F_o(\cos\psi + i\sin\psi) + \varepsilon F_{o-1}(\cos\psi - i\sin\psi)$$

Since the coefficients of sinψ and cosψ terms must be separately equal on the two sides of this equality we obtain

$$A = -\frac{1}{2\omega}\,i(F_{oi} - F_{o-1}) = -\frac{1}{2\pi\omega}\int_0^{2\pi} F_o(a,\psi)\sin\psi\,d\psi$$

and thus

$$\frac{da}{dt} = \Phi(a) = -\frac{1}{2\pi\omega}\int_0^{2\pi} f(a\cos\psi, -a\omega\sin\psi)\,\sin\psi\,d\psi \tag{12}$$

determines the time rate of the amplitude.

THE STABILITY CRITERION

By a theorem related to the averaging principle[5], if $\Phi(a)=0$ has a simple root $a=a_o$, Eq.(7) has a unique periodic solution in the neighborhood of $a=a_o \cos\omega t$ for a sufficiently small $\varepsilon f(Q,\dot{Q})$, and such a periodic solution is asymptotically orbitally stable if $\Phi'(a) < 0$ holds.

Note that

$$\begin{aligned} Q(t-u_i) &= a\cos\left[\omega(t-u_i)+\theta\right] \\ &= a\cos(\psi-\omega u_i) \\ &= a\cos\omega u_i\cos\psi + a\sin\omega u_i\sin\psi \end{aligned}$$

and therefore we can now compute $\Phi(a)$ by making use of Eq.(8) and Eq.(12);

$$\Phi(a) = -\frac{1}{2\pi\omega}\int_0^{2\pi} \omega^2 a\,\cos\psi\,\sin\psi\,d\psi$$

$$-\frac{1}{2\pi\omega}\frac{1}{\Lambda}\int_0^{2\pi}\frac{d}{dt}\{\sum_{m=1}^{\infty} P_o^m\int_0^{\infty} du_1\cdots\int_0^{\infty} du_m\, G_m(u_1,\ldots,u_m)$$

$$\times\exp(a\,U_m\cos\psi + a\,V_m\sin\psi)\}\,\sin\psi\,d\psi$$

$$-\frac{1}{2\pi\omega}\frac{1}{\Lambda}\int_0^{2\pi}\frac{d}{dt}\{\exp(-a\cos\psi)\int_0^{\infty} D(u)$$

$$\times\exp(a\cos\omega u\cos\psi + a\sin\omega u\sin\psi)\ \sin\psi\,d\psi \tag{13}$$

where

$$U_m \equiv \sum_{i=1}^{m}\cos\omega u_i \quad , \quad V_m \equiv \sum_{i=1}^{m}\sin\omega u_i$$

In Eq.(13) the first term vanishes and the remaining terms can be integrated by padrt using $d\psi = \omega dt$ so as to obtain

$$\Phi(a) = \frac{1}{2\pi\Lambda}\int_0^{2\pi}\{\sum_{m=1}^{\infty} P_o^m\int_0^{\infty} du_1\cdots\int_0^{\infty} du_m\, G_m(u_1,\ldots,u_m)$$

$$\times \exp(a\, U_m \cos\psi + a\, V_m \sin\psi) \}\cos\psi\, d\psi$$

$$+ \frac{1}{2\pi\Lambda} \int_0^{2\pi} \{\exp(-a \cos\psi) \int_0^{\infty} D(u)$$

$$\times \exp(a \cos\omega u \cos\psi + a \sin\omega u \sin\psi)\, du\} \cos\psi\, d\psi \qquad (14)$$

Using the equalities

$$\int_0^{2\pi} \exp(p \cos\alpha + q \sin\alpha) \cos\alpha\, d\alpha = \frac{2\pi p}{\sqrt{p^2+q^2}} I_1(\sqrt{p^2+q^2})$$

where I_1 is the modified Bessel's function of the first kind and,

$$a^2(\cos\omega u - 1)^2 + a^2 \sin^2\omega u = 4a^2 \sin \frac{\omega u}{2}$$

Eq.(14) can be put into the form

$$\Phi(a) = \frac{1}{\Lambda} \sum_{m=1}^{\infty} P_o^m \int_0^{\infty} du_1 \ldots \int_0^{\infty} du_m\, G_m(u_1,\ldots,u_m) \frac{U_m}{\sqrt{U_m^2 + V_m^2}} I_1(a\sqrt{U_m^2+V_m^2})$$

$$+ \frac{1}{\Lambda} \int_0^{\infty} D(u) \frac{\cos\omega u - 1}{2 \sin \frac{\omega u}{2}} I_1(2a \sin \frac{\omega u}{2})\, du \qquad (15)$$

The second term can further be simplified by using the identity

$$(\cos\omega u - 1)/\ 2 \sin \frac{\omega u}{2} \quad - \sin \frac{\omega u}{2}$$

Also using the series expansion

$$I_1(x) = \sum_{k=0}^{\infty} \frac{(x/2)^{k+1}}{k!(k+1)!}$$

for the modified Bessel's function; Eq.(15) becomes

$$\Phi(a) = \frac{1}{\Lambda} \sum_{k=0}^{\infty} \frac{(a/2)^{k+1}}{k!(k+1)!}$$

$$\times \left\{ \sum_{m=1}^{\infty} P_o^m \int_0^{\infty} du_1 \ldots \int_0^{\infty} du_m \, G_m(u_1, \ldots, u_m) \, U_m \, (U_m^2 + V_m^2)^{k/2} \right\}$$

$$- \frac{1}{\Lambda} \sum_{k=0}^{\infty} \frac{a^{k+1}}{k!(k+1)!} \int_0^{\infty} D(u) \, (\sin \frac{\omega u}{2})^{k+2} \, du \tag{16}$$

which can be abbreviated as,

$$\Phi(a) = -\frac{1}{\Lambda} \sum_{k=0}^{\infty} \frac{(a/2)^{k+1}}{k!(k+1)!} (\gamma_k - \alpha_k) \tag{17}$$

where

$$\gamma_k = 2^{k+1} \int_0^{\infty} D(u) \, (\sin \frac{\omega u}{2})^{k+2} \, du$$

and,

$$\alpha_k = \sum_{m=1}^{\infty} P_o^m \int_0^{\infty} du_1 \ldots \int_0^{\infty} du_m \, G_m(u_1, \ldots, u_m) \, U_m \, (U_m^2 + V_m^2)^{k/2}$$

As can be seen from Eq.(17) a=0 is a root of $\Phi(a) = 0$ and for this root to be asymptotically stable $\Phi'(0) < 0$ must hold. Furthermore for global asymptotic stability, $\Phi(a) = 0$ should have no other roots.

Since,

$$\Phi'(a) = -\frac{1}{\Lambda} \sum_{k=0}^{\infty} \frac{(a/2)^{k}}{(k!)^2} (\gamma_k - \alpha_k) \tag{18}$$

$\Phi'(0) < 0$ can be satisfied only if $\gamma_o > \alpha_o$ and a sufficient condition for $\Phi(a) = 0$ to have no other roots is $\gamma_k > \alpha_k \ \forall k$. Or if γ_k has a minimum for a finite k, then the latter condition can be replaced with $\min_k \gamma_k > \alpha_k, \forall k$.

Now let us examine the γ_k, α_k terms in Eq.(17). Since it can be shown that $U_m^2 + V_m^2 \leq m^2$ it follows that;

$$\alpha_k \leq \sum_{m=1}^{\infty} P_o^m \, m^k \int_0^{\infty} du_1 \ldots \int_0^{\infty} du_m \, G_m(u_1, \ldots, u_m) \, U_m \tag{19}$$

Also since the feedback kernels are symmetric with respect to the variables u_i, $U_m = \sum_{m=1}^{m} \cos\omega u_i$ can be replaced with $m\cos\omega u_1$ and thus

$$\alpha_k \leq \sum_{m=1}^{\infty} P_o^m \, m^{k+1} \int_0^{\infty} du_1 \ldots \int_0^{\infty} du_m \, G_m(u_1,\ldots,u_m) \cos\omega u_1 \tag{20}$$

holds.

By defining

$$g_m \equiv \int_0^{\infty} du_1 \ldots \int_0^{\infty} du_m \, G_m(u_1,\ldots,u_m) \cos\omega u_1 \tag{21}$$

inequality (20) becomes

$$\alpha_k \leq \sum_{m=1}^{\infty} P_o^m \, m^{k+1} \, g_m \tag{22}$$

On the other hand using the definition of D(u) in Eq.(18)

$$\gamma_k = \sum_{i=1}^{6} a_i \, \gamma_{i,k}$$

can be written, where

$$\gamma_{i,k} = 2^{k+1} \lambda_i \int_0^{\infty} \exp(-\lambda_i u) \left(\sin \frac{\omega u}{2}\right)^{k+2} du \tag{23}$$

Noting that $\gamma_{i,k}$ are the Laplace transforms of the powers of sine functions, it can be shown that there is a positive nonzero γ_k^* such that[3]

$$\gamma_k^*(\omega,a_i,\lambda_i) = \min_k \gamma_k(\omega,a_i,\lambda_i) \tag{24}$$

and that

$$\lim_{k\to\infty} \frac{\lambda_k}{2^{k+1}} = 0 \tag{25}$$

holds.

Through the use of inequality (22) the stability condition $\gamma_k > \alpha_k$, $\forall k$ can now be stated as

$$\gamma_k > \sum_{m=1}^{\infty} P_o^m \, m^{k+1} \, g_m, \qquad k = 1,2,\ldots \tag{26}$$

Here dividing both sides of the inequality with n^{k+1} for a given n and letting $k \to \infty$ it can be seen that the left hand side tends to zero due to Eq.(25) while the terms on the right hand side and with m > n diverge. Therefore in order for the stability condition $\gamma_k > \alpha_k$, $\forall k$ to hold $g_m < 0$, $\forall m > 1$ must be satisfied in which case inequality (26) can be rewritten as;

$$\gamma_k + \sum_{m=2}^{\infty} P_o^m \; m^{k+1} \; |g_m| > P_o g_1; \; k = 0,1,\dots \text{ and } g_m < 0 \; \forall m > 1 \qquad (27)$$

In this last inequality the summation term is the smallest for k=0, hence replacing γ_k with γ_k^* and taking the summation term for k=0 gives us a stricter stability condition which is;

$$\gamma_k^*(\omega,a_i,\lambda_i) + \sum_{m=2}^{\infty} P_o^m \; m \; |g_m| > P_o g_1 \text{ and } g_m < 0, \; \forall m > 1 \qquad (28)$$

This is the generalized stability condition and is seen, in specific applications to be less restrictive than the previous criteria published. Note that the summation term in Eq.(28) is convergent because of the stability condition assumed for the reactivity feedback functional.

γ_k^* in the stability criterion in Eq.(28) depends, among other parameters on ω. However nonlinear systems are in general capable of frequency multiplication, meaning oscillations at a definite frequency ω can, as time proceeds, give rise to oscillations at frequencies that are multiples of ω. Therefore the stability of the nonlinear system should be studied by keeping in mind $k\omega$ multiples of the natural frequency ω. But it can be seen that the $\gamma_k^*(k\omega,a_i,\lambda_i)$ tends to zero as k takes on larger and larger values. In summary, it can be stated that although the delayed neutrons initially have a positive contribution to stability, this contribution vanishes as time proceeds and higher harmonics of the natural frequency oscillations set into the system.

DISCUSSION[5]

The criterion found here depends on the equilibrium power(explicitly through ω and implicitly through the feedback kernels), the delayed neutron parameters a_i and λ_i, the feedback kernels, and Λ. The magnitude γ_k appearing in the criterion depends on the delayed neutron parameters and also on $P_o|G_1(0)|$ and Λ through ω. In applying the criterion to special cases of linear and nonlinear feedback, we observe that it is more general than Welton's criterion[6] due to its dependence on equilibrium power level and delayed neutron parameters and due to its applicability to cases with nonlinear feedback.

For comparing the derived criterion of Eq.(28) with the one obtained by Akçasu and Dalfes[7], the feedback functional can be written as in Akçasu and Akhtar[8]:

$$K_f(P) = \int_0^{\infty} du \; G_1(u) \; P(t-u) + \int_0^{\infty} du \; G_2(u) \; P^2(t-u), \qquad (29)$$

where

$$G_1(t) = A_1 \ \delta(t) + K(t),$$

$$G_2(t) = A_2 \ \exp(-\lambda_{Xe} t), \tag{30}$$

and λ_{Xe} is the decay constant of ^{135}Xe. The constants A_1 and A_2 and the explicit form of the function K(t) are given by Akçasu et al.[1,2] Applying the criterion of Eq.(28) with

$$g_1 = \int_0^\infty G_1(u) \cos\omega u \ du = A_1 + \mathrm{Re}\left[\bar{K}(i\omega)\right]$$

and

$$g_2 = \int_0^\infty G_2(u) \cos\omega u \ du = \frac{A_2 \lambda_{Xe}}{\lambda_{Xe}^2 + \omega^2} \tag{31}$$

where $\mathrm{Re}\left[\bar{K}(i\omega)\right]$ is the real part of the Fourier transform of K(t). It is seen that as $\lambda_{Xe} > 0$, the condition $g_2 < 0$ implies $A_2 < 0$, which was demanded by Akçasu et al[1] for a negative temperature coefficient in order to assure a unique equilibrium state. Using now the criterion of Eq.(28), we obtain

$$A_1 + \mathrm{Re}\left[\bar{K}(i\omega)\right] - \frac{P_o A_2}{\lambda_{Xe}} \cdot \frac{2}{1+(\frac{\omega}{\lambda_{Xe}})^2} < \frac{\gamma_k^*}{P_o} \tag{32}$$

It is seen that in omitting the delayed neutron term, the two criteria that are being compared become somewhat similar in form; however, the criterion of Eq.(32) for the present case gives a larger region of stability[9] as shown in Fig.1.

The special form of the criterion resulting from a liniearized stability analysis, compared with the one of nonlinear analysis with linear feedback, leads to the observation that the criterion of the linearized analysis does not always exclude the existence of possible nonzero stationary amplitudes, whereas such amplitudes are excluded by the nonlinear criterion.

In the preceeding treatment, the parameter ε signifies the ratio of the magnitude of the delayed neutron and other nonlinear feedback terms to that of the linear one [see Eq.(7)], and this significance clarifies the physical conditions corresponding to the smallness of this parameter. The first order treatment has been based on the observation that a unique equilibrium solution that is locally asymptotically stable implies global asymptotic stability. In this order of treatment, the equilibrium solutions for reasonably well-behaved systems differ slightly from the exact solutions and that only quantitatively, but not qualitatively as pointed out by Minorsky[10]. On the other hand, the sufficiency of the first order treatment in describing the system behavior from the stability point of view can be established by a mechanical analogue, used among others, by Welton[11] in obtaining his stability criterion. This analogue, surely valid for physical systems, asserts that if the average work done on an oscillator by any external periodic force is positive during a full period, than in the absence of external the force, the oscillations die out, i.e., the autonomous system would be stable. According to this analogy, the stability criterion obtained from the first order treatment corresponds to the positiveness of the average work done on the system described by the gen-

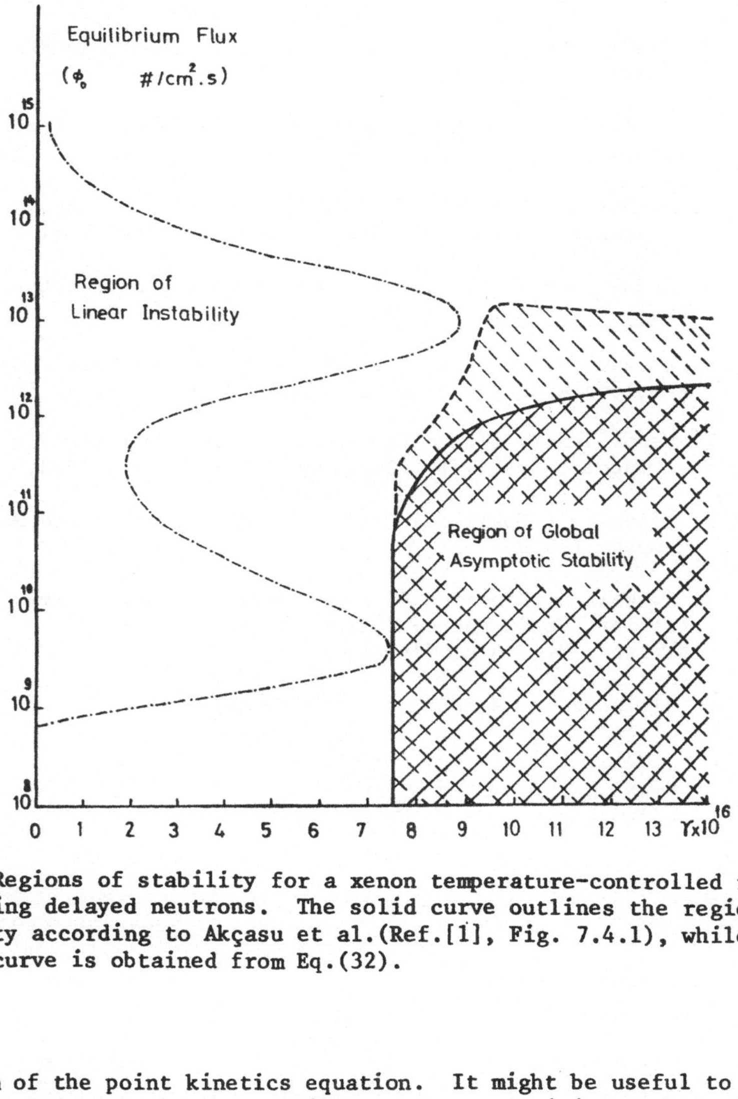

Fig.1. Regions of stability for a xenon temperature-controlled reactor neglecting delayed neutrons. The solid curve outlines the region of stability according to Akçasu et al.(Ref.[1], Fig. 7.4.1), while the dashed curve is obtained from Eq.(32).

eral form of the point kinetics equation. It might be useful to add that for the analogy to hold for nonlinear systems, positiveness of the work done on the system must be assured for external periodic forces of all periods or any period[11]. In this case it is seen that the delayed neutron contribution vanishes due to the higher excitation frequencies that must be considered in this analogy[12].

Acknowledgement

I would like to express my gratitude to research assistant Levent Akin for preparing the manuscript.

REFERENCES

1. Z.Akçasu, G. S. Lellouche, and L. M. Shotkin, "Mathematical Methods in Nuclear Reactor Dynamics", Academic Press (1971).

2. T. B. Enginol, Boğaziçi University Publication, 36/001, (1975); for a summary in English, see also Boğaziçi Univ. J., 3:187, (1975).

3. N. N. Bogoljubov and J. A. Mitropolski, "Asymptotische Methoden in der Nichtlinearen Schwinggungen", Akademie Verlag (1965).

4. M. Urabe, "Nonlinear Autonomous Oscillations", Academic Press, (1967).

5. T. B. Enginol, On The Asymptotic Stability of Nuclear Reactors with Arbitrary Feedback, Nucl. Sci. Eng., 90:231 (1985).

6. T. A. Welton, Systems Kinetics, in "Proc. Symp. Applied Mathematics", vol. XI, American Mathematical Society (1961).

7. Z. Akçasu and A. Dalfes, A Study of Nonlinear Reactor Dynamics, Nucl. Sci. Eng., 8:2 (1960).

8. Z. Akçasu and P. Akhtar, On the Asymptotic Stability of Nuclear Reactors with Arbitrary Feedback, J. Math. Phys. 11:115 (1970).

9. T. Şener and V. Altin, Comparison of Two Criteria for Reactor Asymptotic Stability, Turkish J. Nucl. Sci., 10:3, (1983).

10. N. Minorsky, "Nonlinear Oscillations", Van Nostrand Reinhold Company (1970).

11. T. A. Welton, Kinetics of Stationary Reactor Systems, in "Proc. 1st Int. Conf. Peaceful Uses of Atomic Energy," vol. 5, United Nations, New York, (1956).

12. V. Altin and T. B. Enginol, to be published.

A STOCHASTIC METHOD FOR NUCLEAR POWER PLANT DIAGNOSIS

R. B. Perez* and R. T. Wood*

The University of Tennessee
Knoxville, Tennessee 37916

INTRODUCTION

The value of using neutron noise descriptors, especially the power spectral densities (PSDs) of ex-core detectors, to monitor core barrel motion and fuel element vibrations in pressurized water reactors (PWRs) is well established.[1-6] Indeed, in the mid-frequency range (1-20 Hz), PWR neutron noise is dominated by vibration peaks in the neutron PSD that result from the motion of reactor internals. A qualitative examination of such descriptors from a 1150 MWE Westinghouse PWR plant, revealed significant changes in the plant's noise signature (PSD) as its first fuel cycle proceeded. The purpose of this work is to use a mathematical model of the interaction between small mechanical motions and the core neutronics to provide a tool for quantitative investigation of the PSD structure in the vibrational frequency range.[7]

The purpose of neutron noise analysis is to provide information on plant conditions by the study of power traces from ex-core detectors. Due to the action of the coolant flow through the core and pump induced vibrations, the core barrel, which is bolted at the top of the reactor vessel, performs a pendular motion with a natural frequency determined by the mechanical parameters and constraints of the system.[2-5] In addition, external forces deform the cylindrical shape of the core barrel by inducing vibrational modes which are excited by the turbulence of the coolant flow and by the "table-shaker" effect of the core barrel motion. As a result of the various mechanical motions, the detector response will vary in a manner proportional to the importance function at the location of the disturbance.

A typical ex-core neutron detector normalized PSD for a 1150 MWE PWR is shown in Fig. 1. Below 1 Hz the PSD is dominated by the feedback thermal hydraulic loop effects which evolve with time constants of the order of seconds. Between 3 Hz and 30 Hz, the PSD exhibits a set of resonances arising from the vibrational and pendular modes previously discussed. This disparity in the "natural" frequencies of the system is a fortunate break for the noise analyst as one can perform the modeling of the feedback and mechanical effects in a separate fashion.

IDENTIFICATION OF PSD STRUCTURE

It can be shown[7] on the basis of first order perturbation theory that the neutron detector PSD, $\phi_{11}(\omega)$, is given by the expression:

$$\phi_{11}(\omega) = \sum_{\lambda} \left\{ \frac{\mu_\lambda A_\lambda + (\omega - \nu_\lambda) B_\lambda}{\mu_\lambda^2 + (\omega - \nu_\lambda)^2} + \frac{\mu_\lambda A_\lambda - (\omega + \nu_\lambda) B_\lambda}{\mu_\lambda^2 + (\omega + \nu_\lambda)^2} \right\}, \tag{1}$$

*Oak Ridge National Laboratory, Oak Ridge, Tennessee 37831

where the subindex λ varies over the frequencies of the mechanical vibrations and where we introduce the pole strength and asymmetry factors, A_λ and B_λ, respectively. The result in Eq. (1) shows that the neutron detector PSD is the superposition of a set of resonances at the various frequencies ν_λ. The line shapes are asymmetric with respect to the resonance peak because the term $(\omega - \nu_\lambda)$ changes sign as it goes through the resonance maximum. The second term within the bracket in Eq. (1) is a nonresonant term arising from the reflection of complex poles at negative frequencies.

The parametric model in Eq. (1) has been implemented in the program "CARDIOGRAMA" which performs a least squares fit of the functional relation to measured neutron PSD data. A typical result of the Cardiograma algorithm is shown in Fig. 2, where the data points correspond to the PSD of one of the ex-core neutron detectors at the 1150 MWE PWR SEQUOYAH-1 nuclear power plant. The resonance at 6 Hz has been associated with the pendular motion of the core support barrel (CSB). The peaks at 3.6, 8 and 16 Hz correspond to fuel element vibrations. Two shell mode resonances have been identified: a core barrel shell mode at 12 Hz and a pump-induced shell mode at 20 Hz. The evolution of the neutron PSD as a function of time and power level has been followed for nearly two full cycles at the SEQUOYAH-1 plant.[5] A short-term and several long-term effects have been observed.

The short-term effect observed consisted of the shifting of the fuel vibrational peaks towards lower energies following a decrease in power, while the core barrel and shell model resonance frequencies remain unaltered. Among the various long-term effects, PSDs and CPSDs of the neutron noise exhibit a resonance at frequency of approximately 6.7 Hz. Phases between the cross-core detectors at this frequency are always 180° and the coherence is always greater than 0.8, thereby indicating pendular CSB motion. Another resonance at 8.0 Hz, having the same coherence and phase relationships can also be resolved in the PSDs and CSPDs at the beginning of the first fuel cycle. This resonance occurs at approximately the second mode of fuel assembly vibration (the first mode occurs at 3.6 Hz) as deduced from in-core neutron noise measurements made at SEQUOYAH-1. The fuel assembly fundamental mode decreased in frequency over the first fuel cycle to 3.0 Hz with a corresponding decrease in the second mode resonant frequency. This second mode resonance merged with the CSB resonance (6.7 Hz) to form what appears to be a single broad peak in the 5 to 7.5 Hz range, as shown in Fig. 3. The normalized root mean square (NRMS) of the neutron noise (at nominal full power conditions) in the 5- to 10-Hz frequency range increased by a factor of five from the beginning to the end of the first fuel cycle. At the beginning of the second fuel cycle, the neutron noise amplitude in this same frequency range decreased, but only to a level somewhat higher than at the beginning of the first cycle and resonances could be distinguished only at 3.0 and 7.0 Hz.

RESULTS

In this section, we discuss the application of the present stochastic model to the interpretation of the neutron PSD in the vibrational region. Of the observed long-term effects the most worrysome one is the "disappearance" of the core barrel resonance at the end of the first cycle.

To answer this question, measured PSDs from SEQUOYAH-1 taken at various times during the first and second fuel cycles were fitted using the stochastic model that was developed. Table 1 lists the frequencies and pole-strength factors for the first two fuel element resonances and for the core barrel peak. These results, which were obtained from plant signatures at the beginning, middle, and the end of the first fuel cycle and at the middle of the second fuel cycle, reveal the following features:

(a) the pole strength factors for the first and second fuel resonances increased considerably between April '82 and August '82;

(b) the fuel element peaks shifted towards the low frequency end of the spectrum; and

(c) the pole-strength factor for the core barrel peak remained fairly constant since

April '82, at a level of about one order of magnitude with respect to the value found in April '81.

The large increase in the pole-strength factor and the frequency shift of the second fuel element vibrational mode account for the disappearing act performed by the core barrel peak. However, the ability of the stochastic model to describe the structure of the neutron PSD and to separate the various mechanical motions allowed confirmation that the core barrel motion had not changed. This feature of the model is illustrated in Fig. 4. It can be seen that the broad peak around 7 Hz was interpreted as the contributions arising from the core barrel peak and the second fuel mode vibration. Note that the fitting program would have failed to work properly without the information gathered from earlier data throughout the fuel cycle, where the core barrel resonance was not masked by the resonance at 8 Hz. In fact, the "CARDIOGRAMA" program refused to fit that broad peak with only one resonance. The use of a core barrel contribution, with parameter guesses based on previous experience, was needed to achieve the functional fit shown in Fig. 4.

Examination of Table 1 reveals two plant "aging" effects: (i) the increase of the pole strength function between April '81 and April '82 for the core barrel peak indicates some "loosening" of the core barrel restrictions, and (ii) the shift of the fuel element resonances towards lower frequencies shows a decrease in the "stiffness" of the core's mechanical configuration.

CONCLUSIONS

In summary, first order perturbation theory has been used to relate the response of a neutron detector to small mechanical motions of the reactor internals. By mathematically manipulating the model, an equation for the neutron PSD was obtained that describes each motion in terms of a pole-strength factor, a resonance skewness factor, a vibration damping factor, and a frequency of vibration. This formulation allows each resonance peak to be quantified in terms of four identifiable parameters. The model was fitted to measured PSDs obtained from a PWR at various times during a fuel cycle. The mechanical motion parameters for several resonances were tracked to determine trends that indicated changes in vibrations within the reactor core. In addition, the resonance model gave the ability to separate the resonant components of the PSD after the parameters has been identified. As a result, the behavior of several vibration peaks were monitored over two fuel cycles.

Table 1. Vibrational Frequencies, ν_λ (Hz), and Normalized Pole Strength Factors, A_λ (dimensionless), at Selected Times During the First and Second Fuel Cycles

Date Recorded	Frequency (Hz)			Pole Strength Dimensionless		
	F1[a]	CB[b]	F2[c]	F1	CB	F2
April 81[d]	3.55	7.03	7.82	$4.78 \cdot 10^{-8}$	$8.07 \cdot 10^{-8}$	$9.22 \cdot 10^{-8}$
April 82[d]	3.34	5.88	7.46	$6.91 \cdot 10^{-8}$	$4.31 \cdot 10^{-7}$	$4.48 \cdot 10^{-7}$
Aug. 82[d]	3.08	5.68	7.25	$1.04 \cdot 10^{-7}$	$3.25 \cdot 10^{-7}$	$1.38 \cdot 10^{-6}$
Aug. 83[e]	3.07	5.97	7.26	$2.53 \cdot 10^{-7}$	$2.80 \cdot 10^{-7}$	$1.39 \cdot 10^{-6}$

[a]Fuel element first vibrational mode.
[b]Core support barrel.
[c]Fuel element second vibrational mode.
[d]First fuel cycle.
[e]Second fuel cycle.

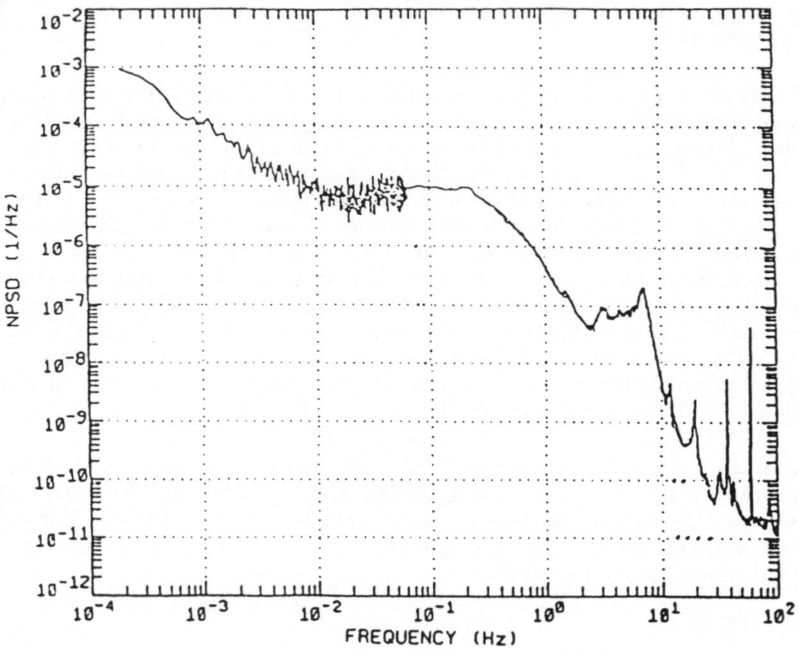

Figure 1. Power spectral density for an ex-core neutron detector at a typical Westinghouse PWR. The data are normalized to the steady state power level.

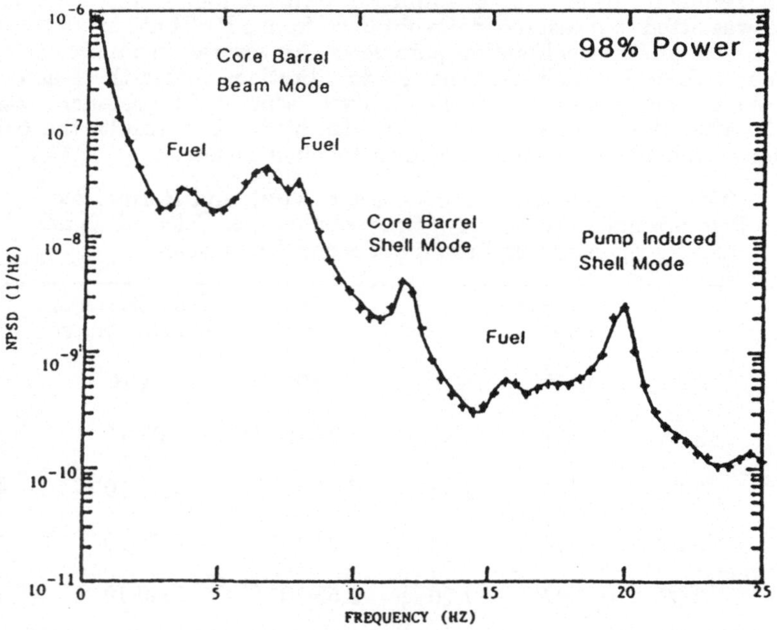

Figure 2. Ex-core neutron detector power spectral density from the Sequoyah-1 Nuclear Power Plant. The points correspond to measured data. The continuous curve is the result of a Cardiograma fit.

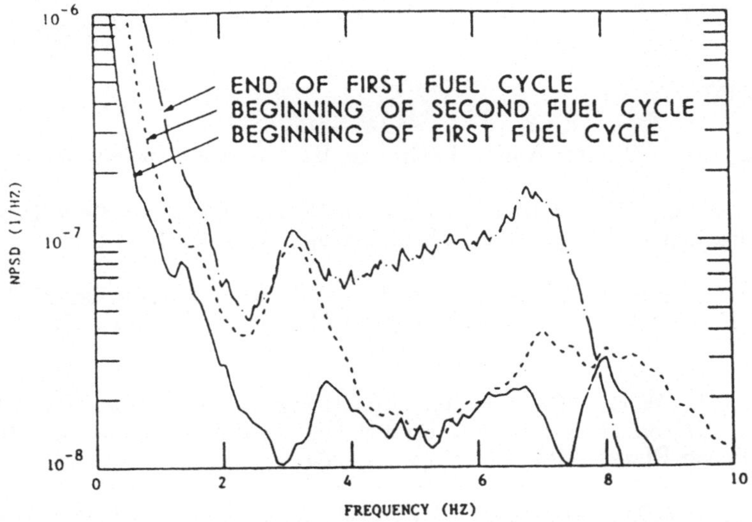

Figure 3. Typical ex-core neutron detector power spectral densities through the first and second fuel cycles at the Sequoyah-1 Nuclear Power Plant.

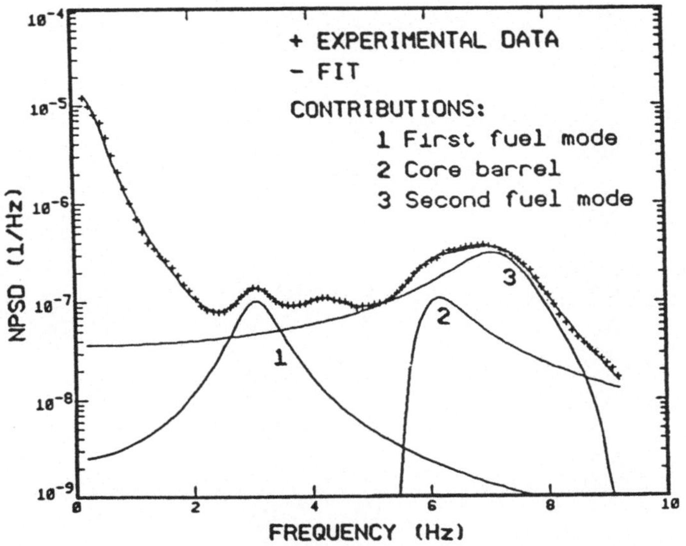

Figure 4. Cardiograma fit to the ex-core power spectral density showing the contributions from the first and second fuel vibrational modes and from the core support barrel pendular motion.

Acknowledgement

Research sponsored by Office of Nuclear Physics, U.S. Department of Energy under contract number DE-AC05-84OR21400 with Martin Marietta Energy Systems, Inc.

REFERENCES

1. J. A. Thie, *Power Reactor Noise*, (LaGrange, IL, American Nuclear Society, 1981).

2. D. N. Fry et al., "Use of Neutron Noise to Detect BWR-4 In-Core Instrument Tube Vibrations and Impacting," *Nucl. Technol.* **43** (1979).

3. W. Bastl, R. Sunder, D. Wach, "On-Line Vibration Monitoring of PWR Internals," *Proceedings of ANS/ENS Topical Meeting on Thermal Reactor Safety* (1980).

4. R. Sunder, D. Wach, "Reactor Diagnosis Using Vibration and Noise Analysis in PWRs," presented at the International Symposium on Operational Safety of Nuclear Power Plants, Marseilles, France, 1983.

5. F. J. Sweeney, J. March-Leuba, C. M. Smith, "Contribution of Fuel Vibrations to Ex-Core Neutron Noise During the First and Second Fuel Cycles of the Sequoyah-1 Pressurized Water Reactor," *Prog. in Nucl. Energy* **15** (1985).

6. F. J. Sweeney, J. P. Renier, "Sensitivity of Detecting In-Core Vibrations and Boiling in Pressurized Water Reactors Using Ex-Core Neutron Detectors," ORNL/TM-8549, Oak Ridge National Laboratory (1985).

7. J. March-Leuba, G. de Saussure, and R. B. Perez *Trans. Am. Nucl. Soc.* **44** (1983).

STOCHASTIC NUCLEAR REACTION THEORY:

BREIT-WIGNER NUCLEAR NOISE

G. de Saussure and R. B. Perez*

Oak Ridge National Laboratory
P.O. Box 2008
Oak Ridge, Tennessee 37831-6354

INTRODUCTION

Our present understanding of neutron-nucleus interaction is largely based on Bohr's statistical compound nucleus model.[1] However, theoretical developments based on quantum corrections[2] to Bohr's model, together with considerable improvements in neutron spectroscopy have predicted and revealed the presence of marked fluctuations in the neutron cross sections of both fissile and fertile nuclei. These fluctuations are wider than the sharp resonances associated with the compound nucleus levels and narrower than the broad structure due to the energy dependence of the neutron penetration coefficients. They, in fact, represent departures from the statistical compound nucleus model in localized energy regions, leading to an intermediate structure which is not predicted by Bohr's model. The observed enhancement of the neutron cross section is due to the presence of doorway states[3,4] in the neutron channel or in the fission channels in fissile nuclei. The understanding and detection of this intermediate structure is of great relevance both in nuclear reaction theory and for the calculation of nuclear reactor parameters.

The purpose of this paper is the application of various statistical tests for the detection of the intermediate structure, which lies immersed in the Breit-Wigner "noise" arising from the superposition of many compound nucleus resonances. To this end, neutron capture cross sections are constructed by Monte-Carlo simulations of the compound nucleus,[5] hence providing the "noise" component. In a second step intermediate structure is added to the Breit-Wigner noise. The performance of the statistical tests in detecting the intermediate structure is evaluated using mocked-up neutron cross sections as the statistical samples. Afterwards, the statistical tests are applied to actual nuclear cross section data.

GENERAL THEORY

The doorway states are special states with a decay width to the continuum. These states have the following properties: (1) their level widths are larger than the average width of the compound nucleus states, (2) they are not eigenfunctions of the total nuclear Hamiltonian, but of a slightly different Hamiltonian which differs from the

* The University of Tennessee, Nuclear Engineering Department, Knoxville, Tennessee 37916

former by a residual interaction, and (3) these special states satisfy the boundary conditions of R-matrix theory.

The introduction of the special levels provides a particularly simple device to interpret the intermediate structure. Endowed with a large level width and with a residual interaction potential, they interact with the compound nucleus levels, which share the strength of the special state. This process leads to enhancements of the cross section in localized energy regions which cannot be explained within the framework of the statistical nuclear model. It can be shown,[6,7] that the intermediate structure is introduced by energy "resonant" reaction widths in the usual Breit-Wigner resonance formula. Hence the Breit-Wigner noise is not additive to the ordered intermediate structure.

MONTE-CARLO SIMULATIONS

Neutron capture cross sections for the ^{238}U nucleus were constructed by sampling from the statistical distribution functions for level widths[8] and level spacings,[9] in accordance to the statistical compound nucleus model. The intermediate structure was simulated by replacing the neutron widths in the $J = \frac{1}{2}, \ell = 1$, ^{238}U compound nucleus state, by their "resonant" counterparts. In every instance the average resonance parameters were adjusted to reproduce the measured average neutron cross section.

STATISTICAL TESTS

Statistical samples were generated by averaging the cross sections over energy bins of variable length, W. Thus, by varying the bin length, W, one obtains sets of cross sections, containing different levels of information.

For instance, by setting the value of W larger than the average spacing of the intermediate structure (a few keV), one erases from the statistical sample both the Breit-Wigner noise and the intermediate structure contributions, leaving only the long range energy dependence of the average cross section on the neutron penetration factors.

The Wald-Wolfowitz[10] (WW) distribution-free statistics test was applied to mocked-up (simulated) ^{238}U capture cross sections with and without the intermediate structure, as well as to sets of measured ^{238}U capture cross sections section data.

The WW test deals with the number of runs, R, of consecutive values which lie above or below a given reference line (in our case the average cross section is taken as the reference line). The WW statistics provides the number of runs, $E(R)$, to be expected from random statistical data, as well as the standard deviation $\sigma(R)$. It can also be shown that the ratio

$$\epsilon_R = \frac{[R - E(R)] - \frac{1}{2}}{\sigma(R)}$$

approximates a normal probability distribution, $P(\epsilon_R)$. Low values for the ratio, ϵ_R, indicate that the statistical sample approximates the statistics of a random sample. The distribution $P(\epsilon_R)$ gives the probability that the tested sample behaves as a random data set.

RESULTS

The WW test for the search of intermediate structure in the ^{238}U neutron capture cross section was applied to two sets of measurements,[8] set I and set II, and for two sets of simulated cross-section data: model I, constructed on the basis of the statistical compound nucleus model, and model II, containing the intermediate structure.

To evaluate the performance of the WW test, this methodology was applied to a set of random data (white noise), as well as to models I and II, and to the experimental data set I and set II. Figure 1 shows the number of runs (R), versus the averaging interval W. For averaging intervals up to 10 keV the results for model I (i.e., the cross section computed according to the statistical compound nucleus model) fall within the error bands of the number of runs expected for white noise. For values of W up to 2 keV, model II as well as sets I and II show deviations from the results for the random set of data. For W=10 keV, the intermediate structure appears to have been washed out and all the tested sets of data behave in a similar fashion. The ratios ϵ_R and the probability $P(\epsilon_R)$ obtained for each cross-section data set are given in Table 1.

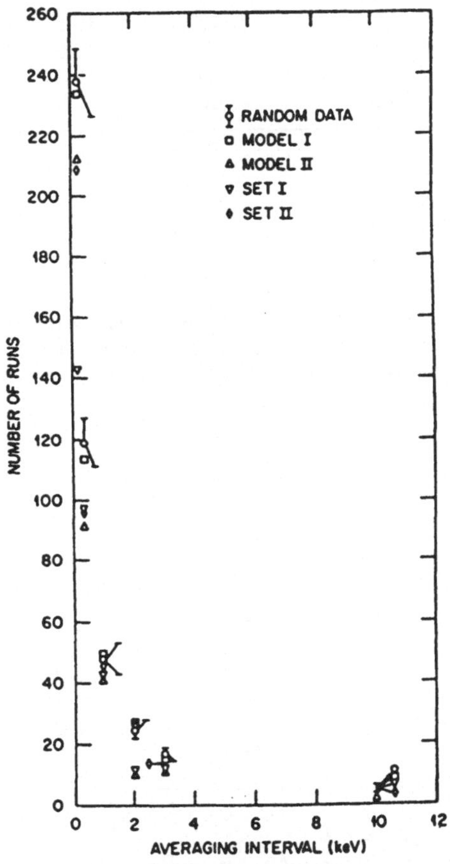

Figure 1. The Wald-Wolfowitz runs test for various simulated ^{238}U neutron capture cross sections and the experimental data, compared with the expected number of runs from a set of randomly distributed objects.

These results indicate that the experimental cross-section data behave similarly to the data constructed on the basis of the statistical nuclear model modified by the inclusion of intermediate structure. For W=400 eV, the significance level for the structure in the ^{238}U capture cross section is equal to that corresponding to about three standard deviations for a normal distribution. The effectiveness of the WW test for the detection of intermediate structure is shown in Table 2.

Table 1. Results of the Wald-Wolfowitz runs test for the ^{238}U capture cross-section measurements (sets I and II) and comparison with the results obtained for model II

Δ[a] (keV)	Set I ϵ_R	Set I $P(\epsilon_R)$ (%)	Set II ϵ_R	Set II $P(\epsilon_R)$ (%)	Model II ϵ_R	Model II $P(\epsilon_R)$ (%)
0.2	8.72	$<10^{-15}$	2.65	0.4	2.20	1.4
0.4	2.81	0.2	2.99	0.1	3.41	0.03
1.0	1.03	15.0	0.42	33.0	1.37	8.0
2.0	2.34	0.9	0.30	38.0	2.49	0.65
3.0	0.98	16.0	0.36	36.0	1.05	14.7
10.0	0.4	34.0	0.40	34.0	1.41	7.8

[a]Energy averaging interval.

Table 2. Results of the Wald-Wolfowitz runs test for the mockup ^{238}U capture cross-section with and without intermediate structure

Δ[a] (keV)	Model I[b] ϵ_R	Model I[b] $P(\epsilon_R)$ (%)	Model II[c] ϵ_R	Model II[c] $P(\epsilon_R)$ (%)
0.2	0.30	38.0	2.20	1.4
0.4	0.74	23.0	3.41	0.03
1.0	0.09	46.0	1.37	8.0
2.0	0.89	18.0	2.49	0.65
3.0	1.08	14.0	1.05	14.7
10.0	0.68	25.0	1.41	7.8

[a]Energy averaging interval.
[b]Mockup cross section without intermediate structure.
[c]Mockup cross section with intermediate structure.

The results pertaining to model I, show that on the average there is a 27% probability of obtaining larger or equal ratios, ϵ_R, from a sample of random data. In the presence of intermediate structure this figure goes down to about 3% for values of the energy averaging interval between 0.2 keV and 2 keV. For larger intervals, $P(\epsilon_R)$ increases again, indicating a progressive washout of the intermediate structure. Hence, the WW test indicates the persistence of intermediate structure effects up to energy intervals as large as 3 keV. For the 400-eV capture cross-section averages, there is a factor of five increase in the ratio, ϵ_R, for the model II case. Clearly, the WW test appears to be a sensitive test for the detection of ordered structures buried in Breit-Wigner noise.

From the present study we draw the following conclusions: (1) the Wald-Wolfowitz runs test is a suitable tool for the detection of intermediate structure in neutron cross-section data, and (2) the application of this test to simulated and measured cross-section data indicates with a high confidence level, the presence of substantial departures from the statistical compound nucleus model.

ACKNOWLEDGEMENTS

Research sponsored by the Office of Nuclear Physics, U.S. Department of Energy under Contract No. DE-AC05-84OR21400 with Martin Marietta Energy Systems, Inc.

REFERENCES

1. C. E. Porter, R. G. Thomas, *Phys. Rev.* **104**, 483 (1956).

2. V. M. Strutinsky, *Nucl. Phys.* **A95**, 420 (1967).

3. H. Feshbach, A. K. Kermann, and R. H. Lemmer, *Ann. Phys. (NY)* **41**, 230 (1967).

4. D. Robson, *Phys. Rev.* **137**, B535 (1965).

5. G. de Saussure and R. B. Perez, ORNL/TM-2599, Oak Ridge National Laboratory Report (1969).

6. H. Weigmann, *Z. Phys.* **214**, 7 (1968).

7. R. B. Perez, G. de Saussure, and M. N. Moore, *Proc. of the Second Intern. Atomic Energy Symposium on Physics and Chemistry of Fission*, Vienna, Austria, 1969.

8. R. B. Perez, G. de Saussure, R. L. Macklin, and J. Halperin, *Phys. Rev. C* **20**, 528 (1979).

9. E. P. Wigner, "Statistical Theory of Spectra Fluctuations," C. E. Porter, Editor, p. 120 in Academic Press, NY 1965.

10. A. Wald and J. Wolfowitz, *Math. Stat XI* **2** (1940).

INDEX